# Goat Science and Production

# Goat Science and Production

Sandra G. Solaiman

WILEY-BLACKWELL
A John Wiley & Sons, Inc., Publication

Edition first published 2010


Blackwell Publishing was acquired by John Wiley & Sons in February 2007. Blackwell's publishing program has been merged with Wiley's global Scientific, Technical, and Medical business to form Wiley-Blackwell.

*Editorial Office*
2121 State Avenue, Ames, Iowa 50014-8300, USA

For details of our global editorial offices, for customer services, and for information about how to apply for permission to reuse the copyright material in this book, please see our website at www.wiley.com/wiley-blackwell.



*Library of Congress Cataloging-in-Publication Data*

Solaiman, Sandra Golpashini, 1952–
Goat science and production / Sandra G. Solaiman.
p. cm.
Includes bibliographical references and index.
ISBN-13: 978-0-8138-0936-6 (alk. paper)
ISBN-10: 0-8138-0936-3 (alk. paper)
1. Goats. 2. Goats–Breeding. I. Title. II. Title: Goat science and production.
SF383.S65 2010
636.3′9–dc22

2009033115

A catalog record for this book is available from the U.S. Library of Congress.

Set in 9.5 on 12 pt Times by Toppan Best-set Premedia Limited



MIX
Paper from responsible sources
FSC www.fsc.org FSC® C013604

# Dedication

To my mother, Mariam, who instilled in me the value of hard work; to my dearest children, Diaco, Dion, and Emeline, who encourage, strengthen, and humble me; to my dearest students who constantly challenge, question, and motivate me:

May your efforts always be sincere, showing commitment to seemingly trivial matters and not devotion only to subjects of high status and importance. In the end, status will fade, yet one can never know the impact these supposedly trivial matters may have on the life of another.

In everything you do, do your best, proceed with integrity, and let your actions be guided by a humane conscience. In doing so, your life portrait will be painted in colors of true fulfillment, unselfish accomplishment, compassionate love, and peace of conscience.

"In all your ways acknowledge him and he will make your paths straight."
Proverbs 3:6

# Contents

# Foreword

Congratulations and commendations to Dr. Solaiman for bringing together a truly international team of experts as contributors to this book addressing the characteristics, management, production, and contributions of goats in our world. These authors have resided and studied in at least 14 countries and traveled extensively throughout many others, molding their perspectives on the roles of goats in global cultures. Sandra, whom I am privileged to know personally, is a consummate professional, more interested in truth than perception and value than appearance. Clearly, these qualities are shared by those selected as contributors.

Goats belong to one of a large number of species regarded generally as small ruminants that occupy special niches throughout the entire world. They vary in size, shape, color, and behavior and supply numerous consumable products. Likewise, the human populations served by goats are diverse. In 2007, the National Research Council of the National Academies published under its Animal Nutrition Series the "Nutrient Requirements of Small Ruminants: Sheep, Goats, Cervids, and New World Camelids." Dr. Solaiman, a member of the writing committee of the NRC issue and a major contributor to the goat sections, determined to expand the application of knowledge of the goat beyond its nutritional requirements to include other biological, social, managerial, and economic considerations. It is to that end and for that purpose that this volume is being published. The codependence of goats and their holders under greatly different circumstances presented special challenges to the authors. Under Dr. Solaiman's editorial leadership, these challenges were met.

Having invested a career working with goats and with colleagues of similar interests, I am grateful for this opportunity to commend this volume to you for your information and enjoyment.

Ed Huston
Professor Emeritus of Animal and Range Science
Texas A&M University
San Angelo, Texas

# Preface

Currently in the U.S., goats have been gaining popularity for their milk and meat products, catching up with popularity they have experienced in other parts of the world. Though more U.S. universities are involved in goat research and extension programs than ever before, few offer courses in goat science and management. The last comprehensive book on goats, "Goat Production," was published more than a quarter of a century ago by Gall (1981), while a more recent publication, "Goats: Biology, production and development in Asia" by Devendra (2007) is focused on Asia. There was a need for a new book covering all aspects of goat sciences and all applicable production practices; therefore, this book was created. This book has been designed to provide a comprehensive, up-to-date compilation of information on goat science and the principles governing their production. It is intended to primarily serve as a textbook for a Junior-Senior level animal science course, which may also be taught as a first-year graduate course in some universities. The depth of scientific information relating to the field of animal and veterinary science presented in this book qualifies it for use in educating veterinary students as well as for use as a resource book for veterinarians, educators, researchers, extension specialists, managers, consultants, and all other goat enthusiasts.

## TOPICS

The topics covered in this book are the collections of my thoughts after reviewing many textbooks and resource books, after conducting more than 20 years of research in the field of goat science and production, both nationally and internationally, and lastly after serving more than 25 years as a teacher in higher education. I would also like to note that many of my contributors have provided me with excellent suggestions that were also incorporated into the contents. I felt that the reader should be introduced to goats in Chapter 1 to understand the main position of this noble animal and its significant contribution to the socioeconomics of subsistence farming and its role in hunger alleviation in the world. Chapters 2, 3, and 4 are focused on major breeds of goats and how to conserve genetic resources in indigenous breeds, especially after introduction of genetically superior breeds, principles of genetics, and breeding. Chapter 5 can serve as a reference and satisfy goat enthusiasts that actively show goats. Showing animals is a common practice in the developed world and goats are no exception as they are commonly included in animal shows. This chapter additionally focuses on ideal meat, dairy, and fiber goat characteristics for proper selection for breeding. Chapter 6 was created mainly for veterinary students and others interested in different functional parts of the goat. This chapter identifies the principles governing common practices, for example, hoof trimming, dehorning, castration, etc., as well as major functional anatomy of different body systems, such as digestive, respiratory, cardiovascular, endocrine, nervous, etc. Chapter 7 focuses on applied reproductive physiology of the goat and covers all of the advances in reproductive technology practiced in goats. Chapters 8, 9, and 10 cover principles governing digestive physiology, nutrient metabolism, unique ingestive behavior of goats, and feeds and feeding practices. Rumen dysfunctions and common problems associated with feeding are also covered in these chapters. The next two chapters, 11 and 12, cover herd health program, diseases and parasites, integrative approaches for prevention and control of parasites, and preferred goat management practices for raising healthy goats. Chapters 13, 14, and 15 focus on goat products such as meat, milk, and fiber including topics such as their uniqueness, quality, and marketability. Chapter 16 explores the unique characteristics of goats and their use in enhancing the environment, reducing fuel load, and preventing fires. Chapter 17 discusses the requirements for housing including a discussion of both shelter and fencing. After properly raising goats, they should be marketable and profitable. Chapter 18 covers basic economic indices based on enterprise budgeting, which can be used to measure profitability and to identify proper marketing channels for goat meat, milk, and fiber.

Reviewing information provided in each chapter, contributors were asked to identify main teaching, research, extension, and outreach activities that are currently lacking. Also by searching the literature and Internet using various search engines, information lacking was identified. Based on information gathered, future needs for teaching, research, extension, and outreach were formulated and briefly summarized in Chapter 19.

## PREREQUISITES

This book is written assuming that the reader has at a minimum a basic understanding and exposure to both biological and animal science. In terms of mathematical and chemical background, it is assumed that the reader has a basic understanding of the topics covered in general chemistry, organic chemistry, and first and second year equivalent mathematics concepts.

# Acknowledgments

I am extremely grateful for a number of individuals who have assisted, encouraged, and motivated me throughout the creation of this book. Specifically, my 90-year-old mother, Mariam, who has instilled in me the value of hard work from a very young age, and my three children, Diaco, his wife Angie, Dianoosh, and Emeline for providing advice, encouragement, and support at various stages of the preparation of this book. I also thank my sister Valentine and my brother-in-law, Dr. Jose Hernandez, for reminding me that there are things in life beyond writing books.

I am deeply indebted to Dr. Denny Marple, my friend and colleague, for his continuous support in providing grammatical corrections and for his advice regarding sections that needed revisions for clarity. His valuable suggestions and comments have significantly shaped the final product you have before you. I am thankful to Dr. Fred Owens, my lifelong advisor and mentor, for reviewing and editing various chapters of this book, and Dr. Ed Huston for believing in me. I am grateful to Dr. Devendra and Dr. Haenlein, my dear mentors, and Dr. Shrestha for their input and support throughout this project and their help with the contents and order of the chapters. I am grateful to all of the contributors for their collaboration and generosity in sharing their information, and a very special thank you to the contributors whose assistance related to their expertise came at times of great need. Specifically, I am grateful for Dr. Olcott, Dr. Dawson, Dr. Kebede, Dr. Gurung, and future Dr. Aviki (my daughter), who extended helping hands in order to make this book a reality. Among others, I am thankful to my dean and friend, Dr. Walter Hill, for his patience and encouragement, to Dr. Luther Williams for his advice, to Dr. Ankumah for his support, and to Dr. Kenneth Andries who contributed to Chapter 19.

One of the special features of this book is the original artwork and illustrations provided by three professional artists and illustrators. I would like to extend a special "thank you" for their contribution: Beth Emery for Chapter 2; Dr. G. Constantinescu for Chapters 5 and 6; and Z. Proctor for Chapter 12.

Finally, I would like to warmly thank Justin Jeffryes, Executive Commissioning Editor, for believing in me and for his encouragement, and Shelby Allen, Senior Editorial Assistant, and the Wiley-Blackwell team for enthusiastic support of this project. I am grateful to the production team at Wiley-Blackwell for their excellent work.

# About the Contributors

The contributors for this book were hand selected from recommendations provided by reputable sources, and by searching the current topics in the literature related to goats. Attempts were made to include authors who represent a wider global demographic distribution in order to provide an international prospective on goats. As indicated in their short biographies, they are experts in their respective fields.

**E.M. Aviki**, BS, MBA. Emeline has a BS in Economics and is an MD/MBA student at Duke University School of Medicine and the Fuqua School of Business, Duke University, Durham, NC. Emeline has 4 years of experience as a teaching assistant for various economics courses including Introductory Macroeconomics and Labor Economics. Her short research experience has focused on health outcomes in transplantation and translational research involving malignant melanoma. She currently has one paper submitted for publication.

**G.F. Bath**, BVSc. Gareth is a Professor of Production Animal Studies, Faculty of Veterinary Science, University of Pretoria. He has more than 40 years of private practice, teaching, research, and outreach activities experience in the areas of sheep and goat nutrition, sustainable parasites control, health systems, and diagnostics. He has mentored numerous graduate and undergraduate students, authored 8 books or book chapters, and has over 150 refereed and non-refereed publications.

**G.M. Constantinescu**, DVM, PhD, mult. Dr.h.c. Gheorghe is a Professor of Veterinary Anatomy, Department of Biomedical Sciences, College of Veterinary Medicine, University of Missouri, Columbia. He has 58 years of teaching and research experience in Clinical Anatomy and the Anatomical Nomenclature. He is one of the five members of the Editorial Board of the 5th edition of the Nomina Anatomica Veterinaria, and an honorary member of the Editorial Board of Experimental Medical and Surgical Research. He has mentored 73 graduate students, and published 25 books, 52 book chapters, and 96 refereed papers. He is also a professional member of the Association of Medical Illustrators.

**I.A. Constantinescu**, MS, DVM. Ileana is a Clinical Assistant Professor of Veterinary Anatomy, Department of Biomedical Sciences, College of Veterinary Medicine, University of Missouri, Columbia. She is one of the eight members of the International Committee on Veterinary Embryological Nomenclature. She teaches Veterinary Anatomy and Developmental Anatomy. She has authored 3 books, book chapters, and more than 20 refereed publications.

**G.H. Crow**, BSc, MSc, PhD. Dr. Crow is an Associate Professor, Department of Animal Science, Faculty of Agriculture and Food Sciences, University of Manitoba, Canada. Gary has 28 years of experience in teaching, research, and outreach working in the areas of quantitative genetics, genetic resources. He has mentored more than 10 graduate students and authored more than 50 refereed and non-refereed publications.

**L.J. Dawson**, BVSc, MS, Dip. American College of Theriogenology. Lionel is an Associate Professor, Veterinary Clinical Sciences, Centre of Veterinary Health Sciences, Oklahoma State University, Stillwater, Oklahoma, and an Adjunct Associate Professor, Research and Extension, Langston University, Langston, Oklahoma. He has 27 years of experience in teaching, research, and outreach in the area of goat nutrition, internal parasites, synchronization and artificial insemination, and male infertility. He has mentored 56 graduate students, interns, and residents, and has authored 6 books or book chapters and more than 120 refereed and non-refereed publications.

**C. Devendra**, B. Agric. Sci., M. Agric. Sci., PhD, DSc. Dev is a Fellow of the Academy of Sciences, Malaysia, and is now an Independent Consultant, with international efforts spanning more than 25 years of experience in animal nutrition and feed resources, animal production systems, and integrated natural resource management. He is on the Editorial Boards for the *Small Ruminant Research*, *Livestock Science*, and *Outlook on Agriculture* journals. He is the author of 17 books and more than 440 publications published internationally. He was Distinguished Visiting Professor to the University of Delaware. His awards include the International Dairy Production (U.S.), Lincoln International Alumni Medal (New Zealand), and the Asian Animal Science (Korea).

**H. Dove**, B. Agr. Sci., Dip.Ed., PhD. Hugh is a Chief Research Scientist, CSIRO Plant Industry, Canberra, Australia. He is a Fellow of both the Australian Institute of Agricultural Science and Technology and the Australian Society of Animal Production, and was awarded the Research Medal of the Nutrition Society of Australia. He has 33 years of experience in the area of diet selection and intake estimation, and crop livestock interaction. He has authored 13 books and book chapters, and more than 200 refereed and non-refereed publications. He is an Associate Editor of the *Journal of the Science of Food and Agriculture* and is on the Editorial Board of the *Encyclopedia of Animal Science*.

**R.A. (Bob) Ebert**, BS, MEd. Bob is an Extension Specialist, Department of Animal Sciences, College of Agriculture, Auburn University, Alabama. He has 25 years of teaching and extension activities in the areas of livestock judging and evaluation, 4-H youth educational programs, and adult education.

**B. Emery**, BS Studio Art, Emphasis in Painting, School of Visual Arts, Florida State University. Beth has participated in many group and solo exhibitions throughout the U.S. and has won numerous awards for painting. Beth likes to take chances with colors and shapes; and committed to visual reality, her work conveys life, energy, and movement. She generously contributed her artistic production to Chapter 2 of this book. For more of her work, please visit www.betemeryart.com.

**S. Galal**, BS, MS, PhD. Salah is a Professor in the Animal Production Department, Faculty of Agriculture, Ain Shams University, Cairo, Egypt. He has more than 45 years of teaching and research experience including working for the FAO in the areas of animal breeding and genetics, population genetics, and animal genetic resources. He is an Associate Editor for the *Small Ruminant Research* journal. He has mentored 25 graduate students, authored 3 books and book chapters, and published more than 100 refereed and non-refereed publications.

**J. Greyling**, BS, MS, PhD. Johan is a Professor and Department Head, Department of Animal Wildlife and Grassland Sciences, Faculty of Natural and Agricultural Sciences, University of the Free State, Bloemfontein, South Africa. He has 30 years of teaching, research, and outreach activities in the areas of reproductive physiology and accelerated reproductive techniques. He has mentored 47 graduate students and authored 3 books and 98 refereed and non-refereed publications. He is currently Associate Editor of the *Small Ruminant Research* journal, sub-Editor of the *South African Journal of Animal Science*, and is on the Editorial Board of the *Livestock Science* journal.

**N.K. Gurung**, BS, MS, PhD, Professional Animal Scientist (PAS), and Hubert Humphrey Fellow. Nar is an Assistant Professor, Department of Agricultural and Environmental Sciences, College of Agricultural, Environmental and Natural Sciences, Tuskegee University, Alabama. He has more than 15 years of research, extension, outreach, and teaching experiences in Animal Nutrition, especially goats, by-products utilization, goat production, and silvopasture systems. He has authored more than 30 refereed and non-refereed publications.

**G.F.W. Haenlein**, Dipl. Ag., DSc., MS, PhD. George is a Professor Emeritus, Department of Animal & Food Sciences, College of Agriculture & Natural Resources, University of Delaware. He has 59 years of research and more than 42 years of teaching and extension experience in the areas of dairy goats, nutrition, milk production, and composition. He is founder and Honorary Editor-in-Chief of the *Small Ruminant Research* journal. He has mentored 20 graduate students and has authored 7 books or book chapters and more than 650 refereed and non-refereed articles.

**E. Kebede**, BS, MS, PhD. Ellene is an Associate Professor, Department of Agriculture and Environmental Sciences, College of Agriculture, Environmental and Natural Sciences, Tuskegee University. She has 15 years of experience in teaching, research, and outreach in the areas of resource management, impact analysis and policy, and participatory research. She has mentored 12 graduate students and authored more than 27 refereed and non-refereed publications.

**C.J. Lupton**, BSc., PhD. Chris is a Professor, Department of Animal Science, College of Agriculture and Life Sciences, Texas A&M University, and Adjunct Professor, Department of Agriculture, Angelo State University, Texas. Chris has 25 years of research and outreach activities in the area of animal fibers (fiber metrology and effects of nutrition, management, selection, and genetics). He has one patent. He has mentored 20 graduate students, and has authored 13 book chapters and more than 250 refereed and non-refereed publications. He is currently serving on the Editorial Board of the *Sheep and Goat Research* journal.

**D.N. Marple**, BS, MS, PhD. Denny is a retired Professor and Department Head of Animal Sciences (2001), Iowa State University. Denny is a past President of the American Society of Animal Science, and a former member of the Boards of Directors of American Society of Animal Science and the Council for Agricultural Science and Technology. He has more than 35 years of experience in teaching, research, and outreach in the area of animal growth and development and was involved in research on swine growth and meat quality as well as beef cattle research on shipping fever and fescue toxicity. He is the author or co-author of 57 refereed journal papers.

**K.W. McMillin**, BS, MS, PhD, Professional Animal Scientist (PAS), Diplomate in the American College of Animal Food Science. Ken is a Professor of Meat Science in the School of Animal Sciences and Department of Food Sciences, the Louisiana State University Agricultural Center at Baton Rouge, Louisiana. He has more than 28 years of experience in teaching, research, and outreach in the area of processing and shelf life of muscle food species, modified atmosphere packaging, and goat meat properties and marketing. He has mentored 14 graduate students, has 2 patents, and has published 9 book chapters, 71 refereed journal articles and conference proceedings, and 64 popular press and technical articles. Currently he serves as Associate Editor of the *Animal Products*, *Journal of Animal Science* and is on the Editorial Board for *Meat Science*.

**J.E. Miller**, DVM, MPVM, PhD. Jim is a Professor of Pathobiological Sciences, College of Veterinary Medicine, and Veterinary Sciences, Louisiana State University Agricultural Center and Adjunct Professor, Department of Animal Science. He has 27 years of research, teaching, and outreach experience in the areas of epidemiology and control of ruminant parasites, host response to parasite infection, and genetics of breed resistance to parasite infection. He has mentored 25 graduate students, authored 5 books or book chapters, and over 100 refereed and non-refereed publications.

**B.M. Olcott**, BS, DVM, MS, MBA. Bruce is an Associate Professor, Department of Veterinary Clinical Science, College of Veterinary Medicine, Louisiana State University. He has 31 years of experience in teaching, research, and outreach in the areas of veterinary production medicine, small ruminant health and production, veterinary science, clinical herd health and production medicine. He has mentored 9 graduate students and has authored 6 book chapters and more than 40 refereed and non-refereed publications.

**F.N. Owens,** PhD, and Professional Animal Scientist (PAS). Fred is a senior research scientist for DuPont-owned Hi-Bred International and a Professor Emeritus of Animal Sciences at Oklahoma State University. He has taught and conducted research on protein nutrition, feed intake, grain processing, rumen function and feed additives for 40 years. He served as president of the American Society of Animal Science and Editor-in-Chief for the *Journal of Animal Science*, mentored numerous graduate students, authored 18 book chapters, and published 460 referred and non-referred articles. In 1996, Fred received the Morrison Award, the most prestigious U.S. research award in animal science.

**Y.W. Park**, BS, MS, M. Div., PhD. D. Min. Young is a Professor at Georgia Small Ruminant Research & Extension Center, Fort Valley State University, Fort Valley, GA, and an Adjunct Professor, Department of Food Science and Technology, University of Georgia, Athens, Georgia. He has more than 30 years of research experience in the area of goat milk and its products. He has mentored more than 20 graduate students. He is on the

Editorial Board for the *Small Ruminant Research* journal. He has published 3 books, 27 book chapters, and more than 240 refereed and non-refereed articles.

**A. Peischel**, BS, MS, PhD. An is an Assistant Professor and Extension Specialist, Tennessee State University/ University of Tennessee—Cooperative Extension Program. She has 25 years of experience in Extension and Outreach in the areas of goat production, vegetation control and environmental enhancement using goats. She has mentored several graduate students and has authored 5 book chapters and numerous refereed and non-refereed publications.

**J.N.B. Shrestha**, BVSc & AH, MS, PhD. Jap is a Research Scientist, at the Dairy and Swine Research and Development Centre, Agriculture and Agri-Food Canada, Sherbrooke, QC, Canada, and also an Adjunct Professor, University of Manitoba, Winnipeg, Manitoba, Canada. He has 35 years of research and outreach activities in animal breeding and genetics, and conservation of domestic animal diversity. He is an Associate Editor for the *Small Ruminant Research* journal. He has authored 20 books and book chapters and more than 200 refereed and non-refereed publications.

**S.G. Solaiman**, BS, MS, PhD, and Professional Animal Scientist (PAS). Sandra is a Professor, Department of Agricultural Sciences, College of Agricultural, Environmental and Natural Sciences; Director of Small Ruminant Research and Education Program, Tuskegee University; and an Adjunct Professor at Department of Animal Sciences, Auburn University, Alabama. She has more than 25 years of teaching, research, and outreach experience in the areas of goat nutrition, trace mineral nutrition, kinetics of digestion and passage, forages and pastures, system nutrient optimization, growth performance, and carcass quality. She was a member of the Small Ruminant Committee of National Research Council, and currently serves as Editorial Board member of the *Small Ruminant Research* journal. She has mentored more than 20 graduate students and has authored 3 books or book chapters and 50 refereed and non-refereed publications.

# Goat Science and Production

# 1
# Perspectives on Goats and Global Production

*C. Devendra, PhD, DSc, FASc and S.G. Solaiman, PhD, PAS*

**KEY TERMS**

Improver breeds—local breeds of goats that have a potential to improve performance of other breeds.
Bezoar—*Capra hircus*, the true goat, one of the five wild ancestors of the domestic goat.
The Silk Road—one of the world's oldest and historically important trade routes.
Agroecological zones—climate, soil, and terrain conditions relevant to agricultural production.
Agropastoralism—mixed crop-livestock systems with extensive grazing in which households are on the move and their livelihoods are involved with the system.
Range-based systems—extensive systems characterized by rainfall of less than 150 millimeters (mm)/year.
4-H Competition—a competitive event important in youth development education.
Rainfed areas—non-irrigated agricultural areas that depend on rain as a source of water.

**OBJECTIVES**

By completing this chapter, the reader should acquire knowledge on:

- Where the goat originated
- How the goat contributes to human culture
- Outlook for goat production in the world
- Outlook for goat production in the United States (U.S.)
- Production systems in the world
- Production enterprises in the U.S.
- Constraints to goat production
- Farming systems research

## INTRODUCTION

The goat species is an important component of animal genetic resources. Together with sheep, and partly because of their size, both are commonly called "small ruminants." Goats have been associated with man since the dawn of agriculture and the domestication of animals. Goats were the first animals to be domesticated by man and continue to hold an important niche particularly in subsistence agriculture in the developing countries, and they support a variety of socioeconomic functions throughout the world.

This chapter provides a comprehensive background and some perspectives on goats and goat production in the world and the U.S. highlighting and discussing the various aspects of goat production. For brevity, discussions are necessarily concise, and readers are encouraged where appropriate to read the references provided when seeking further information.

## EVOLUTION AND DOMESTICATION OF GOATS

Evolutionary biology indicates that the goat was domesticated about 10,000 years ago at the dawn of the Neolithic age. Domestication was associated with three of the oldest civilizations: the Nile in northeast Africa, the Tigris–Euphrates in west Asia, and the Indus in the Indian subcontinent. Archeological investigations of relics from past civilizations show that links between goats and people and their livelihoods were very close. Archeologists indicate that the goat was first domesticated in the "Fertile Crescent" of the eastern Mediterranean. This landmass stretches between the Black and Caspian seas along the coast of Palestine, and curves like a quarter moon toward the Persian Gulf. This is the site where agriculture originated—in the narrow strip bounded by the Euphrates and Tigris rivers.

Domestication of wild goats was evident first in Jericho (Jordon) around 7000 BC as well as in the Zagros Mountains in Gangi Dareh (Iran) around 8000 BC (Zeuner, 1963). Since then, the goat has been involved in many aspects of human culture including religion, tradition, folklore, nutrition, livelihood, and economics (Boyazoglu et al., 2005).

## GOATS IN MYTHOLOGY

Goats, probably more than any other species, have been associated with mythology in many cultures. In ancient mythology, Jupiter was nursed by a goat and Thor's chariot was pulled by a team of goats. The Greek god Pan often is represented as half man and half goat. In Chinese culture for example, the goat spirit Yang Chin is the god of Fan-Yin, the transcendent goat with a white face, horns, a long beard, and a special headdress.

In Chinese astrology, the goat is one of the cycles among the 12 earthly branches of nature. Symbolizing love and happiness, the goat represents Wei, the eighth earthly branch and the embodiment of summer. The goat also is a Mongolian god (Cooper, 1992). The zodiacal sign Capricorn comes from the Latin word *Capra* meaning a goat. In addition, the sign Aries has a mixed etymology. In Umbrian it is a ram, in Greek it is a kid (goat), and in old Irish it is a doe.

## ZOOLOGICAL CLASSIFICATION

The goat is a hollow-horned ruminant that belongs to the mammalian order *Artidactotyla,* suborder *Ruminantia,* family *Bovidae,* and is of either the *Capra* or the *Hemitragus* genera. The distinction between these two genera was based first on horn structure, but the distinction has been confirmed genetically.

**Table 1.1** Zoological classification and ancestry of domestic goats.

| Common name | Species |
|---|---|
| True goat | *Capra hircus*, including the bezoar (*c.h.aegagrus*) |
| Ibexes | *Capra ibex* |
| Caucasian tur | *Capra caucasica* |
| Spanish ibexes | *Capra pyrenaica* |
| Markhor | *Capra falconeri* |

Source: Ellerman and Morrison Scott, 1951.

The domestic goat belongs to the genus *Capra.* It developed from the following five wild ancestors:

1. *Capra hircus*, the true goat including the bezoar (e.g., aegagrus)
2. *Capra ibex*, the ibexes
3. *Capra caucasica*, the Caucasian tur
4. *Capra pyrenaica,* the Spanish ibex
5. *Capra falconeri*, the markhor

See Table 1.1. The Mediterranean breeds also owe some of their distinctive characteristics to the ancestral influence of the extinct *Capra prisca.*

The comparative morphology and breeding experiments indicate that the bezoar of western Asia was the main progenitor of most domestic goats. Both the markhor and bezoar gave rise to the majority of the Indian and central Asian breeds with their distinctive characteristics: long coarse hair, black rather than white, brown, or other colors, and scimitar-shaped horns (Devendra and Burns, 1983).

## IMPORTANCE AND SOCIOECONOMIC RELEVANCE OF GOATS

Goats provide products and services important for man throughout the world. In developed countries goats are valued mainly for their milk, fiber, and meat; while in the developing countries, they are valued mainly for meat, followed by milk, fiber, and skins. Table 1.2 summarizes the products and services from goats in Asia, which will be similar to other regions of the world.

The socioeconomic relevance of goats is greatest in developing countries where they meet socioeconomic, cultural, and recreational needs. Their small size is especially relevant and relates directly to economic, managerial, and

**Table 1.2** Goat products and services in Asia.

| Products | Services |
|---|---|
| Meat (raw, cooked, blood, soup, goat meat extract—"Zeungtang" in Korea)* | Cash income and investment |
| | Security and insurance |
| Milk (fresh, sour, yogurt, butter, cheese) | Prestige in ownership |
| | Gifts and loans |
| Skins (clothes, shoes, water/grain containers, tents, handcraft, shadow play in Indonesia, thongs, etc.) | Religious rituals (e.g., sacrificial slaughter) |
| | Human nutrition—beneficial characteristics of meat and milk |
| Hair (cashmere, mohair, garments, coarse hair rugs, tents, ropes, wigs, fish lures) | Pack transport |
| | Draught power |
| | Medicine |
| | Control of bush encroachment |
| Horns | |
| Bones (handcraft) | Guiding sheep for grazing |
| Manure and urine (crops, fish) | |

*With goats: Total edible proportion is 61%. Total saleable proportion is 82%.

Source: Devendra, 2007a.

biological advantages over other species. The following benefits are noteworthy (Devendra, 1998):

- Income: means to earn supplemental money
- Food: provide animal proteins (milk and meat) for the nutritional well-being of peasants, particularly the undernourished
- Security: form for investment, maintenance of assets, security, and economic stability
- Employment: creation of jobs, including effective utilization of family labor
- Fertilizer: contributes to crop production and farm fertility through the return of dung and urine
- By-product utilization: using nonmarketable crop residues to generate value-added products (for example, meat, fiber, and skins)
- Social values: increases cohesiveness in village activities and religious ceremonies
- Recreation: buck fighting and buck races

The small size of the goat contributes to its popularity. Goats help to meet daily food needs (meat and milk), are easily sold as a source of cash, provide insurance and collateral for various agricultural activities, are valued in religious ceremonies including sacrifices, are gifts during marriage ceremonies, and are even used for sport and recreation. In harsh semiarid and arid environments such as the Sahel, Near East region, and northern parts of the Indian subcontinent to which the species are well adapted, poor and landless farmers often increase the size of their goat flocks to provide greater food and economic security.

## POPULATION SIZE AND GLOBAL DISTRIBUTION

Table 1.3 tabulates the total population size and the global distribution of goats. Several items of note follow:

- The total world population of goats in 2007 was about 850 million.
- The developing countries claim about 41% of the goat population.
- The low-income, food-deficit countries had about 86% of the goat population.
- The largest goat population is in Asia (545 million) followed by Africa (245 million). Combined, these two continents accounted for about 93% of the world's total goat population.
- During the period of 1986–2007, the annual growth of the goat population was 3.5% per year, with growth rate in developing countries of 4.5%. Of the developing nations, only Latin America and the Caribbean had a very low population growth rate for goats.
- In Europe, goats had a negative growth rate. But Europe had 2% of the total world goat population with a sizeable number of breeds (26% of breeds), many of which have been introduced into other countries.
- Oceania had the highest negative growth rate and only 0.1% of the total world goat population. But it is among the highest exporters of goat meat.
- North America had the smallest goat population of about 3 million or about 0.3% of the total goat population, but it has an impressive population growth rate of 3.4% among developed countries.

Table 1.3 shows how goats are distributed very widely globally across a wide variety of agroecological zones (AEZ). Historically, distinct dispersion routes enabled the spread of goats leading to their wide distribution. Such trade routes continue to be used today. Examples would include the early settlement of the West Indies where goats often were carried aboard ships to supply fresh milk during voyages. Thereby, goats from India were introduced to Trinidad, Guyana, and Jamaica. Likewise, goats accompanied immigrants aboard the ship with Captain Cook to Australia in the eighteenth century. Asia, western Asia, and the Indian subcontinent were especially important as

**Table 1.3** Goat populations and their global distribution.

| Region | 1986 (Million) | 2007 (Million) | % of Total (2007) | Average Annual Growth Rate (%/yr) |
|---|---|---|---|---|
| Africa | 153.4 | 245 | 28.8 | 3.0 |
| Asia | 288 | 545 | 64.1 | 4.5 |
| Europe | 20 | 18.1 | 2.1 | –0.45 |
| Americas | 34.6 | 41.1 | 4.8 | 0.9 |
| Oceania | 1.5 | 1.0 | 0.1 | −1.7 |
| North America | 1.8 | 3 | 0.3 | 3.4 |
| Latin America and Caribbean | 32.8 | 38.1 | 4.5 | 0.8 |
| Developing countries | 186 | 351 | 41.3 | 4.4 |
| Food deficit countries | 385 | 728.6 | 85.7 | 4.5 |
| Least developed countries | 126.6 | 246 | 29 | 4.7 |
| Total world | 497.5 | 850.2 | — | 3.5 |

Source: FAO, 2007.

the focal points for goat dispersion along two routes (Devendra and Nozawa, 1976):

1. From Iran, Afghanistan, and Turkistan to Mongolia and northern China along the Silk Road.
2. From the Indian subcontinent to Asia and Europe through the Khyber Pass.

The second route is older, being used by the Indo-Aryan people from the north in the second millennium BC. Mongolia, China, and India received domestic goats from nomadic and seminomadic pastoralists from western and central Asia who used these routes.

Because goats are fully adapted to drier, semiarid to arid AEZ environments, they thrive in sub-Saharan Africa (SSA). Among grassland environments, in the mixed rainfed arid and semiarid grassland systems, and in all mixed systems, goats and sheep outnumber cattle. A similar situation exists in rainfed arid and semiarid Rajasthan in the northern parts of India. In such environments, goats and sheep are critical for the livelihood of the landless and marginal farmers in West Asia, North Africa, north India, north Brazil, and north Mexico.

The principal reason for the wide distribution of the goat species is their intrinsic capacity to adapt to different biophysical conditions and environments. This adaptability is aided by their inquisitive nature and independent habits. This in turn becomes reflected in their adaptations in terms of anatomical, morphological, physiological, feeding behavior, and metabolic modifications under different climatic conditions (Table 1.4).

## GOAT PRODUCTS AND PRODUCTIVITY

Goats produce meat, milk, fiber, and skins. Goat meat is widely consumed locally and may be exported. Milk is of secondary importance, being consumed primarily by the household in developing countries. In developed countries, milk and milk products and notably cheese are sold commercially and consumed widely. Milk and milk products are important contributors to human nutrition (Haenlein, 2004).

Commercially, goat fiber (mohair, cashmere) is the most luxurious fiber in the world and goatskin is a very valuable product with high added value especially in European markets. In developing countries, the value of goatskins often is not recognized due to inadequate knowledge and poor processing methods. India is one exception due to the advances in their tannery function and processing methods. Consequently, the export of skins is a major source of income for India.

Figure 1.1 illustrates global goat meat and milk production from 1980 to 2007. Globally, goat meat and milk production were 5.1 and 14.8 million metric tons (MT) in 2007, respectively. These numbers represent increases of 3.0- and 2.0-fold from 1.7 and 7.7 million MT from those of 1980. The total world meat and milk inventories in 2007 according to the Food and Agriculture Organization of the United Nations (FAOSTAT) were 285.7 and 671.3 million MT (FAOSTAT, 2007), respectively. Goat meat and milk production represents only 2.0 and 2.2% of the global inventory, respectively. Unlike other major meats and milk, goat meat and milk are not widely traded but instead are consumed locally.

**Table 1.4** Characteristics of adaptation by goats to different tropical climates.

| Type of climate | Anatomical | Morphological | Physiological | Metabolism | Feeding behavior/feed utilization | Products |
|---|---|---|---|---|---|---|
| Arid/semi-arid | Large size (30–50 kg), long legs and ears; scrotum shows two distinct sacs | White, black, or brown coat color; shiny surface; white-colored goats absorb less amounts of solar radiation | Panting, sweating, and cooling | Increased mobilization of fat during periods of feed shortage; lower water turnover rate | Browse over long distances; resistance to dehydration; desiccation of feces; increased concentration and reduced urine volume; rumen acts as water reservoir; higher digestive efficiency of coarse roughages; efficient use and retention of nitrogen | Meat/milk/fiber |
| Subtropical | Intermediate size (25–30 kg) | White, black, or brown coat color; less shiny coat | Panting and sweating | Lower water turnover rate | Intermediate | Meat/milk/fiber |
| Humid/subhumid | Small/dwarf size (10–25kg) short legs; small ears | Mainly black or brown coat color; shiny coat | Reduced panting and evaporative cooling due to humidity and also more shade | Low metabolic rate; restricted pituitary function | Reduced walking due to increased availability of forages and crop residues | Meat |

Source: Devendra, 1987.

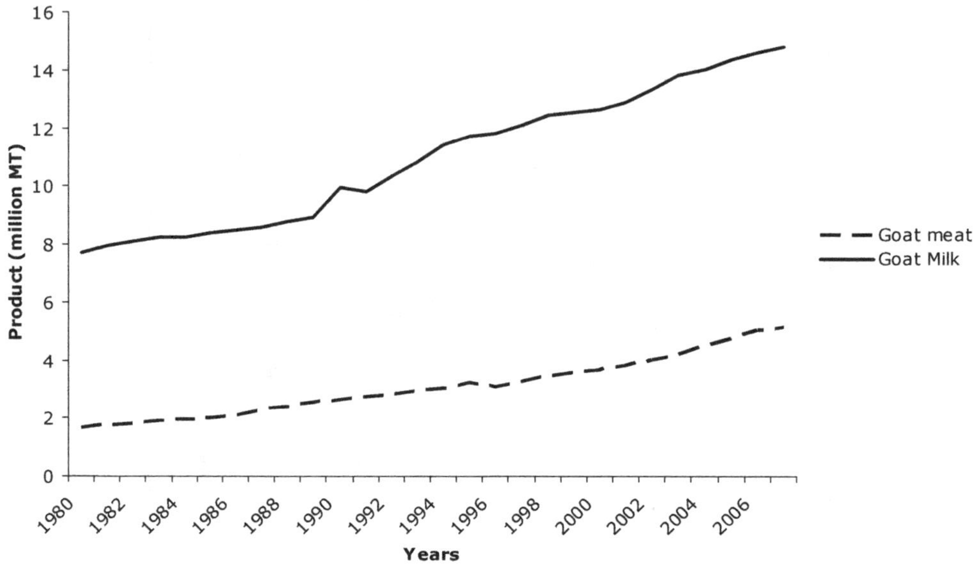


**Figure 1.1** Changes in global goat meat and milk production (FAOSTAT, 2007).

**Table 1.5** Top producers of goat meat.

| Country | Goat meat (MT) | % |
|---|---|---|
| China | 2,253,678 | 43.8 |
| India | 527,000 | 10.2 |
| Pakistan | 350,000 | 6.8 |
| Bangladesh | 195,000 | 3.8 |
| Sudan | 186,000 | 3.6 |
| Nigeria | 148,830 | 2.9 |
| Iran, Islamic Republic | 106,000 | 2.0 |
| Indonesia | 63,410 | 1.2 |
| Greece | 58,000 | 1.1 |
| Mali | 52,820 | 1.0 |
| Others | 1,214,504 | 23.6 |
| World | 5,146,202 | 100 |

Source: FAOSTAT, 2007.

**Table 1.6** Top producers of goat milk.

| Country | Goat milk (MT) | % |
|---|---|---|
| India | 3,823,000 | 25.8 |
| Bangladesh | 2,016,000 | 13.6 |
| Sudan | 1,450,000 | 9.8 |
| Pakistan | 699,000 | 4.7 |
| France | 590,000 | 4.0 |
| Greece | 500,000 | 3.4 |
| Spain | 488,500 | 3.3 |
| Somalia | 393,000 | 2.6 |
| Iran, Islamic Republic | 370,000 | 2.5 |
| China | 268,000 | 1.8 |
| Others | 4,218,152 | 28.5 |
| World | 14,800,534 | 100 |

Source: FAOSTAT, 2007.

Tables 1.5 and 1.6 summarize the contributions of the top 10 countries to global goat meat and milk production. Goat meat and milk are important in most parts of the world particularly in the developing countries. Production parallels the size of goat populations. More than 60% of the total world goat meat was produced by China, India, and Pakistan in 2007. At the same time, almost 50% of the goat milk was produced by India, Bangladesh, and Sudan, and more than 70% of milk was produced in the top 10 countries in goat milk production.

Among exporting countries, Australia leads the world in goat meat exports with 21,952 MT exported in 2006 (the latest data available on the FAOSTAT website in 2009) and 41.1% of the total world market. Ethiopia claimed 25.7% of the export market, followed by China (8.9%), Pakistan (7.4%), and France (5.4%). More than 80% of the goat meat was exported by the top seven countries (Table 1.7).

The top importers of goat meat were the U.S. with 22.6%, followed by United Arab Emirates (16.7%), and China (15.5%). Collectively, these three countries import

**Table 1.7** Top exporters of goat meat.

| Country | Goat meat (MT) | % Total |
|---|---|---|
| Australia | 20,199 | 41.1 |
| Ethiopia | 12,659 | 25.7 |
| China | 4,406 | 8.9 |
| Pakistan | 3,650 | 7.4 |
| France | 2,666 | 5.4 |
| Saudi Arabia | 1,410 | 2.8 |
| New Zealand | 1,132 | 2.3 |
| Others | 3,145 | 6.4 |
| World | 49,137 | 100 |

Source: FAOSTAT, 2005.

**Table 1.8** Top importers of goat meat.

| Country | Goat meat (MT) | % |
|---|---|---|
| U.S. | 9,653 | 22.6 |
| United Arab Emirates | 7,119 | 16.7 |
| China | 6,599 | 15.5 |
| Qatar | 3,013 | 7.0 |
| China, Hong Kong | 2,680 | 6.3 |
| Saudi Arabia | 2,511 | 5.9 |
| France | 1,470 | 3.4 |
| Italy | 1,459 | 3.4 |
| Canada | 1,302 | 3.0 |
| Others | 6,914 | 16.2 |
| World | 42,680 | 100 |

Source: FAOSTAT, 2005.

almost 55% of the total goat meat imports (Table 1.8). Goat milk and skin are not traded according to the FAOSTAT Web site, and trade of goat cheese is minimal.

## THE GOAT INDUSTRY IN THE U.S.

As a contrast with the situation in the developing countries, it seems relevant to examine the goat industry of the U.S. and its outlook.

In the U.S., goats are used mainly for meat, milk, and fiber production, for vegetation control and management, and for reducing the fuel for wildfire.

Meat goat production in the U.S. has increased in recent years because of a growing population of ethnic and faith-based groups that consume goat meat. The national estimates, based on import data only, indicate that in 2007 the U.S. had a deficit of more than 718,000 goats relative to the current demand for goat meat (Solaiman, 2007).

### Status and trends of goat farms and the industry

According to the 2007 U.S. Department of Agriculture-National Agricultural Statistics Service (USDA-NASS), the goat population in the U.S. is a little over 3 million head. Goat numbers doubled from 1985 to 2007 (Figure 1.2). In 2007, more than 83% of goats in the U.S. were meat goats, 10% were dairy goats, and 7% were Angora goats (Table 1.9).

According to the USDA Census (2002), the number of all-goat farms increased more than 19%, and the goat population increased by over 12% from 1997 to 2002; however, the number of farms selling goats increased by over 45%, and goat sales were up by more than 55%. USDA-NASS (2007) sample data indicated that the total number of goats continued to increase, being up 19% from the 2002 USDA Census. The number of meat goat farms increased by 18%, and the number of meat goats increased more than 57% from 1997 to 2002. The number of farms selling meat goats increased by 48% with an increase of more than 108% in meat goat sales. According to the 2007 USDA-NASS sample data, meat goat population was up another 29% from the 2002 USDA Census. The number of dairy goat farms increased by 45%, and dairy goat numbers increased by 52.5% from 1997 to 2002. The number of farms selling dairy goats increased by 71%, and the number of dairy goats sold increased 57%. According to the 2007 USDA-NASS sample data, the number of dairy goats had increased by another 5% from 2002.

During the same period, the number of Angora goat farms declined by over 63% from 1997 to 2002. The number of farms selling Angora goats also declined with 61% fewer Angora goats being sold. Sample data indicated that the number of Angora goats declined another 29% from 2002 to 2007. Although there was a drastic reduction in Angora goat numbers and sales, the increase in the total goat population in the U.S. can be attributed partially to a small increase in the numbers of dairy goats, and a major increase in the numbers of meat goats.

The majority of the meat goats are produced in the South and the Southeast regions of the U.S. In contrast, the states producing dairy goats are located in the Midwest and West, with a small number in the Northeast. The majority of the Angora goats are produced in the South and the Southwest.

Texas leads the nation with the largest number of goats: 44% of the meat goats, more than 70% of the Angora goats, and more than 8% of the dairy goats (Table 1.10).

According to the USDA-NASS 2007 data, Missouri and Florida had the greatest percentage increase in the number

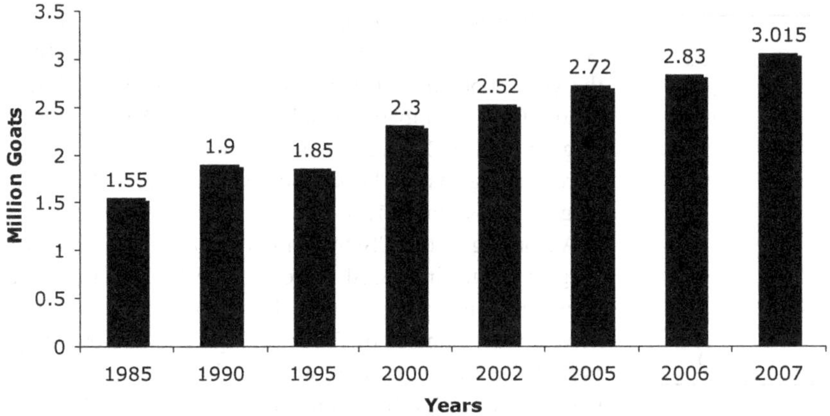


**Figure 1.2** Goat population changes in the U.S. (1985–2007).

**Table 1.9** Changes in goat farming in the U.S. (1997–2007*).

| Category | 1997 | 2002 | 2007 | % of total | % of 1997 |
|---|---|---|---|---|---|
| **All goats** | | | | | |
| Farms | 76,543 | 91,462 | | | 119.5 |
| Number of goats | 2,251,613 | 2,530,466 | 3,015,000 | 100 | 134.0* |
| Farms selling goats | 29,937 | 43,495 | | | 145.3 |
| Number of goats sold | 843,773 | 1,314,310 | | | 155.8 |
| **Meat goats** | | | | | |
| Farms | 63,422 | 74,980 | | | 118.2 |
| Number of goats | 1,231,762 | 1,938,924 | 2,500,000 | 83 | 203.0* |
| Farms selling goats | 24,539 | 36,403 | | | 148.3 |
| Number of goats sold | 532,792 | 1,109,619 | | | 208.3 |
| **Dairy goats** | | | | | |
| Number of farms | 15,451 | 22,389 | | | 145 |
| Number of goats | 190,588 | 290,789 | 305,000 | 10 | 160.0* |
| Farms selling goats | 5,163 | 8,850 | | | 171.4 |
| Number of goats sold | 72,307 | 113,654 | | | 157.2 |
| **Angora goats** | | | | | |
| Number of farms | 5,485 | 5,075 | | | 92.5 |
| Number of goats | 829,263 | 300,753 | 210,000 | 7 | 28.7* |
| Farms selling goats | 1,883 | 1,662 | | | 88.3 |
| Number of goats sold | 238,674 | 91,037 | | | 38.1 |

*Reflects the 2007 USDA-NASS sample data.

of meat goats (213% and 197%, respectively); Iowa experienced an increase in the number of dairy goats (220%); and although the number of Angora goats continued to decline, Colorado, Nebraska, Minnesota, and Missouri experienced more than an 80% increase in the number of Angora goats since 2002.

### Import and Export of Goat Meat

The United States was a net exporter of goat meat before 1990. Exports ceased in 1994 due to an increased domestic demand. This shift reflects increased goat meat consumption nationally. In 2007, the U.S. imported 10,166 MT, nearly 23 million pounds (lb) of goat meat valued at USD

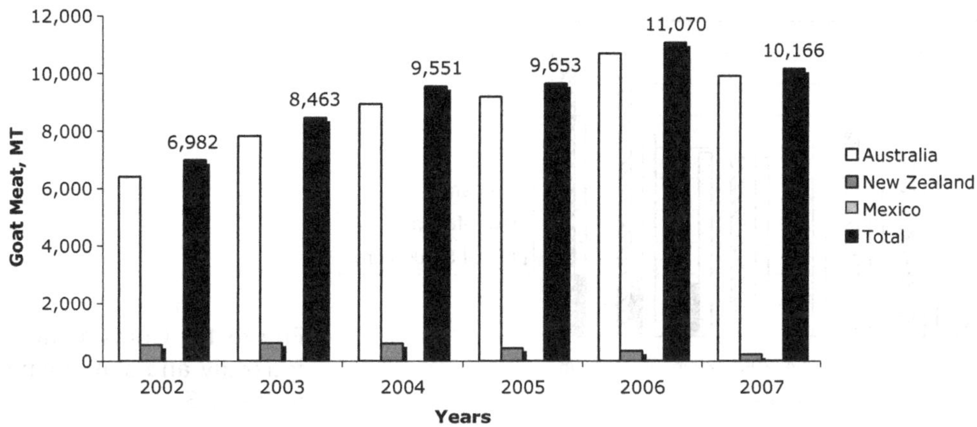


**Figure 1.3** Changes in the amount of goat meat imported to the U.S. Source: USDA-FAS (www.fas.usda.gov/ustrade)

**Table 1.10** The top three states for the numbers of goats in the U.S. (2007*).

| Category | Goat numbers | % of total | % of 2002 |
|---|---|---|---|
| **Meat goats** | 2,500,000 | 100 | 129 |
| Texas | 1,090,000 | 43.6 | 116 |
| Tennessee | 118,000 | 4.7 | 110 |
| California | 100,000 | 4.0 | 163 |
| Georgia | 100,000 | 4.0 | 152 |
| **Dairy goats** | 305,000 | 100 | 105 |
| Wisconsin | 33,000 | 10.8 | 127 |
| California | 30,000 | 9.8 | 80 |
| Texas | 25,000 | 8.2 | 111 |
| **Angora goats** | 210,000 | 100 | 70 |
| Texas | 150,000 | 71.4 | 65 |
| Arizona | 18,000 | 8.6 | 65 |
| New Mexico | 7,500 | 3.6 | 106 |

*Reflects the 2007 USDA-NASS sample data.

$37 million, up 146% from 6,982 MT in 2002; and its price per kilogram (kg) was up 173% from $2.11 in 2002. Based on an average carcass weight of 14.7 kg, an estimated 718,000 goat carcasses were imported. This number of goats imported is a potentially viable value-added enterprise opportunity for the goat industry and increased diversification of small farms in the U.S.

The main exporters of goat meat to the U.S. are Australia and New Zealand with more than 90% of the contribution coming from Australia. Slightly more than 18 MT of goat meat was imported from Mexico in 2007 (Figure 1.3).

### Goats Slaughtered

The number of goats slaughtered at the state and federally inspected plants in 2007 was nearly 830,000, up 3.6-fold from 1999. Meat goat slaughter numbers have increased steadily since 1999 (Figure 1.4). The meat goat industry in the U.S. is in its infancy; therefore, many on-farm slaughters are probably not reported. One can conjecture that for every goat reported, there are at least four (Solaiman, 2007) that are not reported. Also 2006 was the first year that the number of goats slaughtered in state-inspected plants was reported.

### Goat Meat Consumption

Goat meat consumption in the U.S. can be estimated based on imported goat meat and slaughter data at nearly 1.5 million head of goats being consumed annually (2007). This represents an increase of 150% from 2002 and a more than 320% increase from 1999. Total goats imported were estimated from the total weight of goat meat imported and an average carcass weight of 14.7 kg (29.5 kg of live weight is an average slaughter weight reported by the USDA). Goat consumption in the U.S. has increased steadily since 1999 and likely will continue to increase due to current trends in population growth that promote meat goat production (Figure 1.5). The sharp increase in the number of slaughtered goats may reflect the revision to report data from the state-inspected plants that was initiated in 2006.

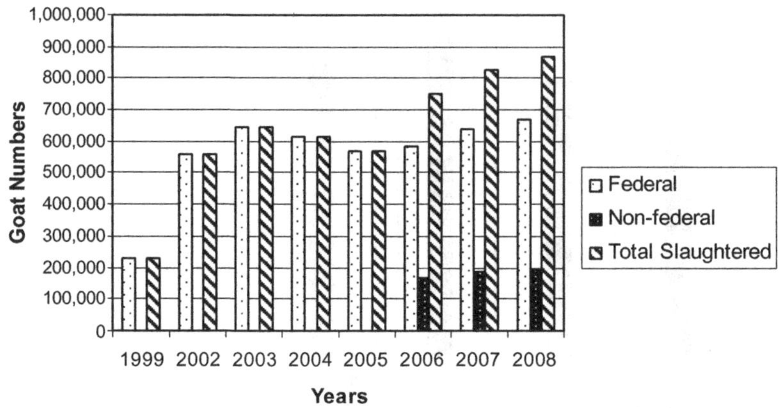


**Figure 1.4** Goats slaughtered in federally and state inspected plants in the U.S.

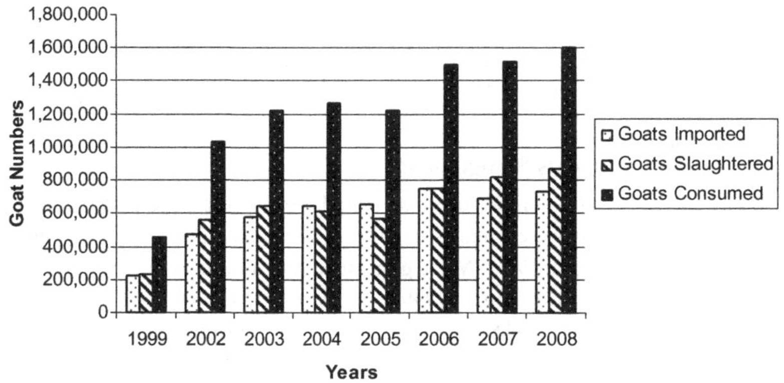


**Figure 1.5** Changes in number of goats consumed in the U.S.

### Factors Affecting Goat Meat Consumption in the U.S.

Population demographic changes have contributed to the increased interest in goat meat production and consumption in the U.S. According to the U.S. 2000 population census, the foreign-born population in the U.S. increased by 57% since 1990, from 19.8 million to 31.1 million, and continues to increase. In 2000, 51.7% of the foreign-born population came from Latin America and 26.4% from Asia. The Hispanic population is rising rapidly and will reach over 100 million or 25% of the population by the year 2050. This group of immigrants has a strong preference for goat meat, and increased consumption of goat meat will stimulate growth in this sector of agriculture.

The population of different cultures in the U.S. has increased. In 2000, over 1 million Buddhists, 1 million Muslims, over 10 million Asians, and over 35 million Hispanics lived in the U.S. (U.S. Census, 2000). Marked increases in these populations from 1990 to 2000 created an opportunity for U.S. agriculture to produce new products to meet food preferences of this ever-increasing population.

Goat cheese also has gained in popularity in recent years. Goats provide healthy meat and milk that fit the designer diets of health-conscious Americans.

## PRODUCTION ENTERPRISES IN THE U.S.

Three major goat enterprises exist in the U.S. today: meat goat, dairy goat, and Angora or fiber goat farms. Meat goat farms predominate with purebred farms, commercial farms, and independent dealers all contributing. These can operate on a local or national basis. Dairy goat and Angora goat production are more established industries than the newly developing meat goat industry.

### Purebred Production Farms

Purebred farms produce a specific goat breed and help to preserve the breed quality and integrity. There are several purebred meat-type, dairy-type, and fiber-type goats in the U.S. Purebred goats usually are more expensive than commercial goats, and they are purchased by producers to improve the quality of their herd. Purebred goat producers also have a potential source of income in the popular 4-H market. As goats gain in popularity, the number of 4-H goat projects, plus the interest in 4-H competition among youngsters, will increase and this will increase the market demand for purebred goats.

### Commercial Goat Production Farms

The majority of meat, dairy, or fiber goat farms in the U.S. produce commercial goats mainly for domestic use. Commercial Angora goat farms usually have a larger number of goats than meat and dairy goat farms, with Angora goats often being raised under range condition. Most of these producers/ranchers purchase their goats from other farms or at auctions, and some produce their own kid crop.

The type of goats sold, their age, weight, sex, and color are important factors for different ethnic groups and must be considered when selling meat goats. Selling brood stock to other producers is another marketing outlet; however, superior stock is required to produce superior-grade goats that will be more expensive than commercial goats. Most commercial farms use crossbred goats and grade animals.

### Multi-enterprise Goat Farming

On multi-enterprise farms, goats that produce meat, fiber, or milk also will serve as a holistic tool to control brush and restore vegetation. Gross margins from combination enterprises with two (meat and fiber), three (meat, fiber, and milk), four (meat, fiber, milk, and brush control), or any combination requires careful economic evaluation. With careful planning, the potential exists to manage a dual purpose or multi-purpose crossbreeding enterprise that produces for fiber, milk, meat, or vegetation management markets. Goat farmers also can produce vegetable crops to further diversify, increase their cash flow, and spread the risk of production.

### Organic Goat Production

Raising goats organically for meat and milk is gaining popularity in the U.S. and can add value and revenue to the farm. Organically raised animals are in high demand; however, according to the USDA Organic Program, livestock produced organically must be managed according to specific USDA organic standards from the last third of gestation. Livestock feeds must be 100% organic. Although feeds such as vitamins and mineral supplements are acceptable, the use of hormones and antibiotics is not permitted in an organically raised animal production unit.

## GOAT PRODUCTION IN EUROPE

The variation in climate increases the diversity of goat production systems in the European countries. Housing and indoor feeding often are required for the cold months, and supplemental feed normally is required. Upland and mountain grazing under marginal conditions in some European countries may require several hectares per animal to sustain production while lowland systems with intensive grazing will have a carrying capacity of several goats per hectare with a small herd size. Many European countries have emphasized development of the dairy industry with superior dairy goat breeds. Goat production systems in Europe are in close harmony with natural conditions in each locality and thus fulfill most criteria of sustainable development in agriculture. More than 75% of goat milk is produced in the Mediterranean region, and most of the goat products (meat, milk and hide) are utilized there. The development from subsistence farming to the marketing of products has led to more specialized production systems. In the goat sector, milk is the main product with meat being secondary in most cases followed by skins and hair. Milk from goats often is regarded as a niche product and is less visible in an open mass-market situation. Goat meat consumption per capita has been fairly stable or declined slightly over the last 40 years (EAAP, 2003). However, ethnic changes in the population may lead to an increased demand for goat meat in countries with large numbers of immigrants.

## GLOBAL GOAT PRODUCTION SYSTEMS

Goat production systems have evolved from the broader livestock systems and vary among the AEZs and types of farming systems (Spedding, 1975; Ruthenburg, 1980). The biophysical environment and the available production resources are especially important in the genesis and intensification of the production systems (Devendra, 2007a). The biophysical environment will determine to a large degree the types of crops that can be grown and the resulting feed resource available for animal production.

In developing countries, the main livestock systems can be divided into four types: landless, mixed crop based, agropastoralist, and range-based. Each of these systems exists in Asia, sub-Saharan Africa (SSA), Central Asia (CA), West Asia and North Africa (WANA), and Latin America and the Caribbean (LAC). While the other systems are common in most of the regions referred to, the term agropastoralism is more specific to Africa and is used to describe agropastoral systems or mixed crop-livestock systems with extensive grazing where households are nomadic and their location and livelihood depend on the system. Agropastoralism has evolved from pastoralism that in the future likely will shift to more intensive mixed crop-animal systems. Range-based systems are essentially extensive systems characterized by rainfall of less than 150 mm/year, chronic droughts, and minimal feed resources. These systems usually are associated with extensive nomadic and transhumant systems. In the Balochistan province of Pakistan for example, 93% of the available 35.7 million hectares (ha) of land is rangeland that supplies 60% of the feed requirements of sheep and goats. The average dry matter yield is about 0.5 MT/ha, and the carrying capacity is about 0.5–1.5 hectares per sheep equivalent. Within the prevailing livestock systems, four categories of goat production systems are identifiable (Devendra, 2007b):

- Rural landless systems
- Extensive systems
- Systems combining arable cropping (tethering, communal and arable grazing systems, and cut-and-carry feeding)
- Systems integrated with tree cropping

Rural landless production systems involve a variety of mitigating practices to overcome the constraints of harsh environments, namely high temperatures, inadequate water, droughts, and chronic feed shortages. These systems are extensive, and are associated with resource-poor nomads, transhumant or agricultural laborers and seasonal migrations with small ruminants, cattle, and camels (Devendra, 1999). Very common in the arid and semiarid regions, notably Pakistan and India, they also exist in the Hindu-Kush Himalayan region in South Asia. Movements follow annual cycles that are triggered by reduced feed and water supplies, and market opportunities. They are also a way of life for the poor. Two problems common to these systems are overgrazing and environmental degradation.

Among the production systems, the extensive systems are the most prevalent in many regions. Integrated systems involving tree crops generally are neglected and underestimated. The decreased availability of arable land in many areas and the need for more food from animals can encourage the development and intensification of this potentially important silvopastoral system. One example is the integration of goat production with production of oil palm (Devendra, 2004).

Ruminant production systems in the developing countries are unlikely to change in the foreseeable future (Mahadevan and Devendra, 1986; Devendra, 1989). However, intensification is likely to increase and a shift, especially from extensive, to systems combining arable cropping, will be induced by population growth. The principal aim should be improved feeding and nutrition, and maximum use of the available feed resources, notably crop residues, low quality roughages, and various leguminous forage as supplements.

Emerging issues associated with the future of these systems need to be addressed in an interdisciplinary fashion. Table 1.11 provides a summary of the broader livestock systems, priority production systems in ruminants, as well as some of the major issues across regions.

## CONSTRAINTS TO PRODUCTION AND THEIR RESOLUTION

Improved goat production requires an understanding of the types and extent of the constraints to improve performance and enhance development and productivity. Both genetic and nongenetic constraints are involved. The latter group is more widespread and includes such examples as nutrient deficiencies, diseases, and poor housing. Constraint analysis also helps to establish priorities for research and development, and in the formulation of strategic programs that are needs based and respond to farm situations.

One essential prerequisite for improved goat production is knowledge and understanding of prevailing production systems, their problems, and constraints that limit production. Systems perspectives and analysis enable one to address these issues, formulate approaches and strategies that can resolve the problems, improve the situation, and increase production. Often the research approach has tended to be top down, whereby perceived and unrealistic constraints are addressed to seek improvements. Failure to respond to needs results in the persistence of the real constraints that limit production. Even within rainfed environments in Asia for example, a systems approach can increase animal production (Devendra, 2000).

The diversity of crops and animals, traditional methods of farming, the role and contribution of animals, multiple crop-animal interactions, and numerous problems of farmers present complex issues that cannot be routinely resolved. Specific problems often cannot be resolved by a single discipline. This has been a major weakness of many research programs involved with natural resource management (NRM) of the past. Farming Systems Research (FSR) and systems methodologies can provide new and very important pathways to promote improvements that can expand meat and dairy production. More importantly, innovative, on-farm linkages that involve farmers,

**Table 1.11** Summary of livestock systems, priority production systems, and major issues across regions.

| Type of livestock systems | Priority production system | Regions: Asia | SSA | CA | WANA | LAC | Major issues |
|---|---|---|---|---|---|---|---|
| 1. Landless | • Peri-urban / urban dairy production | / | / | | / | / | • Surface water contamination |
| | • Peri-urban / urban poultry and pig production | / | / | / | / | / | • Zoonosis<br>• Waste disposal |
| | • Feedlot (cattle or small ruminants) | / | / | / | / | / | • Nutrient flows |
| | • Goat and sheep production | / | / | / | / | / | • Overgrazing |
| 2. Crop based mixed | • Integrated systems with | / | / | / | / | / | • Food-feed systems |
| | • Annual crops (ruminants) and non-ruminants plus fish) integrated systems with perennial crops (ruminants) | / | / | | | / | • All-year-round feeding systems<br>• Nutrient flows / soil fertility |
| | • Beef and dairy production | / | / | / | / | / | • Productivity enhancement |
| | • Goat and sheep production | / | / | / | / | / | • Intensification and specialization<br>• Overgrazing |
| 3. Agro-pastoralist | • Cattle | | / | / | / | | • Feed supplies / drought strategies |
| | • Goat and sheep production | | / | / | / | | • Property regimes<br>• Overgrazing<br>• Trypanosomiasis |
| 4. Range based | • Sheep and goat production | / | / | / | / | / | • Drought strategies<br>• Overgrazing<br>• Property regimes<br>• Marketing |

**Notes:**

(i) SSA - Sub-Saharan Africa, CA-Central Asia, WANA - West Asia and North Africa, LAC - Latin America and the Caribbean

(ii) / indicates that both the production systems and animal species are the most important within the region.

(iii) Major issues *inter alia* are those that currently merit R and D attention. Across regions, the issues are broadly similar as is the case with dairying. Dairy production includes buffaloes and cattle especially in Asia.

Source: Devendra, Morton, and Rischkovsky, 2005.

researchers, community, extension agents, nongovernment organizations, and development agents are necessary. In this context, Eponou (1993) has suggested several elements to integrate agricultural research and technology transfer. These include shared goals between researchers and farmers, synergy, strong leadership for the whole, decision making by consensus, accountability to clients and policy makers, and considering farmers as partners.

### Advantages of Listening to Farmers

Associated with constraint analysis are the very needs and aspirations of individual farmers. Listening to farmers is central for development. Because participatory approaches have many advantages, it is therefore important to invest time and resources to address this issue. Research is meaningless if it does not drive development, and development programs are futile if they do not involve farmers as partners. Major advantages from listening to farmers (Devendra, 2006) include:

- They know the biophysical environment.
- They understand prevailing farming systems.
- They have full understanding of the major constraints and problems.
- Research and development issues are identified.
- They are the target beneficiaries of improved technologies.
- They have deep knowledge of local customs and indigenous and traditional systems.
- Farmers are the agents of change and promotion of wide adoption.
- They have all the local contacts.
- They know current patterns of marketing.
- Farmers are the target beneficiaries of sustainable livelihoods.
- This enables understanding of farmers' perceptions and aspirations about household stability, improved livelihoods, and a better tomorrow.

### Formulation of Participatory Research and Development Programs

Farming systems in regions throughout the world are characterized by a diversity of crops and animals, traditional methods of farming, various contributions of animals, and multiple crop-animal interactions. The numerous problems of farmers present complex issues that cannot be resolved easily. Farming System Research is important to address the needs of farmers.

Farming System Research owes its origin to cropping systems research and methodologies that were developed in Asia and Latin America. The key features of FSR are:

- It seeks to provide a clear understanding of the farming systems and practices.
- It is needs based.
- It has systematic methodology.
- It is multidisciplinary.
- It involves the participation of farmers, researchers, community, extension agents, and development agents.
- It is on-farm.
- It identifies farmers as partners.
- It promotes efficiency in NRM.
- It enables definition of an agenda for promoting development and progress.

The methodology for FSR follows several distinct steps, including:

1. Site selection
2. Site description and characterization (diagnosis)
3. Planning of on-farm research
4. On-farm testing and validation of alternatives
5. Diffusion of results
6. Impact assessment

Figure 1.6 illustrates the steps involved in Farming System Research. Due to the multidisciplinary nature of the activities, other methodologies also are involved.

Table 1.12 provides an example of constraint analysis. For convenience, constraints are identified under the subheadings of environment, socioeconomics, farming

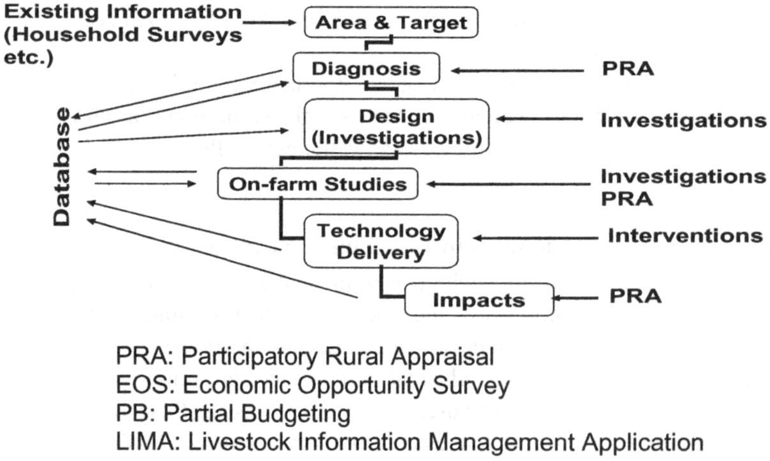


**Figure 1.6** Methodology for farming systems research.

**Table 1.12** Nature and extent of constraints to goat production in typical mixed small-farm systems in humid Southeast Asia.

| Environment | Socioeconomics | Farming systems | Goats | Externalities |
|---|---|---|---|---|
| Poor soil fertility | Low income | Poor access to technology | Inadequate numbers | Poor infrastructure |
| Extreme periods of precipitation (both wet and dry) | Limited assets | Poor crop yields | High cost | Poor marketing facilities |
| Soil erosion | No access to capital | Limited cropping options | Poor performance (e.g., meat and milk yields) | Price fluctuations |
| Pollution | No access to market outlets | Underuse of fodder trees | Poor nutrition | Policies |
| Contamination of surface water | Limited access to knowledge | Inefficient use of production resources | Poor management | No or weak farmer organizations |
| Land subdivision and conversion | Economic isolation | Inadequate feed supply | Diseases (e.g., Nematodes) | No cooperatives |
| Geographic isolation | No access to off-farm labor | Droughts | No extension services | Poor linkages with institutions |
| Soil salinization | | Rising production costs | No vaccines and drugs | Poor rural-urban linkages |

Source: Devendra, 2006.

system, animals, and externalities. Because the list of constraints can be numerous, it is most essential toprioritize these issues to focus on the more important constraints. Analysis and proposed interventions must involve the participating farmers to implement the on-farm activities.

The major challenge for increased resource use by small farm systems in rainfed areas concerns collective and holistic approaches that can shift the level of productivity from a traditional low-input system to an intermediate and market-oriented system. Switching from a sustainable system to an intermediate level must be consistent with agricultural growth, poverty reduction, food security, environmental sustainability, and self-reliance (Devendra, 2000). Figure 1.7 illustrates the different productivity levels and possible shifts. The figure demonstrates that, with various improvements, it is possible to shift productivity from traditional systems to market-oriented systems with maximum productivity.

Pro-poor strategies and social and effective development policies are needed to sustain and increase the contribution from these farms to help feed hungry nations and continents (Devendra, 2007c). Additionally, one must enable an economic environment to spur agricultural development. Hunger is more prevalent where dependence on agriculture and goats is greatest. Every effort must be made to maximize food production of animal origin to provide for sustained nutritional security and improve the socioeconomic well-being of the poor in the future.

## CHALLENGES FOR THE FUTURE

To increase productivity and the contribution of goats, it is urgent to accelerate productivity in tandem with productive potential. Emphasis must be placed on specific areas of research and development, on needs-based delivery of appropriate technologies, and on large-scale development at the farm level. Such development needs to address the following aspects *inter alia:*

- The concept of production-to-consumption that links the owner, producer and consumer
- The importance of needs-based research, in which research is focused on resolving farmers' problems and constraints
- Systems approaches and interdisciplinary efforts that consider whole-farm systems
- Formulation of a research and development agenda that addresses real needs and opportunities for national goat production

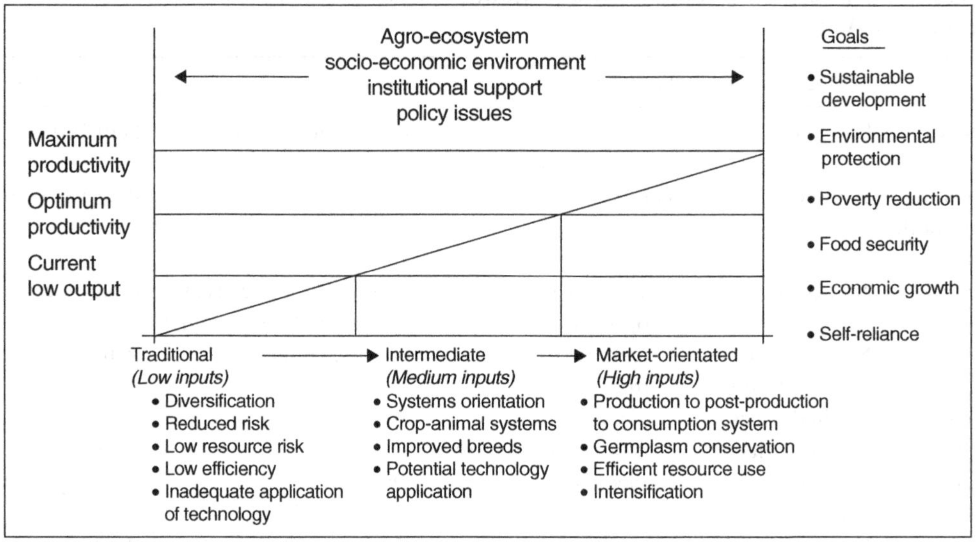


**Figure 1.7** Potential levels of productivity feasible from traditional, market-oriented system and the development of sustainable small farms (Devendra, 2000).

- Large-scale development of the breeds that enhance meat production
- Acceleration of the transfer of technologies on-farm; promotion of wide dissemination of information, and networking

The following chapters provide a comprehensive and up-to-date summation of the science of goat production. The descriptions involve both the developing and developed countries and agroecosystems and a global perspective.

## SUMMARY

The goat is one of the first livestock that was domesticated. About 10,000 years ago, the goat was domesticated by three of the oldest civilizations: the Nile in northeast Africa, the Tigris–Euphrates in west Asia, and the Indus in the Indian subcontinent. The bezoar of western Asia is the main progenitor of most domestic goats. The majority (98%) of the world population of goats is in developing countries where goats are raised under extensive conditions. Goats provide meat, milk, and skins and also contribute to socioeconomic, cultural, and recreational needs of mankind. In developed countries, goat meat, milk, and fiber also have increased in popularity. In the last decade, the demand for goat meat and cheese has increased in the U.S. because population demographics have shifted toward those that traditionally consume goat products. The current domestic supply in the U.S. cannot meet the demand, so the U.S. is the largest importer of goat meat in the world.

Ruminant production systems in the developing countries are not likely to change in the foreseeable future despite some increase in intensification of production and a gradual shift from extensive production to systems that involve cropping of arable land. This change will be driven by population growth. A variety of constraints to goat production exist. Constraints must be identified clearly so that research can provide solutions to improve animal performance, and enhance development and productivity. Use of Farming Systems Research perspectives provide ways to address the issues and to formulate approaches and strategies to resolve the problems and increase animal production to help alleviate world hunger.

## REFERENCES

Boyazoglu, J., I. Hatziminaoglu, and P. Morand-Fehr. 2005. The role of the goat in society: past, present and perspectives for the future. Small Rumin. Res. 60: 13–24.

Cooper, J.C. 1992. Mythological and symbolic significance of goats. In: Symbolic and Mythological Animals. The Aquarian Press, Wellingborough, USA.

Devendra, C. 1987. Herbivores in the arid and wet tropics. In: The Nutrition of Herbivores. Proceedings 2nd International Symposium on the Nutrition of Herbivores. (Eds. J.B.

Hacker and J.H. Ternouth), Academic Press, New South Wales, Australia, pp. 23–46.

Devendra, C. 1989. Ruminant production systems in the developing countries: resource utilisation. In: Feeding Strategies for Improved Productivity of Ruminant Livestock in Developing Countries. IAEA, Vienna, Austria, pp. 5–30.

Devendra, C. 1998. Improvement of small ruminant production systems in rainfed agro-ecological zones of Asia. Annals of Arid Zone, 37:215–232.

Devendra, C. 1999. Goats: Challenges for increased productivity and improved livelihoods. Outlk. on Agri. 28: 215–226.

Devendra, C. 2000. Animal production and rainfed agriculture in Asia: potential opportunities for productivity enhancement. Outlk. on Agric. 29: 161–175.

Devendra, C. 2004. Integrated tree crops–ruminant systems: potential importance of the oil palm. Outlk. on Agric. 33: 157–166.

Devendra, C. 2006. Listening to farmers: participatory approaches for developing goat production. Symp. on Agric. Biotechnology and Environment, Daegu, Korea, pp. 1–19.

Devendra, C. 2007a. Goats: biology, production and development in Asia. Academy of Sciences Malaysia, Kuala Lumpur, Malaysia, 246 pp.

Devendra, C. 2007b. Perspectives on animal production systems in Asia. Livestock Science. 106: 1–18.

Devendra, C. 2007c. Small farm systems to feed hungry Asia. Outlk. on Agric. 36: 7–20.

Devendra, C. and M. Burns. 1983. Goat Production in the Tropics (2nd Revised Ed.). Technical Communication Bureaux of Animal Breeding and Genetics, Commonwealth Agricultural Bureaux, England, 183 pp.

Devendra, C. and K. Nozawa. 1976. Goats in South East Asia—their status and production. Z. Tierz., Zuchtbiol. 93: 101–120.

Devendra, C., J.F. Morton, and B. Rischkovsky. 2005. Livestock systems. In: Livestock and Wealth Creation. Improving the husbandry of animals kept by resource-poor people in developing countries (Eds. E. Owen, A. Kitalyi, N. Jayasuria, and T. Smith). Nottingham, U.K.: Nottingham University Press, pp. 29–52.

EAAP Publication No. 108. 2003. After BSE—A Future for the European livestock Sector. Edited by E.P. Cunningham. Wageningen Academic Publishers, Netherlands, 104 pp. (ISBN 90-7699–823X).

Ellerman, J.R. and T.C.S Morrison-Scott. 1951. Checklist of Palaearctic and Indian mammals, 1758–1946. British Museum, London, U.K., 810 pp.

Eponou, T. 1993. Integrating agricultural research and technology transfer. Public Admin. and Dev., Special issue on Managing Agricultural Research, 13: 307–318.

FAOSTAT. 2007. Food and Agriculture Organization of the United Nations. Rome, Italy.

FAOSTAT. 2005. Food and Agriculture Organization of the United Nations. Rome, Italy.

Haenlein, G.F.W. 2004. Goat milk in human nutrition. Small Ruminant Res, 51: 155–163.

Mahadevan, P. and C. Devendra. 1986. Present and projected ruminant production systems of South East Asia and the South Pacific. Proceedings Forages in South East Asia and the Pacific, ACIAR Proceedings. No. 12: 1–6.

Ruthenberg, H. 1980. Farming systems in the tropics (3rd Ed.). Clarendon Press: Oxford, U.K. 131 pp.

Solaiman, S.G. 2007. Future Outlook of Meat Goat Industry for the U.S. Small Farms. Proc. 4th National Small Farm Conference, Greensborough, NC.

Spedding, C.R.W. 1975. The biology of agricultural systems. Academic Press, London, U.K., 261 pp.

U.S. Census. 2000. U.S. Census Bureau (www.census.gov).

USDA Census. 2002. Census of Agriculture (www.nass.usda.gov/census/census02/).

USDA-FAS. (www.fas.usda.gov/ustrade).

USDA-NASS. National Agriculture Statistic Service. (http://usda.mannlib.cornell.edu/reports/nass/livestock/pls-bban/lsan0305.pdf).

Zeuner, F.E. 1963. A History of Domesticated Animals. Harper and Row, New York, U.S.A., 560 pp.

# 2
# Goat Breeds

*N.K. Gurung, PhD, PAS and S.G. Solaiman, PhD, PAS*

**KEY TERMS**

Herdbook—records and pedigree of a recognized breed of any livestock species.

Breed—a group of animals within a species descended from a common ancestor (similar genotypes) that visibly are similar in most physical characteristics (similar phenotypes).

Dairy breed—breed of goat selected for high milk production, dairy characteristics, large body capacity, and a superior mammary system.

Feral goats—domesticated animals that have been in the wild for over 100 years with no introduction of new genetics.

Meat breed—breed of goat selected for faster gain, muscling, and carcass characteristics.

Fiber breed—breed of goat selected for fiber coat quality.

Companion breed—a group of goats with a mild temperament often used for pets.

Multi-purpose breed—a group of goats that is valued for its combination of traits (milk, meat, and fiber production).

Composite breed—a crossbreed developed from two or more existing breeds that outperform the parent purebreds.

Transborder breeds—breeds that cross the national borders.

**OBJECTIVES**

By completing this chapter, the reader will acquire knowledge on:

- Breed definition
- Morphological characteristics of different breeds
- Breeds commonly used for meat production
- Breeds commonly used for milk production
- Breeds commonly used for fiber production
- Breeds commonly used for pets

## INTRODUCTION

More than 1,153 goat breeds are listed in the Food and Agriculture Organization (FAO) of the United Nations' inventory of domestic animal diversity database (FAO, 2009). According to the FAO (2007), more than 850 million goats exist in the world with more than 95% found in the developing countries. Of this total population, Asia has the highest breed share at 65.3%, followed by Africa with 29.2%, and Central America with 1.3% (Oliver et al., 2005). Only 60% of the breeds are found in the developing countries (Scherf, 2000). Europe is broadest in goat genetic resource (33%) but only 4% of the world's goat population (Galal, 2005). According to Oliver et al. (2005), approximately 31% of the goats are dairy/milk goats in the developed countries compared with only 19% for the developing world. Therefore, worldwide, most goats are valued primarily for their meat.

## BREED AND NUMBERS

In the developing countries, goats are seldom defined in the traditional sense and generally breed societies, breed standards, or herdbooks are lacking. According to the new FAO (2009) database, there are 1,153 goat breeds in the world. However, the World Dictionary of Livestock Breeds (Mason, 1996) lists 565 goat breeds plus 76 varieties (subbreeds) but only 68 herdbooks and/or established breed societies. Such registries are found primarily in Europe and North America. Of 68 herdbooks or breed societies listed, 21 are for the descendents of original breeds. For example, 7 breed societies/herdbooks are listed for the Saanen breed and its descendents in different countries. In developing countries, efforts to establish the purity of breeds or improve existing breeds generally are lacking. Therefore, the majority of goat breeds listed in the FAO database found in the developing countries are not known or recognized outside of their country of origin.

Goat breeds are usually classified by their geographical origin (Epstein, 1969; 1971; 1977), their region and productivity (Devendra and Burns, 1983), or their height at the withers (Devendra, 2007). Some classifications include the shape and size of the horns in conjunction with production and geographical considerations (Mason, 1981) while other classification methods include the primary product of goats and their distribution (The Council of Science and Industrial Research [CSIR], 1970), their morphological characteristics (in India; Acharya, 1982), and hair and coat characteristics (in Pakistan; Hasnain, 1985). Based on these different classification methods, the number of goat breeds in the world has been reported to be 562 by Devendra (2007), 136 by Gall (1996), 565 by Mason (1996), and 1,153 by FAO (2009).

Within developed countries, the majority of goat breeds are classified according to their products. Duplications occur when goats are used for multiple purposes such as meat, milk, fiber, or skin; however, end use has been adopted for goat breed classification in this chapter. Some breeds are mentioned according to their end use, alphabetically under countries of origin for simplification.

## SELECTION OF THE DESIRED BREED OR TYPE

Ideally, a breed should be selected according to its purpose or use, the environmental conditions, and adaptability of the breed (Gipson, 1996). No single breed has all the characteristics desired and is productive under all environmental conditions. Indigenous breeds are adapted to local environments but may not excel in performance. Crossbreeding of local (indigenous) breeds with improved breeds often produces crossbred offspring that have faster gains, higher milk yields, and greater adaptability to different regions (Solaiman, 2007). Hybrids or composite breeds can exploit hybrid vigor. Over 80 composite goat breed populations have been developed in 37 countries in the world (Mason, 1996). Listed below are favorable traits or preferable characteristics usually employed when developing or selecting a breed.

### Adaptability

All goats, regardless of their breed, should readily adapt to the environment where they will grow and perform. Spanish goats raised in hot, dry climates such as west Texas may not adapt favorably to either the humidity of the southeastern U.S. or the irrigated pastures of the western U.S. Improved breeds of goats that are produced under various environmental conditions, such as Boer goats from South Africa or Kikos from New Zealand, may perform differently when produced under environmental conditions that differ from their area of origin. Crossbreeding and selection can enhance environmental adaptability.

### Reproductive Efficiency

Capacity to reproduce is one of the most important production traits for meat, dairy, and fiber goats. High rates of conception, kidding, and weaning are all important. Females (does) that produce more kids per pregnancy are more productive even though kids may have lower birth weights, slower growth, and lower weaning weights. Because goats usually are not managed as individuals, total herd reproductive performance is important. Goat breeds from temperate regions and dairy goats tend to breed only seasonally with the breeding season being from July/August to November/December in the Northern Hemisphere. In contrast, goat breeds from the tropics or meat goats often kid year-round. This trait is desired economically. However, tropical breeds may not have desired meat characteristics and the larger carcasses preferred for meat production.

### Growth Performance

Kid performance usually is evaluated in two segments: average daily gain (ADG) prior to weaning (pre-weaning ADG) and after weaning (post-weaning ADG). For goats sold at weaning (either for Cabrito or meat stock), pre-weaning performance is important and depends on both the kid's ability to grow rapidly and its dam's mothering ability. For kids sold as yearlings or for milk production, both pre-

and post-weaning ADG are important selection criteria. Where resources are available, the choice for larger body size with more capacity to consume feed is desired for milk production. The choice for the size of carcass differs among cultures. In some countries, small carcasses are valued whereas larger, older carcasses are preferred in other cultures.

### Carcass Characteristics

The proportion of muscle to bone is one major criterion for selecting a meat-type goat. Although other traits can be measured on live animals, this trait is measured when animals are slaughtered. The higher the proportion of muscle to bone, the better the animal is rated for meat production. One carcass characteristic often measured is dressing percentage—the proportion of live weight found in the carcass. Dressing percentage for goats usually is 50% or less. Fat content of the carcass increases as the animal ages and develops, but goats usually have a lower proportion of fat in their carcass than other domestic animals. Lower fat content makes goat a desirable product for precooked diet meals. As the animal ages and fattens, the proportion of bone in the carcass decreases while the proportion of lean stays relatively constant.

### Productivity

Productivity of animals in the herd is their yield in terms of meat, milk, or fiber. With accurate records, an individual animals' productivity can be measured to aid in selection for maximum productivity. For milk production, productivity usually is measured as the amount of milk produced and efficiency of milk production. Both the amount and quality of fiber produced are important factors for selection of fiber goats.

When conception rate in the herd increases, litter size will increase, more kids are weaned per doe in a year, and the productivity of the herd increases. But if it takes longer to breed back the doe, or if the doe produces lower-weight kids with lower ADG, a longer time will be required for kids to achieve desirable weaning weights reducing productivity in terms of meat production. In a dairy or fiber-producing enterprise, animal productivity usually is measured in terms of milk or fiber production.

## CHARACTERISTICS OF MEAT GOAT BREEDS

Most of the world's goat population is used for meat production, but in western countries, goats may be used for milk, meat, or fiber production. The dwarf goat breeds of Asia and Africa are exclusively considered as meat animals (Gall, 1981). For instance, in a predominantly Hindu country like India, goat meat provides 47% of the total meat consumed in India because beef consumption is taboo (MLA, 2007). Goat meat often is considered to be healthier than mutton because goat meat is lower in fat and cholesterol, being more comparable to lean meat from chickens. Another unique characteristic of goat meat is the low extent of marbling within muscle tissues (Solaiman et al., 2006).

The major factors involved with selection of meat goat breeds are their growth rate, feed efficiency, total edible and salable percentage of the animal, dressing percentage, and carcass traits including flavor, tenderness, juiciness, and cooking loss. Adaptability to local environmental conditions and feed resources also is important. The type of goat meat preferred will vary with cultural traditions. In some cultures, goat age and sex are important while in other cultures they are not. Similarly, in some cultures, in addition to income from carcasses as the salable product, the economic and religious significance of specific noncarcass components are highly valued (Shrestha and Fahmy, 2005).

According to Devendra (1981), the types of meat produced and consumed in the tropics will include meat from kids (Category 1: usually 8–12 weeks old), meat from young goats (Category 2: 1–2 years old), and meat from old goats (Category 3: 2–6 years old). For the first category, breed selection should be based on pre-weaning growth rate, whereas for the second category, both pre- and post-weaning growth rates need to be considered. For the final category, less emphasis is placed on growth rate than on meat quality.

Leanness of carcass is another universally desired criterion because it reflects the quantity of salable components. For rapid improvement of performance of meat goat productivity, emphasis should be placed on crossbreeding and formation of composite breed populations (Shrestha and Fahmy, 2007). Obviously, emphasis on selection must also consider production levels and resource availability. The ideal type will be different for subsistence-based, market-based, and high input production systems. The most prominent meat goat breeds are listed in this chapter. Although most breeds lack specific standards, where they are available, those standards are listed.

### Black Bengal

Also known as Bengal, this dual-purpose breed is used for both meat and milk production in west Bengal, Bihar, and Orissa regions of northeastern India (Acharya, 1982). Its skin is used extensively for making high quality shoes in India. The Bengal is the predominant breed of Bangladesh. Known for its prolificacy and kidding twice a year with 60% twins and 20% triplets (Sengar, 1976), Bengal are small with adult goats weighing about 13 kg (29 pounds [lb]) with a wither height of 45 cm (18 inches [in.]). Primarily black, individuals may be brown, white, or grey. The breed has a short coat and ears and is bearded.

### Chappar

This multipurpose goat breed originated in the Sind and Las Bela, Baluchistan area of Pakistan and also is known as Jabli, Jablu, Kohistani, and Takru. This breed is used for several purposes including meat, cashmere, hair, and milk production. Its coat color can be black, white, or pied. The breed is of medium size with adult males weighing about 26 kg (57 lb) and adult females weighing 20 kg (44 lb) (Devendra, 2007). Wither height is 60–70 cm (24–28 in.). Daily milk yields average about 0.8–1.0 kg (1.8–2.2 lb).

### Chinese Meat Goat Breeds

Mason (1996) listed 18 meat goat breeds native to China whereas Jiang (1982, 1995) grouped Chinese meat goats into 11 classes, namely the Banjiao, Chengdu Ma, Duan Fuqing, Guizhou White, Haimen, Huai, Leighou, Longlin, White Goat, Matou, and Nanjiang Yellow breeds. Several of these breeds are described below.

#### Huai

This breed originated in the Henan and the Anhui provinces of China and also is known as Huanghuai. Subtypes include Fuyang, Huaipi, and Xuhai. Their color is usually white, and they are medium-sized goats. Adult males weigh 30 kg (66 lb) and adult females 26 kg (57 lb). They have horns and mature early with high prolificacy and rapid growth rate potential (Ying, 1986).

#### Matou

The Matou goat is one of the most prevalent of the Chinese indigenous breeds. According to Moaeen-ud-Din et al. (2008), this breed is extensively distributed in more than 10 provinces (i.e., Sanxi, Henan, and Jiangxi). The breed is very prolific with a kidding rate of 196%. The goat was developed in the border areas of China, in Hunan, Hubei, Sichuan, and Guizhou provinces. The adult male weighs 44 kg (97 lb) while the adult female weighs 34 kg (75 lb). This is a polled breed. Their coat color is white with coarse, medium-length hair. This breed is a dual-purpose breed valued for both its meat and its milk. The average dressing percent is 60%. Skins also are highly valued. The breed is known for its rapid growth rate, big build, and good meat quality.

#### Nanjiang Yellow

A Chinese meat goat breed native to Nanjiang County of the Sichuan province, this breed is well adapted to mountainous areas and is suitable for year-round grazing. The breed has a high rate of growth with high prolificacy potential and a kidding percentage of 198% (Ying, 1986). The adult males weigh 59 kg (130 lb), and females weigh 39 kg (86 lb).

#### South China

This breed resembles the Kambing Katjang (Epstein, 1969) except for having shorter legs. This breed originated in the Yunnan and Kwangtung provinces of China. These animals are relatively small in size with a concave profile and short prick ears. Both sexes have horns and beards. They generally are used for meat, but their skins are valued. The adult male weighs 30 kg (66 lb), and the adult female weighs about 25 kg (55 lb). Wither height is 50–55 cm (23–25 in.).

### Creole

Also known as Reunion Creole, this breed is widely distributed in the West Indies but has spread to 18 different Caribbean and South American countries being a prominent transborder breed (FAO, 2004). A meat breed, Creoles look similar to South American Criollo goats with short ears and horns with colors being black, brown, or pied. Adult males weigh 25 kg (55 lb), and females weigh about 20 kg (44 lb). Wither heights are 60 and 50 cm (24 and 20 in.) for adult males and females, respectively. The reported fecundity rate is 90% with 2.25% prolificacy and 78% viability to weaning (Alexandre et al., 1999).

### Criollo

Criollo goats are distributed across many Latin and South American countries, especially Mexico, Argentina, Bolivia, Peru, and Venezuela (Mason, 1980). They originated with the Spanish conquest and colonization (Lanari et al., 2003). This breed has many characteristics of European dairy breeds. Their color is variable, but the most common color is black. Horns are short. Their main use is as meat, but milk production also is valued. The average adult weighs 39 kg (86 lb). Very hardy, this breed is well adapted for grazing under harsh environmental conditions.

### Kambing Katjang or Pea

A very prevalent meat breed in Southeast Asia, this breed is native to Malaysia and Indonesia. It also is called Kambing, which means goat, and Katjang, which means beans. The breed resembles the goats found in Burma, Thailand, the Philippines, and Taiwan. Medium in size, adult males weigh about 25 kg (55 lb), and females weigh 20 kg (44 lb; Devendra, 2007). Wither height is 60–65 cm (24–26 in.) for males and 56 cm (22 in.) for females. Their hair is coarse, males have beards, and their color usually is black with some patches. It is a prolific breed with twinning being very common.

**Khurasani**

Found in the border regions of Afghanistan and Iran and also in Baluchistan in Pakistan, this dual-purpose breed is valued for both meat and milk. The breed is horned, their size is medium, and their color usually is black but they may be white or grey. Adult males weigh about 25 kg (55 lb), and females produce up to 1.3 kg (2.0 lb) of milk per day. They produce a small amount of cashmere.

**Kiko**

Kiko means "flesh" or "meat" in Maori. This breed is a composite developed in New Zealand from crossing feral goats exhibiting good meat-type qualities with Anglo-Nubian, Toggenburg, and Saanen bucks and is selected for survivability and growth rate in hill country environments (Mason, 1996). Established in 1986, the breed was imported to the U.S. in the 1990s. This breed is hardy, resistant to parasites, produces a lean carcass, and has good mothering ability. In the U.S., Kiko goats and their crosses can be registered by the International Kiko Goat Association either as New Zealand Fullblood Kiko (NZFK), American Premier Fullblood kiko (APFK), Percentage Kiko, American BoKi (cross between NZFK or APFK and Fullblood Boer), American MeatMaker (crosses between NZFK or APFK and American BoKi), and International MeatMaker (crosses between 15/16 Boer and American BoKi). Their color usually is white or creamy. Adult does weigh more than 48.6 kg (107 lb) with some bucks weighing nearly 100 kg (220 lb) (Taneja, 1982). They are very prolific and adapt well to harsh environments. See Figure 2.1.

**Figure 2.1** Kiko (Meat Goat). For color detail, see Appendix A.

**Nepalese Hill**

The predominant breed of Nepal, this breed is found spread across the Siwalik and Mahabharat hill regions of Nepal. The Southern Hill goats are found in the Siwalik range while the Northern Hill goats are spread in the Mahabharat range. Their main use is for meat, but occasionally they may be milked. They usually are brown or black with white markings, or white with black markings. They are named by their coat colors including Khairi, Kali, Seti, Singari, or Dhobini. The Northern Hill goats are heavier than the Southern Hill goats. The male Northern Hill goats weigh about 35–45 kg (77–99 lb), while the females weigh about 24–30 kg (53–66 lb). Wither heights are 60–70 cm (24–28 in.) for adult males while adult females are 50–65 cm (20–26 in.) (Epstein, 1977). Corresponding values for the Southern Hill goats are 12–16 kg (27–35 lb) and 50–55 cm (20–22 in.).

**Small East African**

A very prominent breed across Kenya, Uganda, and Tanzania (Devendra and Burns, 1983), this breed resembles the goats raised by farmers in Rwanda, Malawi, Zimbabwe, and Botswana. Their size is small with short hair and short erect ears, and they vary in color. They are used mainly for meat but also for their skins. They have tassels. They are early maturing, but the twinning rate is low. The types found in Uganda are very longhaired, and they are favored for skins and their long hair. Adult goats of this breed weigh about 30 kg (66 lb).

**Somali**

These goats have different names depending on the country. In Somalia, they are known as Abgal or Modugh; in Ethopia, they are called Ogaden or Benadir; while in Kenya they are Galla or Boran. The breed is mainly used for meat, but skins also are valued. Their coat color is primarily white, and their ears and horns are short. Males are horned, but females may be either horned or polled. The breed is primarily adapted to dry areas (Mason and Maule, 1960).

**South African Meat Goat Breeds**

BOER

The Boer goats were derived from the indigenous goats kept by Namaquo Hottentots and Southern Bantu tribes of

**Figure 2.2** Boer (Meat Goat). For color detail, see Appendix A.

**Figure 2.3** Kalahari Red (Meat Goat). For color detail, see Appendix A.

South Africa although some genes were derived from goats from India and Europe (Casey and Niekerk, 1988). The name Boer means "farmer" in Afrikaans. The South African Boer goats have a white body with a red head and a blaze. This breed is suited for mixed grazing with cattle due to its browsing skills with little impact on grass cover. The adult males weigh between 110 and 135 kg (240 and 300 lb), and females weigh about 90–100 kg (200–225 lb), with an average daily gain of 0.2 kg (0.44 lb) in feedlots and 0.14–0.18 kg (0.3–0.4 lb) on a standard farm.

The Boer goat is a horned breed with lopped ears. The breeding season is extended with does often producing three kid crops every two years. A kidding rate of 200% and weaning rate of 160% is common for this breed. The breed is very prolific with potential for high growth rate and excellent carcass traits.

The Boer goats are distributed widely throughout the world being used to upgrade indigenous goats and improve meat production. They are found in more than 48 countries (FAO, 2004). This breed has been registered in herdbooks in Germany since 1980 with Boer breed societies being recognized in many countries. See Figure 2.2.

#### Kalahari Red

These goats are considered as feral goats of South Africa with the breed being a result of successful crossbreeding programs started in 1990 between feral and African breeds including the Red Boer, the African Boer, and the Nubians. Two improved lines of the Kalahari Red breed exist in South Africa. One line developed exclusively from the red head Boer goat while the other line was developed from unimproved indigenous breeds. Kalahari Red breeders claim that this breed is tougher and more robust than red-headed Boer goats.

This breed is well suited to semidesert and arid conditions because of its hardiness and color that serve as a camouflage from predators. This breed thrives in very harsh conditions. Kalahari Red is very productive, has good mothering abilities, strong teeth, excellent growth rates, large frames, immense goat meat conformation, and a carcass with uniform color. It was imported by Australia in 2000 and crossed with their feral goats. Demand for this breed is increasing in other countries and is prized by North and South American goat breeders (Campbell, 2003). See Figure 2.3.

#### Savanna

Found throughout the African continent, this breed is raised by pastoralists for both milk and meat (Shrestha and Fahmy, 2007). This breed also was developed in South Africa from the crossbreeding of indigenous goats with white Boer goats. Its white coat is a dominant trait over other colors in goats. Demand for these white goats, despite their higher price, has been high for religious ceremonies, for celebration of the birth of a son, and based on the belief that white goats have superior carcass qualities. This breed has high productivity, high muscular development, and strong bones, legs, and hooves. Also, natural selection has resulted in bare skin, horns, and hooves that are totally black and well pigmented (Campbell, 2003). See Figure 2.4.

**Figure 2.4** Savanna Buck (Meat Goat). For color detail, see Appendix A.

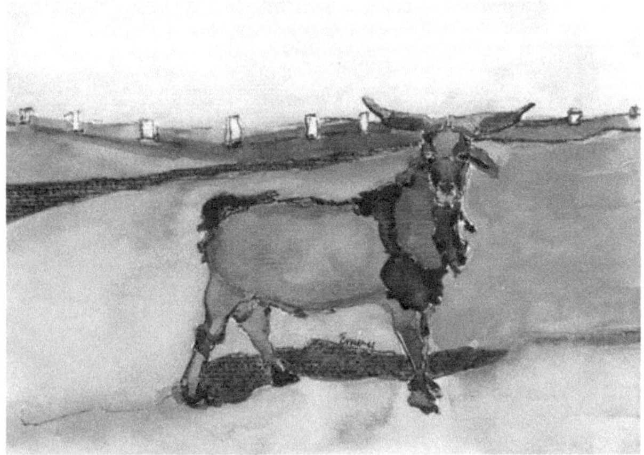

**Figure 2.5** Spanish (Meat Goat). For color detail, see Appendix A.

### Spanish

The Spanish goat in the U.S. originated from Mexican Criollo (Shelton, 1978). Aliases include Woods, Ball Field, Brush, Briar, Hill, and Scrub goats. This breed is distributed throughout the southwestern U.S., particularly Texas, with the major concentration around the Edwards Plateau of central Texas. Not known for rapid growth rates, they are very prolific. Spanish goats are nonseasonal breeders and under proper management can produce three kid crops in 2 years. Mature bucks weigh up to 91 kg (200 lb) and mature does up to 59 kg (130 lb). Until recently, these goats were used primarily to clear brush, but with growing demand, they were crossed with improved imported goats to produce meat of higher quality.

Lacking any particular breed characteristics, they are highly variable in appearance. Due to wide variability among Spanish goats, they offer an excellent opportunity for selection. They are very hardy and excellent foragers because they have survived undomesticated in harsh environments through many years of natural selection. See Figure 2.5.

### Sudanese Meat Goat Breeds

#### Southern Sudan

Small to dwarf in size, this breed is found in the southern region of Sudan. The females do not exhibit any of the signs of the achondroplasia of the West African Dwarf (Mason, 1981). These goats are primarily used for meat. The colors are variable. The males are horned, but females often are polled. Like Somali goats, their horns and ears are short.

#### Sudanese Desert

This breed is found in the semidesert areas of northern Sudan. A subtype in North West Darfur is known as "Zaghawa" (Mason, 1996). This breed is used mainly for meat and milk production. While they usually are white to black in color, they can also be grey. Both sexes are horned. The adult males weigh 40–58 kg (88–128 lb) while females average 33 kg (73 lb). The breed is a nonseasonal breeder with 200% kidding rate.

### Syrian Mountain

This breed is found primarily in Syria, Lebanon, and northern Israel (Mason, 1981). This breed also is known as Black Bedouin, Djelab, Jabali, Jabel or Jebel, Mamber, Mambrine or Membrine, Northern Mountain (Israel), Palestinian, and Syrian black. The hair color is usually black, and ears are long and pendulous. This medium-sized breed has adult males that weigh 60 kg (132 lb), and females weigh about 40 kg (88 lb). The breed is of medium prolificacy with about 130% kidding rate. A multipurpose breed, they are raised for meat, milk, and hair.

### Tennessee Wooden-Leg or Myotonic

One of the few goat breeds indigenous to America, this breed of goats has many names including Tennessee Fainting goats, Nervous goats, Tennessee Wooden Leg, Fall-down goat, Fainting goat, and Stiff-Leg goats. In a frightened state, these goats lock up or fall over and lie very stiff for a few seconds (myotonia). These animals are nonseasonal breeders and are capable of kidding three times in 2 years. Does are easy kidders, good mothers, and milk producers. The breed is medium in size, is very

**Figure 2.6** Myotonic (Meat Goat). For color detail, see Appendix A.

**Figure 2.7** British Alpine (Dairy Goat). For color detail, see Appendix A.

muscular and meaty, and has great potential for crossbreeding by crossing myotonic bucks with new improved breeds such as Boer and Kiko. The fainting trait is recessive, so it is not expressed in crossbreeds (OSU, 2004). This involuntary muscle reaction is reported to increase meat tenderness. See Figure 2.6.

### West African Long-Legged

This breed resembles the West African Dwarf goats except that they are larger in size, and their horns are long and twisted (Mason, 1981). They also are known as Sahelian or Sudan goats, and Fulani. Local names in Mali include Gorane, Niafounke, and Nioro. The breed is found in northern Africa spreading between Chad and Senegal, in the tropical forest, and in thick desert areas. This breed is used for meat and milk purposes. Their faces are either straight or slightly convex, and their ears are short. The coat colors vary from white, black, and brown to pied.

## CHARACTERISTICS OF DAIRY GOAT BREEDS

Dairy goats have been selected primarily for productivity in terms of milk quantity and milk-fat percent, as well as efficiency of production and longevity. Morphological traits closely related to milk production include large body capacity and angularity. Composition of milk is an important criterion for cheese making. The kidding interval is another selection criterion. A longer kidding interval is a desirable trait in temperate climates where there is limited demand for goat meat, while the opposite is true for the tropical countries where goat meat is popular (Devendra and Burns, 1983). Higher reproductive rates and lower generation intervals are of economic importance for dairy goats. Most prominent dairy goat breeds are listed in this chapter: and where available, breed standards are listed.

### Alpine

The Alpine breed originated in Switzerland. Alpines are excellent milkers with well-shaped large udders. Many composite breeds have Alpine ancestry.

The French Alpine breed is descended from Swiss Alpine goats of the Alps. These goats were selected for their uniformity, large size, and high milk production. They have a larger size than the other Swiss breeds. No distinct color pattern has been established; color may range from pure white through shades of fawn, gray, and brown to black. Both sexes have short hair. The males usually have long hairs along the spine. The adult bucks weigh about 77 kg (170 lb) with a height of 85–100 cm (34–40 in.) at the withers. The does are about 61 kg (134 lb) in weight and 75 cm (34 in.) at the withers. They may be either horned or polled. Except for white, all colors are acceptable, but the most common color is chamoisee. Ears are erect and hair is short. They have high milk production and are adapted to most climates (Briggs and Briggs, 1980).

The British Alpine breed has both horned and polled animals. The breed is a medium to heavy milker. The wither height is 95 cm (38 in.) for adult males and 83 cm (33 in.) for females. The Swiss Alpine was imported into Great Britain in 1903. The breed was recognized in 1921, and its herdbook was established in 1925. This breed is ideally suited for temperate climates. See Figure 2.7.

**Figure 2.8** Barbari (Dairy Goat). For color detail, see Appendix A.

### Barbari

This small breed of goats originated in Rajasthan of India and Sandi of Pakistan. These goats have small erect horns and short hair. A white coat with tan spots is common. Adult females are 27–36 kg (60–79 lb) in weight. This breed has been selected for milk production and is very prolific with twins and triplets being common (Devendra, 2007). See Figure 2.8.

### Beetal

Although this breed looks similar to Jamunapari, it is smaller in size. This breed is widely distributed in both Punjab provinces of India and Pakistan along with surrounding areas in Rawalpindi and Lahore. The coat color usually is red but black also is common. The ears are pendulous, and the horns are twisted. The adult male goats weigh about 65 kg (143 lb) while adult females weigh about 45 kg (100 lb) (OSU, 2004). The average lactation yield is about 195 kg (430 lb) in 224 days (Devendra and Burns, 1983). Meat production is remarkable, so this breed is classified as a dual-purpose breed. The dressing percentage is reported to be 46.9% (Bhatnagar et al., 1971). See Figure 2.9.

### Canary Island

This Spanish dairy goat also is used as a meat animal. The breed is very hardy with good mothering ability. Their coat color is variable, and they have saber to twisted horns with a straight face. The ears are drooping, and animals are medium to large in size. They are good milk producers averaging 300–400 kg (661–882 lb) milk with 5% fat when milked once per day (Mason, 1981). The breed is considered very hardy and resistant to disease with a high twinning ability.

**Figure 2.9** Black Beetal (Dairy Goat). For color detail, see Appendix A.

**Figure 2.10** Damascus (Dairy Goat). For color detail, see Appendix A.

### Damascus

This prominent goat breed is spread across the Middle East, especially Syria and Lebanon. They are known as Baladi or Shami, and Aleppo in Turkey (OSU, 2004). This is a dairy breed that contains some Nubian blood. Color usually is red or brown but also may be pied or grey. The hair is long, and males have tassels. Both horned and polled types are produced. See Figure 2.10.

### Golden Guernsey

This breed was developed from local breeds by crossbreeding with Anglo-Nubian and Swiss dairy breeds

**Figure 2.11** Jamunapari (Dairy Goat). For color detail, see Appendix A.

**Figure 2.12** LaMancha (Dairy Goat). For color detail, see Appendix A.

during a period from 1920 to 1950 (Mason, 1996) in the Channel Islands off the coast of Britain. A rare breed, it is recognized as a dairy breed by the British Goat Society. Although not producing as much milk as their Swiss counterparts, the high fat and protein content of milk make it suitable for cheese making. The color varies from cream to brown, and the breed is usually polled.

### Jamunapari

This is the most famous and the largest Indian goat breed (Acharya, 1982). The breed is also known as the Etawah breed, a name that is derived from the Etawah district of Utter Pradesh state. The breed is very tall with long legs, pendulous ears, and a Roman nose. They have short and flat horns. The breed has been exported to as many as eight different countries in the world to improve milk and meat production of local indigenous breeds (FAO, 2004). The adult male weighs around 68–91 kg (150–201 lb), and females weigh about 36–63 kg (79–139 lb) (Devendra and Burns, 1983). Their wither heights are 91–127 cm (36–51 in.) and 76–107 cm (30–43 in.) for adult males and females, respectively. The breed can yield up to 562 kg of milk (1,239 lb) from selected females. Kidding is once a year with a low twinning percentage (10%). See Figure 2.11.

### LaMancha

Developed in Oregon (U.S.), this breed originated in Texas from short-eared Spanish goats that were brought from Mexico by early Spanish explorers and Catholic missionaries. They have distinct short or almost nonexisting ears and come in all colors. The LaMancha goat has a pleasant temperament among dairy breeds. It is a very sturdy animal that can withstand hardships, yet still produce milk. Although selected for milk production, this goat has a lower milk production, but the milk is higher in fat content than the Swiss breeds. The breed society and herdbooks were established in 1969 and 1958, respectively, in the U.S. See Figure 2.12.

### Malabar or Malabari

This dairy breed is found in the southeastern region of Kerala, India, where the climate is hot and humid. Milk yields are about 1 kg (2.2 lb) per day. They are medium-sized animals with adult does weighing about 40 kg (88 lb) with quite variable coat colors. Both sexes are horned, but polled animals also are common (Devendra, 2007). This is a prolific breed with a high twinning ability.

### Nubian

This breed also is called Anglo-Nubian. The breed was developed in England by crossing British goats with African and Indian bucks to form a composite breed in the 1870s. The name originated from Nubia, a region spreading from northern Sudan to southern Egypt. According to Mason (1981), the Nubian is more of a concept rather than a specific breed, because it is difficult to find a Nubian goat in Nubia. This is an all-purpose breed used for meat, milk, and hide. Though not a high milk producer, its milk contains high levels of fat. This breed has a longer breeding season than Swiss breeds. The Nubian goat has been used in as many as 33 countries (FAO, 2004) for upgrading programs for meat and milk

**Figure 2.13** Nubian (Dairy Goat). For color detail, see Appendix A.

**Figure 2.14** Oberhasli (Dairy Goat). For color detail, see Appendix A.

particularly in tropical countries because of its suitability for hot environments. This is a large, graceful dairy breed with droopy long ears and a Roman nose. Mainly horned, they also can be polled. It is always shorthaired. A mature buck usually weighs at least 80 kg (175 lb), and a mature doe usually weighs over 61 kg (135 lb). Composite breeds with Nubians as ancestors include the Zaraibi, Sudan Nubian, Damascus or Shami, Kilis, Mzabite, Mishri, and Anglo-Nubian (Briggs and Briggs, 1980). The U.S. Nubian was derived from the Anglo-Nubian breed of England. See Figure 2.13.

### Oberhasli

This breed originated in the mountains and valleys of Switzerland (OSU, 2004) and is a very energetic and alert breed. It also is known as the Swiss Alpine and was part of Alpine registry. However, in 1979 the Oberhasli breed was established. This breed is medium in size and has chamoisee color with black markings (stripes) on its face, the base of the ear through the dorsal line, belly, and legs below the knee. Oberhasli does are 71 cm (28 in.) in height and weigh around 55 kg (120 lb). See Figure 2.14.

### Saanen

This most reputable dairy goat breed (queen of the dairy goats) is native to the Saanen Valley of Switzerland. Its herdbook was established in 1890 in Switzerland. The breed is a heavy milk producer with average butterfat content. This rugged and vigorous breed has a medium to large frame with does weighing around 68 kg (150 lb) and measuring 75 cm (30 in.) in height at the withers. The adult males weigh 80–91 kg (176–200 lb) and are 90 cm (36 in.) at the withers. It is shorthaired and white or light cream color. They are polled. Along with Nubian goats, this breed has been used frequently to form composite breeds and improve meat breeds (FAO, 2004; Glowatzki-Mullis et al., 2008). This is one of the most widely distributed dairy goat breeds in the world (68 countries) for upgrading local goats to increase milk and meat production (FAO, 2004). Herdbooks have been established in many countries including South Africa (1922), Great Britain (1922), the U.S. (1954), and Italy (1981). See Figure 2.15.

**Figure 2.15** Saanen (Dairy Goat). For color detail, see Appendix A.

**Sable Saanen**

Sables are Saanens that are not white in color; they are rare. These also are called a black variety of Saanen. They come in many colors and combinations and have been a part of the Saanen heritage. The Sable is descended from Saanen goats brought to America by Europeans. In 2005, the Sable was recognized as a separate breed. Usually very dark or black, they can be multicolored.

When two recessive genes for coat color are paired, one from the sire and one from the dam, Sables are created (i.e., both genes must be present) (Dohner, 2001). Sables are recognized by the American Dairy Goat Association. Their weight or height is similar to the Saanen breed.

**Sangamneri**

This breed is native to Pune, Maharastra (Devendra, 2007). Although it is considered a dairy breed, its meat also is valued highly. Colors vary from white to black to brown or pied. The breed is a medium-sized breed with an average weight for females about 29 kg (64 lb). Their wither height is 68 cm (27 in.).

**Stiefelgeiss**

Also known as the Booted goat, this breed originated in St. Gallen, in the northeastern part of Switzerland. This is 1 of the 10 goats in Switzerland with a testified Swiss origin (Glowatzki-Mullis et al., 2008). The breed was almost extinct until the foundation called Pro Specie Rara saved it in the 1980s. The breed is a milk breed well adapted to the extensive system of production in the mountainous regions. They are horned with hair colors that range from grey brown to dark-red brown. They are concentrated in eastern Switzerland with scattered breeding groups also found in the central and western regions of the country (OSU, 2004). The German subvariety is called St. Galler Stiefelgeiss (Mason, 1996). Their coats range from a light-greyish brown to a dark reddish color. The breed is managed by the "Booted Goat Breeders Club of Switzerland."

**Toggenburg**

This breed originated in the Toggenburg Valley of northeastern Switzerland, also known as "Alemu." Among the Swiss dairy breeds, this is the oldest known dairy breed. This breed is medium in size and slightly smaller than other Alps breeds. The doe weighs at least 54 kg (120 lb) while the buck may weigh more than 91 kg (200 lb). The hair is short, and the color is solid from light fawn to dark chocolate. The ears are erect and carried forward. They have white stripes down the face. The height at the withers is at least 83 cm (33 in.) for adult males, and females are 68 cm (27 in.) tall at the withers. The Toggenburg also is known for being very sturdy, vigorous, and alert. They perform best in cooler environments and produce great amounts of milk with average butterfat content. They have been exported to at least 48 countries around the world (FAO, 2004) to upgrade the milking ability of indigenous local goats. As a result, many Toggenburg breeds have developed in different countries. See Figure 2.16.

**Figure 2.16** Toggenburg (Dairy Goat). For color detail, see Appendix A.

## CHARACTERISTICS OF FIBER GOAT BREEDS

Quantity of fiber produced and quality of the fleece have been emphasized in selection of these breeds with less emphasis on body size. Fiber goats generally are smaller in stature than meat and dairy goats. Mason (1996) has listed more than 70 breeds in the world that are used for either hair or cashmere production purposes besides providing meat and milk. Most prominent fiber goat breeds are listed below. Breed standards are listed where available.

**Angora**

This breed originated in the mountainous district of Angora (Ankara, Turkey) in Asia Minor. It is distributed to more than 22 countries in the world (FAO, 2004). The breed society was established in the U.S. in 1900 and in 1921 in South Africa (Mason, 1981). The fiber produced by these goats is called mohair. Though similar to wool, mohair has a much smoother surface and a very thin smooth scale. The Angora goat is a small animal. The average carcass weight is 13 kg (29 lb). This breed is not as prolific as other goat breeds and does lack good mothering ability. Angora goats

**Figure 2.17** Angora (Fiber Goat). For color detail, see Appendix A.

**Figure 2.18** Cashmere (Fiber Goat). For color detail, see Appendix A.

are very delicate and appear more susceptible to internal parasites than other goat breeds. Newborn kids need protection from the cold during their first few days of life. Angora goats have excellent browsing ability and can adapt to areas where sheep would not thrive on land unsuitable for other agricultural uses.

The average amount of mohair produced by these goats is 2.4 kg (5.3 lb) per shearing with animals usually being sheared twice a year in the U.S., South Africa, and Russia but only once a year in Turkey. Mohair does not shed out naturally; therefore, Angora goats must be sheared. See Figure 2.17.

### Atlai Mountain

This fiber (wool) breed of goat originated in Gorno, Atlai, Siberia, Russia, by crossing Don goats (wool goats) with local goats. The breed also is used for production of milk and skins. The coat color usually is black with a grey undercoat. This breed was first recognized as a breed group in 1968 and as a separate breed in 1982. They are well adapted to year-round grazing on pastures under extensive systems in the highlands. They are uniform in color, size, and conformation (Dmitriez and Ernst, 1989). The adult males weigh about 65–70 kg (146–155 lb) and females about 41–44 kg (90–97 lb). The guard hair color is black and true dark. Of very high quality, the fiber is a valuable commodity.

### Cashgora

This is a composite breed that developed in Australia, New Zealand, and Great Britain primarily by crossbreeding Cashmere goats with Angora goats. This is a first cross between Angora goats with feral goats or Angora goats with Cashmere goats. The Angora goats were being used to upgrade feral goats in Australia and New Zealand. There are two foundation breeds developed in Australia (Shrestha, 2005). The fiber has both Mohair and Cashmere qualities.

### Cashmere

The name originated in Kashmir, a mountainous region in northwest India bordering Pakistan. The breed is also known as Pashmina, Changthangi (Kashmir), Mongolian, Alashan Down, Albas Down, Hexi Down, and Tibetan and Xinjian (China). The fiber (luxurious fiber of kings) is a secondary hair shed by goats during the spring and is very fine in diameter. The breed usually is white with long hair. The horns are twisted, and ears are either erect or horizontal. More than 60% of the world's cashmere is produced in China with the remainder in Turkey, Iraq, Iran, Kashmir, Australia, and New Zealand. Cashmere goats were imported from Australia and New Zealand to the U.S. in the 1800s. This breed is easy to raise and needs minimal management. Because they are not agile, they are easy to contain but need shelter due to the insulating nature of their dual coat that is shed in summer. They are sheared once a year, and an adult buck will shear about 1.1 kg (2.5 lb) of fleece, which contains guard hair and about 20% cashmere. See Figure 2.18.

### Kurdi

The Kurdi goat originated in Kurdistan, Iraq, and Iran. The breed is a cashmere-producing breed. Their color usually is white, black, or brown. They also are known as Karadi, Kurdish, Marghaz, Markhaz, Morghose, or Morghoz (Iran). They have a straight or slightly concave face with laterally directed ears with a forward and downward inclination (Mason, 1981) and have twisted horns. Males weigh about 50–60 kg (110–130 lb).

#### Nigora

This is a composite breed developed by crossbreeding Nigerian Dwarfs with Angora goats. Their size is medium to medium small. The American Nigora Goat Breeder Association was established in 2007 in the U.S. According to the breed association, the size of the Nigora breed should be 25% less than either parental breed. Ideally, the breed should be 50% Nigerian Dwarf and 50% Angora goat. The registered Nigoras produce three types of fibers. The type A fiber, similar to mohair, is up to 15 cm (6 in.) in length. The type C fiber is like cashmere, a fine fiber, 2.5–7.5 cm (1–3 in.) in length. The type B fiber is a blend of Type A and C. Friendly, intelligent, playful, and gentle, they are popular as pets.

#### White Himalayan

This is a fiber breed mainly used to produce cashmere, but its meat also is valued. They are found in the northeastern Himanchal Pradesh and northern part of Utter Pradesh, India. They also are known as Kangra Valley, Chamba, Gaddi, or Kashmiri. They are used as beasts of burden. The color is usually white. They have long and twisted horns.

#### Zhongwei

Also known as Chungwei, Zhongwei goats are found in the arid desert steppes counties in the Ningxia Hui Autonomous Region and Gansu Province of China (Cheng, 1984). Besides producing cashmere, they are famous for their pelts from kids slaughtered at 35 days of age. The pelts are white with lustrous staples and attractive curls. Both males and females have horns that twist upward. The adult males weigh about 39 kg (85 lb), and females average 24.5 kg (55 lb). This breed is not very prolific.

## CHARACTERISTICS OF FERAL GOATS

According to the Board of Agriculture (1993), goats known to run wild for at least 100 years with no known introduction of outside genetics are called "feral goats." Some 11 different feral goat breeds are listed in the Atlas of Goat Breeds (Mason, 1996). These include Australian feral goats; Barren Island feral goats (Andaman Islands, India); British feral goats from the mountains of Wales, Scotland, and North England, Desecho Island (off of Puerto Rico), Galapagos (Islands of Ecuador), Guadalupe Island (Lower California, Mexico), Juan Fernandez (Islands off of Chile), Montecristo (Montecristo Island, Tuscany, Italy); New Zealand feral goats (especially Arapawa, an island in Marlborough Sounds, Auckland Forsyth and Great Barrier Islands), San Clemente, and Santa Catalina (Islands off California, U.S.). Feral goats allegedly cause environmental damage. Although considered a menace in Australia, in more recent years feral goats have become sources of revenue for certain regions in Australia from their opportunistic capture and sale (MLA, 2007). Most prominent feral goat breeds are listed in following sections.

#### Australian Rangeland

The Australian feral or wild goats are called "Rangeland Goats" (MLA, 2007). They are thought to be the result of haphazard crossbreeding among Angora, Cashmere, Anglo-Nubian, British Alpine, Saanen, and Toggenburg breeds (Reithmuller, 1998). These are very hardy goats developed through natural selection over many years and are well adapted to harsh environments. They can survive under harsh conditions, maintain high fertility, and produce suitable carcasses. Very popular in Australia, they are exported around the world both as meat and live animals.

#### San Clemente and Santa Catalina Island

These goats are found on two of the eight islands that form California's channel islands along the Pacific coast from Santa Barbara to San Diego (Dohner, 2001). They are believed to be originated from Spanish goats were brought about by explorers. San Clemente goats were nearly eradicated by the U.S. Navy in an attempt to conserve the endangered plants of the islands, but a few hundred goats probably are still alive. These goats are small in size with adult males weighing about 45 kg (100 lb) and adult females weighing about 29 kg (65 lb). Their coat color commonly is red or light brown with black or dark brown markups. The breed is listed as a critically endangered heritage breed by the American Livestock Breeds Conservancy.

## CHARACTERISTICS OF COMPANION (PET) BREEDS

Goats have become very popular as pets in the U.S. They can be handled by children and are used therapeutically with children and adults with physical handicaps and with those that need psychological rehabilitation (Morand-Fehr et al., 2004). A variety of breeds are available as pets. Goats are gregarious animals and enjoy socializing and being in the company of other goats. Dwarf goats make ideal pets because they are small in size and have lower feed, housing, and management requirements. Their different coat colors also add variety. They breed year-round and are very prolific. Selected breeds of pet goats are listed below.

**Australian Miniature**

The Australian miniature goat breed was developed in 1995. These miniature goats became a recognized breed in 2000 by the Australian Miniature Goat Club (Burk's Backyard fact sheet, 2009). The breed is mainly from Angora, Cashmere, and Nubian breeds, combined with Australian feral goats. The breed measures up to 53.3 cm (21 in.) in height for adult females and 58.5 cm (23 in.) for adult males. They are intelligent, gentle, love human companionship, and often form close bonds with their family and other family pets. Their small size makes them easy for members of the family to train and handle. Australian miniature goats are predominantly kept as pets. Some owners use the longhaired types for spinning while others milk their goats.

**Kinder**

This is a composite breed developed by crossing Pygmy goats with Nubian goats. This breed was created in 1988 in the U.S. and is recognized by the American Dairy Goat Association as a dairy goat (OSU, 2004). More muscular than the pygmies, they are popular as pets. This breed also is considered a dual-purpose goat being used for both milk and meat production. The adult does average 52 kg (115 lb). Their milk production averages 680 kg (1,500 lb) in 305 days or less. Animals of this breed grow rapidly.

**Nigerian Dwarf**

The Nigerian Dwarf is a miniature goat of West African origin. They make wonderful pets and great animal projects for young children in 4-H. They are very popular as zoo or park pets also. They have a straight nose and erect ears. Their coat is soft with short to medium hair. Dwarf goats come in many colors but usually are black, or chocolate and gold with random white markings and other color combinations such as red, white, gold, and black.

The mature buck's wither height is 48–50 cm (19–20 in.) with a maximum height of 59 cm (23.6 in.). The adult female height at the withers is 43–48 cm (17–19 in.) with a maximum height of 57 cm (22.6 in.). Dwarf goats are gentle and loveable. Even breeding bucks are handled easily. They breed year-round and are very prolific. They are registerable in three registries: American Goat Society (AGS), International Dairy Goat Registry (IDGR), and Canadian Goat Society (CGS). Dwarf shows are growing in popularity, and dwarfs are becoming more available. Most dwarfs are sanctioned by AGS (OSU, 2004).

**Pygmy**

The Pygmy goat was derived from the Cameroon Dwarf or the West African Dwarf goats of West African countries. The northern, southwestern, and eastern African countries also have similar forms of Pygmy goats. This breed is small with short legs, a large torso, and broad head. Besides being used as pet animals, the breed also is used for meat.

They are very popular as exotic animals. The hair is straight and medium to long. Males have beards, but beards are nonexistent, sparse, or trimmed in females. All body colors are acceptable, the predominant color being a grizzled (agouti) pattern produced by the intermingling of light and dark hairs of any color (National Pygmy Goat Association, 1932).

**Pygora**

The Pygora goat is a composite breed developed by crossing Pygmy goats with Angora goats (Pygora Breeders' Association, 2002). The breed has fleece that is distinct from Angora and the smaller size from Pygmy. This breed was developed in 1987 in Oregon, U.S. The Pygora goats are medium-sized goats. The registered Pygora goat may not be more than 75% registered Angora goat or more than 75% registered Pygmy goat (Jorgensen, 1993). Pygoras are popular as pets because they are very friendly, cooperative, and easy to handle. They are also good show, breeding, and fiber-producing animals. Pygora goats will produce cashmere-like fleece (Classified as Type C), a mohair-like fleece (Type A), or a combination of the two fleeces (Type B).

Adult bucks weigh 34 kg–43 kg (75–95 lb) and reach a minimum height of 58 cm (23 in.) at the withers. Females (does) weigh 29 kg–34 kg (65–75 lb) and reach a minimum height of 45 cm (18 in.). The breed is both horned and polled animals.

## CHARACTERISTICS OF THE MULTIPURPOSE BREEDS

The majority of goats in the world are multipurpose breeds. Mason (1996) has listed more than 29 goat breeds that are used as triple-purpose animals. Products from these goats include meat, milk, fiber, and skin, and they also may be used as pack animals. Most of these breeds are in their original state throughout the world and have not been intensively selected for any products such as meat, milk, or fiber. The application of a multitrait animal model would be ideal to establish optimum breeding objectives to achieve maximum efficiency of multipurpose goats (Shrestha and Fahmy, 2007).

No separate breed categories for multipurpose breeds are provided here to avoid duplications. In the strictest sense, most of the breeds found in the developing countries are used for multiple purposes.

## SUMMARY

The highest diversity among goat breeds of the world is found in the developing countries with goats being raised under extensive systems. Most breeds have not been extensively characterized or described. Consequently, it is difficult to appraise potential strengths and weaknesses of these breeds. This makes it difficult to develop strategies for breed improvement and management programs appropriate for the small farmers who own most of the goats in the world. In addition, a clearer understanding of variability among breeds should provide goat breeders with an opportunity to design effective programs to develop composite mating systems. Although most indigenous goats are used primarily to produce both milk and meat, improved breeds that excel in dairy, meat, and fiber production have been identified and developed and potentially can be used in crossbreeding programs to improve productivity of the environmentally well-adapted indigenous goat breeds.

## REFERENCES

Acharya, R.M. 1982. Sheep and goat breeds of India. FAO Animal Production and Health Paper 30, Food and Agriculture Organization of the United Nations, Rome, Italy, pp. 190.

Alexandre, G., G. Aumont, N. Mandonnet, and M. Naves. 1999. The Creole goat of Guadeloupe (French West Indies): an important genetic resource for the humid tropics. Anim. Genet. Resour. Inform. 26:45–55.

Bhatnagar, D.S., O.S. Tomer, and R. Nagarcenkar. 1971. All India coordinated research project on breeding and management of goats for milk. In: Annual Report, 1971, National Dairy Research Institute, Karnal, New Delhi, India; Indian Council of Agricultural Research 53–54.

Board of Agriculture. 1993. Managing Global Genetic Resources: Livestock. National Research Council, National Academy Press, Washington, DC.

Briggs, H.M. and D.M. Briggs. 1980. Modern Breeds of Livestock. 4th ed. The MacMillan Company, New York.

Burke's Backyard Magazine. 2009. Fact Sheets. http://www.burkesbackyard.com.au/factsheet/Others/Australian-Miniature-Goat/1903.

Campbell, Q.P. 2003. The origin and description of southern Africa's indigenous goats. South African Society of Animal Science, Popular Scientific Papers (4):18–22. http://www.sasas.co.za/scientific/campbell.pdf.

Casey, N.H. and W.A. Van Niekerk. 1988. The Boer Goat I. Origin, adaptability, performance testing, reproduction and milk production. Small Ruminant Research 1:291–302.

Cheng, P. 1984. Livestock breeds of China. Animal Production and Health Paper 46, FAO, Rome, Italy.

Council of Scientific and Industrial Research. 1970. The Wealth of India. Raw Materials VI. Suppl. Livestock (including poultry). Published Information Directorate, CSIR, New Delhi, India.

Devendra, C. 1981. Meat production from goats in developing countries. In: Intensive animal production in developing countries, Thames Ditton, UK; British Society for Animal Production 395–406. Occasional Publication, British Society of Animal Production, 4.

Devendra, C. 2007. Goats: biology, production and development in Asia. Academy of Sciences, Malaysia.

Devendra, C. and M. Burns. 1983. Goat production in the Tropics. Commonwealth Agricultural Bureau International, Wallingford, UK.

Dmitriez, N.G. and L.K. Ernst. 1989. Animal Genetic Resources of the USSR. Animal Production and Health Paper 65 Published by FAO, Rome, pp. 517.

Dohner, J.V. 2001. The Encyclopedia of Historic and Endangered Livestock and Poultry Breeds. Yale University Press, Exhibits Department, New Haven, CT.

Epstein, H. 1969. Domestic Animals of China. Commonwealth Agricultural Bureau, Farnham Royal, Buckinghamshire, UK, pp. 166.

Epstein, H. 1971. The origin of the domestic animals of Africa. Revised in collaboration with I.L. Mason. Vol. I., Vol. II., New York. New York, London, and Munich.

Epstein, H. 1977. Domestic animals of Nepal. Homes and Meier Publishers, Inc., NY.

Food and Agriculture Organization (FAO). 2004. http://www.fao.org/newsroom/en/news/2004/39892/index.html.

Food and Agriculture Organization (FAO). 2007. Food and agriculture organization statistical database. http://faostat.fao.org/site/497/default.aspx.

Food and Agriculture Organization (FAO). 2009. http://dad.fao.org/accessed. Retrieved October 2009.

Galal, S. 2005. Biodiversity in goats. Small Ruminant Research 60:75–81.

Gall, C. 1981. Goat Production. Academic Press, London/New York.

Gall, C. 1996. Goat breeds around the world. CTA, Margraf/FAO, Weikersheim, Deutschland.

Gipson, T. 1996. Breeds and breeding plans. In: Meat Goat Production Handbook. Sponsored by Rural Economic Development Center, Raleigh, NC, and Mid-Carolina Council of Governments, Fayetteville, NC.

Glowatzki-Mullis, M.L., J. Muntwyler, E. Baumle, and C. Gaillard. 2008. Genetic diversity measures of Swiss goat breeds as decision-making support for conservation policy. Small Ruminant Research 74:202–211.

Hasnain, H.U. 1985. Sheep and goats in Pakistan. FAO Animal Production and Health Paper No. 56. Food and Agriculture Organization of the United Nations, Rome, Italy.

Jiang, Y. 1982. Goat breeds and ecological characteristics in China. J. Beijing Agric. Univ. 8: 9–12.

Jiang, Y. 1995. Chinese meat goat breeds and their crosses. Animal Genet. Resources Information, 15:71–81.

Jorgensen, K. 1993. The goat for all seasons/reasons. Fiberfest Magazine, Vol. 1, Issue 1, Fall 1993.

Lanari, M.R., H. Taddeo, E. Domingo, M. Perez Centeno, L. Gallo. 2003. Phenotypic differentiation of exterior traits in local Criollo goat population in Patagonia (Argentina) Arch. Tires., Dummerstorf. 46(4): 347–356.

Mason, I.L. 1980. Sheep and goat production in the drought polygon of northeast Brazil. World Animal Review. No. 34: 23–28.

Mason, I.L. 1981. Breeds. In: C. Gall (Editor) Goat production, Chapter 3, Academic Press, London, England.

Mason, I.L. 1996. A World Dictionary of Livestock Breeds, Types and Varieties. Fourth Edition. C.A.B International, Wallingford, UK.

Mason, I.L. and J.P. Maule. 1960. The indigenous Livestock of Eastern and Southern Africa. Commonwealth Agricultural Bureaux, Farnham Royal, Bucks, England.

Meat and Livestock Australia (MLA). 2007. Goat Farming for the Future. www.mla.com.au.

Moaeen-ud-Din, M., L.G. Yang, S.L. Chen, Z.R. Zhang, J.Z. Xiao, Q.Y. Wen, and M. Dai. 2008. Reproductive performance of Matou goat under sub-tropical monsoonal climate of Central China. Tropical Animal Health Production 40:17–23.

Morand-Fehr, P., J.P. Boutonnet, C. Devendra, J.-P. Dubeuf, G.F.W. Haenlein, P. Holst, L. Mowlem, and J. Capote. 2004. Strategy for goat farming in the 21st century. Small Ruminant Research 51:175–184.

National Pygmy Goat Association. 1932. 149th Ave SE, Snohomish, WA, USA. http://www.npga-pygmy.com.

Oklahoma State University (OSU). 2004. Breeds of Livestock. http://www.ansi.okstate.edu/breeds/goats/.

Oliver, J.J., S.W.P. Cloete, S.J. Schoeman, and C.J. C. Muller. 2005. Performance testing and recording in meat and dairy goats. Small Ruminant Research 60:89–93.

Pygora Breeders Association. 2002. Pygora goat fiber and its uses. The Goat Magazine, Vol. 6 (5), June/July.

Reithmuller, J. 1998. In: "Goat meat for export," Roma, Queensland, Australia. Department of Primary Industries, Australia, pp. 17–19.

Scherf, B.D. (ed.). 2000. World watch list of domestic animal diversity, 3rd ed. Food and Agriculture Organization of the United Nations, Rome, Italy, pp. 726.

Sengar, O.P.S. 1976. Sheep and goat production systems in India. Paper presented Int. Workshop Role of Sheep and Goats in Agric. Develop., Winrock Int. Livestock Res. Training Ctr., Morrilton, AR.

Shelton, M. 1978. Reproduction and breeding of goats. Journal Dairy Science 61:994–1010.

Shrestha, J.N.B. 2005. Conserving domestic animal diversity among composite populations. Small Ruminant Research 56:3–20.

Shrestha, J.N.B. and M.H. Fahmy. 2005. Breeding goats for meat production: a review 1. Genetic resources, management and breed evaluation. Small Ruminant Research 58:93–106.

Shrestha, J.N.B. and M.H. Fahmy. 2007. Breeding goats for meat production: a review. (3) Selection and breeding strategies. Small Ruminant Research 67:113–125.

Solaiman, S.G. 2007. Simply Meat Goats. Tuskegee University Publication No. 115–1006.

Solaiman, S.G., D. Bransby, C. Kerth, R. Noble, B. Blagburn, and C. Shoemaker. 2006. A sustainable year-round forage system for goat production in Southeastern USA. Final Report, southern SARE Project # LS02–141.

Taneja, G.C. 1982. Breeding goats for meat production. Proceedings of the Third International Conference on Goat Production and Diseases, pp. 27–30.

Ying, J. 1986. Some Chinese goat breeds. World Animal Review (58):34–41.

# 3
# Conservation of Goat Genetic Resources

*J.N.B. Shrestha, PhD and S. Galal, PhD*

**KEY TERMS**

Domestic animal—animal that has been bred in captivity for economic profit to the human community.

Feral animal—a domestic animal that has reverted to its wild state following its escape or its release into an environment favorable for propagation.

Breed—a subgroup of domestic livestock and poultry with definable and identifiable morphological characteristics within a species.

Breeds at risk—according to the Food and Agriculture Organization (FAO) of the United Nations, those breeds with <1,000 females, with <20 males, or with a total population of <1,200.

Globalization—the process of unifying people or a local phenomenon into a single society and function as a global phenomena.

Conservation—analysis and protection of biological diversity in the world.

**OBJECTIVES**

By completing this chapter, the reader will acquire knowledge on:

- The most likely dates and places where goats and sheep were domesticated
- The evolution of the goat
- How the goat spread from its place of origin to different parts of the world
- How primitive goats developed to become highly specialized modern breeds
- The degree of diversity among goat breeds, populations, and landraces in the world
- The alarming trend in the loss of domestic animal diversity
- The need to characterize various domestic animals taking into consideration their population size
- The need to conserve those breeds, populations, and landraces that are considered endangered or at risk of extinction
- The social and economic importance of goats in poverty alleviation

## INTRODUCTION

Creative human activity, which began with domestication of goats over 10,000 BC combined with natural causes, has resulted in a colossal amount of inherent potential among 1,153 goat breeds worldwide (FAO, 2009) for economically important morphological characteristics and production performance. Many of the traditional breeds in the industrialized nations and indigenous breed populations in the developing countries are being replaced by commercial breeding stocks derived from a narrow genetic base. These commercial breeds have outstanding productivity and are capable of meeting the demand for quality food and food products of a burgeoning population of humans with increasing disposable income (Shrestha et al., 2008). One outcome of the United Nations Conference on Environment and Development in 1992 held in Rio de Janeiro was the ratification of the Convention on Biodiversity. This promoted national and international attention toward the depletion of biological diversity, a trend that includes goats. To combat the loss of goat genetic resources worldwide, many countries have seized opportunities to initiate in situ and ex situ conservation to preserve the inherent potential of goats and reduce the endangerment or extinction of breeds, populations, and landraces.

Goat provides a sizeable proportion of the human diet in many developing countries. Through owning animals with multifunctionality, small farmers and the landless have survived in the face of hunger, poverty, and the current food crisis (Devendra, 2007). The genetic resources of goats are not unique to each country but instead are affiliated intimately with the economic status, social structure, religious rituals, and most importantly, public policy. Therefore, preservation of animal genetic resources is necessary to sustain current production levels and to address future market demands for commodities, trade, fertilizer, breeding stocks, employment, and recreation. No longer are goats viewed as pests responsible for soil erosion. Instead, goats are being viewed in new roles as potential alleviators of worldwide poverty.

## DOMESTICATION

Domestication is an ongoing process that includes both artificial selection and skillful breeding practice in an environment provided by farmers. Wild species undergo physical changes as they adapt to captivity and are selected for traits of economic importance. Clutton-Brock (1999) defined a domestic animal as "one that has been bred in captivity for purposes of economic profit to a human community that maintains total control over its breeding, organization of territory and food supply." Domestic animals that revert to their wild state following their escape or release into an environment favorable for propagation are known as "feral" animals.

As early as 8500 BC, prehistoric people at the end of the Mesolithic period raised herds of goats and flocks of sheep in the mountainous areas bordering Iran and Iraq (Figure 3.1). This conclusion is based on archeological excavations where goat and sheep bones were uncovered adjacent to human settlements along with radiocarbon dating. Archaeological sites in the Kermanshah Valley of the Zagros Mountains at "Ganj Dareh," Iran (8000 BC) and the Euphrates river valley at "Nevali Cori," Turkey (9000 BC) are believed to be two separate but distinct sites where domestication occurred (Vigne et al., 2005). Other possible sites of domestication include the Indus Basin at "Mehrgarh," Pakistan (7000 BC), "Cayönü," Turkey (8500–8000 BC), "Tell Abu Hureyra," Syria (8000–7400 BC), "Jericho," Palestinian Occupied Territories (7500 BC), and "Ain Ghazal," Jordan (7600–7500 BC). The central Anatolia and the southern Levant also have been proposed as additional sites of early domestication. New research has confirmed that the "Ganj Dareh" site contains the earliest directly dated evidence of livestock domestication (Zeder and Hesse, 2000).

**Figure 3.1** Region where early domestication of goats and sheep occurred.

**Figure 3.2** Alpine ibex (*Capra ibex ibex*) in the Gran Paradiso National Park, Italy. Photo by Marco Festa-Bianchet. For color detail, see Appendix A.

Zoo-archaeological records predating the time of domestication indicate that the ancestors of the present-day domestic goats arose from three genetically distinct maternal lineages rather than one single wild population. Wild goats of the Bezoar or Pasang (*Capra aegagrus aegagrus*), Makhol (*C. aegagrus falconer*), and Ibex (*C. aegagrus ibex*) types are believed to be the ancestors of domestic goats worldwide. Bezoar, Savannah, and Nubian goats, each with its characteristic morphology and distinctive horns, probably were domesticated from these three types of wild goats. Currently descendents of these goats that represent breeds, populations, and landraces are spread in various regions of the world. See Figure 3.2.

Phylogenetic analysis of mitochondrial DNA (mtDNA) from 19 ancient bone remains uncovered at the earliest Neolithic sites in southwest Europe that date back to about 4900 to 5300 BC indicates that two highly divergent goat lineages may have coexisted (Fernández et al., 2006). Genetic sequence analysis of the mitochondrial cytochrome gene supports the concept that two distinct clades of goats existed in the Caucasus with domestication in the Fertile Crescent mentioned earlier. The first farm animals domesticated probably were Bezoar goats (*Capra hircus*). The presence of a sizeable amount of mtDNA diversity from the remains of goats in Europe confirms that goats migrated there, presumably to provide food and fiber during the "Neolithic revolution." Later, humans along with goats traveled to distant lands and in due process established homesteads.

## EARLY MIGRATION ACROSS CONTINENTS

About 7500 BC nomadic and seminomadic pastoralists from the Near East traveled along with their goats across continents. From the West Asian center of civilization, goats spread westward and eastward through the first wave(s) of expansion of agriculture. This was followed by an extensive movement of improved types of goats during the late Neolithic period (with wool sheep) and the expansion of the Roman Empire (with large cattle). With the rise of the Roman Empire, goats with large body size were developed in Europe. This was when hornless goats were first reported. Later, with the demise of the Roman Empire in the 3rd to 6th century, the body size of goats decreased through the "Middle Ages" until the 14th century. Concurrently, the movement of goats was widespread, from north to south in East Africa, east to west in North Africa, and finally spreading across the entire African continent.

Goats with scimitar-shaped horns originated in Egypt in the 5th century BC and migrated into the areas adjoining Syria and Palestine. The Savannah type goats currently are widespread across the African continent, providing meat, milk, and skin to the pastoralists. In North Africa, the Nubian-type goats raised by the sedentary agriculturists originated in India or Iran, and subsequently migrated to Syria and Egypt (Epstein, 1971). The Anglo-Nubian goats are believed to be a cross between the prick-eared goats indigenous to Britain and Nubian-type Zaraibi, Chitral, and Jamunapari goats from Egypt and India.

The rise of the Aryan empire in the 2nd century BC resulted in the movement of goats eastward through the Khyber Pass into the Indian subcontinent. The nomadic and seminomadic pastoralists from Iran and Afghanistan traveled via the Silk Road to Turkistan, Mongolia, and North China. At the same time, Tsinghai and Mongolian goats became established in the Tibetan Plateau. Later, goats moved eastward from western and central Asia to South China, Hainan Island, and the eastern plain of Taiwan. Goats indigenous to central and eastern China may have originated as a cross between the Mongolian goat and the meat goat of South China.

In the 6th century BC, trade between India and South-Eastern Asia was thriving along the maritime route of Burma, Thailand, Malaysia, and Indonesia. During this period, goats migrated from the Indian subcontinent into those countries. Goats from South China possibly were introduced into those countries as well as South Asia. The Kambing Katjang or Pea goats in Malaysia and Indonesia that are raised for meat have a morphology similar to that of indigenous goats in the Philippines, Taiwan, and the islands of southwest Japan. These goats also bear some resemblance to the South China goat and the Black Bengal goat. The Jamunapari and Anglo-Nubian breeds probably share a common heritage to the indigenous goats of Thailand, Malaysia, and Indonesia.

In the last few centuries, explorers and conquerors brought goats during their attempts to find and colonize new lands. Goats were not only easy to adapt and manage, but also were portable, plentiful, and inexpensive. Goats accompanied voyages from Spain and Portugal to the Americas and from France and the U.K. to North America. Goats also accompanied voyages from the U.K. to Australia, New Zealand, and South Africa, and from Africa to the Caribbean countries. The abundance of forage and vegetation, as well as freedom from predators and communicable diseases in sparsely populated regions, resulted in the expansion of goat numbers worldwide. This also contributed to a large feral population from goats that escaped or were released into the wilderness. The export of large numbers of goats into the colonies continued for many years resulting in their expansion worldwide. Goat populations established in distant lands were exposed to both natural selection and skillful breeding practices that resulted in a tremendous increase in phenotypic variability. Specific genetic combinations became established through centuries of evolutionary processes worldwide. Genetic diversity in goat populations has potential merit as an important resource for the future.

## DEVELOPMENT OF BREEDS

The FAO (2000) defines breed as "either a sub-specific group of domestic livestock with definable and identifiable external characteristics that enable it to be separated by visual appraisal from other similarly defined groups within the same species or a group for which geographical and/or cultural separation from phenotypically similar groups has led to acceptance of its separate identity." The Thoroughbred horse first was listed in the General Stud Book in 1791; it included only horses that had won important races. Shire horses, Longhorn cattle, and Leicester sheep followed. The first herdbook in the western world was the Coates' Herd Book for Shorthorn Cattle, established in 1822 in the U.K. This led to the widespread acceptance of a concept developed by Robert Bakewell of Dishley, U.K., that involved combining stocks with similar morphology and production in accordance with the breeders' vision, recording of the pedigree, registration, and maintenance of herdbooks established by breed societies (Lush, 1945). Bakewell's message was "like begets like" and "breed the best to the best." Concurrently, the establishment of rigid breed standards encouraged the selection of goats with uniform morphological characteristics and the requirement for detailed record keeping of pedigreed animals that occasionally included performance measurements. Pure breeds were widely sought and promoted as seed stocks of choice in show rings, exhibitions, and auctions. To ensure the purity of the breed, introduction of any breeding animal that was not registered in the same herdbook was officially prohibited regardless of its genetic merit. The concept of breed was readily accepted and became established in many countries resulting in the development of numerous breeds worldwide.

In the 1920s, South African farmers from the Eastern Cape crossed indigenous goats kept by the Hottentot and Bantu tribes with imported Nubian and Indian goats to produce the Boer breed (Skinner, 1972). In the following years, selection for body size and conformation resulted in goats that excelled in meat production efficiency. In Europe, the Alpine (e.g., Saanen and Toggenburg) and Anglo-Nubian breeds were developed as dairy goats. In the U.S., the Spanish goat, named to be distinct from the Angora and dairy breeds, is of Mexican origin and accounts for most of the goat meat produced in the U.S. The highly variable appearance of these goats may be attributed to natural selection and the indiscriminate crossing with the Nubian (syn., Anglo-Nubian) or Toggenburg breeds.

In India, goats were classified according to both their primary product and their distribution into the following: cashmere-like goats close to the Himalayan mountain range; milk goats in the northern dry areas; black meat goats on the Deccan plateau; and small meat goats adjoining the Bay of Bengal. Many indigenous populations were further classified into some 20 goat breeds on the basis of having morphological characteristics distinct from other breed populations with complementary local names in the vicinity. Pakistan classified goats on the basis of hair and coat smoothness characteristics into 25 breeds. Goats found in the tropical, subtropical, and high mountain climates were exposed to humid, very dry, dry, and humid and cold environments. These goats were classified into breeds according to primary products with 34 for meat, 12 for milk, 8 for prolificacy, and 3 each for pashmina and skin production. According to the FAO (2009), 1,153 breeds of goats currently exist in the world.

## GENETIC RESOURCES: THEIR HABITAT AND MANAGEMENT IN THE WORLD

Goats worldwide are subjected to diverse management practices dependent on husbandry, breeding, productivity, and disease status. Other factors include the interacting influences of diet and environment within the habitat, region, and country, and social and cultural attributes. The present-day goat breeds have diverged considerably from the original characteristics that remain in the wild goats. This divergence is caused by the evolutionary forces of mutation, migration, selection, and genetic drift under varying environmental conditions. It is common to find goats in small or large multiple species herds (cattle and goats) under nomadism and seminomadism in agropastoral production systems and arid rangelands. Goats usually are kept within the perimeter of farms, or in the vicinity of homesteads of the small holder farming systems. These animals have multiple functions and also have been used to control brush and prevent forest fires in North America. Generally, the management practices of goats are traditional, socially tolerant, sustainable, and in harmony with the natural vegetation and local environment. Goats contribute to the efficient use of labor surplus, help meet household needs for food in small holdings worldwide, and help alleviate poverty among the rural poor. In India goats are called the poor man's cow.

In the developing countries of Asia, the poor in rural areas raise a large number of small herds of goats. These people use milk for home consumption and supplement their income by selling goats or meat from surplus animals. Goats survive on vegetation surrounding roads, irrigated pastures and adjoining forest, agricultural by-products, and vegetable toppings from kitchen waste. In Malaysia and the Philippines, goats have been integrated with cropping systems and fish farming, and this has increased production of animal protein (Jalaludin et al., 1992; Libunao, 1990). In contrast, in countries bordering the Himalayan mountain range, the mountain people sustain their livelihood raising goats together with sheep under a transhumance system and nomadic pastoralists. Nomads with their goats and sheep ascend to high mountainous pastures during the spring and descend into the populated areas adjoining croplands and irrigated pastures during the autumn. Farmers raise goats in South China including the Hainan Island and in the western plain of Taiwan in irrigated river basins.

In the Sahel region of Africa, nomadic tribes keep goats to provide milk and meat, and move with them from region to region in search of vegetation around oases, waterholes, and irrigated lands. In the humid zone of West Africa and the southeastern region of Nigeria, small numbers of goats are kept in each household for milk and meat with minimal managerial and labor input.

Countries along the northern coast of the Mediterranean and in northern Europe concentrate on producing milk for cheese production from large herds of goats. This contrasts with the southern coast of the Mediterranean where meat and meat products are the primary end products. Sedentary goat keepers of Tunisia raise a few goats along with other livestock adjacent to oases, receiving income from the sale of meat while using milk for their own consumption. In the eastern and southern regions of Algeria, large herds of the Berber breed are kept for meat while the Arabe breed is raised for producing both meat and milk.

Goats in the eastern U.S. consume diets rich in browse and consume diets based on herbage. In Mexico, goats graze on semiwoody brush rangeland but often are supplemented with concentrate feeds. In the north and west central region of Argentina, large herds of Criollo (syn., Creole) and Anglo-Nubian goats are managed under extensive husbandry (Angel-Neelem and Nellem, 1998). Creole goats in Guadeloupe grazing on irrigated pastures are highly productive under an accelerated kidding program. The transition from goat keeping for sustenance to the large-scale commercial enterprise common in cattle, swine, sheep, and poultry illustrates the potential to increase production efficiency from goats. To date, commercialization has only materialized in the dairy goat industry. However, goats have been used as laboratory animals for research and teaching and for production of biological proteins.

An enormous diversity of goat genetic resources exists. Of the 546 world goat breed populations reported by FAO in 2000, 77 (14%) were found in Africa, 167 (30%) in Asia and the Pacific region, 192 (35%) in Europe, 28 (5%) in Latin America, 77 (14%) in the Near East, and 5 (1%) in North America. This represents a colossal amount of variability in morphological characteristics and potential productivity thereby providing an opportunity to exploit the biological potential of goats through the application of quantitative genetic principles to improve production efficiency. Goat breeds have been selected for meat, milk, prolificacy, pashmina (cashmere), and skins in countries with climatic conditions that range from tropical and subtropical to high mountains in dry, humid, and cold environments as listed in Table 3.1. The wide diversity of these breeds in morphological characteristics and production performance should have immense merit for selection and genetic improvement.

Despite its importance, information about the productivity of breeds of goats is difficult to retrieve. Nevertheless,

**Table 3.1** Goat breeds with potential for genetic improvement.

| Speciality | Breed | Country of origin | Climate | Environment |
|---|---|---|---|---|
| Meat | | | | |
| | Banjiao | China | Subtropical | Humid |
| | Barbari | Pakistan | Tropical | Dry |
| | Black Bengal | India | Tropical | Dry |
| | Black Bengal | Bangladesh | Tropical | Dry |
| | Black Bengal | Pakistan | Tropical | Dry |
| | Boer | South Africa | Subtropical | Dry |
| | Bugri | Pakistan | Tropical | Dry |
| | Chengdu Ma | China | Subtropical | Humid |
| | Cutchi | India | Tropical | Dry |
| | Damani | Pakistan | Tropical | Dry |
| | Du An | China | Subtropical | Humid |
| | Fijian | Fiji | Tropical | Humid |
| | Fuquing | China | Subtropical | Humid |
| | Ganjam | India | Tropical | Dry |
| | Guizhou White | China | Subtropical | Humid |
| | Haimen | China | Subtropical | Humid |
| | Huai | China | Subtropical | Humid |
| | Kaghani | Pakistan | Tropical | Dry |
| | Kail | Pakistan | Tropical | Dry |
| | Katjang | Indonesia | Tropical | Humid |
| | Khasi | India | Mountain | Humid |
| | Kheri | Nepal | Subtropical | Humid |
| | Katukachchiya | Sri Lanka | Tropical | Humid |
| | Lehri | Pakistan | Tropical | Dry |
| | Leizhou | China | Subtropical | Humid |
| | Longlin | China | Subtropical | Humid |
| | Ma'tou | China | Subtropical | Humid |
| | Marwari | India | Tropical | Dry |
| | Osmanabadi | India | Tropical | Dry |
| | Patteri | Pakistan | Tropical | Dry |
| | Sangamaneri | India | Tropical | Dry |
| | Shanzi White | China | Subtropical | Humid |
| | Sirohi | India | Tropical | Dry |
| | Sudan Desert | Sudan | Tropical | Very dry |
| | Tapri | Pakistan | Tropical | Dry |
| | Terai | Nepal | Subtropical | Humid |
| Milk | | | | |
| | Barbari | India | Tropical | Dry |
| | Beetal | India | Tropical | Dry |
| | Black Bedouin | Israel and Egypt | Tropical | Very dry |
| | Damascus | Syria and Lebanon | Subtropical | Dry |
| | Dera Din Panah | Pakistan | Tropical | Dry |
| | Kamori | Pakistan | Tropical | Dry |
| | Jamunapari | India | Tropical | Dry |
| | Jhakrana | India | Tropical | Dry |
| | Malabar | India | Tropical | Humid |

**Table 3.1** *Continued*

| Speciality | Breed | Country of origin | Climate | Environment |
|---|---|---|---|---|
| | Nubian | Sudan | Tropical | Dry |
| | Surti | India | Tropical | Dry |
| | Zaraiby | Egypt | Tropical | Dry |
| Prolificacy | | | | |
| | Barbari | India | Tropical | Dry |
| | Boer | South Africa | Subtropical | Dry |
| | Black Bengal | India | Tropical | Dry |
| | Criollo | South America | Tropical and subtropical | Dry |
| | Malabar | India | Tropical | Humid |
| | Ma'tou | China | Subtropical | Humid |
| | Sudan Desert | Sudan | Tropical | Very dry |
| | West African Dwarf | West Africa | Tropical | Humid |
| Pashmina (Cashmere) | | | | |
| | Kashmiri | Central Asia | High mountains | Cold |
| | Chyangra | Nepal | High mountains | Cold |
| | Singhal | Nepal | High mountains | Cold |
| Skins | | | | |
| | Black Bengal | India | Tropical | Dry |
| | Maradi (Red Sokoto) | Niger and Nigeria | Tropical | Humid |
| | Mubende | Uganda | Tropical | Humid |

Source: Shrestha and Fahmy (2005).

the diversity in production performance based on available reports has been summarized by region of the world in Table 3.2. This represents differences among breeds within environment as well as production performance. Europe has the heaviest breeds with the largest litter size and the highest milk yield. Africa has the largest variation, possibly attributable to diversity among regions and irregularities in recording (Galal, 2005).

Irrespective of its superior genetics and ability to adapt to varying environments, no single breed can meet the requirements of the discriminating consumers worldwide. Indeed, the vast array of food and fiber products available within and among nations testifies to the diversity of needs both among producers and among their respective markets. Goat milk and milk products are well liked in Europe and the Middle East regions, whereas goat meat is considered a delicacy in Asia and the Pacific region. Cashmere production is popular in remote areas adjoining the high-mountain and the Himalayan ranges. Furthermore, the discriminating tastes and preferences of the consumer as well as the changing cultural mosaic of the populations in developed countries have created a demand for meat, cheese, and cashmere from goats. In the future, climatic change, emergence of exotic diseases, development of export markets, and the changing economic status of a country as well as individuals within the country are likely to further alter product availability and preferences. Altered desires and preferences of consumers can be expected to have a profound influence on the choice of goat populations to be kept in specific environments in accordance with fiscal constraints.

## EROSION OF GENETIC RESOURCES

The colossal amount of genetic variability within and among goat breeds provides the raw material necessary to achieve genetic improvement and to meet current and future market requirements. However, genetic variability in goats has steadily decreased at an alarming rate due to natural causes, man-made disasters, loss of habitat, neglect and abandonment of the well-adapted indigenous landraces, breeds, and populations. Also active utilization of few improved breeds to propagate genes necessary for increased production efficiency through skillful human activity has been recurring. According to FAO (2000), nearly 800 farm animal breed populations in the world already have become extinct. Among those remaining, 30% could be considered

**Table 3.2** Diversity in the production performance of goats by region of the world.

| Region/performance | Goats[1] | Mean ± S.D. | Minimum | Maximum |
|---|---|---|---|---|
| **Africa** | | | | |
| Body weight (kg) | | | | |
| Male | 32 | 53 ± 37 | 20 | 130 |
| Female | 32 | 39 ± 22 | 20 | 94 |
| Litter size (kid) | 17 | 1.4 ± 0.33 | 1 | 2.1 |
| Milk yield (kg/lactation) | 3 | 126 ± 116 | 50 | 500 |
| **Asia and Pacific** | | | | |
| Body weight (kg) | | | | |
| Male | 106 | 43 ± 17 | 16 | 130 |
| Female | 106 | 32 ± 12 | 14 | 100 |
| Litter size (kid) | 79 | 1.4 ± 0.37 | 1 | 2.9 |
| Milk yield (kg/lactation) | 63 | 136 ± 109 | 16 | 550 |
| **Europe** | | | | |
| Body weight (kg) | | | | |
| Male | 123 | 66 ± 13 | 35 | 120 |
| Female | 124 | 49 ± 12 | 24 | 120 |
| Litter size (kid) | 28 | 1.6 ± 0.35 | 1 | 2.2 |
| Milk yield (kg/lactation) | 41 | 299 ± 225 | 40 | 775 |
| **Latin America and Caribbean** | | | | |
| Body weight (kg) | | | | |
| Male | 10 | 40 ± 14 | 15 | 70 |
| Female | 11 | 30 ± 13 | 13 | 50 |
| Litter size (kid) | 11 | 1.4 ± 0.3 | 1.1 | 2.0 |
| Milk yield (kg/lactation) | 2 | 63 ± 4 | 60 | 65 |
| **Near East** | | | | |
| Body weight (kg) | | | | |
| Male | 48 | 44 ± 14 | 17 | 75 |
| Female | 47 | 33 ± 8 | 15 | 50 |
| Litter size (kid) | 27 | 1.6 ± 0.43 | 1.1 | 2.5 |
| Milk yield (kg/lactation) | 32 | 150 ± 97 | 35 | 460 |
| **North America** | | | | |
| Body weight (kg) | | | | |
| Male | 5 | 41 ± 21 | 22 | 67 |
| Female | 4 | 35 ± 26 | 13 | 60 |
| Litter size (kid) | 4 | 1.0 | 1 | 1 |

[1]Number of goat breeds with records.
Source: Galal (2005).

**Table 3.3** Risk status in actual number (%) of goat breeds by region of the world.

| Region | Unknown | Not at risk | Extinct | At risk | Total |
|---|---|---|---|---|---|
| Africa | 38 (43) | 50 (56) | 0 | 1 (1) | 89 |
| Asia and Pacific | 29 (20) | 50 (34) | 1 (1) | 67 (46) | 147 |
| Europe | 25 (12) | 86 (43) | 14 (7) | 76 (38) | 201 |
| Latin America and Caribbean | 19 (56) | 15 (44) | 0 | 0 | 34 |
| Near East | 47 (49) | 40 (42) | 1 (1) | 7 (7) | 95 |
| North America | 8 (38) | 3 (14) | 1 (5) | 9 (43) | 21 |
| Total | 166 (28) | 282 (41) | 17 (3) | 160 (27) | 587 |

Source: Galal (2005).

endangered. The number of livestock species in the world deemed at risk of extinction has been estimated at 27% in goats, 22% in sheep, 13% in buffaloes, 29% in cattle, 40% in pigs, 42% in horses, 23% in asses, and 8% in camels (Galal, 2005).

Most traditional breed populations and landraces that have remained in their habitats for centuries, currently are experiencing urban encroachment. The environmental impact from the destruction of habitats compromises the survival of many breed populations. This situation is more critical in the nomadic pastoralist systems due to predators, overgrazing, loss of grazing rights in the mountain and forest pastures, and abandonment of the traditional profession. The majority of the endangered goat breeds, worldwide, rely on a very small number of male and female parents for their propagation; this minimizes the effective population size. There is a general agreement among geneticists that small populations are subject to inbreeding depression and genetic drift that result in the recurring loss of genetic variation and erosion of genetic diversity.

The FAO (1998, 2000) defines "breeds at risk" (critical and endangered) as those with fewer than 1,000 females, with fewer than 20 males, or with a total population size approaching 1,200. The risk increases as the proportion of females declines below 80%. According to this definition goats rank next to buffaloes in percentage of breeds not at risk worldwide (41% versus 64%). At the same time, 3% of the goat breeds have become extinct compared to 12% for sheep breeds and 17% for cattle breeds. Among geographical regions, the highest percentage of extinct goat breeds was in Europe (7%) followed by North America (5%), with the rate of extinction ranging from zero to one in other regions. Yet, goats have the largest percentage of breeds (28%) in the world whose populations are not reported or are unknown compared to other ruminant livestock. This is understandable because goats are distributed in vast areas of the world, and collection of census data is difficult. The risk status for goat breeds by region of the world is presented in Table 3.3. The low rate of extinction can be attributed to goats not being popular in the industrialized countries, thus limiting their exposure to intensification when compared to other livestock and poultry species.

## GLOBALIZATION

Globalization is defined as the process of unifying people or a local phenomenon into a single society and functions as a global phenomenon involving a combination of economical, technological, sociocultural, and political aspects. Globalization involves integration of national economies through international trade, foreign investment, capital flow, migration, and spread of technology. One effect of globalization in agriculture has been the transformation of family operated farms into agribusinesses. Larger farms seek to raise livestock and poultry breeding stocks of superior genetic merit and utilize mechanization and farm buildings or extensive pastures with relatively few employees. This technology has achieved considerable success in increasing food production worldwide to meet the demand of the growing human population.

In the last two decades, national and international attention has been drawn toward the depletion of biodiversity, that includes domestic animals. The erosion of genetic resources that has been occurring at an alarming rate could lead to a critical level in the near future so that the ability of the world communities to develop livestock and poultry to meet current demands and future market requirements for commodities, trade, breeding stocks, employment, and recreation would be impaired.

Many traditional communities worldwide rely on goats and other farm animals for their livelihood, while exporting low value products. This is because animal agriculture ensures food sovereignty, alleviates poverty, and creates wealth in harmony with cultural patrimony. Furthermore, owning goats and other livestock and poultry often has been an indication of wealth and prosperity. The use of animals for religious rituals, insurance against crop failures, family employment, dowry, barter, and payment in kind during conflicts has been widespread and supports sustainable rural development. Small farmers and the landless have, through the ownership of goats and their multifunctionality, survived for centuries in the face of hunger and poverty.

Many of the developing countries, while attempting to increase food production, have followed the path of the developed countries, making substantial strides in increasing production efficiency. However, this rapid pace of development has been achieved at the cost of losing indigenous genetic resources that are not capable of matching the higher productivity of the developed breeds based on present criteria for productivity. Nevertheless, the selection criteria could change in the future as it has in the past. The proportion of goats in developing countries compared to the total world has increased from 96% in 1980 to 98% in 2006; developing countries account for 77% of all of the goat breeds. In the developing countries, the loss of goat breed populations may be greater than typically reported numbers because of the absence of policy and legal instruments to assist in the identification and documentation of all of their domestic animal genetic resources.

In the years following the Second World War, the number of affluent families in the industrialized nations increased rapidly. This created a requirement for a wider range of commodities, facilitating the transformation of wartime technology and economy of scale to help improve production efficiency in agriculture. Research with corn provided scientists with clear evidence of the potential for improvement of productivity from the genetic response to selection and hybrid vigor. This prompted the development and application of quantitative genetic principles for the commercial exploitation of domestic livestock and poultry. Livestock and poultry populations were developed with exceptional productivity and economically important morphological characteristics and production performance. Breeders used a narrow genetic base within a small number of breeds and outstanding animals within each breed to achieve rapid genetic response resulting in populations with exceptional productivity.

Breed populations with superior genetics for specific environments were widely accepted by the commercial breeders. At the same time, the quality and quantity of animal products produced gained consumer acceptance. The multinational corporations seized the opportunity to create wealth from increased production efficiency. Concurrently, all aspects of production were integrated to provide producers with the necessary financing, breeding stocks, animal feed, disease control, insurance, promotion, processing, packaging, marketing, and research findings. The concept of integration was promoted throughout the developed countries and later worldwide, leading to the globalization of animal agriculture. Consequently, advances in quantitative genetics and reproductive technology helped produce more offspring from fewer animals with superior genetic merit that were readily multiplied and capable of supplying the worldwide requirement of breeding stock (Shrestha and Fahmy, 2007a,b). Nearly all genetic resources developed through this process were high-input/high-output breeds and/or crosses.

## CONSERVATION OF BREEDS

Conservation is defined as the analysis and protection of the earth's biological diversity. There is unequivocal agreement among researchers that the colossal amount of genetic variability within domestic animal genetic resources is extremely important to achieve genetic response to selection for current and future market requirements. Many breed populations in the world share a large proportion of genetic information, and it is practically impossible to conserve all the breeds irrespective of their genetic merit or uniqueness. To establish the uniqueness of goat populations, a multitude of events needs to be assessed. These can include complementary historical evidence, cultural heritage, religious rituals, geographic distribution of the habitat, and ultimately, the molecular genetic characteristics of the population. Furthermore, political scrutiny, morphological characteristics, production performance, economic importance, and scientific knowledge need to be addressed.

Historical evidence could range from archaeological findings of animal remains and rock paintings. At the same time, ancestral knowledge and folklore can be important in the absence of written records. Throughout history, there has been extensive migration of pastoralists for better grazing lands, of tribes for economic reasons and food security, settlement following the discovery of new lands, the influence arising from the importation of highly productive exotic animals, and most importantly, natural selection. The Mongols, Romans, Arabs, and Europeans and their subsequent colonization in the world have had significant impacts on the choice, distribution, and method of raising goats and its role in farming. Breed information derived from the goat genome may help establish relationships among goat breed populations worldwide, as well as establish genomic inbreeding. Scientific studies of the genome could contribute to genetic conservation. Genetic principles investigated include polymorphism, gene frequencies, genetic distances, and degree of heterozygosity estimated from blood groups and genetic markers including isozymes, microsatellites and single nucleotide polymorphism (SNP) in the DNA, and sequence variation in the mtDNA (Shrestha et al., 2008).

The majority of endangered goat breed populations worldwide rely on few male and female parents; this results in small effective population sizes that contribute to loss of genetic variation. The best way to reverse this situation is to increase the effective number of parents, particularly by using unrelated male parents. In closed endangered populations, theoretically it is possible to minimize the influence of inbreeding and genetic drift by dividing populations into several complementary breeding groups of one male and one female parent. In the following year, male offspring from the first group may be bred to female offspring from the subsequent group. This would continue in the following years until the male descendent of the first group would be bred to the female descendent of the last group according to a pedigreed breeding structure. This approach may not be practical in most populations of goats because in practice herds maintain about 1 buck for every 20 to 50 does (Shrestha et al., 2008). Nevertheless, breeders with endangered goat populations should attempt to increase the number of bucks that are used as parents of the following generation, or risk loss of genetic variation.

An increase in inbreeding at the rate of 0.1% per generation may be tolerated as long as the loss of genetic variation in commercial herds is prevented (Hill, 1982). This same rate of inbreeding may be considered as the tolerable level in the conservation of endangered breed populations. Furthermore, populations with an effective population size of more than 100 should be able to sustain a genetic response to selection for economically important traits. In the absence of records, a random breeding structure where each male parent has an equal chance of mating with a female parent may be appropriate. The rate of inbreeding can be reduced by 50% with pedigreed breeding compared to random breeding in a small population. Concurrently, in theory, equalizing the number of male and female parents could decrease the variance of progeny numbers per parent, thus increasing the effective population size.

The reestablishment of animal breeds, populations, or landraces in one generation is possible with the cryoconservation of spermatozoa and embryos for potential transfer to an appropriate donor animal. There are reports of closely related species having acted as surrogate dams to reproduce implants from zygotes or clones of endangered domestic animals that are at the verge of extinction. An alternate approach is to backcross the endangered population to a closely related breed for about seven generations resulting in animals with 99% of the original breed. Among others, Alderson (2003) has proposed qualitative criteria as a means for prioritizing breeds of special genetic importance.

Conservation strategies should identify important sources of interruption responsible for the endangerment of the breed population. These may include habitat loss, breed replacement, natural calamity, and loss of genes within individuals. After the sources of interruption have been identified, appropriate policies and action plans to sustain domestic goat diversity need to be developed and implemented while resolving the unique status of the breed population. Conservation breeding strategies for domestic goats under threat of extinction or endangerment can benefit from the following: historical evidence that indicates the unique status of goat populations; existing scientific knowledge of application of quantitative genetic principles; advances in husbandry and disease control measures; and cryoconservation (Shrestha et al., 2008). Conservation of an endangered breed or population can best be realized by "conservation through utilization," (that is, finding ways to effectively use the breed in the prevailing or a modified production systems).

There are 80 composite breed populations of goats that have been developed in 37 countries (Table 3.4). This list is neither comprehensive nor does it necessarily imply that these breeds/populations remain functionally present today. A number of goat breeds with potential for improving efficiency in the production of meat, milk, fiber, and skin under varying agroecological zones (Devendra, 1991) could be used for developing composite breed populations. There is an advantage in saving breeds considered less productive by current standards, especially those that are at risk of extinction that may become of value in the future with changes in production system, environment, and consumer preference for animal and animal products (Shrestha, 2005).

Development of composite breed populations from a combination of an endangered breed with complementary breeds of superior genetic merit for morphological

**Table 3.4** Noncomprehensive list of composite breed populations of goats that have been developed in the world, number of foundation breeds, and the year of origin or year recognized by country of origin.

| Country | Composite breed/ population | No. of foundation breeds[†] | Country | Composite breed/ population | No. of foundation breeds[†] |
|---|---|---|---|---|---|
| Australia | | | Kenya | | |
| | Cashgora | 2 | | Kenya Dual-Purpose | 4 |
| Brazil | | | Kyrgyzstan | | |
| | Branca sertaneja | 2 | | Kirgiz | 2 |
| | Parda sertaneja | 2 | Mongolia | | |
| | SRD | 2 | | Gobi Wool goat | 2 |
| Bulgaria | | | | Unjuul | 2 (1982) |
| | Bulgarian White Dairy | 2 | | Uuliin Bor | 2 (1991) |
| China | | | Morocco | | |
| | Guanzhong Dairy | 2 (1940) | | Fnideq | 2 |
| | Hailun | 3 | Mozambique | | |
| | Hongtong | 2 | | Pafuri | 2 (1928) |
| | Laoshan Dairy | 2 (1919) | Netherlands | | |
| | Nanjiang Yellow | 2 (1960) | | Dutch Pied | 2 |
| Cyprus | | | | Dutch Toggenburg | 2 |
| | Peratiki | 2 | | Dutch White | 2 |
| Denmark | | | Nigeria | | |
| | Danish Landrace | 3 | | Savanna Brown | 2[†] |
| Fiji | | | Norway | | |
| | Fiji | 3 | | Norwegian | 5 |
| France | | | New Zealand | | |
| | French Alpine | 2 (1930)[a] | | Kiko | 2 |
| Germany | | | Pakistan | | |
| | German Improved Fawn | 2 (1928)[a] | | Beiari | 2 |
| | German Improved White | 2 (1928)[a] | | Buchi | 2 |
| Hungary | | | | Jattal | 2 |
| | Hungarian Improved | 2[†] | | Pak Angora | 2 |
| India | | | | Shurri | 2 |
| | Indian Mohair | 3 (1973) | | Sind Desi | 2 |
| | Malabari | 2 | Romania | | |
| | Ramdhan | 2 | | Banat White | 3 |
| Indonesia | | | Russia | | |
| | Peranakan Etawah | 2 | | Altai Mountain | 2 (1982) |
| Israel | | | | Angora-Don | 2 |
| | Israeli Saanen | 2 (1932) | | Dagestan White | 2 |
| | Yaez | 2[†] | | Don-Kirgiz cross | 2 |
| Italy | | | | Russian White | 2 (1905) |
| | Aquila | 4 | South Africa | | |
| | Benevento | 4 | | Boer | 2 (1959)[a] |
| | Campobasso | 4 | Spain | | |
| | Ionica | 2 (1981)[a] | | Barreña | 3 |
| | Potenza | 3 | | Murcia-Granada | 2 (1980)[†] |
| Kazakhstan | | | | Murcian | 2 (1933)[†] |
| | Soviet Mohair | 2 (1962) | | | |

**Table 3.4** *Continued*

| Country | Composite breed/ population | No. of foundation breeds[†] | Country | Composite breed/ population | No. of foundation breeds[†] |
|---|---|---|---|---|---|
| Tajikistan | | | United Kingdom | | |
| | Soviet Mohair | 2 (1962) | | | |
| Tanzania | | | | British Alpine | 2 (1925)[a] |
| | Blended goat | 3 | | British Cashmere | 5 |
| Togo | | | | British Saanen | 2 |
| | Vogan | 2 | | British Toggenburg | 2 |
| Turkey | | | | English Guernsey | 2[†] |
| | Angora | 2 (1900)[a] | | Golden Guernsey | 2[†] (1970)[a] |
| | Bornova | 3 | U.S. | | |
| | Çukurova | 2 | | Kinder | 2 (1988)[a] |
| | Kilis | 2 | | Pygora | 2 (1987)[a] |
| | Taurus | 2 (1973) | Uzbekistan | | |
| Turkmen | | | | Soviet Mohair | 2 (1962) |
| | Soviet Mohair | 2 (1962) | | Uzbek | 2 |
| United Kingdom | | | | Uzbek Black | 2 (1961) |
| | Anglo-Nubian | 4 (1910)[a] | | | |
| | British | 3 (1896)[a] | | | |

[a]Year of origin or year recognized.

[†]Year breed's society, association, or stud book was established.

Sources: Mason (1996); Shrestha (2005).

characteristics and production performance could be beneficial in conservation through utilization (Shrestha, 2005). Conservation may be in situ (that is, in the location where the breed population naturally exists) or ex situ, where the animals are translocated to different areas as in the case of ecofarms, zoos, reserves, and historical sites. The ex situ conservation of germplasm (frozen gametes, embryos, or somatic tissues) has achieved considerable success in goats. The subsequent propagation of the breed population should occur without disrupting the cultural integrity, religious rituals, traditions, political affairs, and food security among nations.

The conservation of domestic livestock and poultry that first began in the U.K. with the "Rare Breeds Survival Trust" has spread to a number of countries in the world. The FAO instigated a global initiative for the conservation of domestic animals and published a World Watch list of domestic animal diversity (FAO, 2000). Furthermore, the Nordic Gene Bank of Farm Animals is one of the most advanced conservation institutions. Following the ratification of the Convention on Biodiversity, many countries in the world have realized the importance of conserving domestic animal diversity and started identifying and characterizing indigenous and established domestic animal populations, as well as preserving their genetic resources. Active participation in their preservation is being achieved in cooperation with conservationists, biologists, geneticists, archaeologists, and breeders as well as nongovernmental organizations.

## SUMMARY

Goat is among the earliest livestock that was domesticated around 9000 to 10,000 BC at the border between Iran and Iraq. The zoo-archaeological records that predate the time of domestication indicate that the ancestors of the present-day domestic goats started from three genetically distinct maternal lineages rather than a single wild population. Wild goats of the Bezoar or Pasang (*Capra aegagrus aegagrus*), Makhol (*C. aegagrus falconer*), and Ibex (*C. aegagrus ibex*) types are believed to be the ancestors of domestic goats worldwide. Pastoralists moved with their goats in different directions spreading goats to different continents. More recently colonization and the discovery of new continents carried the goat to further lands. Today, 1,153 breeds of goats exist, but this wealth of diversity is exposed to erosion due to intensification of the production system and excessive dependence on a few high-input/high-output breeds. Diversity needs to be maintained to develop breeds with potential merit for current use and for future unknown situations. To sustain the diversity among goat breeds, we must characterize the breeds, enumerate their populations, conserve populations at risk of extinction, especially those breeds that possess unique characteristics. It has been long said that "the goat is the poor man's cow." In more current terminology, the goat has been recognized as an important tool for poverty alleviation particularly among the rural poor and the landless.

## REFERENCES

Alderson, L. 2003. Criteria for the recognition and prioritisation of breeds of special genetic importance. Animal Genetic Resources Information 33: 1–10.

Angel-Neelem, M. and M.A. Nellem. 1998. Goat production in Argentina. Example of the Santiago del Estero province. Capricorne 11, 2 and 11–16.

Clutton-Brock, J. 1999. A Natural History of Domestic Animals. Cambridge University Press, UK.

Devendra, C., 1991. Breed differences in productivity in goats. In: Maijala, K. (Editor) World Animal Science. B8. Genetic resources of pig, sheep and goat. Elsevier, Amsterdam, The Netherlands, pp. 431–440.

Devendra, C. 2007. Small farm systems to feed hungry Asia. Outlook on Agriculture, 36: 7–20.

Epstein, H. 1971. The Origin of the Domestic Animals of Africa, Vol. II. Africana Publishing Corporation, New York, 719 pp.

FAO. 1998. Secondary Guidelines: management of small populations at risk. Food and Agriculture Organization of the United Nations, Rome, Italy, 215 pp.

FAO. 2000. World Watch List of Domestic Animal Diversity. Third Edition, Rome, Italy, 726 pp.

FAO. 2009. http://dad.fao.org/accessed. Retrieved October 2009.

Fernández, H., S. Hughes, J.D. Vigne, D. Helmer, G. Hodgins, C. Miquel, C. Hänni, G. Luikart, and P. Taberlet. 2006. Divergent mtDNA lineages of goats in an Early Neolithic site, far from the initial domestication areas. Proceedings of the National Academy of Sciences 103: 15375–15379.

Galal, S. 2005. Biodiversity in goats. Small Ruminant Research 60: 75–81.

Hill, W.G. 1982. Prediction of response to artificial selection from new mutations. Genetic Research 40: 255–278.

Jalaludin, S., Y.W. Ho, N. Abdullah, and H. Kudo. 1992. Strategies for animal improvement in southeast Asia. In: Proceedings of the 25th International Symposium on Tropical Agricultural Research, Tsukuba, Japan, 24–25 September 1991. Tropical Agriculture Research Series 25: 67–76.

Libunao, L.P. 1990. Goat/fish integrated farming in the Phillippines, Ambio 19, 8 and 408–410.

Lush, J.L. 1945. Animal Breeding Plans. Iowa State University Press, Ames, Iowa, 443 pp.

Mason, I.L. 1996. A World Dictionary of Livestock Breeds, Types and Varieties. Fourth Edition, Commonwealth Agricultural Bureaux International, Wallingford, Oxon., UK, 273 pp.

Shrestha, J.N.B. 2005. Conserving domestic animal diversity among composite populations. Small Ruminant Research 56: 3–20.

Shrestha, J.N.B. and M.H. Fahmy. 2005. Breeding goats for meat production: a review. (1) Genetic resources, management and breed evaluation. Small Ruminant Research 58: 93–106.

Shrestha, J.N.B. and M.H. Fahmy. 2007a. Breeding goats for meat production: a review (2) Crossbreeding and formation of composite population. Small Ruminant Research 67: 93–112.

Shrestha, J.N.B. and M.H. Fahmy. 2007b. Breeding goats for meat production: a review (2) Selection and breeding strategies. Small Ruminant Research 67: 113–125.

Shrestha, J.N.B., S. Galal, A. da S. Mariante, A. Kotze, J.V. Delgado, and C. Devendra. 2008. Impact of globalization on livestock and poultry genetic resources in the world. In: Proceeding of the 7th RBI Global Conference of Conservation of Animal Genetic Resources, Hanoi, Vietnam, pp. 1–6.

Skinner, J.D. 1972. Utilization of the Boer goat for intensive animal production. Tropical Animal Health Production 4: 120–128.

Vigne, J.D., D. Helmer, and J. Peters. 2005. The First steps of Animal Domestication: New Archaeo-zoological Approaches. Oxbow Books, Oxford, UK.

Zeder, M.A. and B. Hesse. 2000. The initial domestication of goats (*Capra hircus*) in the Zagros mountains 10,000 years ago. Science 28: 2254–2257.

## ADDITIONAL SUGGESTED READINGS

Alderson, L. and I. Bodo. 1992. Genetic Conservation of Domestic Livestock. Vol. II, Commonwealth Agricultural Bureaux International, Wallingford, UK, 282 pp.

Balter, M. 2007. Seeking agriculture's ancient roots. Science 316: 1830–1835.

Bruford, M.W., D.G. Bradley, and G. Luikart. 2003. DNA markers reveal the complexity of livestock domestication. Nature Reviews—Genetics 4: 900–910.

Hodgson, R.E. 1961. Germplasm Resources. American Association for the Advancement of Science, Washington D.C., USA, Publ. No. 66, 381 pp.

Kotzé, A., J. Bester, U.S. Kusel, and I. Plug. 1994. Early domesticated animals of South Africa. In: Proceedings of the Third Global Conference on Conservation of Domestic Animal Genetic Resources, Canada, pp. 338–341.

Mariante, A. da S., A.A. do Egito, M. do S.M. Albuquerque, and U.G.P. de Abreu. 1998. Iberian livestock breeds in Brazil: Almost 500 years of natural selection. Proceedings of the Fourth Global Conference on Conservation of Domestic Animal Genetic Resources, Nepal, pp. 109–112.

Oldenbroek, J.K. 1999. Genebanks and the conservation of farm animal genetic resources. DLO Institute of Animal Science and Health, The Netherlands, 119 pp.

Rodero, A., J.V. Delgado, and E. Rodero. 1992. Primitive Andalusian livestock and their implications in the discovery of America. Archivos de Zootecnia (extra), 383–400.

Simon, D.L. and D. Buchenauer. 1993. Genetic diversity of European livestock breeds. EAAP Publ. No. 66, Wageningen Press, Wageningen, The Netherlands, 591 pp.

# 4
# Breeding and Genetics

*J.N.B. Shrestha, PhD and G.H. Crow, PhD*

**KEY TERMS**

Mendelian genetics—the study of underlying principles governing inheritance of morphological characters according to the laws of Segregation and Independent Assortment.

Population genetics—the area of genetics that deals with genetic composition of biological populations and patterns of genetic changes as a result of evolution and natural selection.

Quantitative genetics—the study of genetic principles governing inheritance of continuous measures of performance determined by many genes and an environmental component.

Breeding—application of quantitative genetic principles and skillful practice in selection and mating systems to improve the efficiency of animal production.

Selection—disproportionate contribution of the parents to their offspring as a result of differences in fitness.

Crossbreeding—crossing of two or more divergent breeds to produce offspring.

**OBJECTIVES**

By completing this chapter, the reader will acquire knowledge on:

- The underlying principles governing inheritance of morphological characteristics and production performance from parents to their offspring
- How forces of evolution and natural selection cause gene frequency changes in biological populations
- That quantitative traits in an individual follow an extension of the basic Mendelian principles of inheritance
- The role of genetics and environment in determining the phenotype of the individual
- Many specialized goat breeds that remain in the world have potential genetic merit to respond to current and future market requirements
- The inherent potential for performance transmitted from parents to their offspring is additive for many traits of economic importance
- The role of nonadditive genetic variation associated with the crossing of complementary breeds
- That real improvement of production efficiency can be achieved with the application of quantitative genetic principles to exploit biological potential
- The socioeconomic values, fiscal constraints, religious rituals, responsiveness to indigenous knowledge and traditional skills in improving production efficiency of goats

## INTRODUCTION

Long before any knowledge of the underlying principles governing inheritance was established, domestic animals were chosen according to the perception of the breeder. In 1865 and 1866 (rediscovered in 1900), Gregor Mendel, also known as the father of genetics from Brno (now in the Czech Republic), published the principles of segregation and independent assortment governing transmission of morphological characters from parents to their offspring. This early work established the theoretical basis of modern genetics. Although these studies were conducted in plants, the nature of inheritance was in the same way applicable to all animal species.

Mendel studied characteristics of peas in which there were only two alleles for each trait in a population, and one allele was dominant to a recessive allele. The inheritance of wattles in goats is an example of such a characteristic (Lush, 1926). Wattles are loose appendages of skin on the neck of some goats. The alleles for the presence and absence of wattles can be denoted as *W* and *w*, respectively, with *W* dominant to *w*. The three possible genotypes are the homozygous dominant *WW*, the heterozygote *Ww*, and the homozygous recessive *ww*. Both *WW* and *Ww* possess wattles and *ww* individuals do not. In a dihybrid mating *Ww* × *Ww*, the expected outcome is 1 *WW*: 2 *Ww*: 1 *ww* in terms of genotypes and 3 Wattle: 1 Abscence of wattle in terms of phenotype.

The dominance-recessive model does not fit all situations. There are some instances where the heterozygote is distinguishable from the two homozygotes. When the heterozygote is intermediate in appearance to either homozygote, the gene action is additive. If the heterozygote is more similar to one of the homozygotes, gene action is termed incomplete dominance.

In a population of animals, there may be more than two alleles for a characteristic. Any one animal may possess only two, but in the population as a whole, many alleles may exist. The *Agouti* gene is one of several genes affecting coat color in goats. According to Adalsteinsson et al. (1994), there may be as many as 11 alleles at this locus, when considering many breeds. Also these alleles produce effects as varied as a white coat, to grey and red.

In the years following Mendel's work, the nature of gene action was discovered to include additive and dominance effects as well as concepts of epistasis, pleiotrophy, and linkage. Population genetics soon followed, and concepts such as Hardy Weinberg equilibrium and inbreeding were developed. The important role of mutation, migration, selection and genetic drift as forces of evolution and natural selection became apparent. The present day goat can be attributed to the effects of many of these forces along with skillful breeding in the premises of the breeder.

Improvement of productivity from genetic response to selection and hybrid vigor was consistent leading to the development and application of quantitative genetic principles for the commercial exploitation of domestic livestock and poultry. Breeding methodologies such as selection, crossbreeding, and formation of composite populations achieved considerable success in exploiting the biological potential of domestic animals. The outcome has been the development of livestock and poultry with increased production efficiency that provide a reasonable economic return over investment. At the same time, the quality of animal products produced has been widely accepted by the consumer.

## POPULATION GENETICS

Population genetics is important in animal production since herds that represent the production units are in fact populations. Genetic improvement of a herd of goats thus involves management of the genes in a population. In the 1920s and 1930s, the principles of Mendelian genetics with Darwinian natural selection were successfully integrated by R.A. Fisher, J.B.S. Haldane, and Sewall Wright leading to the development of population genetics. Population genetics came to occupy an important role in evolutionary theory as well as in the practical improvement of farm animal populations.

### Genotype and Gene (Allele) Frequency

In population genetics, the genetic composition of a population is described in terms of the frequency of the individual's genotype, for example, *BB*, *Bb*, and *bb*. Gene frequencies are calculated from these genotypic frequencies by considering that the homozygotes possess two copies of an allele and heterozygotes possess only one copy of a given allele. The frequencies of *B* and *b* alleles in the population are the proportion of all *B* or *b* alleles counted across both the homozygous and heterozygous individuals and together add to 1.0. When allele frequencies are known for a population, the frequencies of offspring of each genotype can be determined from the frequencies of mating of their parents. Let the frequency of the *B* allele be equal to $q$ and those of the *b* allele be equal to $(1 - q)$. In a population where these alleles have the same frequencies in bucks and does, the mating will result in the following frequencies of genotypes (Table 4.1).

The sum of the frequencies of the three genotypes is equal to $q^2\ \{BB\} + q(1 - q)\ \{Bb\} + q(1 - q)\ \{Bb\} + (1 - q)^2\ \{bb\} = 1$. If $q = 0.6$ as in the numerical example shown

**Table 4.1** Frequency of alleles from bucks and does and genotypic frequency of their offspring.

| | | Alleles from bucks | |
|---|---|---|---|
| | Allele | $B$ ($q$) | $b$ $(1 - q)$ |
| Alleles from does | $B$ ($q$) | $BB$ ($q^2$) | $Bb$ $\{q(1 - q)\}$ |
| | $b$ $(1 - q)$ | $Bb$ $\{q(1 - q)\}$ | $Bb$ $\{(1 - q)^2\}$ |

Note: $q$ and $(1 - q)$ are the frequencies of $B$ and $b$ alleles, and $BB$, $Bb$ and $bb$ are genotypes.

above, the genotype frequencies in offspring are expected to be $F_{BB} = q^2 = 0.36$, $F_{Bb} = 2q(1 - q) = 0.48$ and $F_{bb} = (1 - q)^2 = 0.16$.

**Hardy-Weinberg Equilibrium**

A large random mating population of goats in the absence of mutation, migration, selection, or drift will have gene and genotype frequencies that remain unchanged from generation to generation. This state is referred to as Hardy-Weinberg equilibrium.

An application of the relationship that exists between allele frequencies and genotype frequencies is as follows: consider *black coat color* to be recessive to *white coat color* in goats. A large herd of *white* goats gives birth to one *black* kid out of every 100 kids. The genotypic frequency of the *black color* is expected to represent $(1-q)^2 = \frac{1}{100} = 0.01$. It is important to note that this is true in the absence of mutation, migration, selection or drift acting to change gene frequencies. The allelic frequency can thus be calculated as the square root of the frequency of the recessive genotype as

$$(1-q) = \sqrt{(1-q)^2} = \sqrt{0.01} = 0.10$$

It is then possible to calculate the frequency of the allele for *white color* of the kid $q = 1.0 - 0.1 = 0.9$. Similarly, frequency of the *white color* ($BB$) will be $q^2 = 0.9 \times 0.9 = 0.81$ or 81%, heterozygote ($Bb$) will be $2q(1 - q) = 2 \times 0.1 \times 0.9 = 0.18$ or 18%, and those of the *black color* ($bb$) will be $(1 - q)^2 = 0.1 \times 0.1 = 0.01$ or 1%.

**Forces That Change Gene Frequencies**

MUTATION

There are a large number of detrimental and lethal conditions controlled by a single recessive gene with a mutation. In some cases when the goat suffers from an inherent disorder, the animal is able to survive and reproduce but the condition is detrimental (for example, polled bucks with hypoplasia of the testes). True lethal conditions result in death before or shortly after birth, delayed lethal condition is expressed later in the life, and partial lethal condition becomes fatal under certain conditions. Hutt (1964) has described a number of lethal and semilethal conditions in chickens.

Muscular hypertrophy in cattle, also known as double muscling, is an example of a gene that can have detrimental effects as well as favorable effects. Calves homozygous for the gene have an increased incidence of difficult calving and thus increased calf mortality at birth but possess up to 20% more muscle mass than normal animals that are homozygous. The molecular basis for a lethal condition arising from an inherited disorder in more than 135 traits of animal species other than the human and the mouse can be found in a database of genes available online describing Mendelian Inheritance in animals (Nicholas, 1987).

Mutation rates are generally very small and cause little change in the population unless its effects are accumulated over many generations. Mutation in combination with other forces acting on gene frequencies, however, can be a very important force. Selection, for example, can rapidly increase the frequency of a rare mutant allele.

In breeding populations where bucks are to be widely used to produce large numbers of kids across many herds it is important to establish if the animal is a carrier of alleles for lethal or deleterious conditions. This can be achieved by test mating to mates known to carry a recessive allele for a condition, or to close relatives such as daughters. DNA tests are available for a few conditions. When an abnormal condition conforming to a well-known syndrome appears, the condition may be genetic in origin.

When there is a genotype × environment interaction, the condition would occur in specific environments. The detrimental condition may be environmental in origin. The condition may be a result of lack of amino acid, vitamin, or mineral and in some cases, from the consumption of toxic plants. Detrimental conditions introduced by environment do not recur when the environment is changed.

MIGRATION

In a large population, in each generation the proportion of migrants and natives are $m$ and $(1 - m)$, respectively. Let the frequency of the allele in the natives and migrants be $q_0$ and $q_m$, respectively, and after one generation those of the mixed (migrants and natives) population be $q_1$. The frequency in the mixed population written as the proportion of migrants and their allelic frequency together with the proportion of natives and their allele frequency can be described as: $q_1 = mq_m + (1 - m)q_0$. Consider a population of 100 goats consisting of 10 migrant and 90 native goats. Let the frequency of alleles in the migrant ($q_m$) and native ($q_0$) populations be 0.5 and 0.01, respectively. The allelic frequency of the population in one generation ($q_1$) following the introduction of migrant goats will be:

$$q_1 = mq_m + (1-m)q_0 = \frac{10}{100} \times 0.5 + \frac{90}{100} \times 0.01 = 0.059.$$

The expression for the frequency of allele in the mixed population after one generation can be simplified as:

$$\begin{aligned} q_1 &= mq_m + (1-m)q_0 = mq_m + q_0 - mq_0 \\ &= mq_m - mq_0 + q_0 = m(q_m - q_0) + q_0. \end{aligned}$$

The change in allelic frequency from one generation to another ($\Delta q$) is the difference in the frequency of the alleles before ($q_0$) and after migration ($q_1$). $\Delta q = q_1 - q_0 = m(q_m - q_0) + q_0 - q_0 = m(q_m - q_0) = 0.10 \times (0.5 - 0.01) = 0.049$.

The rate of change in allelic frequency depends on the proportion of migrants and the difference in allelic frequencies of migrant and native populations. Therefore, the change in gene frequency of the population as a result of migration is 0.049.

### Selection

When goats in a population contribute unequally to the next generation, selection occurs. The reason for this unequal contribution is that goats differ in their reproductive fitness, and some leave no offspring while others leave many. Consider fitness to be influenced by a single locus represented by two alleles, *B* and *b*, with allele frequencies of $q_0$ and $(1 - q_0)$ in generation 0. Table 4.2 describes a situation where selection is against the *bb* genotype, with a proportion, *s*, of the *bb* individuals unable to reproduce.

The term $(1 - s)$ is equal to the proportion of the *bb* genotype parents which reproduced in each generation, also known as their fitness. The proportion of the total population in generation 0, which survives to reproductive age is $= 1 - s(1 - q_0)^2$. The genetic contribution of the offspring shows a loss of $s(1 - q)^2$ as a result of selection.

The frequency of alleles in survivors of generation 0 is the same as the allele frequencies in their offspring at birth in generation 1. Individuals selected to become parents for the next generation result in offspring that have a new gene frequency. Let the new allelic frequency of *b* following one generation of selection be $(1 - q_1)$. This is equal to the genetic contribution from all the recessive individuals (*bb*) and one half of the heterozygotes (*Bb*) as follows: $(1-q_1) = \frac{(1-s)(1-q_0)^2 + q_0(1-q_0)}{1-s(1-q_0)^2}$. The change in allelic frequency following one generation of selection is the difference between the previous and the new allelic freq uencies as follows:

$$\Delta(1-q) = (1-q_1) - (1-q_0).$$

Substituting $(1 - q_1)$ from the earlier equation, $\Delta(1 - q)$

$$= \frac{(1-s)(1-q_0)^2 + q_0(1-q_0)}{1-s(1-q_0)^2} - (1-q_0) = \frac{sq_0(1-q_0)^2}{1-s(1-q_0)^2}$$

The change in allelic frequency is dependent on the initial gene frequency $(1 - q_0)$ and the intensity of selection($s$).

Fitness of genotypes may be described in terms of complete, partial, no, and overdominance. With incomplete or partial dominance a heterozygous individual has a phenotype intermediate to the two homozygous. The

**Table 4.2** Genotype of goats and their frequencies, relative fitness, genetic contribution, and genotype frequency in survivors.

| Genotype of goats | BB | Bb | bb |
|---|---|---|---|
| Genotypic frequency | $q_0^2$ | $2q_0(1 - q_0)$ | $(1 - q_0)^2$ |
| Relative fitness | 1 | 1 | $(1 - s)$ |
| Genetic contribution | $q_0^2$ | $2q_0(1 - q_0)$ | $(1 - s)(1 - q_0)^2$ |
| Frequency in survivors | $\frac{q_0^2}{1-s(1-q_0)^2}$ | $\frac{2q_0(1-q_0)}{1-s(1-q_0)^2}$ | $\frac{(1-s)(1-q_0)^2}{1-s(1-q_0)^2}$ |

Note: $q$ and $(1 - q)$ are the frequencies of *B* and *b* alleles; $(1 - s)$ is the proportion of the *bb* genotype parents which reproduced in each generation.

change in gene frequency as a result of selection under different levels of dominance is described in Table 4.3.

Example: In a random breeding population of Angus cattle suppose all *red calves* are destroyed. If 100 *red calves* are born each generation out of a total of 10,000 births, the distribution of the three genotypes are shown in Table 4.4.

$$q = 0.9 \text{ and } (1-q) = 0.1, \text{ and } s = 1$$

Therefore:

$$\Delta(1-q) = (1-s)(1-q)^2 + \frac{sq(1-q)}{1-s(1-q)^2} = 0.09$$

It is important to note that with selection against homozygous recessive genotypes (*bb*), the allelic frequency of *b* changed by 0.01, from 0.1 to 0.09. This indicates the difficulty associated with selection against recessive genotypes, such as lethal genes, color, and markings.

**Table 4.3** Genotypes and their frequency, and genetic contribution with various levels of dominance for fitness.

| Genotype | *BB* | *Bb* | *bb* | $\Delta(1-q)$ |
|---|---|---|---|---|
| Genotypic frequency | $q^2$ | $2q(1-q)$ | $(1-q)^2$ | $\frac{-sq(1-q)^2}{1-s(1-q)^2}$ |
| Complete dominance (selection against *bb* genotypes) | 1 | 1 | (1 – s) | |
| No dominance (selection against *b* allele) | 1 | 1 – (½)s | (1 – s) | $\frac{-\frac{1}{2}sq(1-q)}{1-s(1-q)^2}$ |
| Complete dominance (selection against *B* allele) | (1 – s) | (1 – s) | 1 | $sq(1-q)^2$ |
| Overdominance (selection against *BB* and *bb* genotypes) | $(1-s_1)$ | 1 | $(1-s_2)$ | $\frac{q(1-q)\{s_1q-s_2(1-q)\}}{1-s\{1-(1-q)^2\}}$ |

Note: $q$ and $(1-q)$ are the frequencies of *B* and *b* alleles, (1 – s), 1 – (½)s, $(1-s_1)$ and $(1-s_2)$ are the proportion of parents of the specific genotype which reproduce in each generation.

**Table 4.4** Genotypes and their initial frequency and number, genotypic frequency, fitness, and genetic contribution.

| Genotype | *BB* | *Bb* | *bb* | Total |
|---|---|---|---|---|
| Initial frequency | $q^2$ | $2q(1-q)$ | $(1-q)^2$ | |
| Initial number | 8,100 | 1,800 | 100 | 10,000 |
| Genotypic frequency | 0.81 | 0.18 | 0.01 | 1.0 |
| Fitness | 1 | 1 | 0[1] | |
| Genetic contribution | $q^2$ | $2q(1-q)$ | $(1-s)(1-q)^2$ | $\frac{-sq(1-q)^2}{1-s(1-q)^2}$ |

Note: $q$ and $(1-q)$ are the frequencies of *B* and *b* alleles, $(1-s)$ is the proportion of the specific *bb* genotype parents which reproduce in each generation.

[1]Since (s = 1).

Selection with dominance will change allelic frequency by acting against the recessive allele $(1 - q)$. This will decrease the frequency of the recessive allele $(1 - q)$ and the genotype $(1 - q)^2$ and increase the frequency of the dominant allele $q$ and genotype $q^2$. The change will be gradual and steady with the greatest change occurring at intermediate frequencies (Table 4.3).

Genetic Drift

Dispersive forces of evolutionary significance influence gene frequencies during the assortment of chromosomes at meiosis, and in other processes involved in the transmission of genes from one generation to the next. The theoretical basis of genetic drift is the sampling of families from populations, individual animals from families, and genes within those individuals that cause the gene frequencies to vary in either direction from their initial values in each generation. The effect of genetic drift is most pronounced in small populations.

In closed populations with a limited number of parents, the sampling variance increases homozygosity above that in the initial population. This effect is a result of the effective number of parents ($N_e$), which is a function of the number of breeding bucks $N_m$ and does $N_f$ in the population. When mating is random, each parent has an "equal chance" of contributing to the next generation. It has been shown, however, that the number of kids per parent follows a Poisson distribution in many cases. The effective number of parents ($N_e$) will be $\frac{4N_mN_f}{N_m + N_f}$ which can be rewritten as $\frac{1}{N_e} = \frac{1}{4N_m} + \frac{1}{4N_f}$ (Wright, 1931). In a random breeding population the change in inbreeding level ($\Delta F$) per generation depends upon the effective population size and is equal to $\frac{1}{2N_e}$ which can be rewritten as $\frac{1}{8N_m} + \frac{1}{8N_f}$. This change in inbreeding level, $\Delta F$, is a measure of the increase in homozygosity in the population.

Example: In a population of goats consisting of 10 bucks and 100 does that produces kids, the effective population size ($N_e$) will be $\frac{4N_mN_f}{N_m + N_f} = \frac{4 \times 10 \times 100}{10 + 100} = 36.4$. Also $\Delta F = \frac{1}{2N_e} = \frac{1}{2 \times 36.4} = 0.0138$ per generation, indicating that 1.38% more will become homozygous for a given locus in this generation.

When $q$ is the gene frequency of the favorable allele and $N_e$ is the effective number of parents, the change in gene frequency has a variance equal to $\frac{q(1-q)}{2N_e}$. The change in mean performance from one generation to the next has variance due to genetic drift ($\hat{\sigma}_d^2$) which can be estimated as $\frac{\hat{\sigma}_g^2}{N_e}$, where $\hat{\sigma}_g^2$ is additive genetic variance. This can be rewritten as $\hat{\sigma}_d^2 = 2\Delta F \hat{\sigma}_g^2$.

**Extension of the Hardy-Weinberg Equilibrium**

The Hardy-Weinberg equilibrium also applies to situations where there are more than two alleles in a locus, more than a single locus, and sex-linked loci in the absence of mutation, migration, selection and genetic drift. Under equilibrium conditions with a large random mating population, the genotype and gene frequencies will remain unchanged from generation to generation. These extensions are beyond the scope of the present chapter, but reference to them can be found in a text such as Falconer and MacKay (1996).

Sex-linked Loci

There are two sex chromosomes in mammals denoted by $X$ and $Y$. The sex chromosome makeup of females is $XX$ referred to as the homogametic sex, and of males is $XY$ the heterogametic sex. Of genes found on the sex chromosomes, most are associated with the $X$ chromosome, very few with the $Y$ chromosome. In referring to sex-linked loci it is usually loci on the $X$ chromosome that are being considered, and that is the case here. Note that females have two $X$ chromosomes, and the behavior of a sex-linked character in females will be the same as for characters affected by genes on autosomes (that is, the genotype will be composed of two alleles). Gene action such as dominance or additivity can contribute to the phenotype. With males, on the other hand, they possess only one $X$ chromosome, so the phenotype will be the result of only one allele, and gene action such as dominance will play no role. This difference between males and females thus affects the behavior of gene and genotype frequencies. Suppose a sex-linked character has two alleles in a population, $B$ and $b$, with frequencies $p$ and $q$, respectively. At Hardy-Weinberg equilibrium, females will have genotypes, $X^BX^B$ ($p^2$), $X^BX^b$ ($2pq$) and $X^bX^b$ ($q^2$). Males in the population will be $X^BY$ ($p$) and $X^bY$ ($q$).

Transmission of genes from male and female parents to male and female offspring is also different. Females receive one $X$ chromosome from each parent, while males receive their $X$ chromosome only from their female parent. Gene frequencies may differ between males and females in a population not in equilibrium. It is because of this unequal mode of transmission of sex-linked genes to male and female offspring that if the equilibrium is disturbed, the time required to reach equilibrium will usually be more than one generation.

## QUANTITATIVE GENETICS

Quantitative traits observed in the phenotype of the animal are affected by many factors; some are genetic in origin while others are environmental in nature. Many such traits are measurable on some numeric scale (for example, weights, milk yield, or body dimensions). Some quantitative traits may have binary outcomes, such as survival or death, but the outcome for an individual animal is due to an underlying scale of risk, which has genetic and environmental effects. An individual past a certain threshold on the scale of risk dies. The genetic effect on a quantitative trait is, in most practical situations, a result of many genes. The number and effects of genes are often unknown, although in recent times, more and more genes and their effects on phenotype are being discovered. The purpose of this section is to illustrate how genotype translates into phenotype for a quantitative trait, using a simple one-locus model for most of the development. Concepts such as breeding value, important for quantitative traits, will be developed here.

### Phenotypic Value, Genotypic Value, and Population Mean

In the present development, phenotypes are observed for individuals of different genotypes, and environmental effects are considered to be not important. In a population represented by three genotypes, *bb*, *Bb*, and *BB*, with measures of average phenotypic performance as $P_{bb}$, $P_{Bb}$, and $P_{BB}$, respectively, the midpoint of the two homozygotes, $m = (P_{bb} + P_{BB})/2$. Defining $u$ as follows: $u = P_{bb} - m$. Now let the genotypic value of genotypes, *bb*, *Bb*, and *BB* be $P_{bb} - m = -u$, $P_{Bb} - m = au$ and $P_{BB} = u$, respectively. Note that with this formulation, the genotypic value increases by $u$ for every addition of a *B* allele to the genotype (depending on degree of dominance), and the heterozygote is allowed to show varying degrees of dominance. With the level of dominance $a = 0$, for example, the heterozygote is exactly intermediate in value to the two homozygotes, denoting gene action known as codominance. When $a = 1$, the heterozygote *Bb* is equal to the homozygote *BB*, denoting complete dominance.

Consider the *dwarf* gene in goats which is recessive (*b*) to normal size (*B*) and responsible for a decrease in body size (Table 4.5).

**Table 4.5** Genotypes and their frequency, phenotype, genotypic value, and sum.

| Genotype of goats | *BB* | *Bb* | *bb* |
|---|---|---|---|
| Genotypic frequency | $q^2$ | $2q\,(1-q)$ | $(1-q)^2$ |
| Phenotype | $P_{BB}$ | $P_{Bb}$ | $P_{bb}$ |
| Genotypic value | $u$ | $au$ | $-u$ |
| Sum | $uq^2$ | $2auq(1-q)$ | $-u(1-q)^2$ |

Note: $q$ and $(1 - q)$ are the frequencies of *B* and *b* alleles, $u$ is the genotypic value for the *b* allele, and $a$ is the level of dominance.

**Table 4.6** Average effect of a gene.

| Genotype of parents | *BB* | *Bb* | *bb* |
|---|---|---|---|
| Genotypic frequency | $q^2$ | $2q\,(1-q)$ | $(1-q)^2$ |
| Effect of adding b to the genotype of offspring | BB × b ↓ Bb | Bb × b ↓ Bb, bb | bb × b ↓ none |

Note: $q$ and $(1 - q)$ are the frequencies of *B* and *b* alleles.

The population mean is the sum of the product of the frequency and the genotypic value for all classes of genotypes $= uq^2 + 2auq(1 - q) - u(1 - q)^2$. Note that the midpoint of homozygotes should be added to this result to give the actual mean. The population mean is calculated as a weighted average of genotypic values.

### Breeding Value and Selection

Parents transmit genes to their offspring. Therefore, it is desirable to obtain a measure of the effects of these genes on their offspring. This is known as the "transmitting ability" of the individual and is equal to one-half of a goat's breeding value. The average effect of a gene will help explain this concept (Table 4.6).

Shown in Table 4.6 is the effect of adding the *b* allele to a population of *BB*, *Bb*, and *bb* individuals. The average value of the *b* allele can be derived from the product of genotypic frequencies and their value.

The average effect of an allele can be defined in a random breeding population where individuals receive an allele from one parent while the allele received from the other parent will be at random (Table 4.7). The average effect is expressed as a deviation from the population mean. It is important to note that the average effect of a gene substitution is greater when the frequency of the unfavorable gene increases.

**Table 4.7** Genetic contribution of parents to their offspring, average effect of genotypes and breeding value.

| Genotypes produced in the offspring | | Gametes produced by parents | | | |
|---|---|---|---|---|---|
| | | *B* allele | | *b* allele | |
| Genotype | Value | Frequency | Sum | Frequency | Sum |
| *BB* | $u$ | $q$ | $uq$ | | |
| *Bb* | $au$ | $(1-q)$ | $au(1-q)$ | $q$ | $auq$ |
| *Bb* | $-u$ | | | $(1-q)$ | $-u(1-q)$ |
| *Sum* | | | $uq + au(1-q)$ | | $auq - u(1-q)$ |
| *Average effect* | | $\alpha_1 = u(1-q)\{1 + a(1-2q)\}$ | | $\alpha_2 = -uq\{1 + a(1-2q)\}$ | |
| Genotype | | *BB* | | *Bb* | *bb* |
| Average effect of alleles | | $2\alpha_1$ | | $\alpha_1 + \alpha_2$ | $2\alpha_2$ |
| Breeding value | | $2(1-q)\alpha$ | | $(1-2q)\alpha$ | $-2q\alpha$ |

Note: $q$ and $(1-q)$ are the frequencies of $B$ and $b$ alleles, $\alpha_1$ and $\alpha_2$ are average effect resulting from the substitution of $B$ and $b$ alleles, and $\alpha$ is the average effect of the substitution = $\alpha_1 - \alpha_2$.

Breeding value is derived from the mean value of an animal's offspring, and is calculated as twice the average deviation of the offspring mean from the population mean. The deviation is multiplied by two because the individual only transmits one-half of its genes to the offspring. The remaining one-half is from a random sample of genes in the population. The actual value is the sum of the breeding value and population mean.

Breeding value defined in terms of average effects of genes is equal to the sum of the average effects of all genes that it carries as shown in Table 4.7.

### Inbreeding and Relationships

Inbreeding is used to denote the mating of parents that are more closely related to each other than the average of the population. Related parents are more likely to transmit the same genes to their offspring than unrelated parents. Inbreeding has an influence on the genotype frequency of the offspring such that there will be an increase in homozygosity and reduction in heterozygosity. The term, "inbreeding," is used in another way in the "inbreeding coefficient," which is used to describe the level of homozygosity resulting from the mating of related animals.

The degree of relationship between individuals can be measured by the coefficient of relationship ($R_{XY}$). Relationship is expressed in terms of the probability that two individuals have the same alleles by descent because of common ancestry (Figure 4.1). Full sibs are related by 50% because both the parents are common. On average 50% of the genes are common whereas the remaining 50% differ.

When offspring $X$ and $Y$ receive 50% of their genes from the buck, the probability that both offspring receive the same gene from the buck is $0.5 \times 0.5 = 0.25$. If offspring $X$ and $Y$ receive 50% of their genes from the doe, then the probability that both offspring receive the same gene from the doe is equal to $0.5 \times 0.5 = 0.25$. In full siblings, offspring $X$ and $Y$ have the same buck and doe as parents. Therefore, the probability of receiving the same genes from both the buck and doe parents is the sum of the two frequencies which is $0.5 = (0.25 + 0.25)$ or 50%. In half siblings, offspring $X$ and $Y$ have one common parent. Therefore, the probability of receiving the same genes will be 0.25 or 25%. The relationship between the parent and offspring is 0.5 or 50%. Therefore, the probability of receiving the same genes will be 0.25 or 25%.

Inbreeding is a measure of increase in homozygosity in individuals based on the relationship between the sire and dam. It is denoted as $F_\times$, which is known as Wright's coefficient of inbreeding and is one-half of the relationship between parents. Inbreeding of an individual can be estimated from the following equation:

$$F_\times = \sum\left[\left(\frac{1}{2}\right)^{n+n'+1}(1+F_A)\right]$$

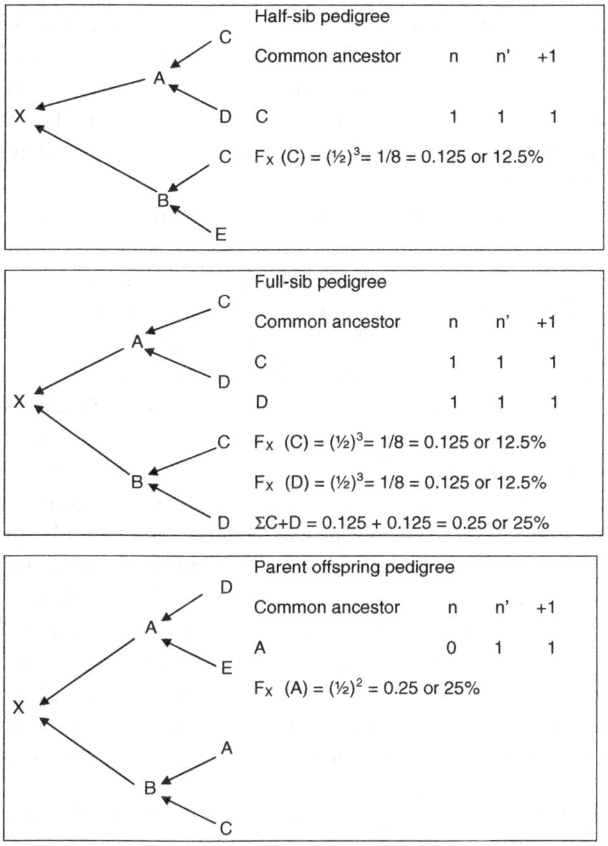


**Figure 4.1** Coefficient of inbreeding for full-sib, half-sib and parent-offspring pedigree.

Where n = number of generations from the sire to the common ancestor, n′ = number of generations from the dam to the common ancestor, Σ = summation of the various paths (there may be more than one common ancestor, and thus more than one path connecting the sire or dam to the common ancestor), 1/2 = probability of receiving one allele or the other at each segregation, and $F_A$ = inbreeding of the common ancestor.

As noted earlier, it is possible to compute the inbreeding rate ($\Delta F$) per generation in a closed population from the effective number of breeding males $N_m$ and females $N_f$ in the population, as follows: $\Delta F = \frac{1}{8N_m} + \frac{1}{8N_f}$. $\Delta F = \frac{1}{8N_m}$, when $N_f > 12$ because $\frac{1}{8N_f}$ is negligible and can be ignored. It is possible to estimate the increase in inbreeding per generation from the effective number of sires used as parents. The estimate of increase in inbreeding per generation will be 0.1250 for 1 sire, 0.0625 for 2 sires, 0.0250 for 5 sires, and 0.0125 for 10 sires.

The detrimental influence of inbreeding has been demonstrated in a number of studies. The consequence is a reduction in the overall vigor of the offspring, with a greater influence on traits associated with fitness, such as conception, fertility, and survival.

### Phenotypic, Genetic, and Environmental Variation

In a population, the variance is a measure of the average squared deviations from the mean for any measurable morphological characteristic and production performance. In goat breeding there is interest in the components of variance starting with the phenotypic variance ($\sigma^2_P$) which constitutes the sum of the genetic ($\sigma^2_G$), environmental ($\sigma^2_E$), and genotype × environment interaction ($\sigma^2_{G\times E}$) variance components. The genetic variance ($\sigma^2_G$) can be further subdivided into additive genetic ($\sigma^2_A$), dominance ($\sigma^2_D$), and epistatic ($\sigma^2_I$) variance components. This has been described in the following:

$$\sigma^2{}_P = \sigma^2{}_G + \sigma^2{}_E + \sigma^2{}_{G\times E}$$

$$\sigma^2{}_G = \sigma^2{}_A + \sigma^2{}_D + \sigma^2{}_I$$

In selection programs, the genetic effect of interest is based on additive genetic variance ($\sigma^2{}_A$) arising from the average effects of genes, which determines the transmitting ability from the parents to offspring. The nonadditive genetic variance comprises dominance variance ($\sigma^2{}_D$) associated with effects due to the combination of alleles at a locus, and epistatic variance ($\sigma^2{}_I$) associated with combinations of alleles at more than one locus. Allele combinations that produce dominance and epistatic effects on traits are lost at meiosis, when only one allele of each pair for each gene forms a gamete. Thus offspring do not have the same allele combinations as their parents. It is possible, however, for breeding programs to take advantage of the nonadditive genetic effects through the use of specialized mating systems.

There is clear evidence to suggest environment is a significant source of variation in many morphological characteristics and production performance. The environmental variation is associated with nongenetic sources of variation arising from differences among years, age of doe, season, sexes, types of birth and rearing, management, diets, housing, and pens. The influence of year and season is random in nature; therefore, it is difficult to predict their performance. Sex is of course a genetic effect, but it can be considered as a hormonal and developmental environment in which a trait is expressed. For example male kids grow more rapidly than female kids and attain a heavier body weight at maturity. Type of birth and rearing can be important; for example, single-born kids grow more rapidly than those born as twins, which in turn grow more rapidly than triplets. Part of this effect is due to care and milking ability of the dam divided amongst the one or several offspring with the limiting factor being the milk production ability of the dam. The impact on performance of goats from management, diet, housing and pen are important. Therefore, every attempt should be made to minimize their influence. To obtain a clear estimate of the genetic component, it is important to ensure the environmental effects on production performance are similar for a group of animals.

### Heritabilities, Repeatabilities, and Correlations

The heritability for any measurable morphological characteristic and production performance indicates the ability of both the parents to transmit superior productivity to their offspring. Heritability in the broad sense ($H^2$) is the ratio of the genetic ($\sigma^2_G$) to total phenotypic variance ($\sigma^2_P$), which is given by $H^2 = \frac{\sigma^2_G}{\sigma^2_P}$. The genetic variance is not a useful measure in selection programs because not all genetic effects are transmitted from parents to their offspring as mentioned previously.

Heritability ($h^2$) in the narrow sense is the ratio of the additive genetic variance ($\sigma^2_A$) to total phenotypic variance ($\sigma^2_P$), which is given by $h^2 = \frac{\sigma^2_A}{\sigma^2_P}$. Heritability can be calculated in a number of ways. One method uses phenotypic measures for the same trait in parents and their offspring. The slope of the line relating the performance of offspring to that of their parents or midparent average gives a measure of heritability in the narrow sense. Heritability estimates have also been derived from regression using other types of relatives (daughter-dam, son-sire, and offspring-parents), and from intra-class correlation (using full siblings or half siblings). The choice of the procedure depends, in part, on the nature of the trait (for example, daughter-dam regression for milk production because milk is only expressed in the female sex). Modern techniques and computer programs based on mixed models (Boldman et al., 1995) are now the methods of choice and make use of information from all types of relatives in a population to give estimates of heritability for a trait. Estimates of heritability for reproductive traits, body weight and daily gain in goats are presented in Table 4.8.

Some traits can be repeatedly expressed over an animal's lifetime. In making culling decisions, it is useful to know whether an animal's performance measure is likely to be repeated. This is a slightly different type of decision than the selection of parents in a breeding program, but it is important in the management and operation of breeding and commercial herds.

Repeatability ($R$) is a measure of the similarity of performance measurements on the same trait over the life of an individual. For example, with milk production, a doe will perform in a similar way from one lactation to the next, not only because the same genes with all of the same genetic effects are expressed, but also because there will be some environmental effects such as presence of mastitis that are permanent in nature that affect all lactations. However, lactations from the same doe will differ due to temporary environmental effects. A simple way to estimate the repeatability of milk production in goats would be to calculate the correlation between first and second lactation milk yield measurements for a sample of does. Though there are better ways to determine repeatability, this demonstrates what it represents. In terms of the variance components discussed previously, repeatability is defined as the ratio of genetic ($\sigma^2_G$) and permanent environment ($\sigma^2_{PE}$) variance for a trait to total phenotypic ($\sigma^2_P$) variance as follows: $R = \frac{\sigma^2_G + \sigma^2_{PE}}{\sigma^2_P}$. Examples of traits that are expressed repeatedly in a goat's life include number of kids born, milk production, and weight of mohair. Estimates of repeatability for multiple births and daily gain in goats are presented in Table 4.8.

The phenotypic correlation ($r_P$) is an estimate of association between two traits measured on an individual, denoted by $r_P = \frac{\sigma_{XY}}{\sqrt{\sigma^2_X \sigma^2_Y}}$. Where, $\sigma_{XY}$ is an estimate of the phenotypic covariance between traits $X$ and $Y$, and $\sigma^2_X$ and $\sigma^2_Y$ are estimates of their respective phenotypic variances.

The phenotypic correlation is due to genetic effects that are in common for the two traits, as well as environmental effects that affect both traits. There are thus genetic ($r_G$) and environmental ($r_E$) correlations between the two traits that play a role in the observed phenotypic correlation. As mentioned, the genetic correlation between two traits is, in part, due to genes that have effects on both traits. This phenomenon is known as pleiotropy. If two traits are affected by many of the same genes then the genetic correlation will be high (+1 or −1). If different genes affect

**Table 4.8** Estimates of heritabilities and repeatabilities for reproductive traits, body weight and daily gain, and direct and maternal heritability and their correlation.

| Trait/Breed | Heritability | Repeatability |
|---|---|---|
| Age at first kidding | | |
| Alpine × Beetal, Beetal | 0.56–0.48 | |
| Litter size | | |
| Beetal, Black Bengal | 0.15–0.09 | |
| Alpine × African common | 0.02 | |
| Multiple births | | |
| Egyptian Baladi | 0.25 | 0.29 |
| Beetal | 0.15 | 0.22 |
| Black Bengal | 0.09–0.17 | 0.15 |
| Birth weight | | |
| Black Bengal | 0.07–0.40 | |
| Boer | 0.18–0.36 | |
| Jamunapari | 0.46–0.55 | |
| Body weight at 12 months | | |
| Black Bengal | 0.32 | |
| Jamunapari | 0.13–0.31 | |
| Boer | 0.31 | |
| Daily gain (birth to 150 days) | | |
| West African Dwarf | 0.38–0.63 | 0.21–0.38 |

| | Heritability | | Correlation |
|---|---|---|---|
| Breed/Trait | Direct | Maternal | Direct × Maternal |
| Birth weight | | | |
| Boer | 0.16–0.33 | 0.14–0.36 | −0.31 |
| Turkish Angora | 0.02 | 0.10 | 0.18 |
| Weaning weight | | | |
| Boer | 0.18–0.27 | 0.05–0.60 | −0.15 |
| Turkish Angora | 0.03 | 0.10 | 0.91 |
| Yearling fleece weight | | | |
| Turkish Angora | 0.06 | 0.04 | −0.91 |
| Litter size | | | |
| Turkish Angora | 0.06 | 0.04 | −0.07 |

Source: Shrestha and Fahmy (2007).

the two traits, then the genetic correlation will be low (near zero). Another reason for genetic correlations to occur is when genes affecting two traits are linked. When this is the case, the genetic correlation will decay over generations as a result of recombination, which breaks up the specific allele combinations responsible for the correlation. Genetic correlations can be estimated from a procedure similar to that for heritability, but now two traits are measured on parents and offspring (or other sets of relatives). For definition and practical purposes, the genetic correlation is usually calculated using additive genetic effects, denoted by $r_A = \frac{\sigma_{AXY}}{\sqrt{\sigma^2_{AX}\sigma^2_{AY}}}$. Where, $\sigma_{AXY}$ is an estimate of the additive genetic covariance between traits $X$ and $Y$, and $\sigma^2_{AX}$ and $\sigma^2_{AY}$ are estimates of their respective additive genetic variances of the traits.

An example applying knowledge of genetic correlations is in selection for early growth in goats. Body weights at different ages (for example, birth, weaning, and maturity) tend to have genetic correlations that are positive and medium in size. This means that a selection program to increase early growth will tend to result in a breeding population with a larger mature size in future generations. Another example is that growth rate of young goats has a negative genetic correlation with feed conversion (measured as feed intake per unit gain). This means that goats selected for rapid growth require less feed per unit gain.

A number of traits expressed early in the goat's life are affected both by the goat's own genetic capability, as well as effects from the maternal parent in the form of milking ability or nurturing ability. These two effects are referred to as direct and maternal effects. Each of these is heritable, and there is also a genetic correlation between the direct and maternal effects. These are shown in Table 4.8 for body weights at birth and weaning, yearling fleece weight, and litter size. A negative correlation between direct and maternal effects of −0.15 for weaning weight in the Boer breed indicates goats that have a genetic capability for rapid early growth tend to become does that lack the inherent potential of providing a complementary environment for early growth to their offspring. This means that it would be somewhat difficult to find goats that are good for both abilities. It is a fairly weak correlation, however, so it is not of great concern.

The environmental correlation between two traits can be estimated from a procedure similar to that mentioned above for genetic correlations and is defined as $r_E = \frac{\sigma_{E_{XY}}}{\sqrt{\sigma^2_{E_X}\sigma^2_{E_Y}}}$. Where, $\sigma_{E_{XY}}$ is an estimate of environmental covariance between traits $X$ and $Y$, and $\sigma^2_{E_X}$ and $\sigma^2_{E_Y}$ are estimates of their respective environmental variances.

Estimates of genetic and environmental correlations are necessary to develop selection indexes based on multiple traits that are useful in practical goat breeding. This is because the breeder is interested in improving many traits (for example, fertility, lean muscle growth, milk yield, milk composition, and hardiness) all useful in increasing productivity.

## GOAT BREEDING

Goats were chosen according to the perception of the breeder for centuries until the early 18th century when the Coates' Herd Book for Shorthorn Cattle and herd societies were established in England. The concept developed by Robert Bakewell of combining stocks with similar morphological characteristics resulted in the registration of animal pedigree and the maintenance of herdbooks that restricted introductions, contributing to the development of a large number of pure breeds (Lush, 1945). This was followed by the development of the Alpine, Angora, Saanen, and Toggenburg breeds. Goat populations expanded rapidly across many parts of the world as a result of favorable climatic conditions, abundance of vegetative growth, religious rituals, ease in transportation, and absence of communicable diseases.

Major advances achieved in the field of animal breeding following the application of quantitative genetic principles have contributed to the improvement of livestock and poultry. These include breed evaluation, crossbreeding, and new breed formation as well as simultaneous estimation of breeding values of the sires, dams, and their offspring, identification of superior individuals with potential merit for genetic improvement to be used as parents to produce the following generation, and Marker-Assisted selection. Quantitative genetic methodologies have shortened the time required for rapid improvement of economically important traits, all known to influence the income and profitability from raising goats. Any program for genetic improvement of goats must provide a comprehensive and technically sound assessment with sufficient flexibility to meet the needs of a diversity of environmental and managerial situations. At the same time, every effort should be made to maintain a uniform test environment and minimize the influence of the environment.

### Selection

Selection is the process of choosing individuals in a population to be used as parents to produce the next generation of offspring. This process creates differential rates of reproduction because selected individuals produce offspring and culled animals do not. The breeder is interested in improving a number of morphological characteristics and production performance, but limited resources usually result in the choice of a few important measurements for improvement.

The selection differential for a measurable characteristic or performance is the superiority of goats selected as parents compared to the average of all candidates available in that generation. This can be computed as the difference between the average of selected goats ($\overline{X}_S$) and the overall population average ($\overline{X}$), given by $\overline{S} = \overline{X}_S - \overline{X}$.

Selection is a continuous process having relevance in giving all goats in a population an opportunity to express their genetic potential. In practice, selection differentials are calculated separately for the buck and doe, and then averaged. This is because fewer bucks are selected compared to does resulting in greater selection pressure from the bucks. A simple average of the selection differentials of buck and doe is used because the male and female groups each contribute equally to their offspring.

When selection is applied to a population, genetic progress occurs in the form of change in performance of individual goats from one generation to the next. This change ($\Delta G$) depends upon the selection differentials of the selected does ($\bar{X}_{SF}$) and bucks ($\bar{X}_{SM}$), as well as the heritability of the trait. For simple selection on phenotypic performance of goats, genetic progress per generation for each trait is predicted as $\Delta G = h^2\left(\frac{\bar{X}_{SF} + \bar{X}_{SM}}{2}\right)$.

Selection can occur among individuals, family (full or half siblings), or a combination of both. In a closed population of goats, selection of individual goats based on their phenotypic measurements is often referred to as mass selection. In general, traits with medium to high heritabilities, such as carcass measures and growth rate, are best selected on the basis of performance of the individual goat. Family selection becomes important when selecting for traits that are low or moderate in heritability such as reproductive traits. In this case, families are selected based on the mean performance of a half- or full-sibling group. With pedigree selection an individual goat is selected based upon the presence of important ancestors or high-performing goats in its pedigree. This procedure is still used in breeding champion horses and dogs to improve performance. However, care must be taken not to place too much emphasis on the presence of an ancestor with outstanding performance in a pedigree when it is more distant than a parent. A grandparent, for example, has only 25% of its genes in common with its grand-progeny. Progeny testing is common in many species and is valuable for sex-limited traits such as milk production. However, this procedure can be expensive because of the need to retain animals until the superior individual is identified, thus extending the generation interval.

The choice of the method is dependent on the expected annual genetic response to selection and the associated costs. It must be noted that long-term genetic progress is dependent on the influence of several factors. These include reproductive rate, number of traits under selection, heritability, genetic correlation, inbreeding and effective population size, consistency of selection goals, uniform test environment, generation interval, and genotype-environment interaction.

**Multiple Trait Selection**

Economically important traits can be selected singly or in combination. There are several ways of doing this. With Tandem selection, one trait is selected for one or more generations until the desired improvement has been achieved, and then a second trait is selected until required improvement has been achieved with it. This process can be continued until all of the traits of interest are improved. Tandem selection is useful for three or fewer traits that are not correlated, otherwise the efficacy of selection will be reduced.

Another method of selection for multiple traits makes use of Independent Culling levels for each trait of interest. Threshold levels would be established for each trait, and goats exceeding the threshold in the desired direction for all traits would be selected for use as parents. The advantage of this procedure is that selection can occur at various stages, for example, during birth, weaning, and market weight reducing, the number of goats that have to be retained, before the final stage of selection with the last trait measured. This procedure is most effective when the number of traits does not exceed three and the intensity of selection is high. Both the Tandem and Independent Culling level of selecting parents have an operational advantage in that they can be implemented without knowing genetic parameters for the traits or their exact economic worth. It would be desirable, however, that the choice of traits and the selection emphasis placed upon them be based on knowledge of their economic worth and expectation of genetic progress (heritability).

The most efficient procedure in selecting for multiple traits is the Selection Index. This is a function of phenotypic measures chosen according to the breeding objective together with weights derived from genetic parameter estimates and their economic values. The Selection Index is given by $\Sigma X_i b_i$, where $X_i$ represents phenotypic traits ($i$) such as fecundity, lean meat yield, milk production, etc., and $b_i$ are coefficients derived from additive genetic variance and covariance of the measurement along with their economic worth (for methodology see Cameron, 1997).

In the past, the high cost of computer processing and slow turnaround time were disadvantages with index selection. Easy access to high-speed computers, however, has made the Selection Index an important tool for prediction of the breeding worth of multiple traits. One problem with index selection is with situations where information is collected on animals over a period of time. Here, it is important to retain all of the animals until the evaluation is complete. This may contribute to financial loss from the need to keep less productive animals until culling is complete. Nevertheless, the ability to identify animals with high genetic merit is greater with the Selection Index as long as genetic parameter estimates of performance traits included in the selection criterion and their economic values are accurate.

Best Linear Unbiased Prediction (BLUP) or Selection is a mixed model methodology for predicting the breeding value of individuals that are candidates for selection. A mixed model has fixed and random effects (see Boldman et al., 1995) that account for the environmental and genetic effects that are parts of each phenotypic expression of a trait by an animal. In this procedure, a system of equations based on fixed (often environmental effects such as herd, year, and age of the animal) and random effects (including animal breeding values, permanent environment effects), affecting one or more traits is used. This procedure requires genetic parameter estimates and makes use of the relationship among individuals to give the best possible estimates of the animal's breeding values, given the information that is available.

In the U.S., G. Wiggins has contributed toward the development of national programs for evaluation of dairy goats. National goat evaluation programs have used animal models to predict the breeding values of bucks, does, and their offspring for milk yield and composition, as well as linear body measurements. Estimated transmitting ability (ETA) or predicted transmitting ability (PTA) of each parent is an estimate of the genetic merit that is transmitted to their offspring, which is equal to one-half the breeding value. Genetic comparisons are made among individuals in a contemporary group, which may include goats of the same age group or lactation, within year, herd, and breed. Furthermore genetic comparisons are made among individuals not in a contemporary group relative to a rolling genetic base. Animals of a given breed with superior genetic merit can be found in different herds and years, because of widespread use of artificial insemination especially in dairy goats. Similarly, most probable producing ability (MPPA) provides an estimate of the milk production potential in subsequent lactations. Often body conformation measurements have been included in the evaluation of dairy goats resulting in a production type index (PTI).

### Quantitative Trait Loci and Marker-Assisted Selection

In the 1990s, considerable strides were achieved in mapping the goat genome based on highly polymorphic microsatellite markers and mitochondrial DNA. Furthermore, single nucleotide polymorphisms (SNP) and DNA chips have been developed to rapidly screen the genome. Quantitative trait loci (QTL) are regions in the genome closely linked to the genes associated with a quantitative trait. Breeding strategies based on QTLs require phenotype, genotype and pedigree records from a large number of individuals that vary in performance. In practice, a QTL refers to a region on a chromosome with an effect large enough to be detected and mapped in the genome. This may be considered as an early step in identifying and sequencing the actual gene that could be genotyped for selection purposes. Marker-Assisted selection exploits major genes by genotyping animals for genetic markers linked to QTLs.

There are several types of markers. An indirect marker is closely linked to the gene that has the effect of interest. A direct marker is the gene itself, or the portion of the gene that contains the mutation producing the effect of interest. The value of Marker-Assisted selection depends on the nature of traits associated with increased production efficiency as follows: when marker information is available to combine with phenotypic measurements to give an estimate of breeding value for an animal, the marker information can improve the accuracy with which the breeding value of an animal is estimated. The value of the marker information depends upon the heritability of the trait. With lowly heritable traits, the marker information improves the accuracy of the estimated breeding value, whereas with highly heritable traits the marker information is less valuable. The use of markers can also be important with sex-limited traits, for example where dairy bulls are selected initially on the basis of the value of their parents. A more accurate selection of these young bulls could be made using marker information on the males themselves. Another situation where marker information may prove valuable is if the trait can only be measured after the slaughter of the animal. In this case, marker information could be used instead to estimate the genetic merit of potential breeding animals for carcass traits.

### Crossbreeding

Crossbreeding is a common practice used in the commercial production of meat from many species of farm livestock. This may be due to the fact that efficient commercial meat production requires the superior performance in reproductive traits of adult animals as well as superior performance of young animals reared to market age. Getting superior performance in these different abilities generally requires the use of more than one breed, as well as use of a genetic phenomenon called hybrid vigor or heterosis. On the other hand, with dairying in cattle and other species, crossbreeding is much less common. Often

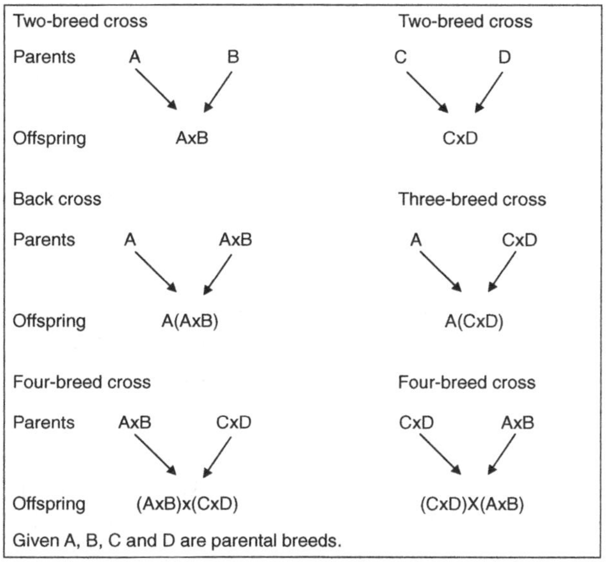


**Figure 4.2** Crossbred offspring derived from two, three, and four breeds.

there is only one breed that is superior to all others in lactation performance, and crossbreeding may not improve efficiency of lactation production. Examples of various crossbreeding systems are shown in Figure 4.2.

There are several approaches to crossbreeding with considerable merit for improving productivity in goats. Crossbreeding with more than two breeds depends on the number of parental breeds or populations available for crossing, and among parents the order of mating pure breeds and their crosses. Initially single cross offspring to be used as parents are produced from the mating of two purebreds. This is followed by crossing selected male or female crossbred parents with a purebred or crossbred of the opposite sex. There is clear evidence to suggest the genetic basis of average superiority of three-breed cross or multiple-breed cross over their constituent single crosses is due to the complementary breed effects in the crossbred animal, maternal effects due to the breed combination of the female parent, all supplemented with heterosis for individual performance as well as maternal ability.

In goats, crossbreeding strategies for meat production may involve crossing bucks of meat-type sire breeds with does of the fecund-type breeds to produce kids with increased growth rate and carcass quality, while benefiting from the reproductive rate and the maternal influence of the doe. In general, locally adapted does of a fecund-type dam breed exhibiting a high frequency of multiple births are mated with bucks of a meat-type sire breed to produce two-breed cross offspring. Operationally, it is possible to allocate one-third of does in the parental breed to produce purebred offspring for herd replacement. Crossbred kids for market may be produced by mating the remaining two-thirds of the does to pure- or crossbred bucks that differ in breed composition.

In practice, does from an established breed in the region or indigenous population may be maintained on the farm. Requirements for bucks may be met either by purchasing bucks with growth potential from a reputable breeder, or using semen from meat-type breeds. An alternate breeding strategy involves breeding a proportion of indigenous goats to produce purebred offspring for herd replacements. The remaining indigenous goats may be crossed with bucks of an improved dairy breed (Saanen, Toggenburg, Nubian, Damascus or Jamunapari) to produce crossbred does for milk production. Finally, the mating of crossbred does to bucks of an improved meat breed (Boer) could benefit from potential genetic merit in the specific three-breed cross offspring.

Backcross offspring can be produced by mating the single cross does to bucks of the sire breed. In practice, purebred offspring for herd replacement can be produced from one-third of the female parents. Two-breed cross offspring are produced by mating the remaining two-third of the female parent with bucks of the alternate breed purchased from an outstanding breeder. Selected crossbred offspring become does that are subsequently mated to bucks of the sire breed to produce the backcross offspring. In this procedure, benefit due to heterosis can occur from maternal performance of the crossbred female parent and their backcross offspring. Yet, the full complement of heterosis and genetic superiority associated with the parental breeds will not be realized. In many developing countries, the need to purchase bucks of one breed to sire both the single cross and backcross offspring has made this method popular. The choice of breeds for crossbreeding and the adaptability of their offspring to the environment where the offspring will be raised are pertinent to avoid any detrimental influence.

Crossbreeding studies in a number of countries corroborate performance of offspring derived from single- and backcross mating surpassed those of their contemporary purebred offspring. In India, matings based on Alpine and Saanen bucks with Beetal and Malabari does, Saanen bucks with Saanen × Malabari does, as well as Angora bucks with Sangamaneri and Angora × Sangamaneri does were utilized to compare single- and backcross does with their contemporary purebreds. Both types of crossbreds

**Table 4.9** Means for body weight at birth (kg) and at various ages, age at first kidding, kidding interval, service period, litter size, live weight at slaughter, hot carcass weight and dressing percentage for goat breeds and their crosses[1].

| Breed | | Body weight at | | | | |
|---|---|---|---|---|---|---|
| Sire | Dam | Birth | 3 mo | 6 mo | 9 mo | 12 mo |
| Beetal | Beetal | 2.9 (0) | 7.7 (0) | 12.2 (0) | | 21.8 (0) |
| Alpine | Beetal | 3.2 (10) | 10.3 (34) | 13.8 (13) | | 40.1 (84) |
| Saanen | Beetal | 3.4 (17) | 10.4 (35) | 14.3 (17) | | 26.9 (23) |
| Malabari | Malabari | 1.7 (0) | 5.7 (0) | 9.3 (0) | 11.1 (0) | 15.2 (0) |
| Alpine | Malabari | 1.9 (12) | 6.3 (11) | 8.9 (−4) | 12.2 (10) | 17.6 (16) |
| Saanen | Malabari | 2.3 (35) | 6.1 (7) | 10.2 (10) | 13.3 (20) | 17.8 (17) |
| Saanen | Saanen × Malabari | 2.7 (17) | 5.9 (−3) | 9.7 (−5) | | |
| Sangamaneri | Sangamaneri | 1.9 (0) | 7.3 (0) | 10.6 (0) | 13.5 (0) | 17.3 (0) |
| Angora | Sangamaneri | 2.1 (11) | 7.3 (0) | 10.6 (0) | 13.0 (−4) | 16.0 (−8) |
| Angora | Angora × Sangamaneri | 2.2 (5) | 8.2 (12) | 11.2 (6) | 13.4 (4) | 15.4 (−11) |

| Breed | | Age at first | Kidding | Service | Litter size (%) | | |
|---|---|---|---|---|---|---|---|
| Sire | Dam | kidding (day) | interval (day) | period (day) | Single | Twin | Triplet |
| Beetal | Beetal | 534 (0) | 313 (0) | 173 (0) | 41 | 51 | 9 |
| Alpine | Beetal | 495 (−7) | 323 (3) | 201 (16) | 59 | 36 | 5 |
| Saanen | Beetal | 546 (2) | 300 (−4) | | 67 | 29 | 5 |
| Malabari | Malabari | 700 (0) | 295 (0) | 141 (0) | 55 | 41 | 4 |
| Alpine | Malabari | 685 (−2) | 329 (−12) | 184 (31) | 65 | 35 | |
| Saanen | Malabari | 585 (−16) | 407 (38) | 260 (84) | 31 | 63 | 6 |

| Breed | | Age at slaughter | Live weight at | Hot carcass | Dressing |
|---|---|---|---|---|---|
| Sire | Dam | (mo) | slaughter (kg) | weight (kg) | percentage |
| Beetal | Beetal | 9 | 15.2 (0) | 7.7 (0) | 50 (0) |
| Alpine | Beetal | 9 | 18.3 (20) | 9.5 (23) | 52 (4) |
| Saanen | Beetal | 9 | 18.8 (24) | 9.4 (22) | 50 (0) |
| Alpine | Beetal | 5 | 14.1 (−7) | 5.3 (−31) | 38 (−24) |
| Sangamaneri | Sangamaneri | 9 | 11.5 (0) | 5.1 (0) | 44 (0) |
| Angora | Sangamaneri | 9 | 12.9 (12) | 4.4 (−14) | 38 (−14) |

Sources: Acharya et al., 1982; Acharya, 1988; Shrestha and Fahmy, 2007.
[1]Percent deviation from dam breed in parenthesis.

were superior to the purebreds in body weight of kids from birth to 12 months of age; reproductive performance of does manifested in age at first kidding, kidding interval, service period, and proportion of single, twin, and triplet; and hot carcass weight and dressing percentage of kids (Table 4.9).

The influence of heterosis on performance may vary according to the breeds involved in crossing together with the environment where the offspring is raised. The mating of bucks of the Beetal breed with does of the Sirohi breed and Beetal × Sirohi cross failed to demonstrate heterosis (Table 4.10) in the single- and backcross combinations compared to contemporary purebreds derived from their female parental breed for the body weights of kids from birth to 12 months of age, kid mortality, feed efficiency and dressing percentage (Taneja, 1982). Lack of heterosis

**Table 4.10** Means for body weight (kg) of single-born kids at birth and at various ages by breed and their crosses[1].

| Breed | | Body weight at | | | |
|---|---|---|---|---|---|
| Sire | Dam | Birth | 6 mo | 9 mo | 12 mo |
| Year: 1978–79 | | | | | |
| Sirohi | Sirohi | 2.8 (0) | 13.6 (0) | 17.1 (0) | 21.3 (0) |
| Beetal | Sirohi | 3.1 (11) | 14.3 (5) | 17.1 (0) | 22.3 (5) |
| Year: 1980 | | | | | |
| Sirohi | Sirohi | 2.9 (0) | 12.5 (0) | 17.0 (0) | |
| Beetal | Sirohi | 3.1 (7) | 13.3 (6) | 17.6 (4) | |
| Beetal | Beetal × Sirohi | 2.9 (−6) | 12.4 (−7) | 16.2 (−8) | |

| Breed | | Body weight at | | | |
|---|---|---|---|---|---|
| Sire | Dam | Birth | 1 mo | 2 mo | 4 mo |
| Black Bengal | Black Bengal | 1.2 (0) | 2.1 (0) | 3.8 (0) | 4.4 (0) |
| Jamunapari | Black Bengal | 1.4 (17) | 2.5 (19) | 4.4 (10) | 5.5 (25) |

| Breed | | Body weight at | | |
|---|---|---|---|---|
| Sire | Dam | Birth | 6 mo | 12 mo |
| Black Bengal | Black Bengal | 1.8 (0) | 11.4 (0) | 16.5 (0) |
| Beetal | Beetal | 2.5 (0) | 12.9 (0) | 19.0 (0) |
| Beetal | Black Bengal | 1.9 (6) | 12.4 (9) | 18.7 (13) |
| Black Bengal | Beetal | 2.3 (−8) | 12.7 (−2) | 18.2 (−4) |

| Breed | | Reproductive traits (days) | | | |
|---|---|---|---|---|---|
| Sire | Dam | Age at conception | Age at kidding | Service period | Kidding interval |
| Black Bengal | Black Bengal | 307 (0) | 451 (0) | 151 (0) | 294 (0) |
| Beetal | Black Bengal | 395 (29) | 543 (20) | 167 (11) | 314 (7) |

| Breed | | Body weight at | | |
|---|---|---|---|---|
| Sire | Dam | Birth | 4 mo | 9 mo |
| Year: 1982–85 | | | | |
| East African | East African | 2.0 (0) | 8.8 (0) | 11.9 (0) |
| Toggenburg | East African | 2.2 (0) | 9.7 (10) | 14.2 (19) |
| Anglo-Nubian | East African | 2.2 (10) | 9.5 (8) | 13.0 (9) |
| Galla | Galla | 2.6 (0) | 9.1 (0) | 13.1 (0) |
| Toggenburg | Galla | 2.6 (0) | 10.2 (12) | 13.9 (6) |
| Anglo-Nubian | Galla | 2.8 (8) | 10.4 (14) | 15.1 (15) |

**Table 4.10** *Continued*

| Breed | | Body weight at | | | Mortality to 56 days |
|---|---|---|---|---|---|
| Sire | Dam | Birth | 56 days | Market | |
| Barki | Barki | 2.1 (0) | 6.2 (0) | 12.6(0) | 24(0) |
| Zaraibi | Zaraibi | 2.1 (0) | 6.3 (0) | 10.9(0) | 31(0) |
| Damascus | Damascus | 3.1 (0) | 9.1 (0) | 17.5(0) | 35(0) |
| Zaraibi | Barki | 2.6 (24) | 7.1 (15) | 13.2 (5) | 25 (4) |
| Damascus | Barki | 2.6 (24) | 7.8 (26) | 15.6 (24) | 33 (38) |
| Barki | Zaraibi × Barki | 2.2 (−15) | 7.5 (6) | 14.7 (11) | 34 (36) |
| Barki | Damascus × Barki | 2.3 (−12) | 7.5 (−4) | 14.8 (−5) | 21 (−36) |

| Breed | | Litter weight at weaning | Milk intake per 1 kg gain | Total 16-wk milk yield |
|---|---|---|---|---|
| Sire | Dam | | | |
| Barki | Barki | 7.85 (0) | 9.17 (0) | 80.0 (0) |
| Zaraibi | Zaraibi | 9.21 (0) | 7.43 (0) | 116.5 (0) |
| Damascus | Damascus | 11.92 (0) | 6.78 (0) | 146.0 (0) |
| Zaraibi | Barki | 10.15 (29) | 7.70 (−16) | 117.7 (47) |
| Damascus | Barki | 10.41 (33) | 7.70 (−16) | 136.8 (71) |
| Barki | Zaraibi × Barki | 7.94 (−22) | 8.73 (13) | 93.0 (−21) |
| Barki | Damascus × Barki | 10.35 (−1) | 8.20 (6) | 123.0 (5) |

Taneja, 1982; Singh et al., 2002; Singh et al., 2000; Ruvuna et al., 1988; Abdelsalam et al., 1994; Abdelsalam et al., 2000; Shrestha and Fahmy, 2007.

[1]Percent deviation from dam breed in parenthesis.

may be a consequence of marginal differences in the mature weight of the parental breeds. In contrast kids derived from bucks of the heavier Jamunapari breeds and does of the dwarf Black Bengal breed were significantly heavier than the contemporary purebred kids of their female parent. Singh et al. (2002) reported that kids derived from the larger Beetal breed exceeded the dwarf Black Bengal and their reciprocal crosses in body weights from birth to 12 months of age demonstrating 3–4% heterosis at 6- and 12-month body weights. However, they found no difference between does of the Black Bengal breeds and Beetal × Black Bengal crosses in age at first conception, while the latter showed a significant increase in age at first kidding, service period, and kidding interval.

In Kenya, Ruvuna et al. (1988) carried out a crossbreeding study of several breeds and reported that the body weight of kids from birth to 9 months of age were lowest for East African, heaviest for Anglo-Nubian × Galla, and intermediate for Galla, Toggenburg × East African, Toggenburg × Galla, and Anglo-Nubian × East African (Table 4.10). Kids derived from Toggenburg and Anglo-Nubian buck and East African does exceed contemporary kids derived from their female parental breed by 9–19% for body weight from birth to 9 months of age. In another study, Ruvuna et al. (1992) concluded Toggenburg and Anglo-Nubian–sired kids had similar growth rate, slaughter weight and carcass composition. Galla dams produced kids that were heavier and leaner than those from contemporary East African dams.

In Egypt, Abdelsalam et al. (1994) evaluated crosses among bucks of the Zaraibi and Damascus breeds with does of the Barki breed, and bucks of the Barki breed with does of Zaraibi × Barki, and Damascus × Barki crosses resulting in single- and backcrosses and contemporaries derived from their purebred parents (Table 4.10). In the Zaraibi × Barki cross heterosis of 12–24% was observed for body weights from birth to market and 9% for mortality to 90 days. Corresponding estimates of heterosis for the Damascus × Barki cross varied from 0–4% for body weights and 12% for mortality. Furthermore, Abdelsalem et al. (2000) reported Barki-sired does of Damascus × Barki cross compared to those of Zaraibi × Barki cross produced more milk for a longer lactation, had heavier litters, and a more efficient milk conversion ratio. Relative to contemporary Barki does, crossbreds produced more milk and heavier litter weights.

There has been extensive research (Flamant and Morand-Fehr, 1982) directed toward genetic improvement of goat breeds from continental Europe. Studies suggest increasing milk yield in the female parent results in increased growth rate of their crossbred offspring. In Mexico, Sánchez et al. (1994) used a stall-fed system to evaluate the grading-up of local goats derived from high (≥7/8) and low (<7/8) proportions of the Alpine, Granadina, Nubian, Saanen, and Toggenburg dairy breeds from continental Europe. The authors concluded that crossbred kids from high and low proportions of the Alpine, Saanen, and Toggenburg breeds were similar to the Nubian breed but heavier than the Granadina breed. Concurrently, Montaldo et al. (1995) reported crossbred kids from high and low proportions of the Alpine, Saanen, and Toggenburg breeds were more productive compared to their contemporary local goats in milk production, lactation length and efficiency, litter size, and weight.

The mating of crossbred bucks to purebred or crossbred does can capitalize on paternal heterosis and contribute toward libido and sperm production with possible influence on fertilization and embryonic survival (Notter, 1987). Lopez-Perez et al. (1998) showed crossbred offspring from Boer × Spanish bucks and Spanish does exceeded the Spanish breed in body weights at birth and 90 days by 11 and 16%, respectively. There is also the potential for lower performance due to rearrangement or segregation of genetic combinations between the chromosomes of the crossbred parent as well as the inability to benefit in performance from maternal heterosis and favorable genes in the crossbred doe. Nevertheless, crossbred bucks may be suitable in environments that may be stressful, due to unfavorable climate, grazing condition and disease.

The mating of bucks from crosses between two meat-type breeds with does from crosses between two fecund-type breeds results in a four-breed cross offspring. The resulting cross can capitalize on the full potential of maternal, paternal, and individual heterosis. The drawback is the need to use four breeds. However, in large scale commercial production of goats there is the possibility of maintaining one or two parental breeds or their crosses within the farm to produce does for use as female parents (Figure 4.2). At the same time, crossbred bucks derived from alternate breeds not used to produce the crossbred doe may be purchased as required for use as male parents.

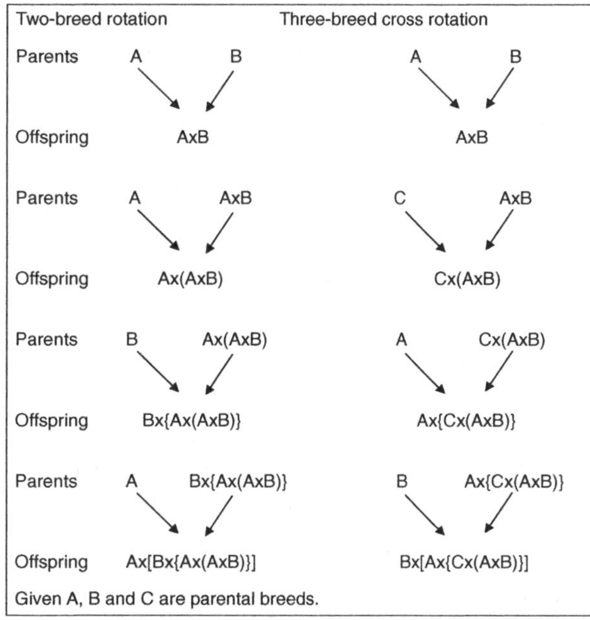


**Figure 4.3** Rotational crosses based on two and three parental breeds.

The crossbreeding of two or more breeds in rotation involves mating does from a specific breed with bucks of an alternate breed (Figure 4.3). In this procedure, bucks may be purchased from reputed breeders. Crossbred offspring from the previous mating are bred to bucks of an alternate breed not used previously to sire the female parent in succession. The drawback arises from large differences in performance of offspring from the different sire breeds. This can contribute to a lack of uniformity in the marketing of animal products to the consumer. In theory, the performance of two- or three-breed rotational cross is reduced due to lower heterozygosity, by one-third or one-seventh the average difference in performance between single crosses and parental breeds, respectively. There is no special benefit from breed differences in maternal versus individual performance. Also further loss in performance can occur in the offspring from interbreed recombination in the gametes as a result of contribution of genes from the dam and maternal granddam.

In the U.K., regional segmentation of sheep production involves raising the Hill breeds in the mountains, and the Down breeds in the lowland followed by their crossbreeding to produce market lambs. Despite the practical constraints of having to use more than two breeds for crossbreeding, substantial gain can be achieved from average breed superiority and heterosis among complementary breeds. In goats, there is no practical evidence of any serious attempt to exploit potential genetic merit among multiple breeds. The crossing of does of the indigenous breeds with bucks of the more productive dairy breeds to produce highly efficient crossbred does in more remote (or tribal) areas for use in terminal crosses under more intensive production close to urban markets demonstrates considerable opportunity.

### New Breed Development

Breeders expect the desirable qualities of two or more breeds to be combined into a composite population to approach the level of production from crossbreeding. In addition to the operational advantage of having to maintain a single breeding population, there is no need to purchase new animals, minimizing the risk of introducing diseases. Although many composite populations were developed for specific objectives, methodologies today differ from those used in the past principally in their intensity and deliberate application of greater knowledge of quantitative genetics. In theory, the genetic potential for performance in a newly formed composite population when compared to specific or rotational crosses involving the same number of breeds is lower due to the reduced level of heterozygosity. Heterosis in a composite breed based on two breeds is two-thirds of that in single crosses, while in composites made of three or more breeds heterosis can be three-fourths or more of that in single crosses. One problem with the use of a single composite population is that it is not capable of exploiting breed differences in maternal versus individual performance. Concept of the composite populations, however, has become an integral part of breeding strategies exploited by breeding companies for commercial production of livestock and poultry.

Grading-up of indigenous breeds has occurred with the Saanen breed in Korea; the Saanen and Toggenburg breeds in the Russian Federation, Germany, and the Czech Republic; the Angora breed in India; and feral goats in Australia and New Zealand resulting in adequate productivity leading to the development of 80 composite breed populations in 37 countries (Shrestha and Fahmy, 2007). In developing countries, the replacement of indigenous breeds through unplanned or accidental introduction has contributed to the development of composite populations. In Algeria, the introduction of the Murcia and Maltese breeds of goats for crossbreeding with the indigenous Berber, Makatia, and Arabia was followed by further introduction of the imported Alpine, Toggenburg, and Saanen breeds. The interbreeding of a number of goat breeds in isolated regions has resulted in the development of small meat breeds in Sri Lanka, India, Papua New Guinea, and Fiji; various Criollo (syn. Creole) breeds of Latin America and the West Indies; Spanish goats of southwest U.S.; and Sem Raça Definida goats of Brazil.

Composite populations with considerable genetic merit are the Anglo-Nubian, Boer, French Alpine, Kilis, LaMancha, and Peranakan Etawah breeds. The Boer breed having excelled in meat production efficiency has been exported to a number of countries including Australia, Canada, China, France, Germany, Israel, New Zealand, Sri Lanka, and the U.S. to meet the growing demand for goat meat (Skinner, 1972; Malan, 2000).

### Breeding Strategies

In a closed nucleus population, with all genetic evaluations occurring in the nucleus herd, offspring with the highest genetic merit in the nucleus are selected as parents to continue improving the nucleus herd. Nucleus herds could accelerate genetic response from the use of artificial insemination, and multiple ovulation and embryo transfer (MOET). Selected bucks and does of superior genetic merit in the nucleus herd are placed in multiplier herds to increase the number of goats with potential genetic merit for transfer to commercial herds. The commercial herds are exclusive in producing kids to provide animal products for the customer. In this procedure, it may take a generation or two to transfer the genetic improvement from the nucleus to the multiplier and finally the commercial herds.

In an open nucleus population, where evaluations occur in the nucleus and multiplier herds, offspring with superior genetic merit in the multiplier herds are used as parents in the nucleus herd. Unlike the above procedure with a closed nucleus where transfer of genetic merit occurs in one direction from nucleus to multiplier, exceptional offspring in the multiplier herd contribute genetic merit to the nucleus herd and can improve the rate of genetic change in the nucleus. These breeding strategies have been modified in various countries according to segmentation of the commercial goat industry.

## CHALLENGES

There are numerous (about 1,153) breeds of goats in all continents demonstrating the availability of a colossal amount of genetic resources for increasing production efficiency. The evaluation of breeds presents an opportunity to characterize their qualities and shortcomings. There is always the possibility that the inherent potential for specific morphological characteristics and production performance necessary for the improvement of established goat breeds may be lacking in a country or region.

Developing countries share more than 90% of the world's goat population. Breeders of nomadic and seminomadic origin on average receive a substantial portion of their income from fiber and meat while milk is used for home consumption. These producers are unaware of scientific accomplishments in the field of genetics, nutrition, and husbandry that have helped other livestock and poultry species become more productive.

In breed evaluation studies, imported breeds may underperform because of poor test environments relative to their nutritional requirements resulting in inconsistent conclusions. This may be due to the situation where goats are often reared on marginal lands not suitable for full expression of the genetic potential of the crossbred offspring. Developing breeding strategies for goats that are raised under sedentary, nomadic and seminomadic management present a real challenge. This is only possible by adding sufficient value to their products, as well as recognizing their significant role in maintaining cultural heritage and sustaining biodiversity while helping alleviate poverty.

## SUMMARY

Evaluation of breeds for specific objectives can identify breeds with potential genetic merit for morphological characteristics and production performance in the selection criteria. Benefit from the use of synchronization of estrus, super ovulation, artificial insemination, batch breeding and timing of kidding according to requirements for the marketing of goats and their products are apparent. Novel reproductive technologies using superovulation, embryo transfer, and cloning have potential merit for genetic improvement. These reproductive technologies combined with year-round kidding of complementary goat breeds demonstrate considerable merit in increasing the number of kids marketed per doe annually. Furthermore, there is opportunity for large-scale commercial production of meat and milk from goats close to markets in large urban populations.

Prospects for commercial goat production can be enhanced by complementing breeds that have demonstrated considerable genetic merit for the performance traits of economic importance with those established from the indigenous breeds. Fiscal constraints and requirements for quarantine have in the past permitted only the importation of a small number of animals, mostly males. Currently, the ease with which semen and embryos can be transported provides an opportunity to sample a large number of unrelated animals and establish foundation herds with as broad a genetic base as possible.

More recently, the promise of molecular methodologies has been enhanced due to considerable advances in microsatellite marker identification and the development of very comprehensive gene maps. There is potential merit in the application of Marker-Assisted selection to genetic evaluation when large numbers of markers are identified improving the accuracy of estimation of breeding values. In practice these already exist for bulls with superior genetic merit.

## REFERENCES

Abdelsalam, M.M., A.E. Haider, A.M. Aboul-Naga, I.S. El-Kimary, and M. Eissa. 1994. Improving performance of desert Barki kids by crossing with Zaraibi and Damascus goats. Egyptian Journal of Animal Production 31:85–97.

Abdelsalam, M.M., M. Eissa, G. Maharm, and A.I. Haider. 2000. Improving the productivity of Barki goat by crossbreeding with Damascus or Zaraibi breeds. Alexandria Journal of Agricultural Research 45:33–42.

Acharya, R.M. 1988. "Goat breeding and meat production." In: Goat meat production in Asia, edited by C. Devendra, pp. 14–29. Proceedings of Workshop in Tando Jam, Pakistan, International Development Research Centre, Ottawa, Canada.

Acharya, R.M., R.K. Misra, and V.K. Patil. 1982. Breeding strategy for goats in India. Indian Council of Agricultural Research, New Delhi, India, 111 pp.

Adalsteinsson, S., D.P. Sponenberg, S. Alexieva, and A.J.F. Russell. 1994. Inheritance of goat coat colors. Journal of Heredity 1994:267–272.

Boldman, K.G., L.A. Kriese, L.D. Van Vleck, C.P. Van Tassell, and S.D. Kachman. 1995. A Manual for Use of MTDFREML. A Set of Programs to obtain Estimates of Variances and Covariances [DRAFT]. U.S. Department of Agriculture, Agricultural Research Service, Washington D.C., Lincoln, NE, USA, 114 pp.

Cameron, N.D. 1997. Selection indices and prediction of genetic merit in animal breeding. Commonwealth Agricultural Bureau International, Wallingford, UK, 216 pp.

Falconer, D.S. and T.F.C. McKay. 1996. Introduction to Quantitative Genetics, Fourth Edition. Longmans Green, Harlow, Essex, UK.

Flamant, J.C., and P. Morand-Fehr. 1982. Milk production in sheep and goats. In: "World Animal Science." C. Production-system approach. 1. Sheep and Goat production, edited by I.E. Coop. Elsevier, Amsterdam, The Netherlands, pp. 275–295.

Hutt, F.B. 1964. Animal Genetics. The Ronald Press Company, New York, USA, 546 pp.

Lopez-Perez, D., S.D. Lukefahr, and D.F. Waldron. 1998. Comparison of crossbred Boer × Spanish and purebred Spanish breed-types for kid growth and litter size traits. Sheep and Goat Research Journal 14:144–147.

Lush, J.L. 1926. Inheritance of horns, wattles and color in grade Toggenburg goats. Journal of Heredity 17:73–91.

Lush, J.L. 1945. Animal Breeding Plans. Iowa State University Press, Ames, Iowa, USA, 443 pp.

Malan, S.W. 2000. The improved Boer goat. Small Ruminant Research 36:165–170.

Montaldo, V.H., A. Juárez, J.M. Berruecos, F. Sánchez. 1995. Performance of local goats and their backcrosses with several breeds in Mexico. Small Ruminant Research 16:97–105.

Nicholas, F.W. 1987. Veterinary Genetics. Oxford Science Publishers, Clarendon Press, Oxford, UK, 600 pp.

Notter, D.R. 1987. The crossbred sire: Theory. Journal of Animal Science 65:99–109.

Ruvuna, F., J.F. Taylor, M. Okeyo, M. Wanyoike, and C. Ahuya. 1992. Effects of breed and castration on slaughter weight and carcass composition of goats. Small Ruminant Research 7:175–183.

Ruvuna, F., T.C. Cartwright, H. Blackburn, M. Okeyo, and S. Chema. 1988. Gestation length, birth weight and growth rates of pure-bred indigenous goats and their crosses in Kenya. Journal of Agricultural Science, Cambridge 111:363–368.

Sánchez, F., H. Montaldo, and A. Juárez. 1994. Environmental and genetic effects on birth weight in graded-up goat kids. Canadian Journal of Animal Science 74:397–400.

Shrestha, J.N.B. and M.H. Fahmy. 2007. Breeding goats for meat production: a review (2) Crossbreeding and formation of composite population. Small Ruminant Research 67:93–112.

Singh, D.K., N.S. Singh, and L.B. Singh. 2000. Genetic studies on reproductive trait of goats. Indian Journal of Animal Science 70:1255–1257.

Singh, S., Z.S. Rana, and D.S. Dalal. 2002. Genetic and non-genetic factors affecting growth performance in goats. Indian Journal of Animal Research 36:12–16.

Skinner, J.D. 1972. Utilization of the Boer goat for intensive animal production. Tropical Animal Health and Production 4:120–128.

Taneja, G.C. 1982. Breeding goats for meat production. Proceedings of the 3rd International Conference on Goat Production and Disease, Tucson. Arizona, U.S.A., pp. 27–30.

Wright, S. 1931. Evolution in Mendelian populations. Genetics 16:97–159.

## ADDITIONAL SUGGESTED READINGS

Devendra, C. 1982. Breed differences in productivity in goats. In: "World Animal Science." B8. Disciplinary approach, edited by K. Maijala. Elsevier, Amsterdam, The Netherlands, pp. 431–440.

Devendra, C. and M. Burns. 1983. Goat Production in the Tropics. Commonwealth Agricultural Bureau International, Wallingford, U.K., 183 pp.

Quartermain, A.R. 1991. Evaluation and utilization of goat breeds. In: "World Animal Science." B8. Disciplinary approach, edited by K. Maijala. Elsevier, Amsterdam, The Netherlands, pp. 451–469.

Shrestha, J.N.B. 2005. Conserving domestic animal diversity among composite populations: a review. Small Ruminant Research 56:3–20.

Shrestha, J.N.B. and M.H. Fahmy. 2005. Breeding goats for meat production: a review (1) Genetic resources, management and breed evaluation. Small Ruminant Research 58:93–106.

Shrestha, J.N.B. and M.H. Fahmy. 2007. Breeding goats for meat production: a review (3) Selection and breeding strategies. Small Ruminant Research 67:113–125.

# 5
# Animal Evaluation

*R.A. Ebert, MS and S.G. Solaiman, PhD, PAS*

**KEY TERMS**

Dairy type—goat breeds selected for the combination of traits that make them suitable for milk production.
Fiber type—goat breeds selected for the combination of traits that make them suitable for fiber production.
Meat type—goat breeds selected for the combination of traits that make them suitable for meat production.
Steep shoulders—refers to the slope of the scapula or shoulder blade.
Steep pasterns—refers to the slope of the pastern.
Base-narrow—animals that lack natural muscling or thickness, boldness, or spring of rib.
Knock-kneed—a defect in which an animal has knees close together.
Cluster teats—a grouping of three or more teats in which all, some, or none may be functional.
Fishtail teats—a teat with two functional milk channels with slight separation between the channels.
Senior does—show ring term for does in milk production or more than 24 months of age.
Junior does—show ring term for does not in milk production.

**OBJECTIVES**

By completing this chapter, the reader will acquire knowledge on:

- The differences in the three types of goats
- The external anatomical terminology of goats
- The relation between structural correctness and functionality of the animal
- Determining age, based on dental examination
- Body Condition Scores (BCS)
- About goat shows

## INTRODUCTION

Goat owners and producers need a "working knowledge" of external caprine anatomical terms for selection of phenotypically superior animals.

The conformation of an animal is the correctness of structure, musculature, and body proportions in relation to each other, as evaluated visually. Different breeds of goats have differing roles in the goat industry. All animals, whether selected for meat, milk, or fiber need to be structurally sound. Evaluation of animals is important for two primary reasons, namely the need for animals to be critiqued on their merits as replacement animals for the herd and for the critical show-ring evaluation by a professional judge.

## FACTORS AFFECTING ANIMAL EVALUATION

Many factors may be involved in evaluating an animal for replacement herd, sales and dispersal, or superiority in show-ring judging. A few main factors will be briefly discussed in the following sections. However, more detailed information can be found on different goat breeders' association or university extension websites.

### Size

The size is relative and, in large part, is based on the environment in which the animals are managed and breed. The environmental factors such as rainfall, forage availability, temperature, and the level and skill of management dictate the production scheme in which animals are maintained. Skeletal frame size or weight, and bone and muscle development are indicators of size. There is a significant variation among breeds in regards to size. Frame is commonly referred to as the height of the animal at the hips. Larger-framed animals may not always be better animals. Smaller animals tend to be more efficient. Larger animals tend to be leaner and have a faster growth rate. In all cases, the frame size of the animal is determined by the environment or production goals. Larger animals tend to be preferred in show-ring judging.

### Volume or Body Capacity

Much emphasis should be placed on the internal dimension, volume, or capacity of an animal. Volume is important as it indicates an expanded internal capability for lungs (breeding capacity), rumen (feeding capacity), and utero function (reproductive capacity—easy birthing). Volume is three-dimensional (for example, length, depth, and width). Animals with expanded spring or arch of ribs (broad-chested), long-sided (from shoulder to hip), and deep-bodied or deep-flanked are preferred.

### Sex Character

#### Male

Bucks should appear rugged, stout, and masculine. Ruggedness is associated with stoutness of bone, identified by a larger circumference of cannon bone and size of the foot. Width between the eyes and prominence of jawbone are other indicators of masculinity. Masculinity is extremely important in evaluation of bucks in show-ring judging. Bucks must have two large testicles that are well shaped, equally sized, firm to the touch, and hang evenly in a single scrotum. Bucks with only one testicle have a condition known as cryptorchidism—a condition in which the testes fail to descend into the scrotum. Bucks with one testicle or testicles that hang unevenly, vary in size, are hard to the touch, or are soft and spongy may be lacking in fertility or may be subfertile, if fertile at all. An indicator of fertility is the scrotal circumference (SC), measured midway of the scrotum with both testicles descended. This measurement (larger is more desirable) is an indicator of the buck's ability to produce semen and is also an indicator of early puberty in both male and female progeny. The Small Ruminant Pocket Guide (2006) recommends SC to be a minimum of 30 cm at 8 months of age, while the Kentucky Governor's Office of Agricultural Policy recommends SC of 25 cm or more at the same age.

#### Female

Does should be feminine and more refined in their features. Femininity is typically associated with a longer head and neck coupled with a more refined head and muzzle. The udder should be soft, smooth, well shaped, and balanced. The female external genitalia should be well developed and properly structured. An infantile vulva often indicates a lowered rate of fertility. Femininity, especially in dairy-type does, is extremely important in evaluation of does in show-ring judging.

### Breed Character

For those breeders with a purebred or seed-stock breeding program, special emphasis should be placed on enhancing the breed characteristics that make the chosen breed unique. Specific breeding plans and goals must be considered. Being familiar with the characteristics that make each breed unique should be researched and followed. Structural correctness and ability to adapt to the environment should not be overlooked. The ability to grow, for meat breeds, and the ability to produce milk, for dairy breeds, must also be evaluated. Judges in the show ring place tremendous emphasis on breed character, because the show ring is the best opportunity for breeders to showcase their breeding program. Breed character is also important for maintaining the characteristics that make each breed unique and distinct from other breeds.

#### Meat-type Goats

When the primary purpose of goat ownership is to provide meat, the amount of muscling becomes important (Figure 5.1). Muscling should be apparent throughout the body of the animal. The two primary indicators of muscling are the back of the animal and the stifle and thigh regions. Animals with an abundance of natural muscling are heavy muscled or "thick topped." This is evidenced by the bulging of muscle along the topline of the animal, over the rack, through the loin, and out the rump to the tail head. The loin eye muscle, or longissimus dorsi, yields one of the most tender, highest priced, and most valuable cuts of the goat carcass. Another indicator of muscling is the bulge of the

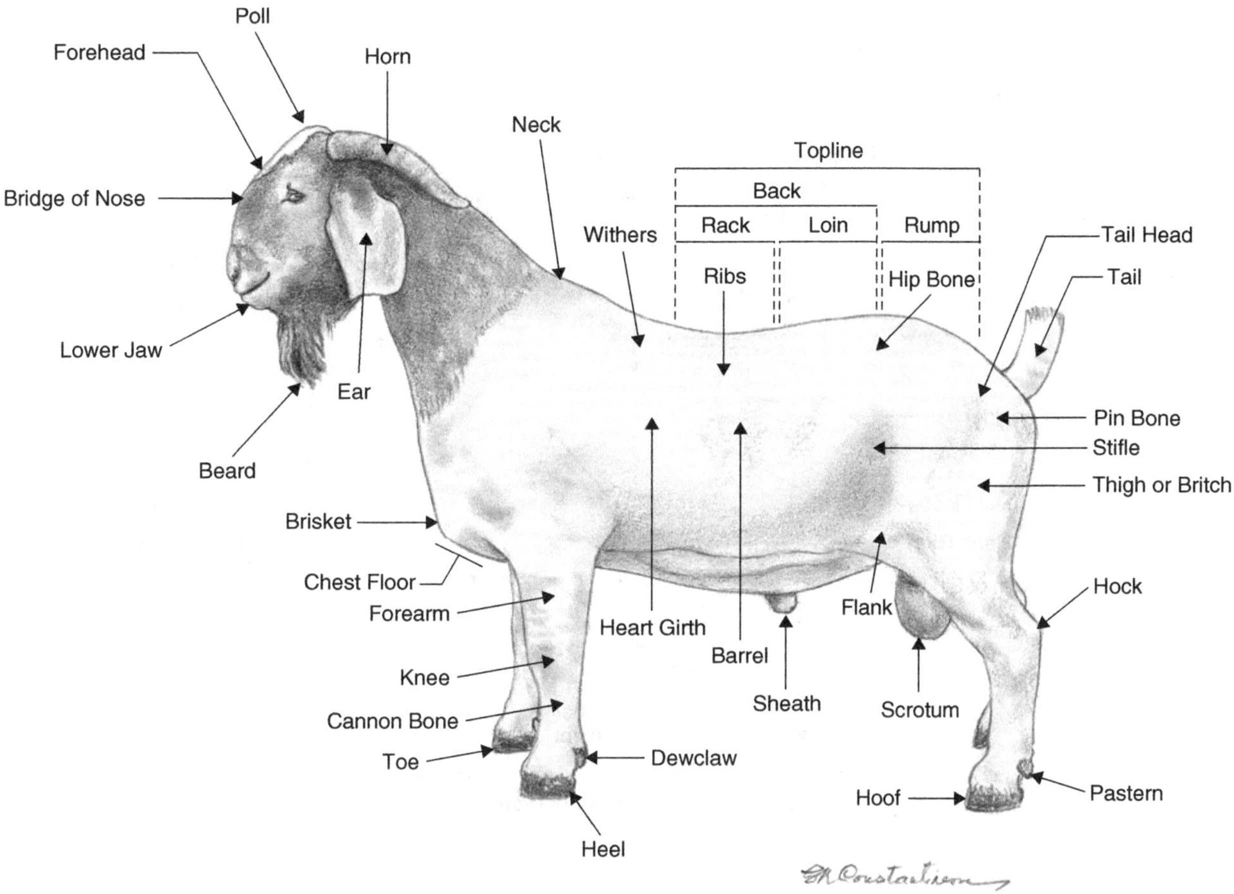


**Figure 5.1** Meat-type goat.

stifle muscle, viewed as the animal walks. The wider the animal stands and walks, usually the heavier muscled the animal is. The thickest part of a meat animal should be the width from stifle to stifle. The shoulders should be muscular but not coarse or out of proportion with the rest of the body and other muscle indicators. Another indicator of heavy muscling is the circumference of the cannon bone and size of the foot or hoof. Traditionally, the bigger the foot, the bigger the bone, and the more muscle mass. For more information on meat goats, please refer to Chapter 13.

### Dairy-type Goats

If milk production is the primary goal of the breeding program, animals are selected for dairy characteristics (Figure 5.2). Dairy characters are denoted by angularity of body shape, general trimness or refinement, and lack of meatiness. Dairy breeds are characterized by long, refined necks, sharpness over the withers, a deep heart girth, and a mammary system that is indicative of problem-free milking. Other important characteristics include a wide chest floor, width between the pins, and structural correctness of feet and legs. Dairy breed type refers to the characteristics of a particular breed as it relates to color, head and ear size and shape, and frame size.

In dairy goat evaluation, the mammary system of the doe in production must be evaluated. The udder must be strongly attached with a strong and smooth fore and rear udder attachment and a strong medial suspensory ligament. The udder should be balanced and symmetrical in shape, and soft and pliable to the touch. A pendulous, ill-shaped, unbalanced, hard, and leathery udder is objectionable. A prominent milk vein, located in front of the udder is desirable. Major emphasis in the show-ring judging (approximately 35%) is placed on the mammary system of does in production (senior does; Table 5.1). For more information on dairy goats, please refer to Chapter 14.

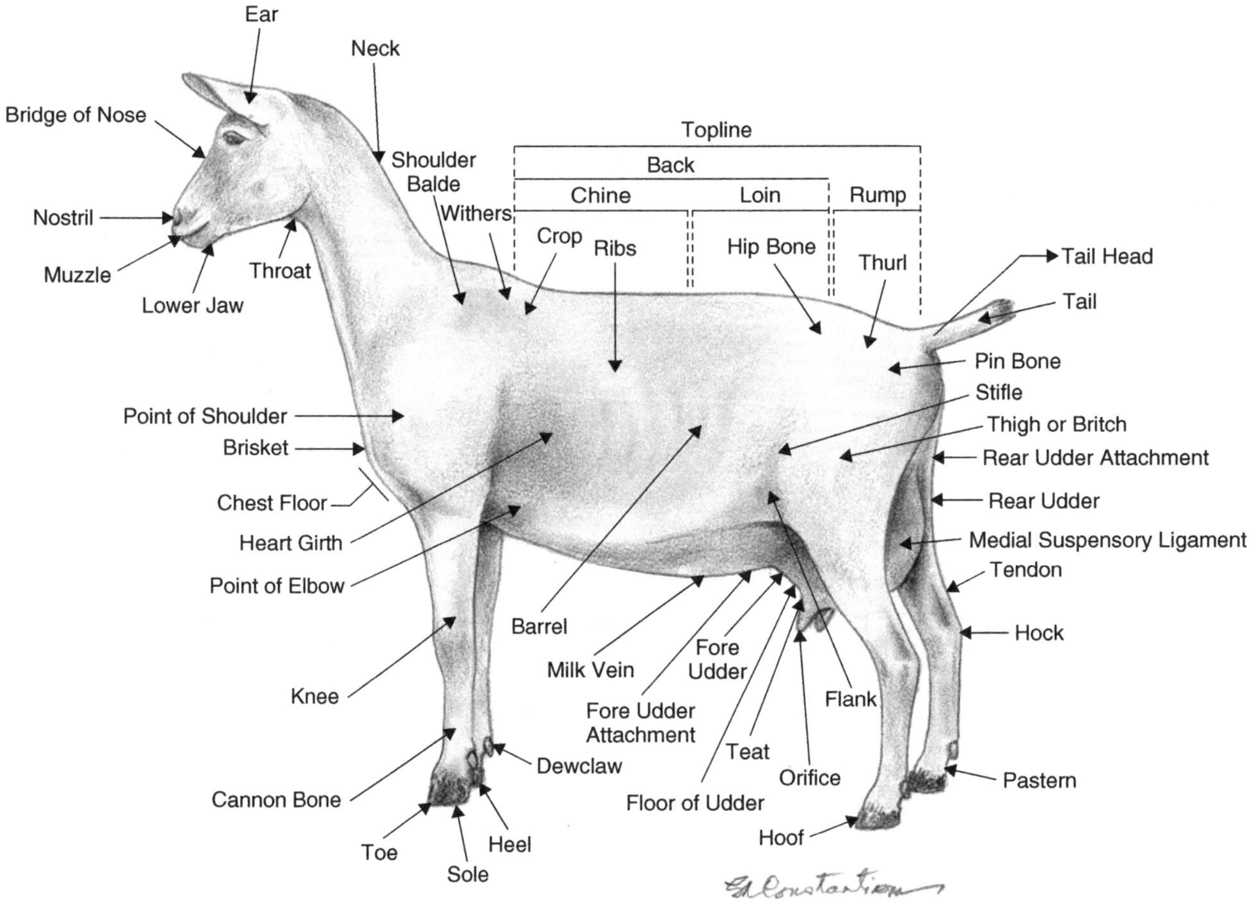


**Figure 5.2** Dairy-type goat.

**Table 5.1** Suggested scorecard for evaluating different types of goats for various traits.

| Goat Type Trait | Dairy, Senior Doe | Dairy, Junior Doe | Dairy, Buck | Angora | Market Meat Goat |
|---|---|---|---|---|---|
| General Appearance | 35 | 55 | 55 | 40 | 75 |
| Dairy Character | 20 | 30 | 30 | | |
| Body Capacity | 10 | 15 | 15 | | 25 |
| Mammary System | 35 | | | | |
| Fleece | | | | 60 | |
| TOTAL | 100 | 100 | 100 | 100 | 100 |

FIBER-TYPE GOATS

Angora goats produce mohair (Figure 5.3). Selection of goats for mohair production should be based on three fiber production traits: length of fiber, diameter of fiber, and uniformity (length and diameter) of fleece. The show-ring scorecard suggests that 60% of the emphasis in evaluation of fiber goats be placed on the fiber or fleece character (Table 5.1). Structural correctness, as discussed earlier in this chapter, is important to longevity of production for fiber goats. See Chapter 15 for a complete review and discussion of fiber-type goats.

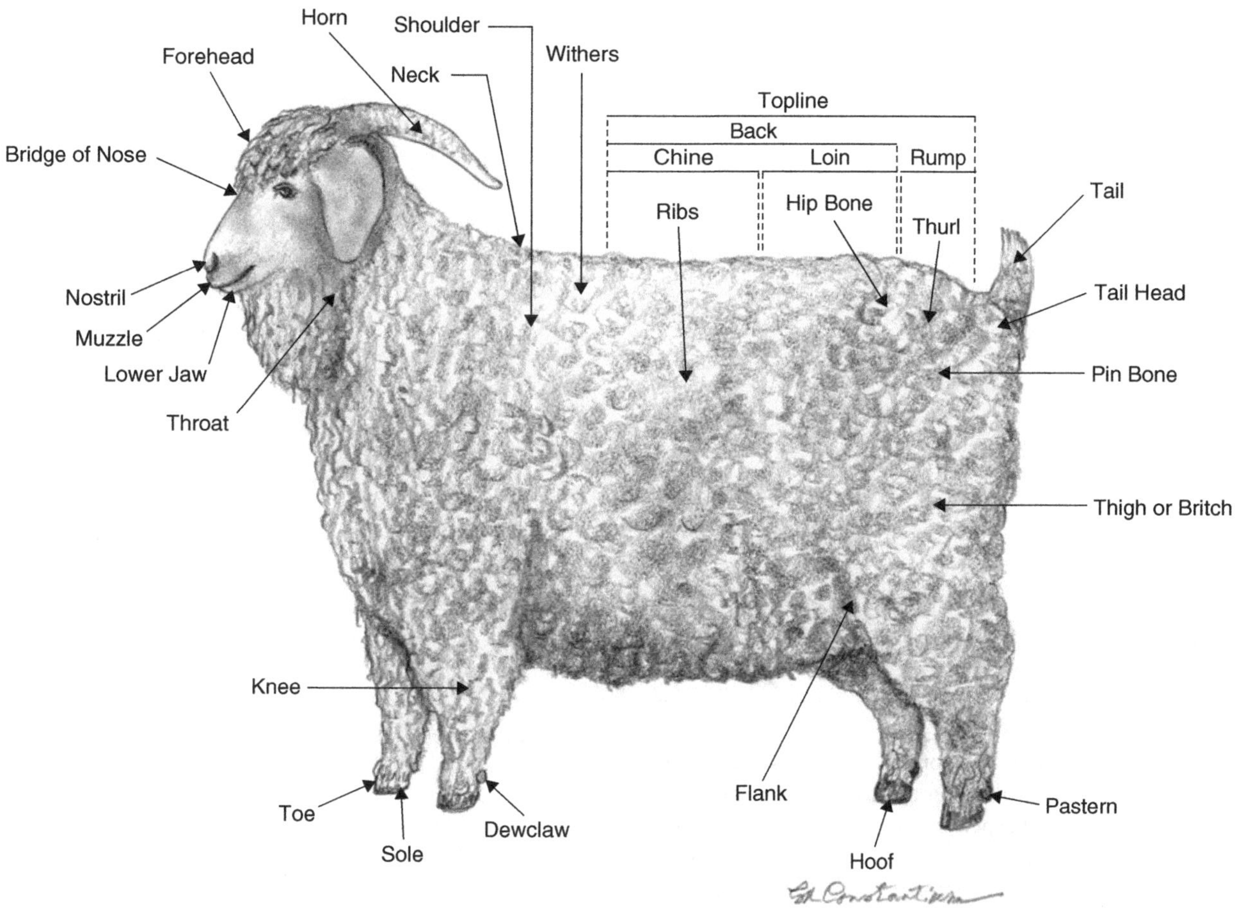


**Figure 5.3** Fiber-type goat.

## DETERMINATION OF AGE

Other methods of evaluation include the ability to estimate the approximate age of the goat by observing the teeth development. Goats should have a specific weight at a certain age. Age and size proportions are important in evaluating an animal. Although diet and health can influence the growth of teeth in goats, the age of a goat can be estimated based on their eight lower incisors. At birth, kids are born with four pairs or eight temporary incisor teeth. At approximately 1 year of age, the loss of the first two "baby teeth" occurs. Annually, two baby teeth are replaced with permanent teeth, until all baby teeth are replaced, at approximately 4 years of age. The amount of wear, after 4 years of age, is helpful in determining age. Some goats may be missing one or more teeth and are referred to as broken mouthed. Some older goats are called "gummers" when the teeth are worn down to the gum of the jaw. See Figure 5.4 for examples of what teeth in a kid, a yearling, a 2-year-old, a 3-year-old, and a full mouth of teeth for a 4-year-old goat should look like.

## STRUCTURAL CORRECTNESS

Structure is the foundation of the animal, and correctness of structure is important to build upon for better animal productivity. The basis for "Form to Function" in any specie of livestock is related to the angles of bone structure and the ability to move or travel with ease. The animal's ability to convert feed and forage to a usable protein product (meat, milk, or fiber) can be compromised by structural problems and difficulties. If an animal is unsound and not capable of performing or functioning at an optimum level because of structural problems, then emphasis on selection for any measurable trait is for naught. Production (meat, milk, or fiber) is directly influenced by structural

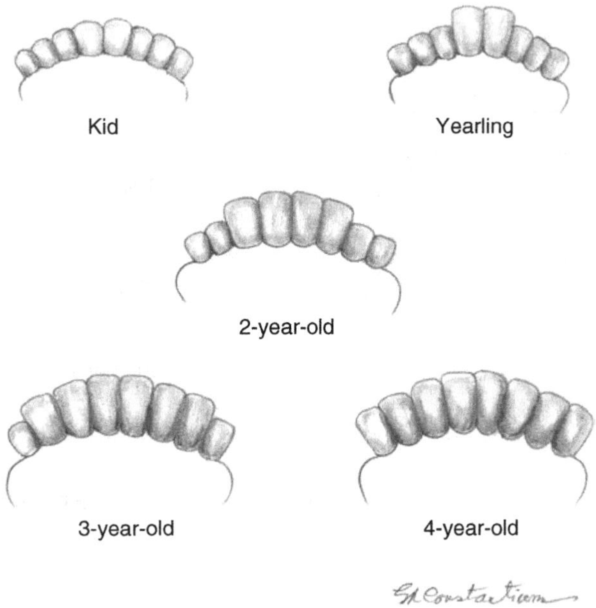


**Figure 5.4** Determination of age based on dental growth.

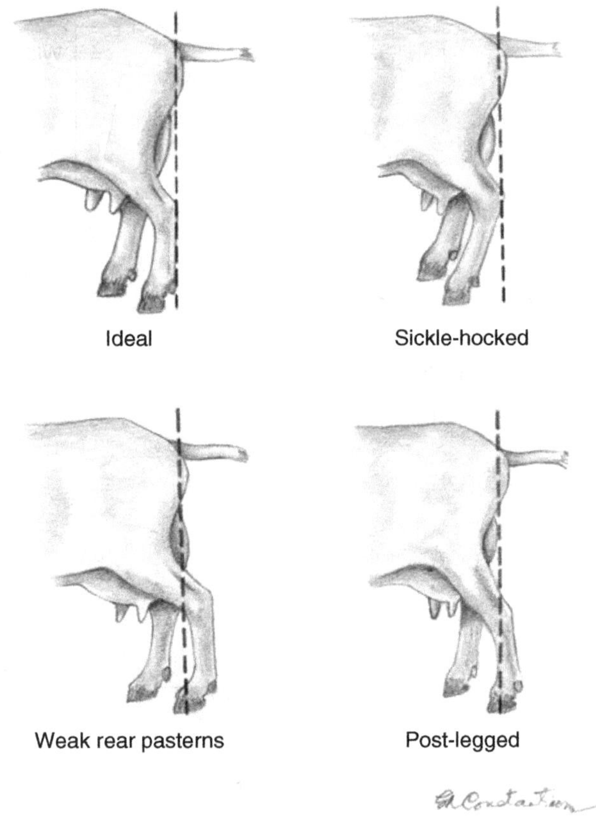


**Figure 5.5** Side-view rear leg evaluation.

soundness or lack thereof. For this reason, selection for sound and functional breeding animals is very important to goat producers and especially to seed-stock or purebred breeders. On the judge's show-ring scorecard, structural correctness is the major component used to assess general appearance.

### Side-View Evaluation

The best way, visually, to evaluate structural correctness of an animal is from the side view. The length and depth of the body and the volume (to a lesser degree) of the animal, as well as both front and rear leg structures can be evaluated at the same time from the side view. For longevity of service, it is imperative that bucks be especially sound in rear leg and hip structure. Movement or ability to travel can be evaluated by observing the length of stride or step; placement of the rear foot in relation to the front foot; and the fluidity of motion.

#### Rear Legs

The side view of rear leg evaluation is presented in Figure 5.5. A goat that has the "ideal" angulations at its rear legs will be approximately straight down from the pins to the hock and the dewclaw. The sickle-hocked condition is when an animal has too much angulation to the set or angle of the hocks. This trait is often associated with "weak rear pasterns." Animals that are "post-legged" have little or no angulations to the curvature of the hind legs. Post-legged animals are prone to suffer stifle-joint problems and disorders. Usually this trait is associated with steep or straight pasterns. Animals that are "post-legged" are prone to become unsound much more quickly than animals that are "sickle-hocked."

#### Front Legs

The side view of front leg evaluation is presented in Figure 5.6. A goat that has the "ideal" angulations at its front legs will be approximately straight down from the shoulder to the knee and the dewclaw. Common front leg structural problems include the condition of "buck-kneed" or "over at the knees," a condition when the knees are pitched forward. This is a serious defect and is associated with steep shoulders and steep pasterns and, quite frequently, the post-legged condition, which limits the ability to travel and graze efficiently. Steep shoulders refer to the slope of the scapula or shoulder blade. Too steep of an angle places

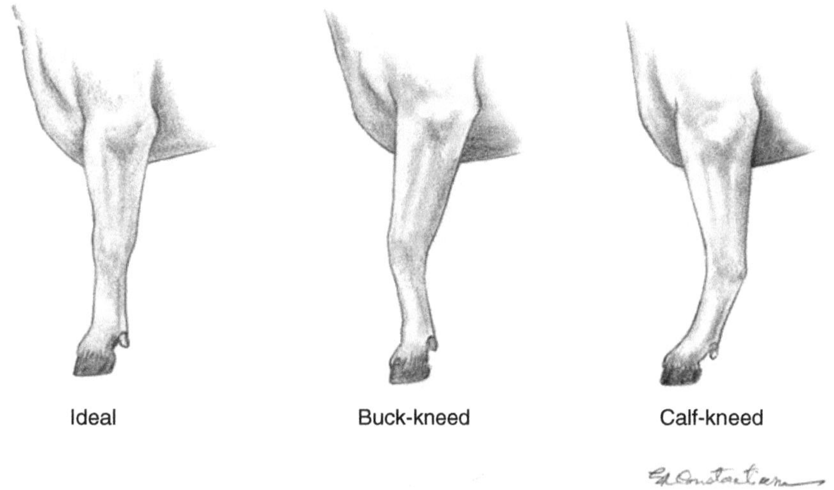


**Figure 5.6** Side-view front leg evaluation.

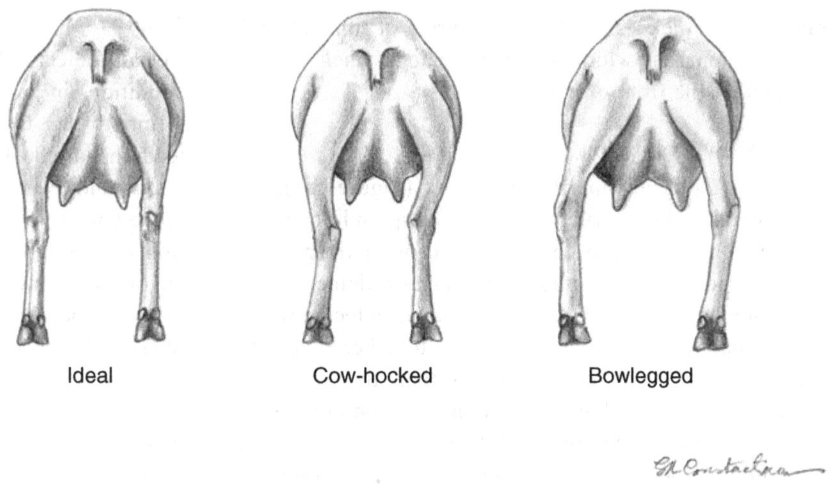


**Figure 5.7** Rear-view leg evaluation.

stress on the skeleton because the shock-absorbing ability of the forelimbs is reduced. Steep pasterns refer to the slope of the pastern. Too steep of an angle places stress on the skeleton because the shock-absorbing ability of the foot is reduced. Both of these conditions can reduce the animal's ability to travel. "Calf-kneed" or "back at the knees" is a less serious defect.

### Rear-View Evaluation

In viewing an animal from the rear (Figure 5.7), the "ideal" legs should come out of the center of the hindquarters and go straight down to the ground. The most common structural problem viewed from the rear is the condition known as "cow-hocked," also known as narrow or close at the hocks. Larger or longer outside toes magnify this condition. "Cow-hocked" is fairly common and seldom is a hindrance to the form and functionality of an animal. "Bowlegged," or wide at the hocks, occurs less frequently and is magnified by larger or longer inside toes. This condition is more serious and is frequently associated with animals that are base-narrow and/or have swollen or puffy hocks. Base-narrow refers to animals that lack natural

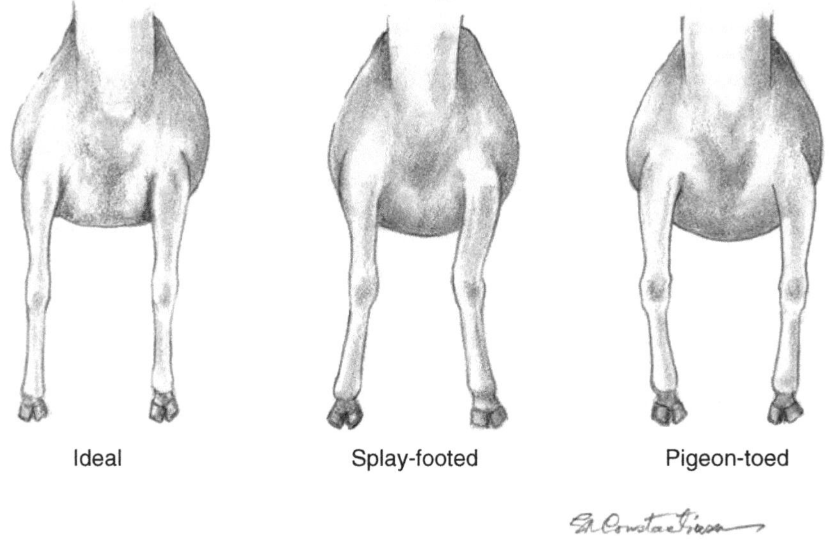


**Figure 5.8** Front-view leg evaluation.

muscling or thickness, boldness or spring of rib, and walk with no apparent width between the legs when viewed from either the front or the rear.

### Front-View Evaluation

In viewing the animal from the front (Figure 5.8), the correct (ideal) knee structure is straight up and down and in-line with the forearm, the knee, the cannon, the pastern, and the foot. "Splay-footed" is a common defect and is associated with legs being knock-kneed—a defect in which an animal possesses knees close together and the lower forelimb and foot are wider than the knees, or turned-out—a general term describing any combination of knock-kneed and splay-footed. "Pigeon-toed" animals are not common, but this trait is a serious defect because it is often associated with other defects such as being wing-shouldered—extreme width of the shoulders, often causing the front legs to be bowed, base-narrow, or bowlegged.

## OTHER EVALUATION PARAMETERS

In goats, other anatomical parts should be considered for soundness or defects. Eyes, ears, mouth, teats, and hooves are important. Animals that are blind or cannot hear should not be used in any production scheme and should be culled.

Foot structure and hoof size should be in proportion to the bone structure of the animal. Cracked or split hooves can be a concern for longevity. Animals with colored hooves tend to be preferred over animals with lighter-colored hooves.

### Mouth

Certain defects of the jaw should be observed. Different conditions in jaw alignments are presented in Figure 5.9. A mouth that is "ideal" in correct alignment has incisor teeth that are flush with the pad of the upper jaw, providing a true and even bite. The trait known as "parrot mouth" is the condition of an undershot jaw (bottom jaw) or overbite (upper jaw). "Monkey mouth" is the condition of an overshot jaw (bottom jaw) or underbite (upper jaw). Both of these conditions lack coordination between the teeth and the dental pad, thus can hinder prehension or mastication.

### Teats

Teat structure is of critical importance. Does should have no more than two functional teats for each side of the udder for meat-type goats and no more than one functional teat for each side for dairy-type goats. It is important that the structure of the udder is such that the offspring can nurse unassisted or the milking machine fits properly. Animals with pendulous udders, bulbous or over-sized teats should be culled. Different teat structures are presented in Figure 5.10. One functional teat (with its milk channel) and one nonfunctional teat, or two functional teats are acceptable for meat goat type. Split teats (teats with two distinctly separate milk channels) are not preferred. Cluster teats, a grouping of three or more teats in which all, some, or none may be functional, and fishtail teat, a teat with two functional milk channels with slight separation between the

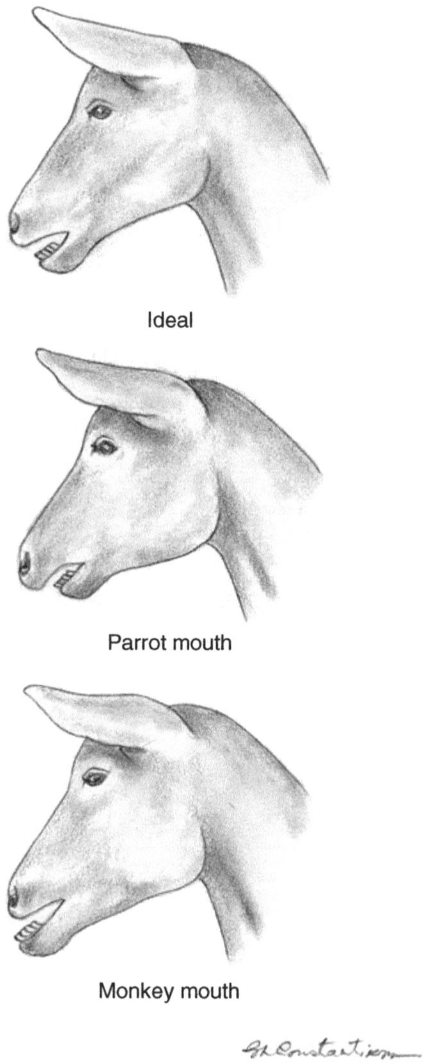


**Figure 5.9** Different conditions in jaw alignment (mouth).

channels, should be discriminated against. Retaining offspring for production purposes of bucks or does with teat or udder defects may not be in the best interest of the breeding program. Evaluation of the teat structure of the buck's offspring (and his dam) cannot be overlooked because the buck will have a significant genetic influence on the does retained for replacement purposes.

## BODY CONDITION SCORE

Body Condition Score (BCS) is another tool used in the evaluation and management of goat herds. It is based on live animal visual determination of body fat deposition. Scores range from 1 to 5, with a score of 1 being very lean, thin, emaciated, or fat deficient, and a score of 5 being extremely fat or obese. The full range of BCS is presented in Figure 5.11. The ideal BCS is between 2 to 3, with an average of 2.5. Richardson (2004) found that does with a BCS of less than 1.5 were nine times more likely to abort compared to other does. They concluded that to maintain pregnancies, BCS at mating time should be above 2. A high-energy diet producing a higher BCS at mating time will help overcome a lowered conception rate.

## THE SHOW-RING JUDGING AND GOAT SHOWS

In the United States, for more than 150 years, the show ring has been a "major influencer" of breeding programs in all species of livestock. The opportunity to showcase the best animals that an exhibitor owns or produces, and having the animals evaluated by a competent and unbiased professional judge compared with contemporaries in the show ring, is an important part of the livestock show scene of animal agriculture in the U.S. and around the world.

A goat show is an event usually associated with a fair or livestock exposition where goat breeders and exhibitors display their animals for evaluation and comparison by a professional judge. Goat shows can have three distinct and different components—breeding animals, market animals, and showmanship where the exhibitor is judged.

The majority of shows are for breeding animals. Judges use specific scorecards to evaluate animals (Table 5.1). Scorecards are numerical guides used by professional judges to evaluate animals in the show ring. They are especially important for evaluation of dairy- and fiber-type goats. These shows are for does and bucks, and the animals are evaluated on their merits as replacement animals for the goat herd to be placed into production as meat, milk, or fiber animals.

Some goat shows are for market animals for sales and dispersals at which the animals are evaluated on their merit as being ready for harvest and consumption. Traditionally wethers (nonintact males) make up the majority of animals in these shows, although other animals often are allowed to compete.

The remaining component, showmanship, is a sport and is traditionally limited to youthful exhibitors (usually members of 4-H, Future Farmers of America [FFA], or junior members of national breed registries). In this

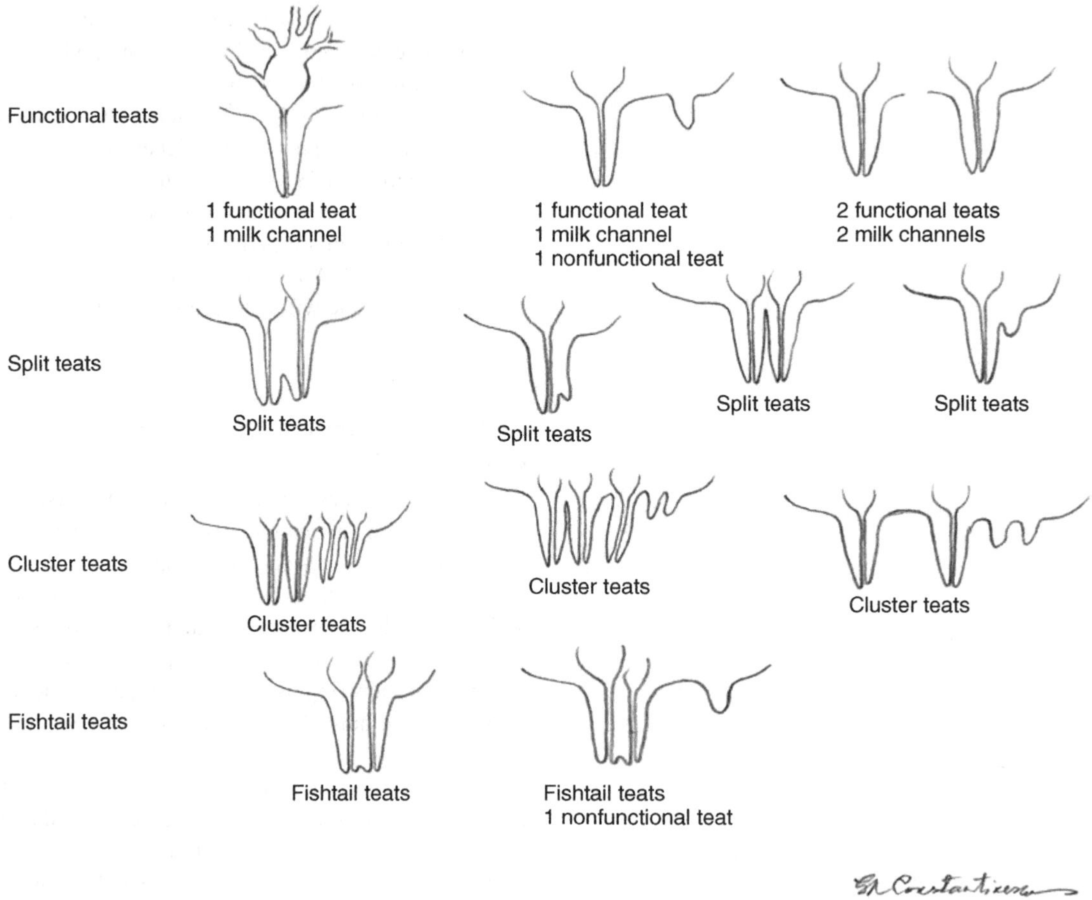


**Figure 5.10** Different teat structures.

competition, the exhibitors are evaluated on the ability to showcase their animals' strongest traits, without drawing attention to or showcasing themselves in the show ring. Other criteria considered for evaluation of the youthful exhibitors are the abilities to handle, manage, and control the animal while being exhibited, as well as mannerisms and courtesies displayed in the show ring to the judge and other exhibitors. Often, the judge will ask questions of the exhibitor to ascertain the working knowledge and degree of involvement the youthful exhibitor has in their "project" animal. In showmanship, the conformation of the animals is not evaluated, only the exhibitors and their ability and knowledge. The animal's behavior (in the show ring and under the control of the exhibitor) does however play an important role in the partnership of animal and the exhibitor that is evaluated in showmanship.

## SUMMARY

This chapter summarizes the three types of goats and the importance of knowing and understanding correct and appropriate terminology of the external parts of goats. It is important to know and understand phenotype or visual selection criteria to identify superior animals. Topics covered are size, volume (body capacity), sex characteristics, and phenotype characteristics of meat, dairy, and fiber types of goats. Points about mouth, teats, and body condition scoring are also covered. In addition, the show-ring aspect of goat production is briefly discussed.

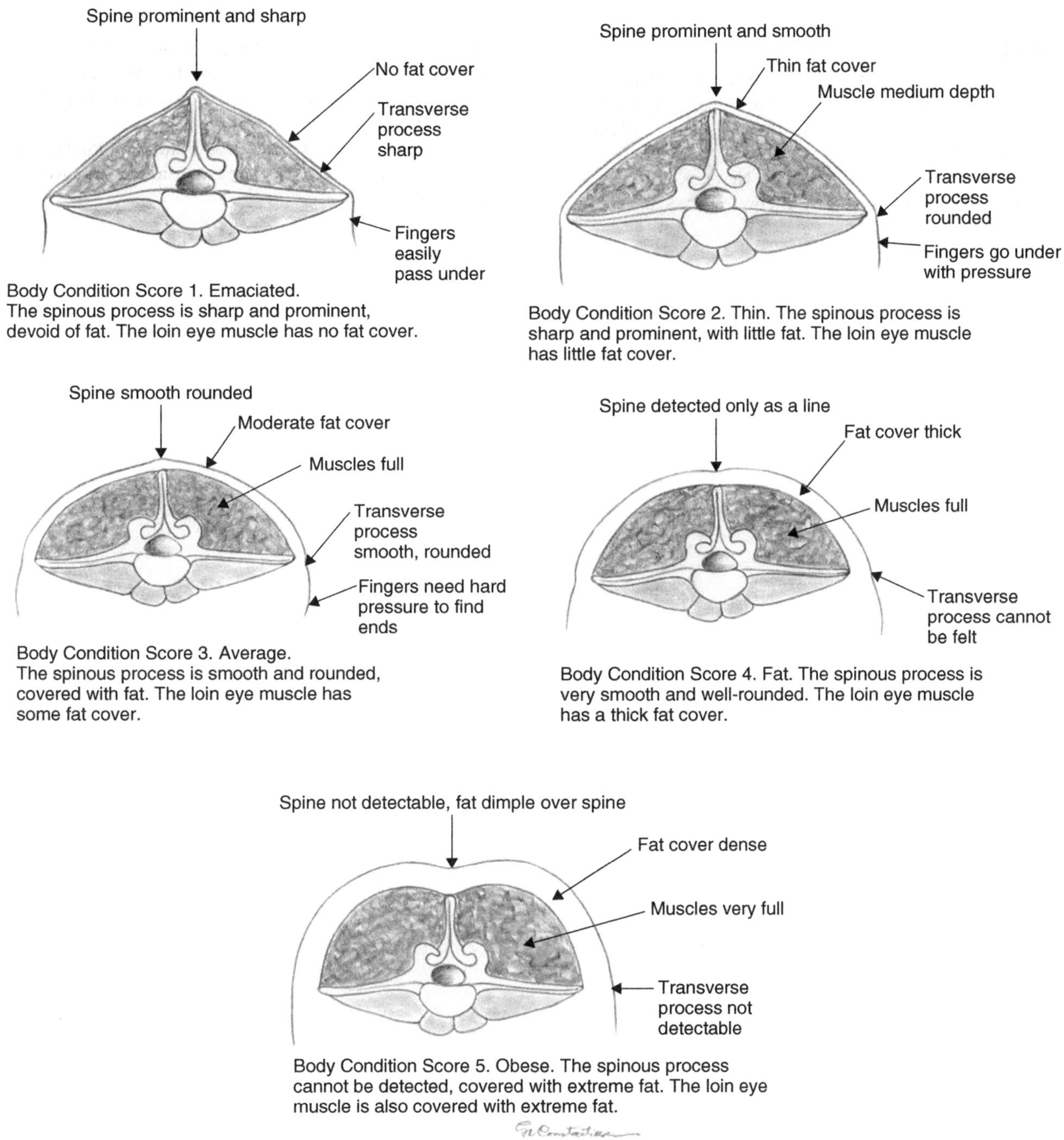


**Figure 5.11** Body condition scores from 1 to 5 for goats.

## REFERENCES

Kentucky Governor's Office of Agricultural Policy. Goat Diversification Program. http://agpolicy.ky.gov/funds/documents/goat_amendment8_040220.shtml (Accessed May 6, 2008).

Richardson, C. 2005. Reproduction News: Kidding rates of goats affected by low body energy reserves under range conditions of northern Mexico. Ontario Ministry of Agriculture Food & Rural Affairs. http://www.omaf.gov.on.ca/english/livestock/goat/news/srs0501a2.htm. (Accessed May 12, 2008).

Small Ruminant Pocket Guide. 2006. Extension Publication. AN-1296. Alabama Cooperative Extension System, Auburn University, AL.

# 6
# Functional Anatomy of the Goat

*G.M. Constantinescu, DVM, PhD, mult. Dr.h.c. and I.A. Constantinescu, DVM, MS*

**KEY TERMS**

Median plane—divides the body in two symmetrical halves.
Sagittal (paramedian) planes—parallel to the median plane.
Transverse planes—at right angles to the longitudinal axis of the body, limbs, or any other organ or part.
Dorsal planes—parallel to the back and corresponding surfaces of head, neck, tail, dorsal aspect of the carpus, metacarpus, and forelimbs' digits, tarsus, metatarsus, and hind limbs' digits.
Ventral planes—opposite to dorsal planes.
Medial—toward the median plane.
Lateral—opposite to the medial, away from the median plane.
Cranial—toward the head, for the limbs proximal to carpus and tarsus.
Caudal—toward the tail, for the limbs proximal to carpus and tarsus.
Rostral—toward the nose, applied to the head only.
Palmar—the caudal aspect of carpus, metacarpus, and digits.
Plantar—the caudal aspect of tarsus, metatarsus, and digits.
Proximal—near the origin in the limbs, tail, and for the penis, the attached end.
Distal—away from the origin in the limbs, tail, and for the penis, the free end.
Axis—the central line of the body or any anatomical structure.
Axial—toward the axis; in referring to the digits, the axis of the limb passes between the third and fourth digits, therefore, the axial aspect of each digit faces the axis.
Abaxial—away from the axis; in referring to the digits, the abaxial aspect of each digit faces away from the axis.

**OBJECTIVES**

By completing this chapter, the reader will acquire knowledge on:

- That there is an international terminology for all anatomical structures called Nomina Anatomica Veterinaria
- The differences between the several types of bones of the skeleton, and the role of different groups of muscles
- The parts of the horn and of the hoof
- The differences between the four compartments of the heart and between arteries and veins
- The role of the respiratory passages and the respiratory organs
- The anatomy and functionality of the prediaphragmatic and the postdiaphragmatic digestive systems
- The reproductive system in association with the urinary system, reproductive organs including essential organs, excretory passages, tubular genital organs, external genital organs, and annex glands
- The structure and function of the placenta and mammary glands
- The difference between the central, peripheral, and autonomic nervous systems
- The location and role of the endocrine glands
- The difference between the sense organs
- The role of the immune system

## INTRODUCTION

Etymologically, "anatomy" comes from the Greek "*anatome*," dissection (*ana*, apart + *tome*, a cutting). Anatomy is that branch of the biological sciences that deals with the body parts. The "functional anatomy" is the anatomy studied in its relation to function. It is our perception that functional anatomy is the most appropriate term for this book. Anatomy can be studied from several perspectives, therefore, it is named accordingly, such as descriptive, topographic, regional, functional, systematic, etc.

Functional anatomy includes the following systems: skeletal, articular, muscular, cardiovascular, respiratory, digestive, urinary, reproductive, nervous, and immune systems. In addition, the horn and the digital organ, the mammary gland, the endocrine glands, and the sense organs are covered. Before the skeletal system, only illustrations depicting the anatomical regions of the goat are shown.

Considering that anatomy is the basic knowledge in morphological studies, the decision was made to restrict the content of this chapter to the gross anatomical structures. Those interested in histological and ultrastructural aspects of different tissues are invited to consult adequate books (for example, Eurell and Frappier, 2008).

The anatomical regions of the goat are illustrated in Figures 6.1 to 6.6.

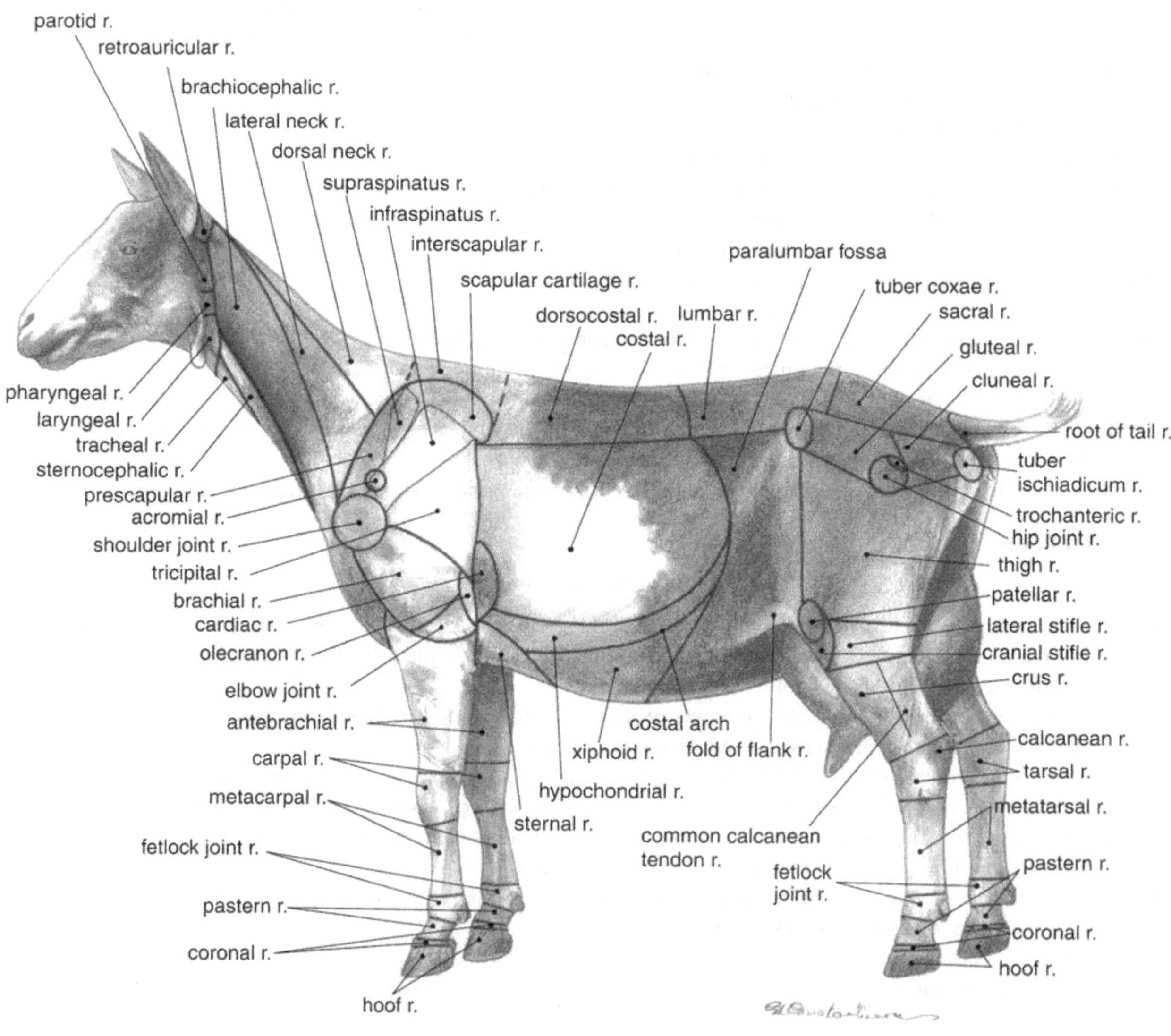


**Figure 6.1** Anatomical regions, lateral aspect of the body.

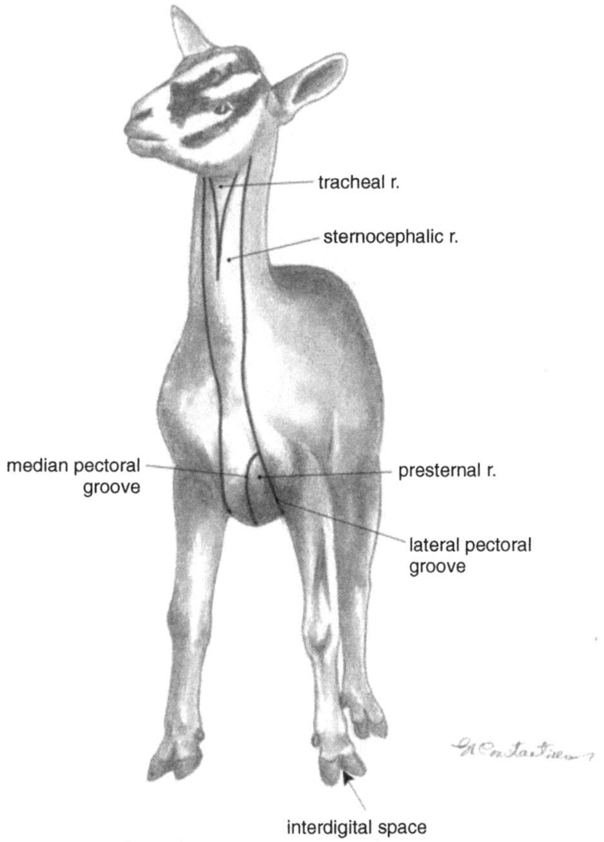


**Figure 6.2** Anatomical regions, cranial aspect of the body.

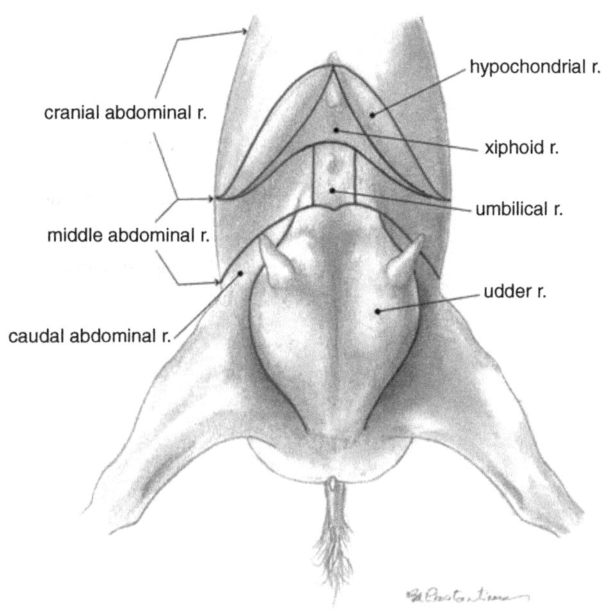


**Figure 6.3** Anatomical regions, ventral aspect of the body, female goat.

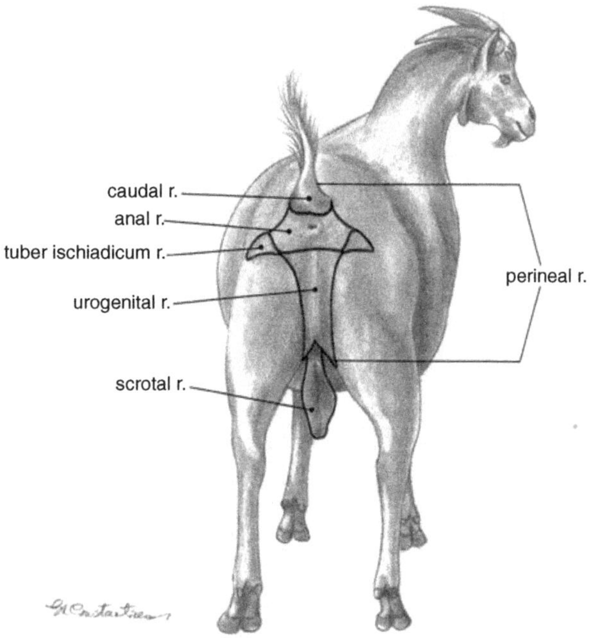


**Figure 6.4** Anatomical regions, caudal aspect of the body, male goat.

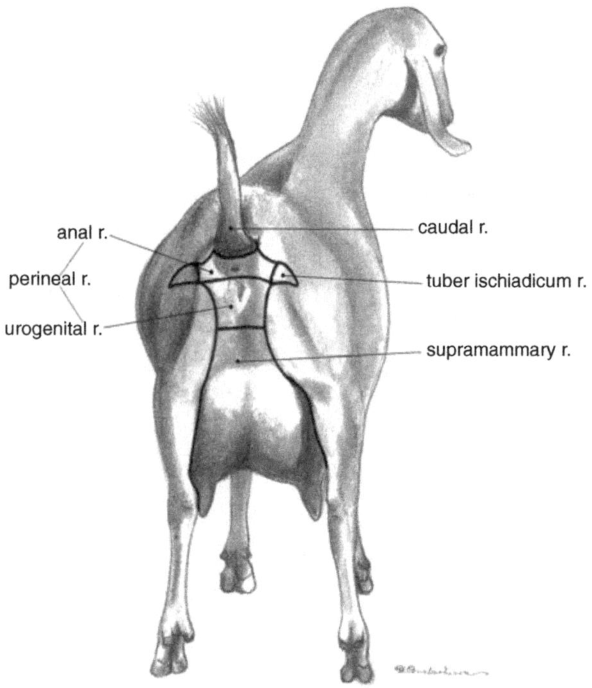


**Figure 6.5** Anatomical regions, caudal aspect of the body, female goat.

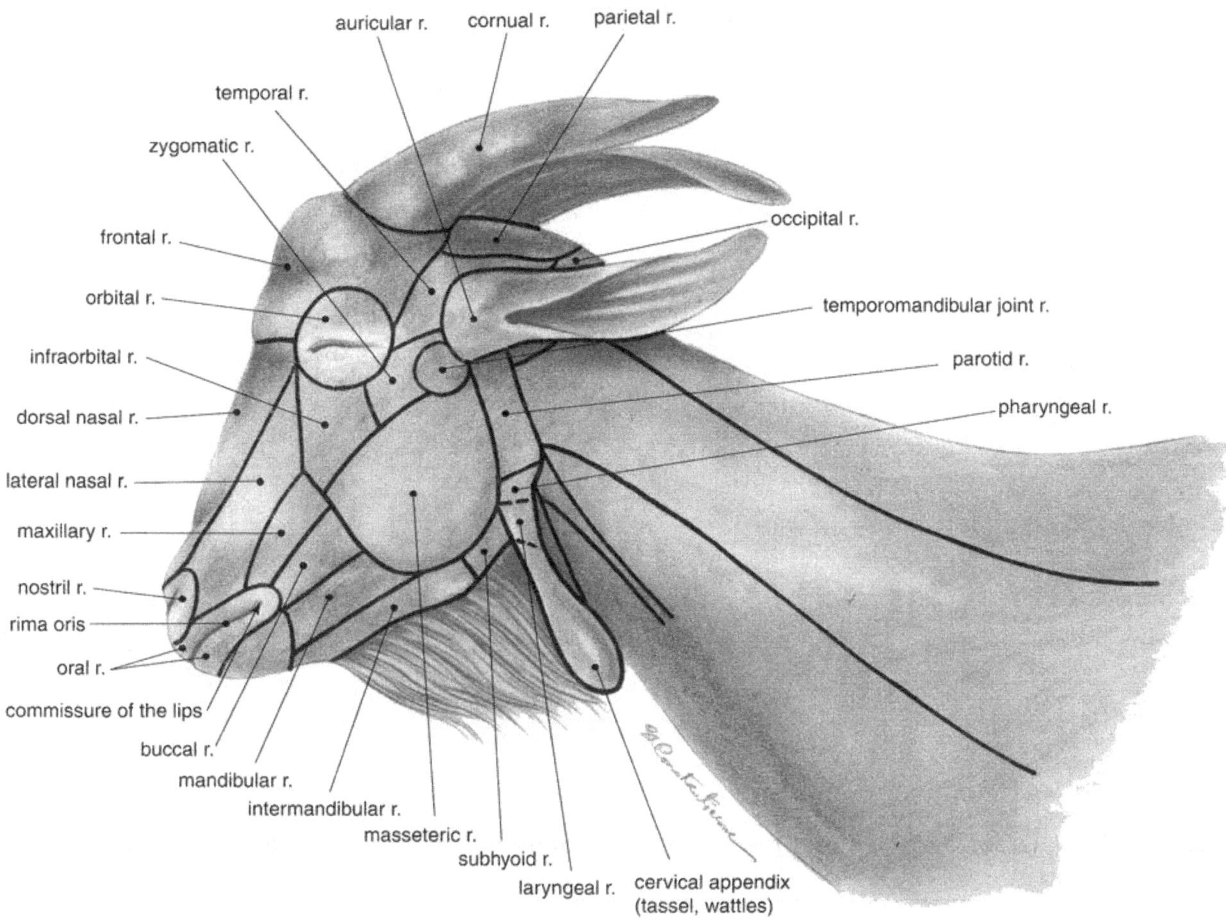


**Figure 6.6** Anatomical region of the head (lateral side).

## THE SKELETAL SYSTEM (THE BONES)

The bones are rigid yet elastic structures that articulate with other bones or cartilages to form the skeleton. They are covered by a connective tissue coat called the periosteum. At the periphery all bones are compact. There is a lamellar spongy tissue inside the bones. There are three major types of bones: long, short, and irregular. The long bones (the majority of the bones of the limbs) have a body and one or two articular extremities, and are centered by a medullary canal. The short bones (the carpal and the tarsal bones) do not have a medullary canal. The irregular bones (the vertebrae) do not fit in any geometrical shape.

The skeletal system of the goat, as of all of the domestic mammals, consists of the axial skeleton and the appendicular skeleton.

### The Axial Skeleton (the Skull, the Vertebral Column, the Sternum, and the Ribs)

#### The Skull

As a mammalian type, the skull of the goat consists of several bones fused with each other, except the mandibles (usually called "mandible"), the only bones with mobility against the rest of the skull. All bones of the skull are symmetrical, except the vomer, ethmoid, sphenoid, and occipital bones. In addition, there is a hyoid apparatus, consisting of small bones and located between the tongue, the pharynx, and the larynx. The skull of the goat is different from that of the large ruminants and even the sheep. As a common characteristic of a ruminant skull, the goat lacks the superior incisive teeth. The skull of the goat (except the mandible; see Figure 6.25) is illustrated in Figures 6.7

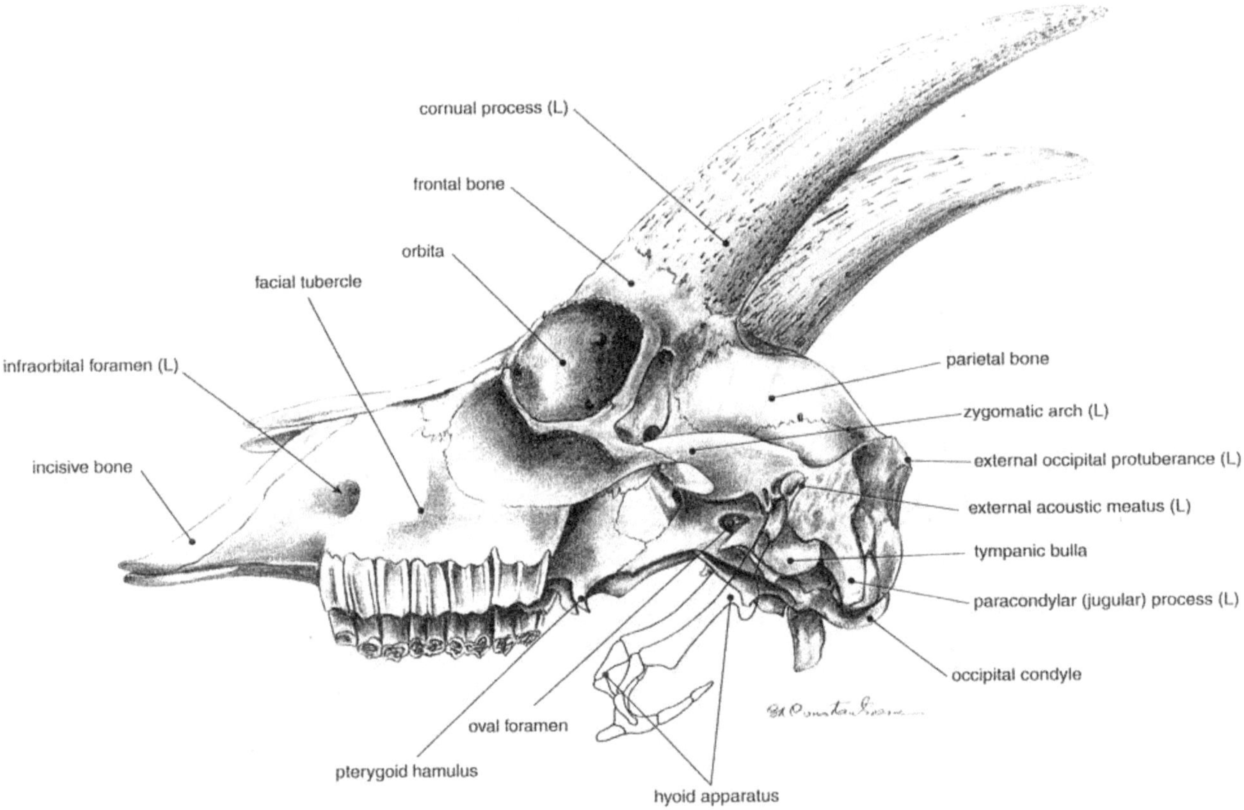


**Figure 6.7** Skull, lateral view (without mandible). (L) = landmark for physical examination and/or clinical approach.

to 6.9. The structures marked with an "L" are landmarks for clinical purposes.

The functional anatomy of the skull is much more complex than that of the other bones, because it protects the brain and two of the endocrine glands, the initial segment of the respiratory system, and the initial segment of the digestive system. Also, some of the prominences and foramina are landmarks for clinical purposes.

### THE VERTEBRAL COLUMN

The vertebral column has a multiple role in the normal functioning of the body: it protects the spinal cord (within the vertebral canal), contributes to the roofs of the thoracic and abdominal cavities, and provides attachments for ligaments and muscles. Also as part of the functional anatomy, the vertebral column provides landmarks for epidural and paravertebral anesthesia. Each vertebra consists of a body and the arch, both enclosing the vertebral canal. There is an intervertebral disc between the bodies of the vertebrae, with a fibrous ring surrounding a gelatinous center. The functional role of the intervertebral discs is to allow a certain degree of mobility of the vertebral column as a whole. There are five regions of the vertebral column: cervical, thoracic, lumbar, sacral, and caudal. The cervical vertebrae in all domestic mammals are much longer than in humans, but the total number is the same for all mammals. The thoracic vertebrae form part of the roof of the thoracic cavity. The lumbar vertebrae form the roof of the abdominal cavity, whereas the sacrum is the roof of the pelvic cavity. Using symbols, the following are the numbers of vertebrae for each region: C7, T13, L6, S5, Cd11–14. Their profiles are illustrated in Figure 6.10 and Figure 6.11.

### THE STERNUM

The sternum is a unique structure, consisting of seven sternebrae as shown in Figure 6.10, and is the bony floor of the thoracic cavity.

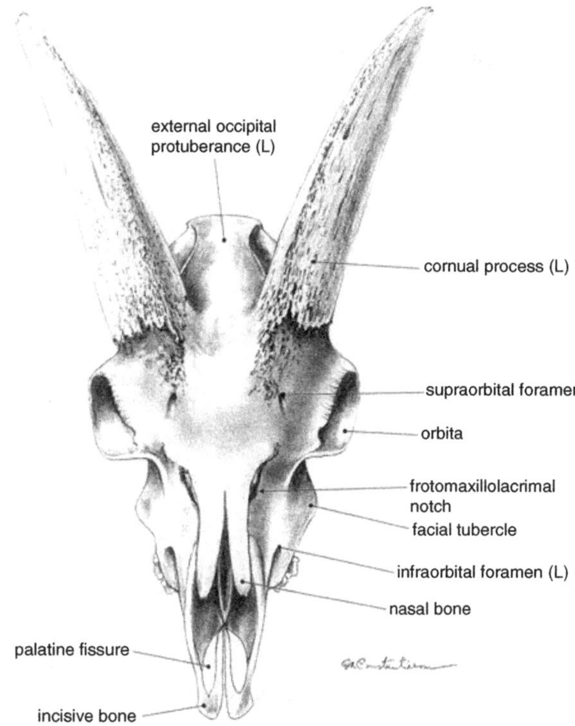


**Figure 6.8** Skull, frontal view. (L) = landmark for physical examination and/or clinical approach.

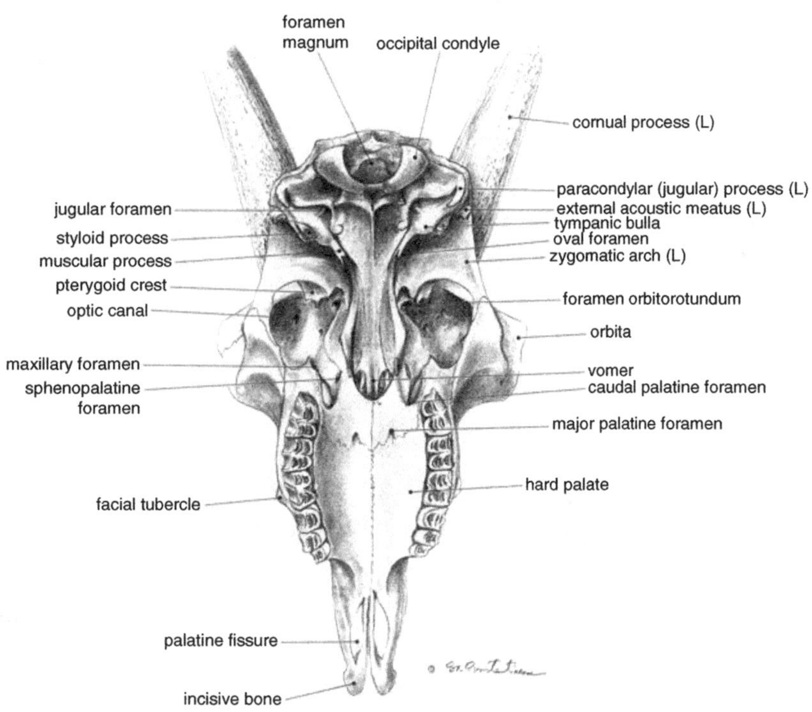


**Figure 6.9** Skull, ventral aspect (without the mandible). (L) = landmark for physical examination and/or clinical approach.

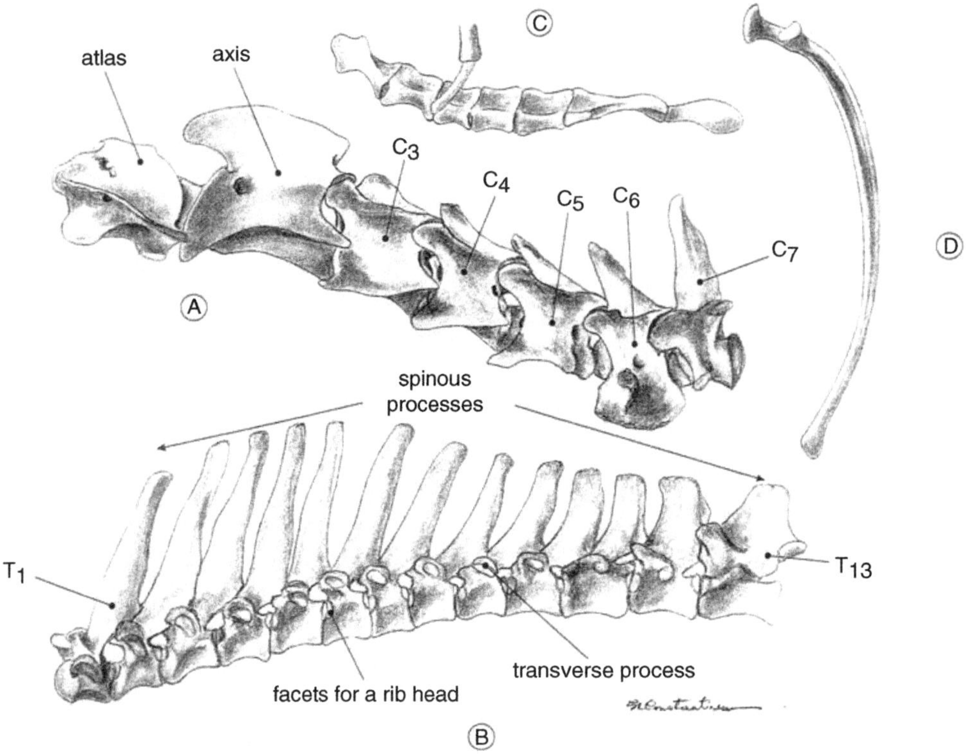


**Figure 6.10** Bones of the axial skeleton: A. Cervical vertebrae, lateral view; B. Thoracic vertebrae, lateral view; C. Sternum; D. Rib.

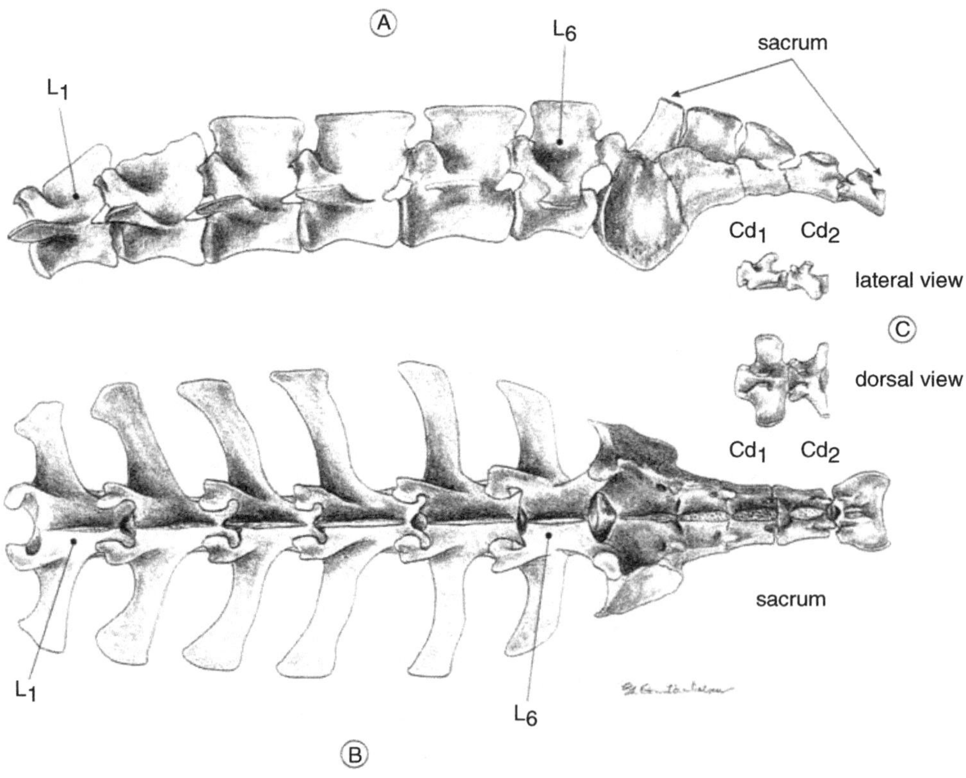


**Figure 6.11** Bones of the axial skeleton: A. Lumbar vertebrae, sacrum, lateral view; B. Lumbar vertebrae, sacrum, dorsal view; C. Caudal vertebrae.

#### The Ribs

The ribs are paired structures, and the numbers of them correspond to the numbers of thoracic vertebrae (×2). Each rib has a bony part and a cartilage at the distal end. The bony part articulates proximally with two adjacent thoracic vertebrae and the intervertebral discs, and the cartilage articulates with the sternum. Depending upon the direct or indirect attachment to the sternum, the ribs are divided into two categories: sternal and asternal. The sternal ribs are eight pairs, whereas the asternal ribs are only five pairs. The cartilages of the latter form together a cartilaginous arch called costal arch. One rib is illustrated in Figure 6.10.

The functional anatomy of the thoracic vertebrae, sternum, and ribs is important for the normal activity of the respiratory and cardiovascular systems. They build up the thoracic cavity for the protection of the lungs and the heart with the major vessels and nerves, lymph nodes, and the thoracic thymus, and facilitate the act of respiration.

### The Appendicular Skeleton

#### The Thoracic Limb

The thoracic (fore) limb consists in a proximo-distal direction of the scapula (the bone of the shoulder), the humerus (the bone of the arm, or brachium), the radius and ulna (the bones of the forearm, or antebrachium), the carpal bones (the bones of the knee), the metacarpal bones (the bones of the metacarpus), and the phalanges with the sesamoid bones (the bones of the digits). They are illustrated in Figure 6.12. Similar to the skull, the structures marked with an "L" are landmarks for clinical purposes.

#### The Pelvic Limb

The pelvic (rear) limb consists in a proximo-distal direction of the coxal (the bone of the rump), the femur and patella (the bones of the thigh), the tibia (the bone of the crus, or leg), the tarsal (the bones of the hock), the metatarsal (the bones of the metatarsus), and the phalanges with the sesamoid bones (the bones of the digits). A specific note for the coxal bone: it is formed by the fusion of three bones—the ilium, the pubis, and the ischium. By the pubis and ischium, the symmetrical bones are articulated with each other. The bones of the pelvic limb are illustrated in Figure 6.13. Similar to the skull and the thoracic limb, the structures marked with an "L" are landmarks for clinical purposes.

The functional anatomy of the appendicular skeleton refers to the following roles: they connect with the body; support the body weight; are related to each other by joints; are the levers that mobilize the body and produce displacement by the contraction of muscles; and some of the prominences are landmarks for clinical purposes. The prominences and depressions of the surface of the bones were modeled during thousands of years of evolution by pulling the ligaments and contraction of the muscles, to respond to a species-specific life.

## THE ARTICULAR SYSTEM (THE JOINTS)

There are three types of joints: mobile, semimobile, and immobile. The mobile joints (the joints of the limbs) allow the entire spectrum of movements, while the semimobile joints (the intervertebral joints) allow very reduced movements. The immobile joints immobilize the bones of the skull, for example, in the adult life after the brain developes fully.

The joints of the vertebral column provide for the bodies and arches of the vertebrae. Intervertebral discs are interposed between two adjacent vertebral bodies. The joints of the ribs include ligaments between the heads of the ribs and the bodies of vertebrae and the intervertebral discs (dorsally), and the cartilages and the sternum (ventrally). The joints of the sternum provide for the sternebrae.

The joints of the limbs consist of joint capsules and ligaments. The joint capsules are fibrous externally, and synovial internally. The fibrous joint capsule is the continuation of the periosteum from one bone to another. The synovial membrane, a thin and delicate structure, contains synovial fluid, which allows a smooth gliding of the bones against each other.

Most of the joints of the skull are sutures, with the exception of the temporomandibular joint (symmetrical), which is mobile. From the functional point of view, it allows the movements of the mandible against the rest of the skull, during the prehension of food and the mastication process. The movements of the mandible open or close the mouth, and move laterally during the mastication (the latter is a species-specific feature for all ruminants).

## THE MUSCULAR SYSTEM (THE SKELETAL MUSCLES)

There are three types of muscles in the body: skeletal, visceral, and cardiac. Functionally, the skeletal muscles produce the movement of bones, and contribute to the standing position and to the displacement of the body. They are called striated muscles because of a specific intimate structure. The visceral muscles mobilize the viscera

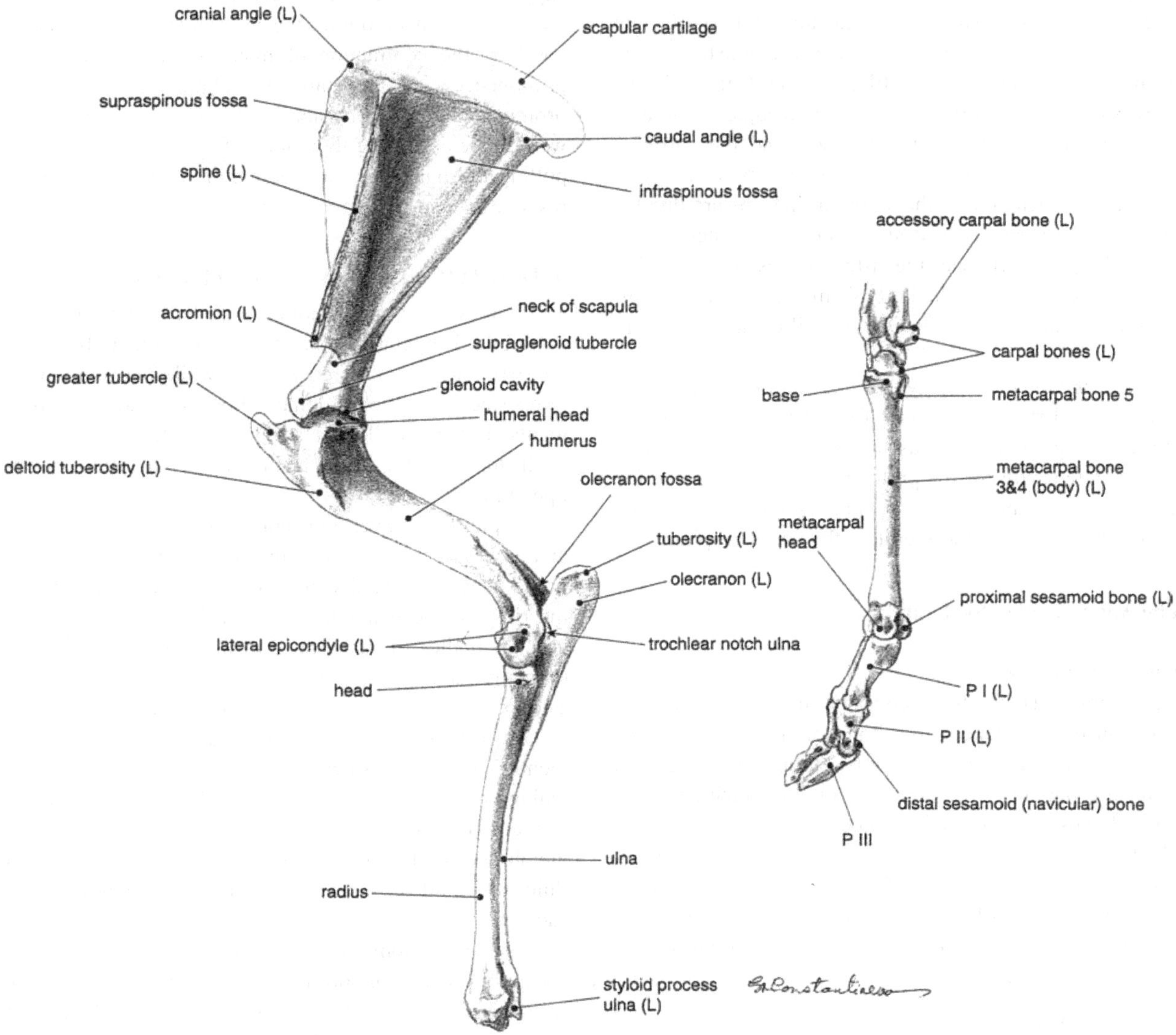


**Figure 6.12** The skeleton of the thoracic limb, lateral aspect. (L) = landmark for physical examination and/or clinical approach.

(internal organs) to move the ingesta from the stomach to the rectum, and are structurally smooth. The cardiac muscle is responsible for the contraction of the heart, and it is a combination of the two precedent muscles.

The tendons are important adnexa of the skeletal muscles. They facilitate the insertion (attachment) of most of the muscles on bones. A widened tendon in the form of an expanded sheet is called an aponeurosis. The superficial muscles are covered by a connective tissue layer called fascia, which does not allow the muscles to exceed a certain volume during the contraction. The superficial fasciae, as well as the deep fasciae send septa separating the muscles from each other. Each muscle has at least one origin and one insertion. The origin is attached to the bone, which doesn't move, whereas with the insertion, the muscle pulls the bone and provides movement.

The following sections discuss the skeletal muscles of the goat selected by regions.

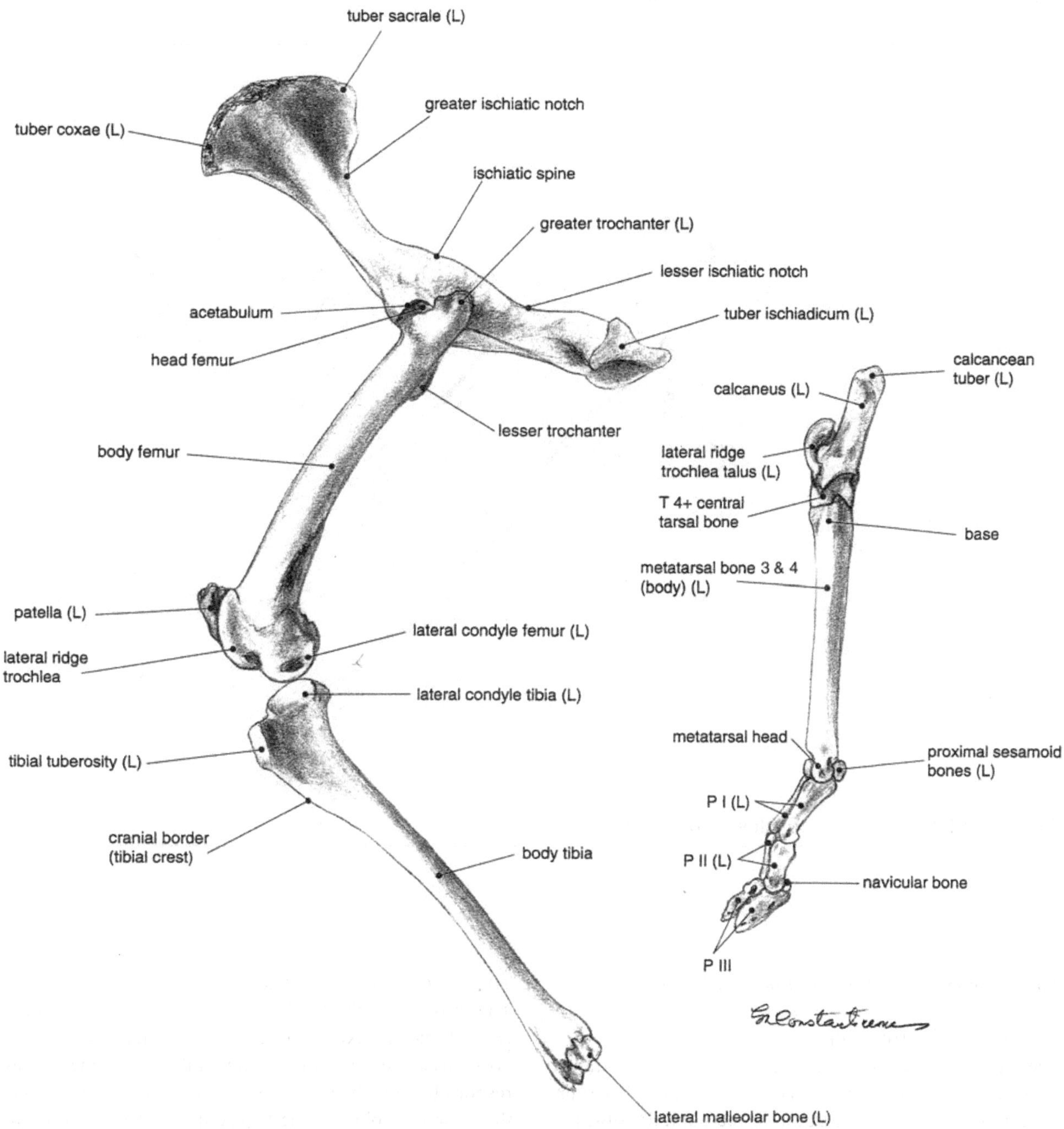


**Figure 6.13** The skeleton of the pelvic limb, lateral aspect. (L) = landmark for physical examination and/or clinical approach.

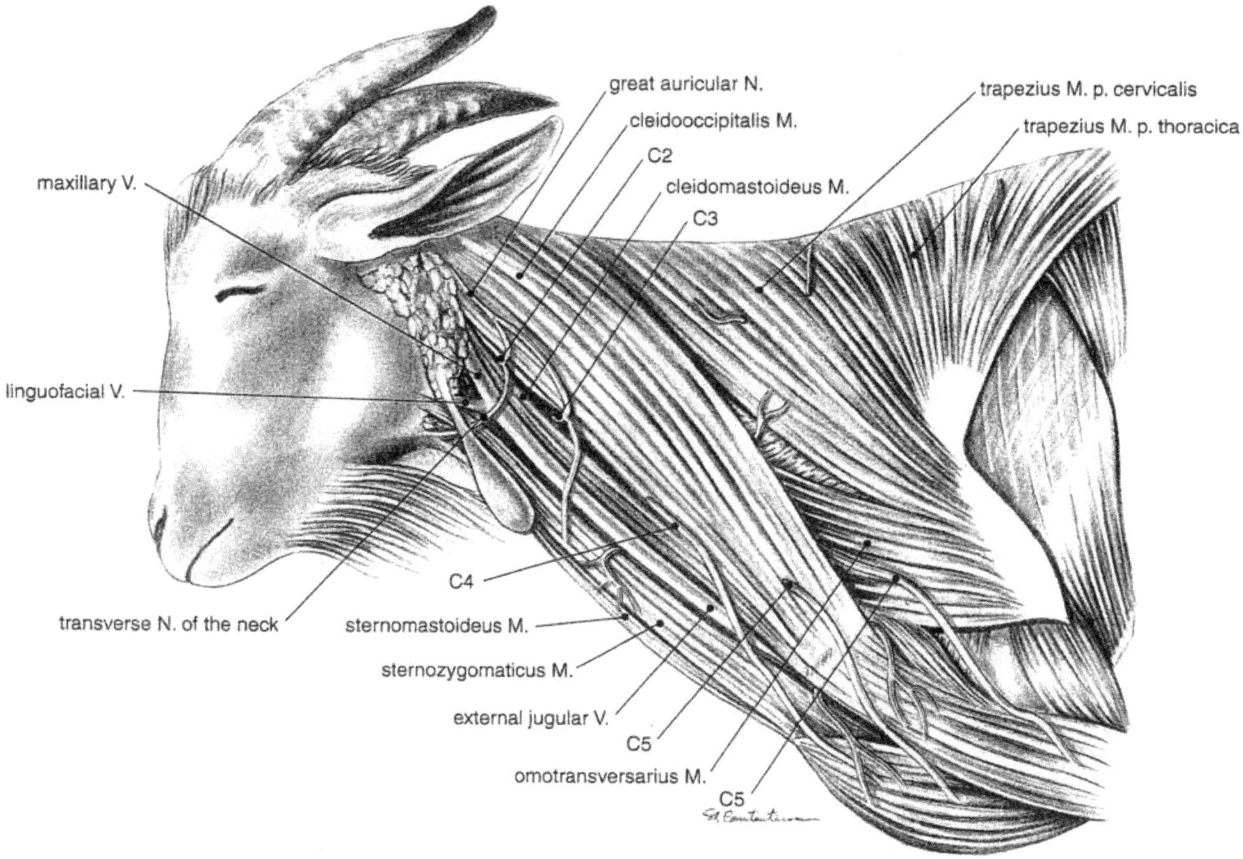


**Figure 6.14** Superficial muscles of the neck.

## The Muscles of the Neck, Trunk, and Abdomen

### The Muscles of the Neck

The muscles of the neck are located dorsal and ventral to the vertebral cervical column. They originate and insert on the cervical vertebrae, the skull, the sternum, the ribs, the thoracic limb, and the thoracic vertebrae. They contract either on one side of the body, or symmetrically, to produce all kinds of movements. The ventral muscles of the neck protect the esophagus, the trachea, and major vessels and nerves. The superficial cervical muscles are illustrated in Figure 6.14.

From the functional standpoint, the muscles of the neck are holding the weight of the head, and making dorsal, lateral, and ventral movements of varying degrees of the neck and head.

### The Muscles of the Trunk

The muscles of the trunk are divided in two categories: epaxial and hypaxial. The epaxial muscles attach to the dorsal aspect of the thoracic and lumbar vertebrae and are responsible for (their primary function is) the extension of the vertebral column. The hypaxial muscles attach to the ventral aspect of the same vertebrae and do the flexion of the vertebral column (which is their primary function). At the same time, all muscles of the trunk can perform lateral movements and limited rotations of the vertebrae in their long axis. The muscles of the trunk also include the pectoral (chest) muscles.

### The Abdominal Muscles

The abdominal muscles are divided in two categories: dorsal and ventral. The dorsal abdominal muscles cover the ventral aspect of the lumbar vertebrae inside of the abdominal cavity, whereas the ventral muscles contribute to the ventral and lateral walls of the abdomen. The function of the dorsal abdominal muscle is to flex the thoracolumbar vertebrae, whereas the function of the ventral abdominal muscles is to support the weight of the abdominal viscera (internal organs within the abdominal cavity) and to act in respiration, defecation, micturition (the act of urination), and parturition (birth). There are four ventral abdominal muscles: the external and internal abdominal obliques, the transversus abdominis, and the rectus abdominis muscles. The aponeurosis of the external abdominal oblique is also involved in the passage of the spermatic cord from the abdominal cavity to the scrotum (see the Reproductive System), in the descent of testicles, and in the suspensory apparatus of the udder (see the Mammary Gland). Due to the considerable weight of the abdominal viscera, an additional structure helps in supporting them. This is the tunica flava abdominis, or the yellow tunic, because it has abundant elastic fibers in its texture fibers that give a yellow appearance. The tunica flava is attached firmly to the ribs and surrounds the ventral abdominal muscles. It contributes also to the suspensory apparatus of the mammary gland (see the Mammary Gland). This elastic structure is necessary to suspend the abdominal organs in a passive way and to prevent the abdominal muscles from being under constant load (and fatigue).

#### The Muscles of the Thoracic Limb

There are specific muscle actions that should be introduced before covering the muscles. The flexion is that action which reduces the angle between two bones in their long axis. The extension is opposite to the flexion (increases the angle). The adduction is the movement toward the median plane. The abduction is opposite to adduction (movement away from the median plane).

The muscles of the shoulder, the muscles of the arm, the muscles of the forearm, and the muscles of the metacarpal region are grouped by position and function, as follows:

- There are lateral and medial muscles of the shoulder. As function, the lateral muscles are extensors and abductors of the arm. The medial muscles are adductors and flexors of the arm.
- There are cranial and caudal muscles of the arm. As function, the cranial muscles are flexors of the forearm, whereas the caudal muscles are extensors of the forearm.
- The muscles of the forearm are disposed on the cranial and on the caudal aspects of the radius and ulna, and both groups have two different functions.
- There are cranial antebrachial muscles that extend the carpus and muscles that extend the digits. As for the caudal antebrachial muscles, those attached to the carpal bones flex the carpus, whereas the muscles attached to the phalanges flex the digits.
- There is only one muscle in the metacarpal region, which is mostly tendinous. Its name is the interosseous muscle. It plays the role of suspensory apparatus for the digits.
- The muscles of the thoracic limb from the lateral and from the medial perspectives are illustrated in Figure 6.15 and Figure 6.16.

#### The Muscles of the Pelvic Limb

Two terms, propulsion and rearing, should be explained here. Propulsion is the action of driving or propelling the body forward, whereas rearing is the action of pulling the body off the ground with the pelvic limbs still on the ground.

The muscles of the pelvic limb consist of the muscles of the rump, the muscles of the thigh, the muscles of the crus, and one muscle of the metatarsal region.

The muscles of the rump function as abductors and extensors of the thigh. One of them is also a propulsor, and helps in the action of rearing.

There are three groups of muscles of the thigh: cranial, caudo-lateral, and medial muscles. As function, the cranial muscles are extensors of the crus by means of the patella. As function, the caudolateral muscles are the most powerful muscles of the body, and they perform flexion and extension of the crus, are propulsors, and act in rearing. The rearing is a very important act for the goats, as browsers, who in search for food reach considerable heights in trees (bipedal stand) to access leaves. As to function, the medial muscles of the thigh are all adductors. At the ventral aspect of the coxal bones, a paired muscle originates from the so-called symphiseal tendon, involved in the suspensory apparatus of the udder (see the Mammary Gland).

Similar to the antebrachial muscles, the muscles of the crus can be grouped into cranial and caudal muscles. Some cranial muscles are flexors of the tarsus, and some are extensors of the digits. There are two subgroups of caudal muscles: superficial and deep. The superficial muscles are either extensors of the tarsus, or flexors of the digits (only one muscle). One deep muscle is a flexor of the crus, while the others are flexors of the digits. The muscles of the pelvic limb are illustrated in Figure 6.17 and Figure 6.18. There is an interosseous muscle in the pelvic limb, too.

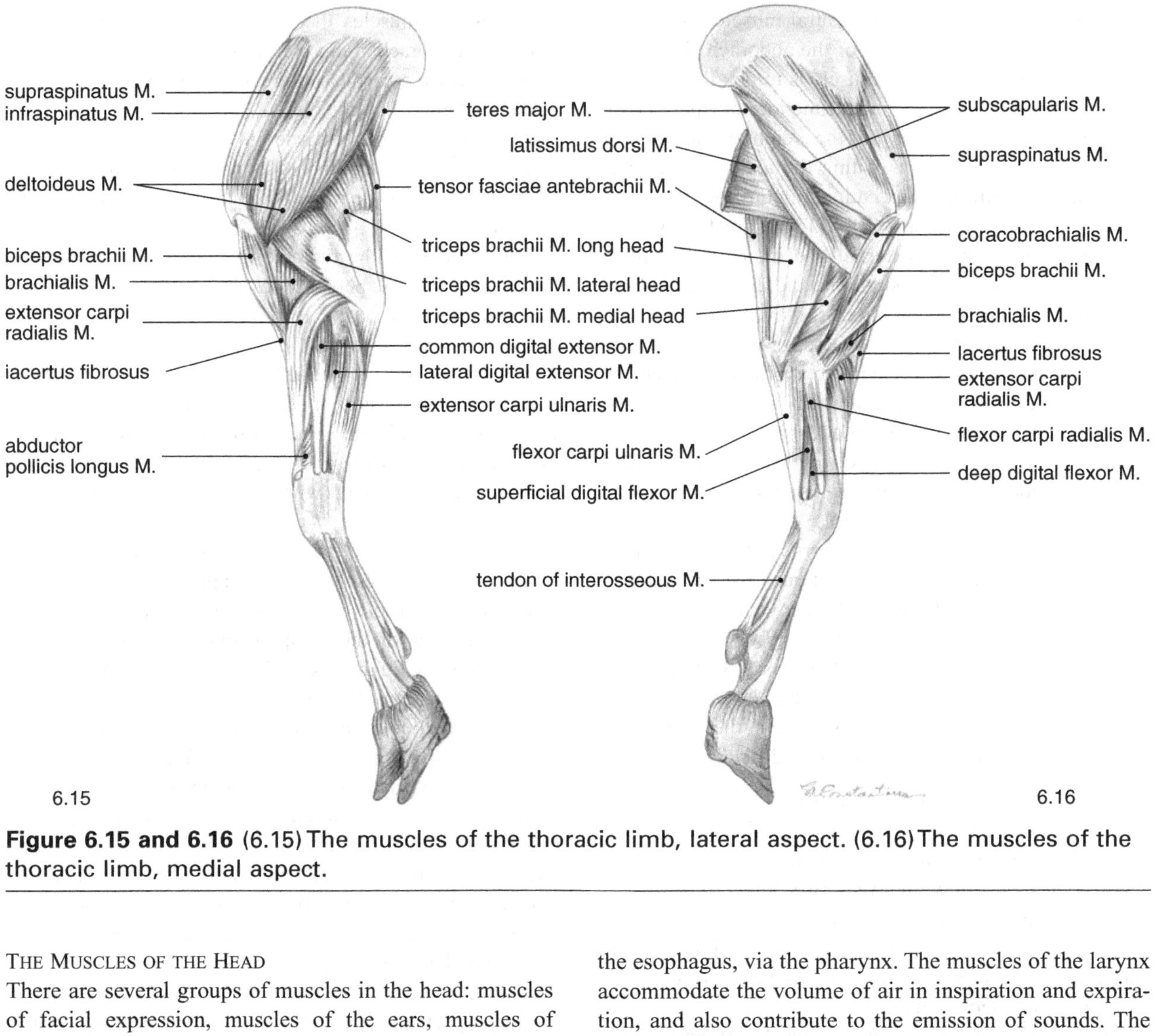


**Figure 6.15 and 6.16** (6.15) The muscles of the thoracic limb, lateral aspect. (6.16) The muscles of the thoracic limb, medial aspect.

THE MUSCLES OF THE HEAD

There are several groups of muscles in the head: muscles of facial expression, muscles of the ears, muscles of mastication, muscles of hyoid apparatus, and lingual, pharyngeal, and laryngeal muscles (the visceral muscles of the head).

The function of these muscles is different from one group to the others. The muscles of facial expression speak for themselves. They are attached to the surface of the skull, around the entrance of the mouth, and around the eyes. The muscles of the ears move them in all directions. The muscles of mastication move the mandible against the rest of the skull. The muscles of the hyoid apparatus help in deglutition. The lingual muscles move the tongue in all directions and are involved in the mastication, insalivation, and formation of the alimentary bolus. The muscles of the pharynx help the alimentary bolus to pass from the oral cavity into the esophagus, via the pharynx. The muscles of the larynx accommodate the volume of air in inspiration and expiration, and also contribute to the emission of sounds. The superficial muscles of the head are illustrated in Figure 6.19.

## THE HORN AND THE DIGITAL ORGANS

### The Horn

Generally speaking, both male and female goats bear horns. Goats may be horned or polled. Homozygous polled goats have a high incidence of infertility. The base, the body, and the apex are the parts of the horn, and they are supported by a symmetrical bony cornual process of the frontal bone. There is an epithelium of the horn externally, which is homologous to the epidermis of the skin. The horn is composed of fine, closely packed epidermal tubules, which appear on the surface as longitudinal striations. The

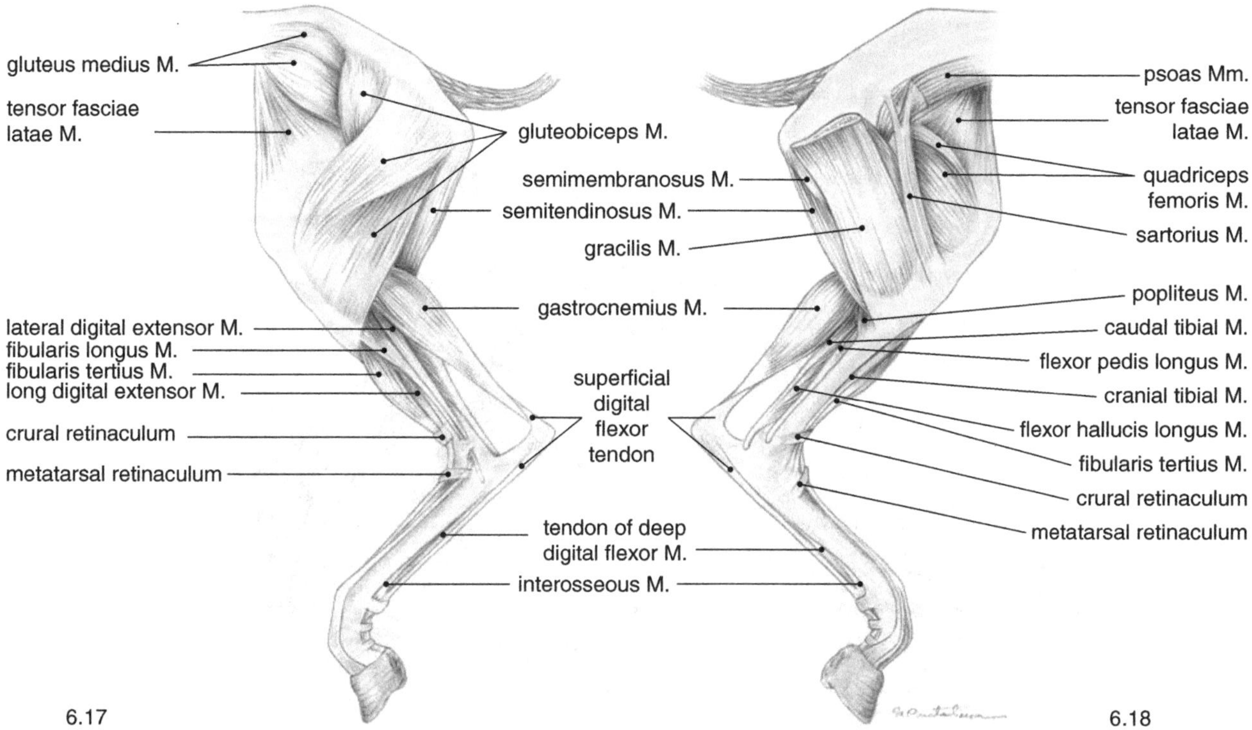


**Figure 6.17 and 6.18** (6.17) The muscles of the pelvic limb, lateral aspect. (6.18) The muscles of the pelvic limb, medial aspect.

base of the horn is surrounded by a zone of transitional epidermis called epiceras (epikeras), with the same relation to the horn as the periople to the hoof, and with the same functionality. The dermis of the horn is the continuation of the dermis of the skin. It is interposed between the epidermis of the horn and the osseous corneal process, and is very vascular. The dermis consists of dermal papillae that penetrate into conical epidermal depressions. The epidermal tubules are produced by the dermal papillae. The rostral part of the horn is in relationship to the frontal sinus, whereas sometimes the caudal part of the horn may be in contact with the frontal plate or calvaria (the roof of the brain cavity). Therefore, one must proceed carefully during the dehorning of a goat.

The cornual artery provides blood supply to the horn. The branches of it are accompanied by the cornual nerve and by the infratrochlear nerve, both branches of the ophthalmic nerve (of the trigeminal nerve). Therefore, both nerves should be blocked before dehorning a goat. The goats also have horn glands (see the Touch Organ—Common Integument with the Sense Organs).

### The Digital Organ

The digital organ includes the main hooves (claws), two for each limb, which are the continuation of the epidermal skin. They cover and protect live structures such as bones, joints, tendons, vessels, nerves, the digital cushion, and the dermis. For each limb, there are two additional accessory hooves, located on the palmar/plantar aspect of the fetlock joint. The fetlock joints are illustrated in Figure 6.1. The accessory hooves are called dewclaws and are not provided with all anatomical structures as the main/functional hooves are.

The third phalanx, the distal half of the second phalanx, and the distal sesamoid bone are the bones included in each hoof. Several ligaments join the phalanges and the distal sesamoid bone with each other. Axial and abaxial tendons of the digital extensor muscles on the dorsal aspect and the tendon of the deep digital flexor muscle on the palmar/plantar aspect are the only tendons included in the hoof.

The digital cushion is a pad of fibroelastic tissue in intimate contact with the deep digital flexor tendon, and

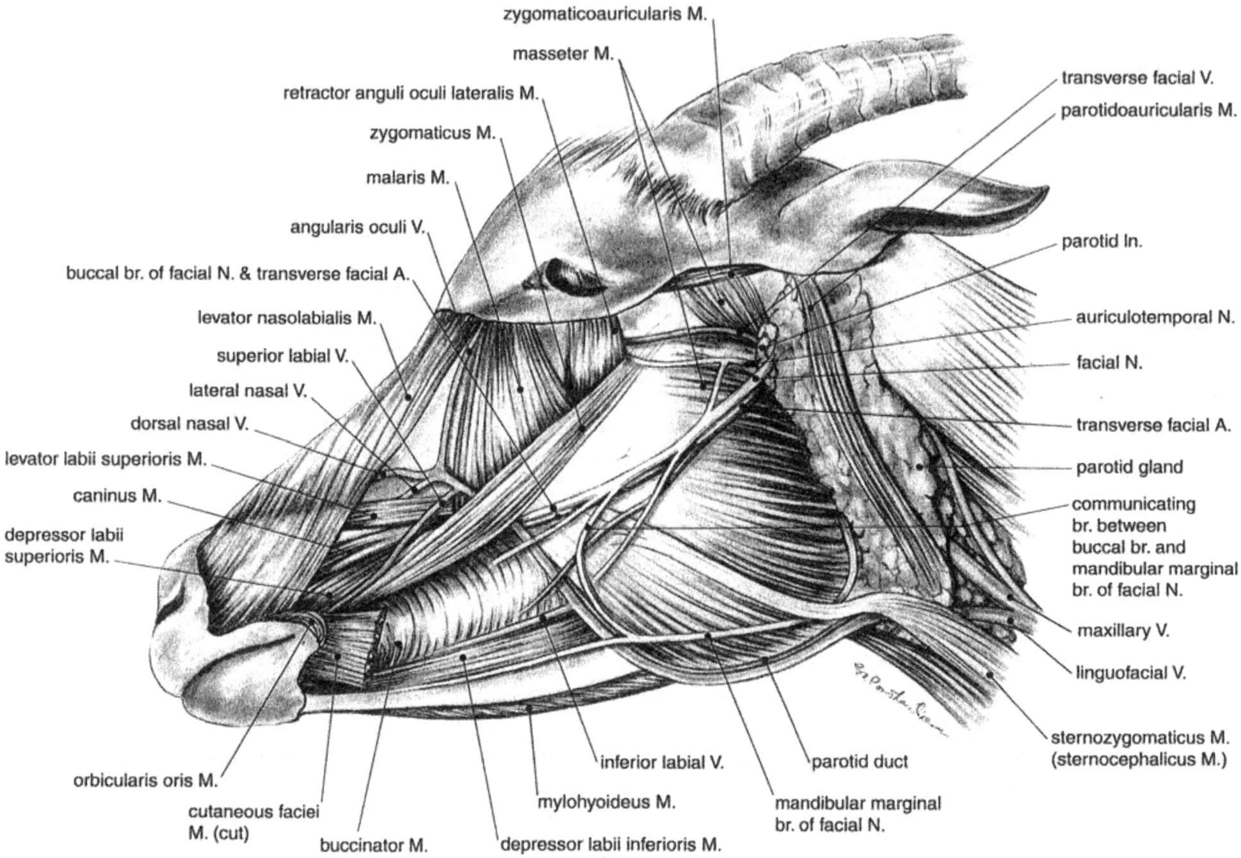


**Figure 6.19** Superficial structures of the head.

its function is to absorb shocks when the hoof touches on uneven ground.

There are two groups of dermal structures corresponding to the wall of the hoof and to the sole, and they are collectively called "corium" (pl. coria). The corium does not attach to the bones. It lies on a connective tissue pad interposed between the dermis (the corium) and the bone (the third phalanx). The structures corresponding to the wall of the hoof are in a proximodistal order—the perioplic corium, the coronary corium, and the laminar (lamellar) corium—on both axial and abaxial aspects. The dermal corium of the sole is called solear corium. With the exception of the lamellar corium, the other coria are provided with papillae (Figure 6.20 e, f, g).

The hoof itself (sometimes referred to as the hoof capsule) is a horny structure, consisting of the wall, the sole, and the bulb of the hoof.

With the foot on the ground, the wall shows both axial and abaxial aspects. The abaxial aspect can be arbitrarily divided into three zones: the toe (in the dorsal position), the quarter in the middle, and the heel (in the palmar/plantar position). In the goat, the sole is reduced in size, and it does not touch flat ground while the goat is standing; therefore, the weight of the body is bearing by the wall and the bulb of the hoof. If the ground is not flat upon contact with the sole, the digital cushion absorbs part of the shock. There is a narrow zone of periople (a thin and shiny protective layer of transitional epidermis) outside, and in a proximal position of the wall. Distal to the periople is the hard wall of the hoof. Toward the palmar/plantar aspect of the hoof, the periople covers the bulb of the hoof.

The sole is pointed dorsally and touches the irregular bulb of the hoof palmarly/plantarly. There is a "white zone" between the sole and the wall of the hoof. The bulb of the hoof is an irregular horny structure that continues with the bulb of the heel (Figure 6.20 a, b, c, d). The bulbar horn (epidermal) covers and protects the bulbar corium (dermal). In the goat, the bulbar horn is soft.

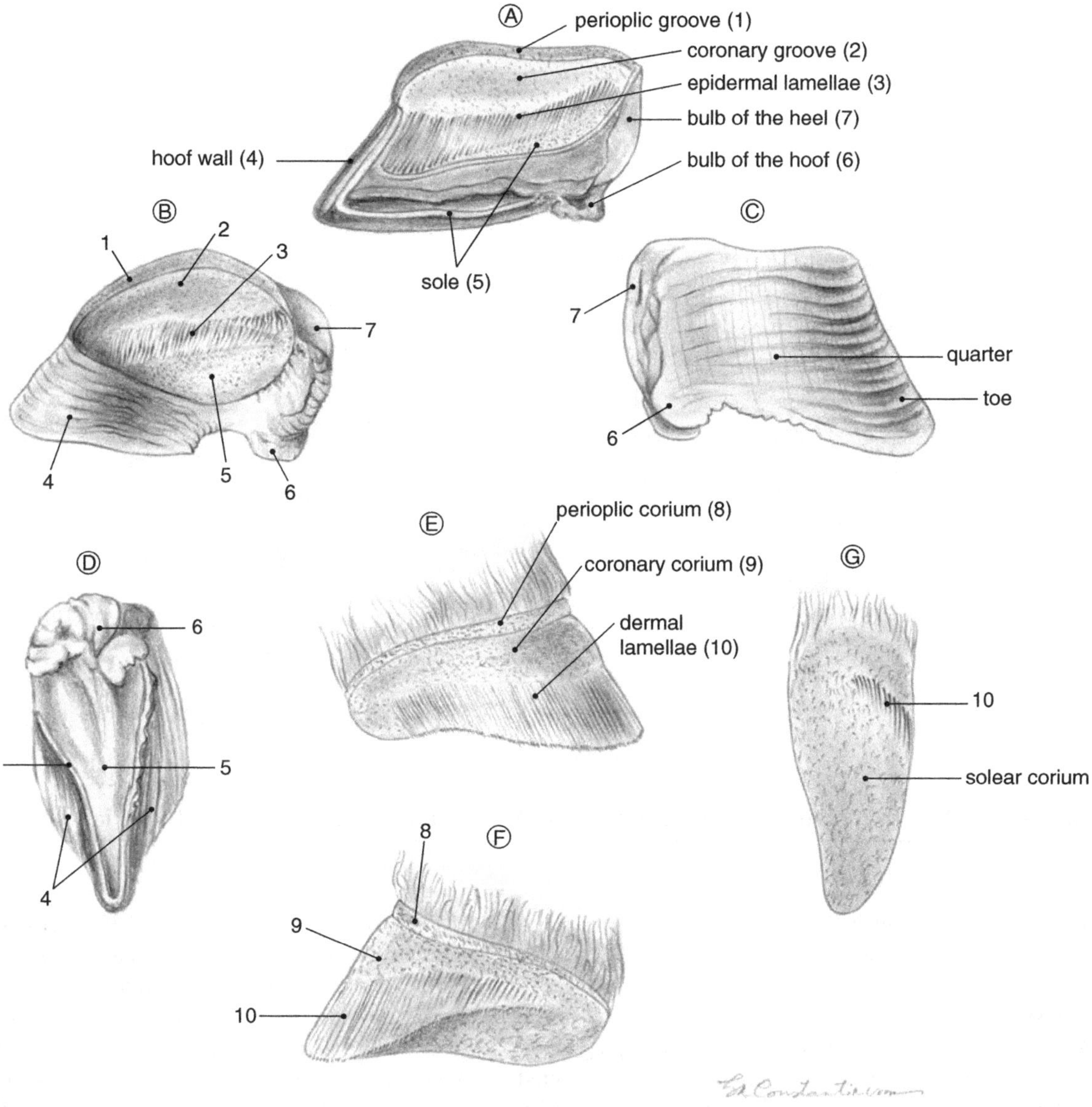


**Figure 6.20** The hoof and the dermal coria: A. Median section; B. Inner aspect of the hoof; C. The hoof wall; D. The solear aspect of the hoof; E. Dermal corium, abaxial aspect; F. Dermal corium, axial aspect; G. Dermal corium, solear aspect.

The inner surface of the hoof wall corresponds to the structures of the parietal corium, which are protected by the wall. In a proximodistal direction, they are the perioplic groove, the coronary groove, and the horny lamellae (laminae). The latter are extremely numerous (between 550 and 700). The horny lamellae are white in color regardless of hoof wall color, and are parallel with each other. They end in small papillae and will show up on the solear aspect of the hoof between the sole and the wall of the hoof as the "white zone" (see Figure 6.20). The digital arteries and veins and the digital nerves supply heavily the live tissues of the hooves.

## THE CARDIOVASCULAR SYSTEM

The cardiovascular system consists of the heart, the blood vessels (arteries, veins, and capillaries), and the lymphatic system (lymph nodes and lymph vessels). The spleen is associated with the cardiovascular system, but it will be

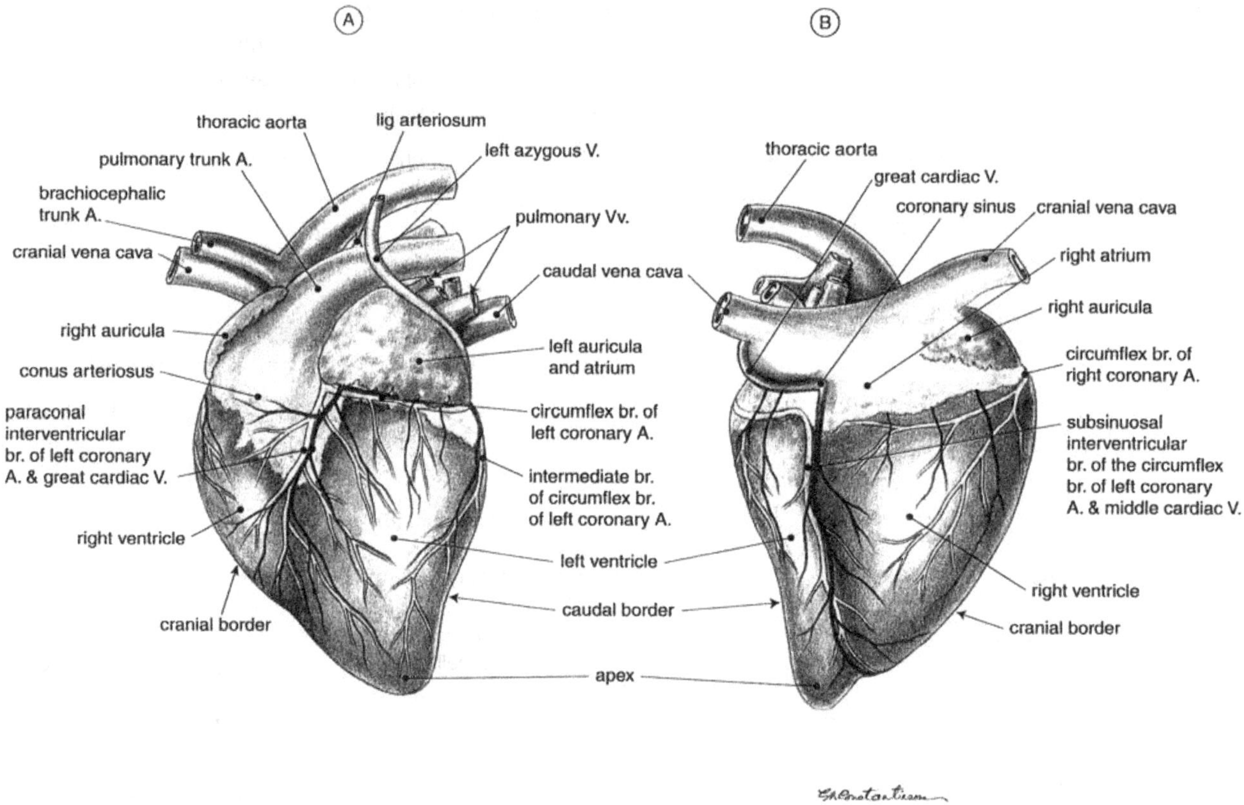


**Figure 6.21** External configuration of the heart: A. Left (auricular) side; B. Right (atrial) side.

described with the digestive system, because of its topographic relations to the stomach.

### The Heart

As an aspiro-repellent pump receiving and pumping blood into the body, the heart is a muscular organ protected within the thoracic cavity. It is intimately surrounded by a double layer pericardium, and is located between the two lungs. From the outside, one can estimate the position of the heart between the ribs III and VI.

#### External Configuration

The heart has a base oriented dorsally and a tip oriented ventrally. The base consists of two atria (sing. atrium), right and left, on top of two ventricles, right and left. The base is associated with the great vessels of the heart and is much smaller than the ventricles. The atria are separated from the ventricles by a horizontal groove called the coronary groove, which protects the coronary arteries and the great cardiac vein.

The atria end in the auriculae (sing. auricula), on both sides of the origin of the pulmonary trunk. The atria are visible on the right side of the heart, whereas the auriculae are on the left side of the heart. The two ventricles are separated from each other by two vertical grooves that protect arteries and veins nourishing the heart. The two aspects of the heart and the major vessels are illustrated in Figure 6.21.

#### Internal Configuration

The right atrium communicates with the right ventricle, whereas the left atrium communicates with the left ventricle. The communications are provided with atrioventricular (AV) valves, complex structures allowing the blood to pass from the atria to the ventricles, but not back to the atria. They consist of three (for the right AV valve) and two (for the left AV valve) leaflets anchored to strong papillary muscles by tendinous cords. The pulmonary arterial trunk originates from the right ventricle, while the aorta originates from the left ventricle. Both are provided with three-valvule valves. The venae cavae (cranial and caudal) open into the right atrium, whereas the pulmonary veins open into the left atrium. Uneven muscle fibers with distinct design can be identified inside of the atria and the ventricles. All prominences inside of the heart are covered by a delicate membrane called the endocardium. The latter will continue inside of the vessels starting from, and ending at, the heart, as endarteries and endoveins, respectively. Around the origin of the aorta, two cardiac bones can be detected.

### The Arteries and the Veins

The blood is a red-colored circulating fluid in the arterial and venous trees, whose physical, chemical, and biological characteristics can be found in corresponding books.

The heart's function is complex and can be demonstrated by the systemic circulation and the functional circulation. The systemic circulation starts in the left ventricle, which pumps the oxygenated blood into the whole body through the aorta, arteries, arterioles, and capillaries up to the finest structures. From the tissues, through capillaries, venules, and veins, the blood with carbon dioxide empties into the right atrium through the cranial and caudal venae cavae. From here, the blood is pumped into the right ventricle. The functional circulation starts from the right ventricle, which pumps the venous blood into the lungs, via the pulmonary trunk, pulmonary arteries, and arterioles. At the level of pulmonary alveoli, the blood yields the carbon dioxide and accumulates the oxygen through the phenomenon of hematosis. From the lungs, the oxygenated blood passes via pulmonary venules and veins into the left atrium, and from there it is pumped into the left ventricle, and the cycle continues. The contraction of the ventricles and the atria is called "systole" and the relaxation "diastole." The complete cycle of cardiac contraction and relaxation is called the "cardiac cycle." During the diastole, the corresponding chamber (atrium or ventricle) is filled with blood, while during the systole, the chambers contract and expel blood.

The venous blood from the gastrointestinal tract and the pancreas is collected in the portal vein. This vein starts from capillaries, and ends by capillaries, in the liver. The blood from the portal vein is processed in the liver in species-specific nutrients. From the liver, the blood travels by way of the caudal vena cava to empty into the right atrium.

The walls of the arteries up to arterioles and the veins (except the venules) consist of an outer layer, a middle layer, and an internal layer. The outer is fibrous, the middle is muscular and elastic, and the internal is the continuation of the endocardium. The veins of the limbs and ventral abdomen are provided with valvules, which orient the blood flow in one direction only, toward the cranial and the caudal venae cavae.

### The Lymphatic System

The lymphatic system consists of lymph nodes and lymph vessels. The lymph nodes receive the lymph via afferent lymphatic vessels, and lymph runs away from the lymph nodes via efferent lymphatic vessels. The lymph is that fluid, clear and transparent, which is collected from the body tissues, and flows in the lymphatic vessels passing through lymph nodes.

The lymph nodes are located under the skin, deep between muscles, in the head and in the natural cavities of the body (the thoracic and abdomino-pelvic cavities). The lymph nodes in the natural cavities receive lymph from the internal organs. The lymph reaches finally the venous system through the cranial vena cava.

There are lymphatic structures (tissues) scattered throughout the body, such as the tonsils, whose function is to guard against the passage of infection deep into the respiratory, digestive, and/or reproductive systems. They are found in the pharynx, larynx, intestine, prepuce, vagina, etc.

Some organs do not have lymph vessels, such as the bone marrow, the articular cartilages, the alveoli of the lungs, the renal cortex and medulla, the placenta, the splenic pulp, the epidermis, the central nervous system, and the meninges.

The lymph nodes and the other lymphatic structures are involved in the immune system (see the Immune System).

## THE RESPIRATORY SYSTEM

The respiratory system has an essential role in the body, bringing in outside air with oxygen necessary for all physiological processes, and exhaling air with carbon dioxide as the result of these processes. The respiratory system is responsible for the gas exchanges between the blood and the surrounding air. It consists of two categories of anatomical structures: the air (respiratory) passages and the essential organs.

### The Air (Respiratory) Passages

#### The Nasal Cavity

The nasal cavity consists of the nostrils, the nasal cavity proper, and the choanae. The nostrils are the rostral openings of the nasal cavity, and are sculpted in an area called the nasal plane. They lead to the entrance of the nasal cavity called the nasal vestibule. The roof, lateral walls,

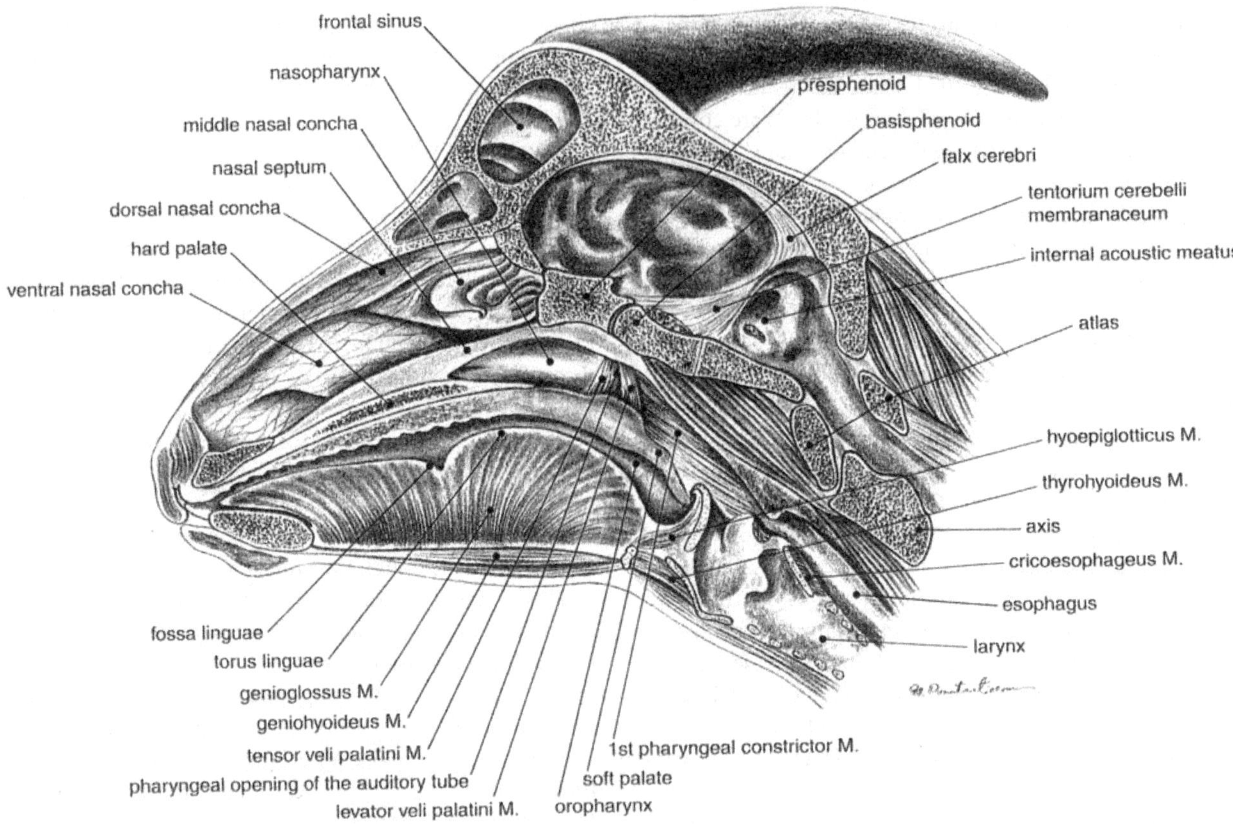


**Figure 6.22** Median aspect of the head, right half.

and floor of the nasal cavity proper are outlined by the bones of the skull. The floor is the hard palate, which separates the nasal cavity from the oral cavity. The vomer bone, which lies on the hard palate, supports the nasal septum, a cartilaginous blade separating the nasal cavity into two symmetrical halves. Inside of the nasal cavity, three pairs of fine bony scrolls called nasal conchae are separated by spaces called meatuses (Figure 6.22).

The nasal cavity including the conchae is intimately lined by a mucosal coat, which has respiratory characteristics in the rostral two-thirds of the cavity and olfactory characteristics in the caudal third of the cavity. The latter is responsible for transmitting smell impulses to the brain.

There are several cavities within bones of the skull called paranasal sinuses that are associated with the nasal cavity. They are lined by respiratory mucosa and communicate among them and with the nasal cavity. Their names correspond to the bones that protect them, such as frontal, lacrimal, maxillary, palatine, and conchal sinuses. A symmetrical extension of the frontal sinus penetrates within the base of the horn and is known as the cornual diverticulum (pl. diverticula). The function of the paranasal sinuses is not yet fully understood, but the frontal sinuses with the cornual diverticula protect the brain from the heavy blows during the fight of male goats.

### The Pharynx

The next segment of the air passages is the pharynx. This is a very important organ, with three compartments: the nasopharynx related to the respiratory system, the oropharynx belonging to the digestive system, and the laryngopharynx, where the respiratory pathway and the digestive pathway cross with each other.

The nasopharynx has a roof (the sphenoid bone of the skull), a floor (the soft palate), and lateral walls (see Figure 6.22). On the lateral walls, the symmetrical openings of the auditory tubes can be seen. The soft palate is the musculo-mucosal continuation of the hard palate and has a definite role in deglutition (see the oropharynx in the next section). The laryngopharynx is the smallest part of the pharynx, located dorsal to the larynx.

#### The Larynx

The larynx is an air passage, a musculo-cartilaginous organ between the pharynx and the trachea, and is also involved in phonation. It consists of paired and unpaired cartilages, joints, muscles, and a mucosal membrane. The muscles move the cartilages between each other and the entire organ rostrally and/or caudally. Some of the muscles regulate the amount of air entering the larynx, and others produce sounds. The sound/voice is produced in the middle of the larynx, by the vibration of the vocal cords.

#### The Trachea

Consisting of incomplete cartilages, the trachea looks like a pipe, allowing the air to flow toward the lungs and from the lungs, and yet is collapsible. The cartilages are connected by annular ligaments. On the dorsal aspect, where the two ends of the cartilages face each other, the tracheal muscle is shown on the inner surface of the trachea. The trachea travels throughout the neck, ventral to the cervical vertebrae and the esophagus, and enters the thoracic cavity. Still with the esophagus on top of it, the trachea bifurcates into the primary bronchi, dorsal to the heart. The primary bronchi enter the lungs. Before the bifurcation, the trachea sends an accessory bronchus to the right lung.

### The Essential Respiratory Organs (the Lungs)

The lungs are located within the thoracic cavity. The latter is a space outlined dorsally by the thoracic vertebrae, ventrally by the sternum, and laterally by the ribs and the intercostal muscles. The thoracic vertebrae are lined in the first half by a muscle, whereas the sternum is covered by a ligament and a muscle. The entrance into the thoracic cavity (the thoracic aperture) is outlined by the first pair of ribs, the first thoracic vertebra, and the first sternebra. The caudal wall of the thoracic cavity is the diaphragm, a vital organ, which functions as the primary muscle of inspiration. The diaphragm consists of peripheral muscles, which send toward the center intermingled aponeuroses collectively called the central tendon (the former phrenic center). The diaphragm is pierced by three holes for allowing important organs to pass from the thoracic cavity into the abdominal cavity and vice versa. In a dorso-ventral order, they are the following: the aortic hiatus (a large foramen [opening]), the esophageal hiatus, and the foramen of the caudal vena cava.

The thoracic cavity is divided into two halves by a connective tissue partition called mediastinum. The thoracic cavity is lined by a connective tissue called endothoracic fascia and a serous membrane called pleura. The pleura associated with the walls of the thoracic cavity, on both sides of the mediastinum and covering the pericardium, is called parietal pleura. The pleura covering the lungs is called visceral or pulmonary pleura. The two categories of pleurae are in continuation with each other at the level where the two primary bronchi branch from the trachea and are heading toward the lungs.

Each lung has an apex (toward the thoracic aperture), a base (lying against the diaphragm), a costal surface and a mediastinal surface, and borders. Both lungs are divided into lobes separated by fissures. On the mediastinal aspect of each lung, there is an indentation called the hilus, where the primary bronchi and pulmonary vessels connect with the lungs. The primary bronchi and the pulmonary vessels form the roots of the lungs. The following are the pulmonary lobes: cranial and caudal right and left lobes (the right cranial lobe is divided into a cranial and a caudal parts); the right lung is provided by a middle lobe; and on the ventral aspect of the caudal lobe, an accessory lobe can be identified on the right lung. The accessory lobe is separated from the right caudal lobe by the passage of the caudal vena cava. On the surface of the cranial lobes and the middle lobe, a polyhedral design may be seen, due to the abundance of connective tissue separating the pulmonary lobules (Figure 6.23 and Figure 6.24).

The internal configuration of the lungs shows a segmentation of the primary bronchi (lobar and segmental bronchi, bronchioles, and respiratory bronchioles), ending into the alveolar sacs. The alveolar mucosa supports a vascular rete (network). Here the gas exchange takes place between the venous blood leaving the carbon dioxide replaced by the oxygen and becoming arterial blood, a process called hematosis. The functional anatomy of the lungs is self explanatory.

## THE DIGESTIVE SYSTEM

In all species, the digestive system provides nutrients from food and eliminates the wastes. However, in all ruminants including the goat, the digestive system is adapted to allow a quick storage of forage, which is digested slowly in safe locations, to avoid predators in the wild. Therefore, the digestive organs are different from those of the horse, pig, and carnivores; and their function will be detailed in Chapter 8, Digestive Physiology and Nutrient Metabolism. The digestive system consists of a succession of structures starting from the oral cavity and ending at the anus. To

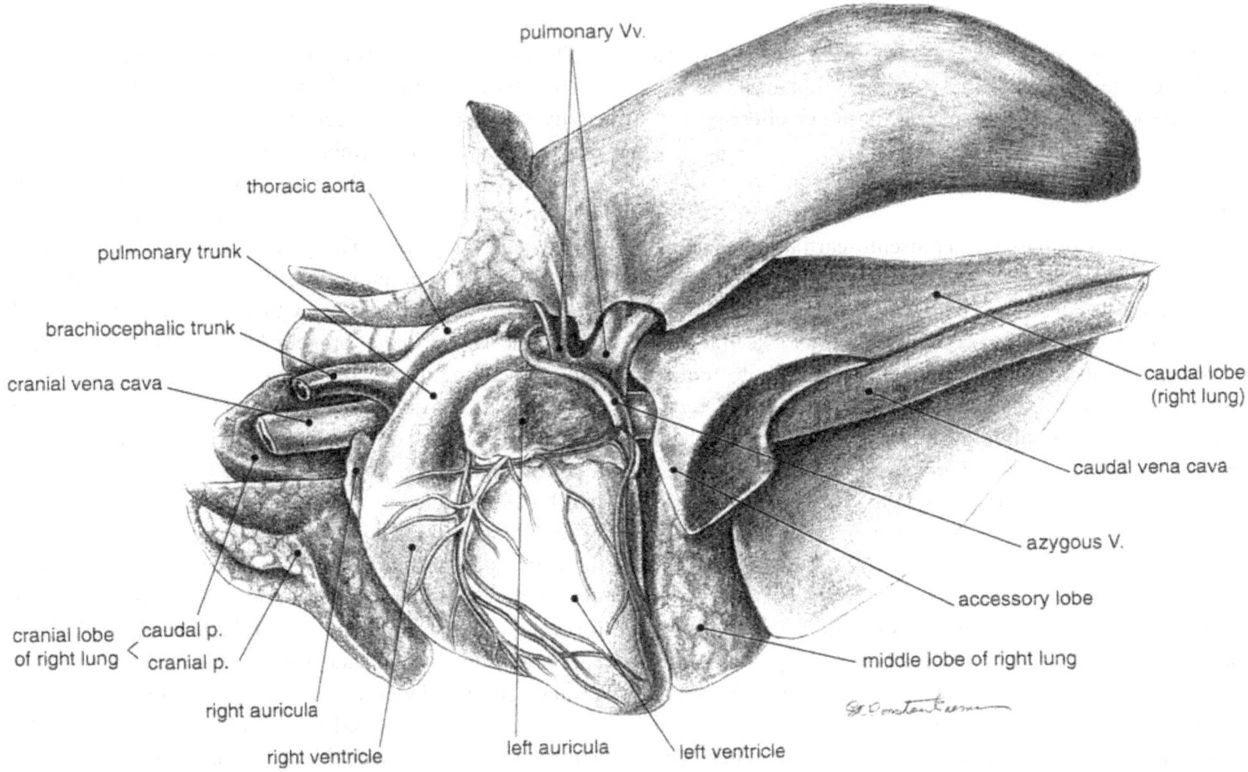


**Figure 6.23** The ventral aspect of the lungs and the auricular side of the heart.

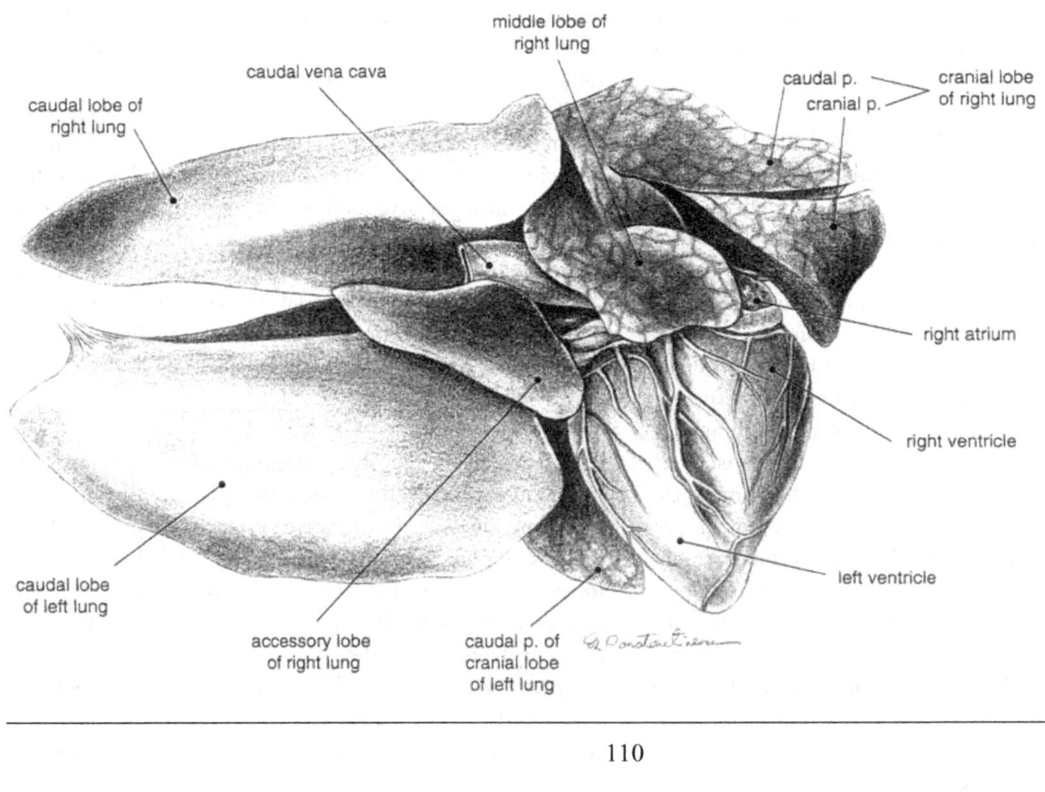


**Figure 6.24** The ventral aspect of the lungs and the atrial side of the heart.

ease the understanding of the functional anatomy of the digestive system, it may be divided in two major groups of structures: the prediaphragmatic and the postdiaphragmatic digestive systems. FunctioFnally, the former is adapted to prehension, mastication, insalivation, and deglutition of the alimentary bolus, whereas the latter is responsible for the digestion of food and elimination of wastes.

### The Prediaphragmatic Digestive System

#### The Oral Cavity

The oral cavity, or the mouth, is located in the head, under the nasal cavity. It starts with the superior (upper) and inferior (lower) lips and continues caudally with the oropharynx. The roof of the oral cavity is represented by the hard palate covered by a mucosa. The floor is rostrally bony and caudally muscular. The lateral walls (the cheeks) are muscular (Figure 6.22).

Goat lips are very mobile. The upper lip lacks the philtrum (present in the sheep), which favors the grasping and tearing of browse (Smith and Sherman, 1994). The lips and the incisive teeth are responsible for the prehension of food.

The hard palate is covered by a mucosa with transversal palatine ridges, having the mechanical role of holding the alimentary bolus within the oral cavity (notice that while grazing, the head is oriented toward the ground). At the level of the first ridge, the incisive papilla can be identified. In front of the latter, the mucosa of the hard palate is very thick and called the dental pad. This structure replaces the upper incisive teeth. The hard palate continues caudally with the soft palate. The bony floor of the oral cavity is the incisive part of the body of the mandible, whereas muscles support the tongue. The lateral walls of the oral cavity are muscular, including the masseter muscle, involved in mastication.

The mucosa of the oral cavity is provided with papillae of different colors, sizes, and shapes, which play a similar mechanical role as the palatine ridges, holding the alimentary bolus within the oral cavity. One symmetrical elevation of the mucosa in the upper oral vestibulum is the parotid papilla, where the parotid duct opens. On the floor of the oral cavity proper, a symmetrical sublingual caruncle can be seen, like a mucosal flap protecting the openings of the mandibular and major sublingual salivary ducts. The mucosa of the oral cavity is supplied by the trigeminal nerve.

The tongue and the teeth are the major structures inside the oral cavity. When the mouth is closed, the space between the teeth and the lips and the lateral walls is called the oral vestibulum. The space surrounded by the teeth is called the oral cavity proper, where the tongue lies. As annex glands of the prediaphragmatic digestive system, the salivary glands will be described later. They are located around the oral cavity, but their product, the saliva, is emptied into the oral cavity.

#### The Tongue

The tongue is a very mobile organ; however, it is not used in prehension of food. A root, a body, and an apex can be distinguished with ease. On the surface of the body, an elevation called torus linguae is characteristic for all ruminants. The tongue is capable of a wide variety of movements, and it is covered by a mucosa provided with papillae.

The lingual papillae are filiform (with a cornified tip), fungiform (mushroom-shaped), and vallate (mushroom-like papillae embedded into the lingual mucosa and surrounded by a circular groove). The filiform papillae have a mechanical role, whereas the other two types bear taste buds and are involved functionally as the primary stage in the perceiving of taste (through the gustatory system; see the Gustatory Organ in the Sense Organs). The filiform papillae may be lens-shaped (lenticular papillae) or conical-shaped (conical papillae). The filiform papillae are scattered on the surface of the tongue, the fungiform papillae are numerous along the borders of the apex, and the vallate papillae, in numbers of 12–18, form two irregular rows on the torus linguae.

The muscles moving the tongue originate from the hyoid apparatus. The narrow symmetrical space between the tongue and the mandible is called the lateral sublingual recess. The tongue is supplied by the hypoglossal nerve (to the muscles), by the trigeminal, facial, glossopharyngeal, and vagus nerves (for the taste buds) and the mucosa for tactile, pain, and temperature stimuli.

#### Teeth

There are three categories of teeth: incisors, premolars, and molars. The premolars and molars are collectively called "cheek teeth." All teeth are implanted into dental alveoli—bony sockets in some bones of the skull. There are only four inferior incisors on each side in all ruminants. From the midline to the periphery, they are called central, first intermediate, second intermediate, and corner incisors. The latter are considered canine teeth that have migrated rostrally. The superior incisors are replaced by the mucosal dental pad. Three superior and inferior premolars and three molars are on each side. There are deciduous teeth, which appear immediately after birth. Some of them are replaced by permanent teeth. Therefore, there are

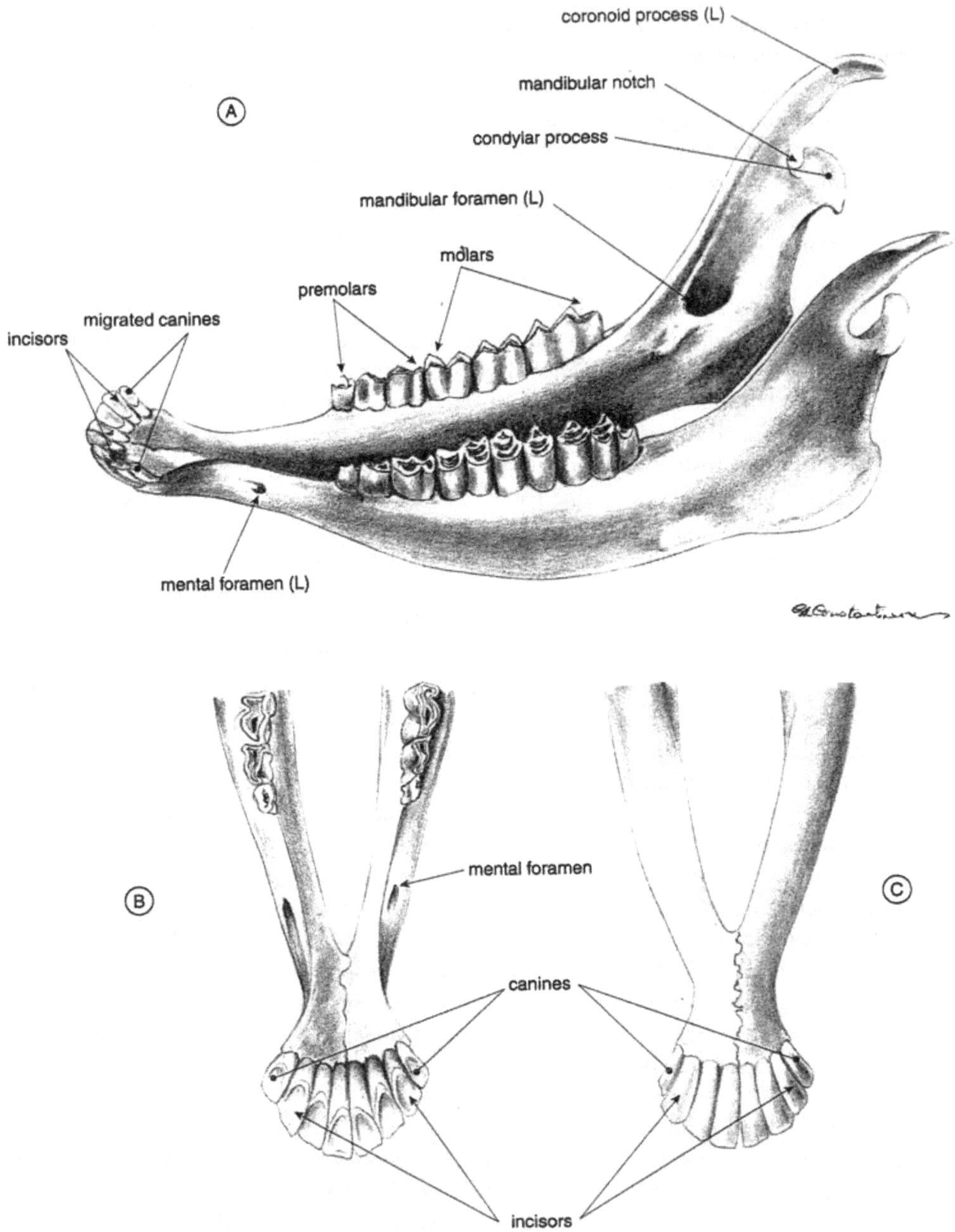


**Figure 6.25** The mandible: A. Laterodorsal view; B. Dorsal view, rostral half; C. Ventral view, rostral half. (L) = landmark for physical examination and/or clinical approach.

two dental formulae. The formula for the deciduous (milk) dentition is:

$$2(\text{Di } 0/4; \text{Dc } 0/0; \text{Dp } 3/3;) = 20$$

The formula for the permanent dentition is:

$$2(\text{I } 0/4; \text{C } 0/0; \text{P } 3/3; \text{M } 3/3) = 32$$

in which D = deciduous tooth, I(i) = incisor, C = canine, P(p) = premolar, M(m) = molar. Because the teeth on one side are in the same number as on the other side, multiply the number of teeth of one side by 2.

The shapes of the premolars and molars are very much alike, whereas the shapes of the incisors are very different from the other group of teeth (Figure 6.25). Brachygnathism (abnormal short mandible) and prognathism (abnormal forward projection of the mandible) are common in goats, and they adversely affect feeding behavior (Smith and Sherman, 1994). The teeth are held in the alveoli by dental ligaments. In ruminants, the ligaments of the incisors are loose, looking like old teeth, with a varying degree of mobility.

The mucosa of the vestibulum and of the oral cavity proper attach to the teeth under the name of gingiva.

### The Salivary Glands

The salivary glands, as annex structures of the oral cavity, empty the saliva inside of the oral cavity. The role of saliva is to mix with the forage during the mastication and to contribute to the formation of the alimentary bolus. There are two categories of salivary glands: scattered (minor) and compact (major). The labial, buccal (under the cheeks' mucosa), and lingual are the scattered salivary glands. The compact, major salivary glands are the parotid, mandibular, and sublingual salivary glands.

#### *The Parotid Gland*

See Figure 6.19. Located between the base of the ear, the Wing of Atlas, the caudal border of the mandible, and the linguofacial vein, this is the largest salivary gland of the goat. The parotid duct accompanies the facial vein along the rostral border of the masseter muscle and opens in the oral vestibulum opposite the fourth superior cheek tooth (the first superior molar).

#### *The Mandibular Gland*

This gland is partially covered by the parotid gland, the external jugular vein, and its branches. Only the caudoventral part of the gland is apparent. It is triangularly shaped, and the mandibular duct travels rostrally to open under the sublingual caruncle.

#### *The Sublingual Salivary Glands*

There are two sublingual salivary glands: monostomatic and polystomatic. The monostomatic gland (with only one excretory duct) is as long as the line between the first and the last lower cheek teeth and parallels the mandibular duct. The major sublingual duct opens with the mandibular duct under the protection of the sublingual caruncle. The polystomatic gland (with many salivary ducts) is located caudodorsal to the previous gland and has numerous short excretory ducts that open into the lateral sublingual recess (the narrow spaces under the tongue).

### The Pharynx

The second compartment of the prediaphragmatic digestive system, the pharynx, consists of three parts: the oropharynx, the nasopharynx, and the laryngopharynx.

The oropharynx is apparently a part of the oral cavity proper, but they are separated by a symmetrical mucosal fold, the palatoglossal fold (glossal = lingual), where the entrance into the pharynx is considered. The oropharynx continues to the epiglottis and is outlined by the tongue ventrally, the soft palate dorsally, and the palatoglossal folds laterally. The caudal border of the soft palate continues dorsally and laterally with two symmetrical mucosal folds called palatopharyngeal folds, which outline the transition between the nasopharynx and the laryngopharynx. Within the oropharynx, the ventral side of the soft palate shows two slit-like openings, the entrances into the palatine tonsils. The nasopharynx and the laryngopharynx were described in The Respiratory System. This is the place to consider the pharynx as a whole and to talk briefly about its conformation and functional anatomy. The soft palate is a mobile structure, movable by contraction of three muscles: one is the tensor, another one is the elevator, and the last shortens the soft palate. All of them help during deglutition. The muscles of the pharynx are constrictors and dilators. The constrictor muscles are grouped in the first, second, and third constrictor muscles. There is also a muscle that dilates the pharynx. The names of these muscles are related to the bones or cartilages at their respective origin. Their insertion is on the lateral and dorsal walls of the pharynx.

The deglutition is the physiological reflex act following the formation of the alimentary bolus. It starts with the bolus pushing the soft palate dorsally, concomitant with the contraction of the elevator of the soft palate, and of the first constrictor of the pharynx. At the same time, the soft palate is shortened. The first wave of contraction is followed by the second and the third waves, under the action of the second and the third constrictor muscles of the pharynx. At the end of the deglutition, the pharynx dilates, and the tensor of the soft palate restores the resting position of the soft palate.

### The Esophagus

The caudodorsal opening of the laryngopharynx leads into the esophagus. There are three segments of the esophagus: the cervical part, the thoracic part, and a very short abdominal part. In the cervical region, the esophagus lies on the dorsal aspect of the trachea and is slightly oriented to the left. In the thoracic cavity, it lies on the dorsal aspect of the trachea, until the trachea bifurcates into the two bronchi. In the rest of the thoracic cavity, the esophagus runs between the aorta (dorsally) and the caudal vena cava (ventrally). Passing through the esophageal hiatus, the esophagus opens on the dorsal aspect of the reticulum by an orifice called cardia. Cardia can be approximated from outside at the level of the eighth left intercostal space.

### The Postdiaphragmatic Digestive System

This consists of segments of the gastrointestinal tract and two annex glands, the liver and the pancreas, all located within the abdominal cavity. The abdominal cavity is a natural component of the body wall, outlined by the diaphragm cranially, the entrance into the pelvic cavity caudally, the lumbar vertebrae covered by dorsal abdominal muscles dorsally (the roof), and ventral abdominal muscles on the lateral walls and on the ventral wall (the floor). The whole cavity communicates with the pelvic cavity and are both lined by a connective tissue layer called transverse fascia and the parietal peritoneum. The latter sends several peritoneal folds to connect/suspend different organs by name of ligaments, mesentery, omenta (singular omentum), etc., before surrounding the viscera, altogether known as the visceral peritoneum. The pelvic cavity is outlined between the sacrum dorsally and the two coxal bones laterally and ventrally. Toward the end of the pelvic cavity, the peritoneum ends by reflecting itself from the roof to the floor of the pelvic cavity and surrounding the pelvic viscera. There is a certain amount of connective tissue and muscle behind the peritoneum, holding the viscera in place.

#### The Stomach

The stomach of the goat is pluricompartmented and consists of four parts: rumen, reticulum, omasum, and abomasum. The entire stomach is surrounded by visceral peritoneum. The walls are provided with powerful muscles lined by a specialized mucosa.

##### *The Rumen*

The rumen, with a capacity of 13–30 liters, fills almost the entire left side of the abdominal cavity along with the reticulum and part of the abomasum (Figure 6.26).

The external configuration of the left side as well as of the right side of all four compartments removed from the abdominal cavity are illustrated in Figure 6.27 and Figure 6.28.

Two lacelike sheets of peritoneum collectively called the walls of the greater omentum originate from both the left and right sides of the rumen, from the longitudinal grooves. They surround the rumen and switch the direction from the left to the right side of the body like a hinge for supporting most of the digestive viscera. Thus, the left wall becomes superficial, and the right wall becomes deep, in relation to one another.

The surface of the rumen is provided by two longitudinal grooves (right and left), which meet in the cranial and caudal grooves, respectively. They separate the rumen into two divisions called sacs, dorsal and ventral. Caudally the ruminal sacs end by two blind sacs, whereas cranially the dorsal sac continues with the ruminal atrium, and the ventral sac ends in the ventral recess. The blind sacs are separated from each other by the caudal groove, and from the dorsal and ventral ruminal sacs by the coronary grooves (dorsal and ventral, right and left). The ruminal atrium is separated from the ruminal recess by the cranial groove. The rumen is separated from the reticulum by the rumino-reticular groove. The internal configuration of the rumen shows strong muscles grouped in pillars, and a mucosa with one fold (rumino-reticular) and papillae. The pillars and the fold correspond to the outside grooves (longitudinal, cranial, caudal, and rumino-reticular grooves, respectively). The ruminal content consists of ingesta, water, bacteria, and protozoa to start microbial digestion, which continues also in the reticulum and the omasum. The protozoa scavenge bacteria (primarily facultative anaerobes), a process that helps keep the bacterial symbiont healthy.

The primary function of the rumen is to mince the ingesta, to mix it with water, and to start the microbial digestion. As a result, methane and carbon dioxide accumulate above the ingesta in the rumino-reticular.

##### *The Reticulum*

The reticulum has a capacity of 1–2 liters and lies on the floor of the abdominal cavity, in intimate contact with the diaphragm. It can be seen on both sides. It looks like an elongated ball and has a honeycomb-like mucosa. The reticulum is separated from the ruminal atrium by the rumino-reticular groove, that from inside corresponds to the rumino-reticular fold. Its function is to squeeze the ingesta by contractions that mix and move ingesta throughout the rumen and create an explosion of contents to the omasal canal during the second of two reticular contractions that initiate the reticulo-rumen contraction cycle.

##### *The Omasum*

The omasum has a capacity of less than 1 liter and is mainly located on the right side of the abdominal cavity. It looks like a flattened ball, and inside it has 35 folds or laminae that are very thin and of different sizes (only three sizes instead of four as in the other ruminants—the shortest is absent). As a function it continues the process of squeezing the ingesta and water absorption, and the omasal canal forces the content into the omasal body. The folds of the omasum absorb much of the fluid from the content and, at regular intervals, the content is passed into the abomasum.

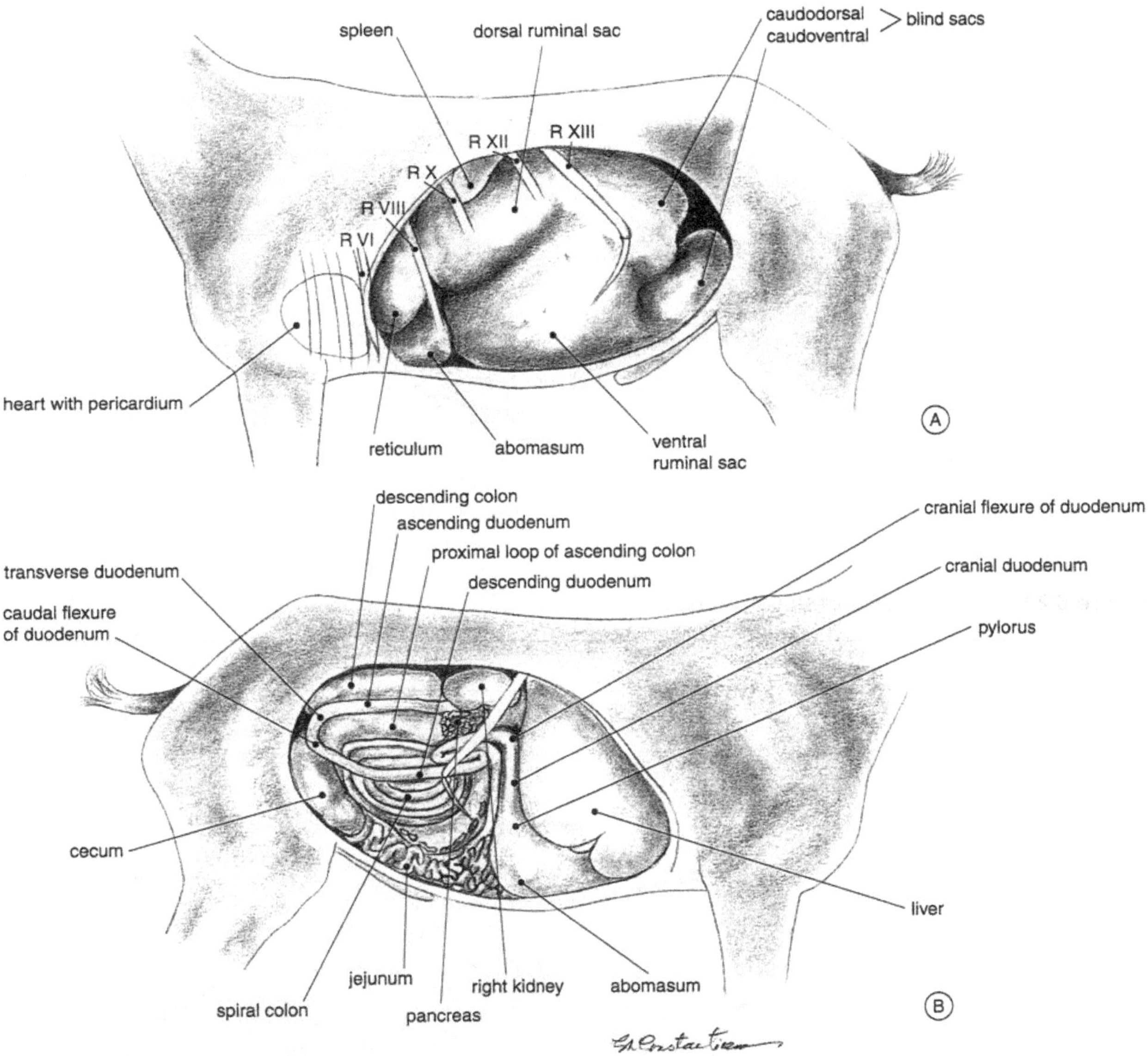


**Figure 6.26** Topography of abdominal viscera: A. Left side; B. Right side.

Starting from the cardia, a *gastric groove* is formed, running on the walls of the reticulum and omasum, and ending in the abomasum. A certain amount of already-shredded forage is pushed back from the reticulum into the oral cavity by an antiperistaltic wave of the esophagus, is remasticated, reinsalivated, and redeglutited as the mericic bolus. Peristaltic versus antiperistaltic waves will be explained in Chapter 8. This bolus bypasses the rumen, drops into the reticulum through the gastric groove, then passes into the omasum to be squeezed of water as much as possible. This is the role of the muscles and configuration of mucosal folds of the reticulum and omasum. Mixed with as little water as possible, the bolus passes into the abomasum and is submitted to a chemical digestion process.

In suckling youngsters, the gastric groove is transformed into a gastric duct, which allows the milk to pass from the esophagus into the abomasum, the only compartment of the stomach capable of digesting milk. The normal functioning of the gastric groove is as important in adulthood as it is for the suckling kids (baby goats).

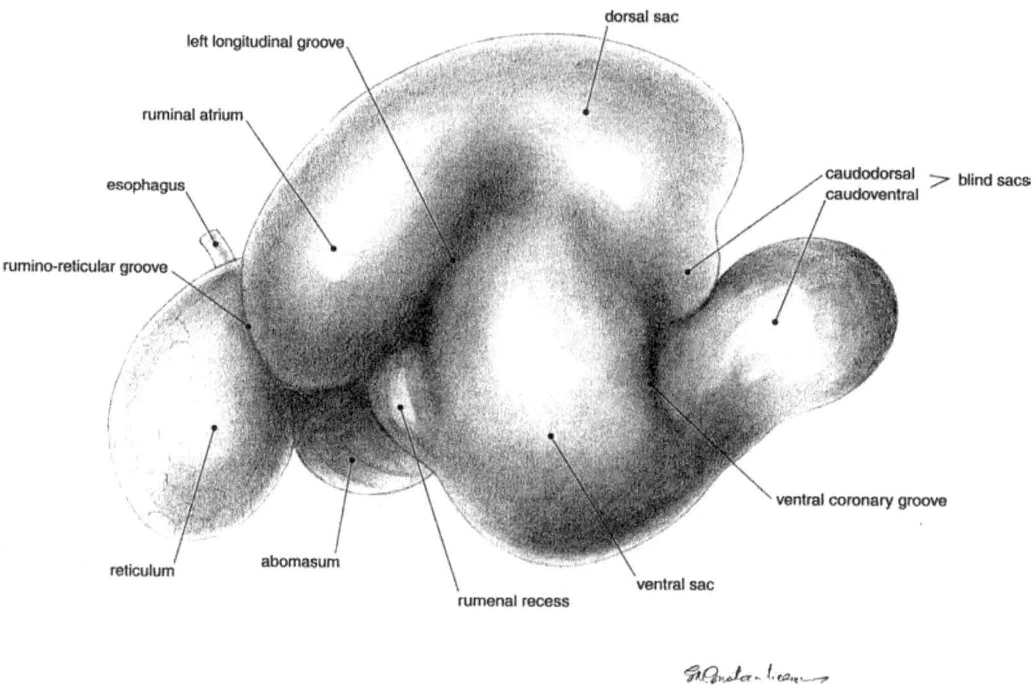


**Figure 6.27** The left aspect of the stomach.

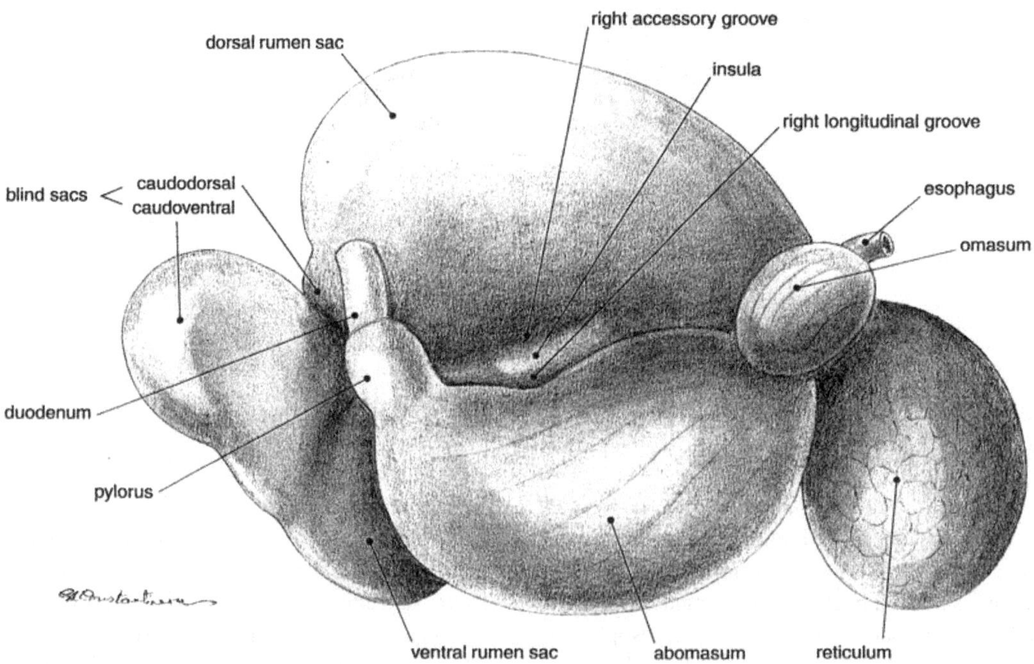


**Figure 6.28** The right aspect of the stomach.

*The Abomasum*

*The abomasum* has a capacity of between 2 and 3 liters and lies on the floor of the right side of the abdominal cavity; sometimes it can be detected also on the left side. It is pear shaped and is located inside with spiraled folds of the mucosa. It starts from the communication with the omasum and ends by the communication with the small intestine called pylorus. There a powerful sphincter (a circular muscle) is present, allowing the transit of the ingesta to pass from the abomasum into the small intestine only by a reflex act. In the goat, as in all ruminants, the pyloric sphincter forms a protuberance called torus pyloricus. The function of the abomasum is to perform the chemical digestion (the HCl hydrolysis). Most of the volatile fatty acids (the short-chain fatty acids) like acetic, butyric, and propionic acids have been absorbed at this point. The advantage of the ruminants is that all the bacterial protein that enters the abomasum provides an adequate protein source for the host animal. The volatile fatty acids are energy substrates produced during microbial digestion and they provide nearly 80% of the ruminant's energy needs.

The goat stomach has a mechanical and a chemical role in the digestion process, which will be detailed in Chapter 8, Digestive Physiology and Nutrient Metabolism.

### The Intestines

There are small and large intestines in all mammalian species. The total length of the goat intestines is 20–40 meters, with the small intestine being between 20 and 30 meters. The average size of the large intestine is 7 meters long. Small and large intestines (with the blood supply) are shown in Figure 6.29. All intestinal segments have a lumen and a similar constitution: the external coat is the peritoneum, the middle coat is represented by smooth muscles, and the internal coat is a specific type of mucosa.

*The small intestine* consists of three segments: duodenum, jejunum, and ileum. The lumen of all of the segments is the same, and is approximately 2 cm in diameter. Functionally, they facilitate most of the digestion.

The duodenum travels on the right side of the body, makes several flexures, and continues with the jejunum. A small elevation called major duodenal papilla can be seen inside of it, where the bile duct and the pancreatic duct open on its tip. The jejunum is the longest segment of the small intestine. A fold of the peritoneum that is called mesentery suspends the jejunum. The ileum is the shortest segment of the small intestine, and it joins the cecum, the first part of the large intestine.

*The large intestine* consists of three segments as well: cecum, colon, and rectum. The main function of the large intestine is the reabsorption of ingesta water. Also, there is a microbial fermentation in the cecum and proximal colon and much less in the distal colon, as the contents become dehydrated. The cecum is much larger in diameter than the small intestine. It is almost horizontally positioned, on the right side of the abdominal cavity. The colon has three segments called ascending, transverse, and descending colons. The ascending colon starts from the cecum as a proximal loop. Then it is coiled making several inflexions called the spiral loop, and ends as the distal loop. The transverse colon is very short, in a transverse position from the right to the left, and continues with the descending colon. The rectum, followed by the anal canal is the last segment of the gastrointestinal tract and is located within the pelvic cavity. The anus is the end and the caudal opening of the digestive system.

The jejunum, ileum, cecum and ascending colon lie on the greater omentum, in the space called supraomental recess.

The role of the intestinal mass is to continue the digestion process of the ingesta, and to eliminate the waste material. The entire physiology of digestion will be detailed in Chapter 8, Digestive Physiology and Nutrient Metabolism.

## The Annex Glands

The *liver* is the largest gland of the body, whose complex role related to digestion will be discussed in Chapter 8, Digestive Physiology and Nutrient Metabolism.

Situated between the diaphragm and the stomach, the liver has an oblique position from right to left and from dorsal to ventral, because of the permanent pressure of the stomach during the developmental life. The parietal surface lies against the diaphragm, whereas the visceral surface comes in contact with the stomach and the intestines (Figure 6.30).

The liver is suspended from the diaphragm by several ligaments, and is connected to the stomach and the duodenum by the lesser omentum. The liver is incompletely divided by lobes. In all ruminants, it is provided by the fewest lobes: left, quadrate, right, and caudate. The gallbladder, where the bile is temporarily stored, is insinuated between the quadrate and the right lobes. It collects the bile produced by the liver, leads it through the cystic duct (the duct of the gallbladder), to empty as the common bile duct into the duodenum, on the tip of the major duodenal papilla.

The *pancreas* makes contact with several viscera, such as the rumen, omasum, spleen, liver, duodenum, right kidney, and jejunum. It consists of the body, the right lobe, and the left lobe. The portal vein is surrounded by a pancreatic notch. The pancreatic duct opens into the major duodenal papilla, together with the bile duct.

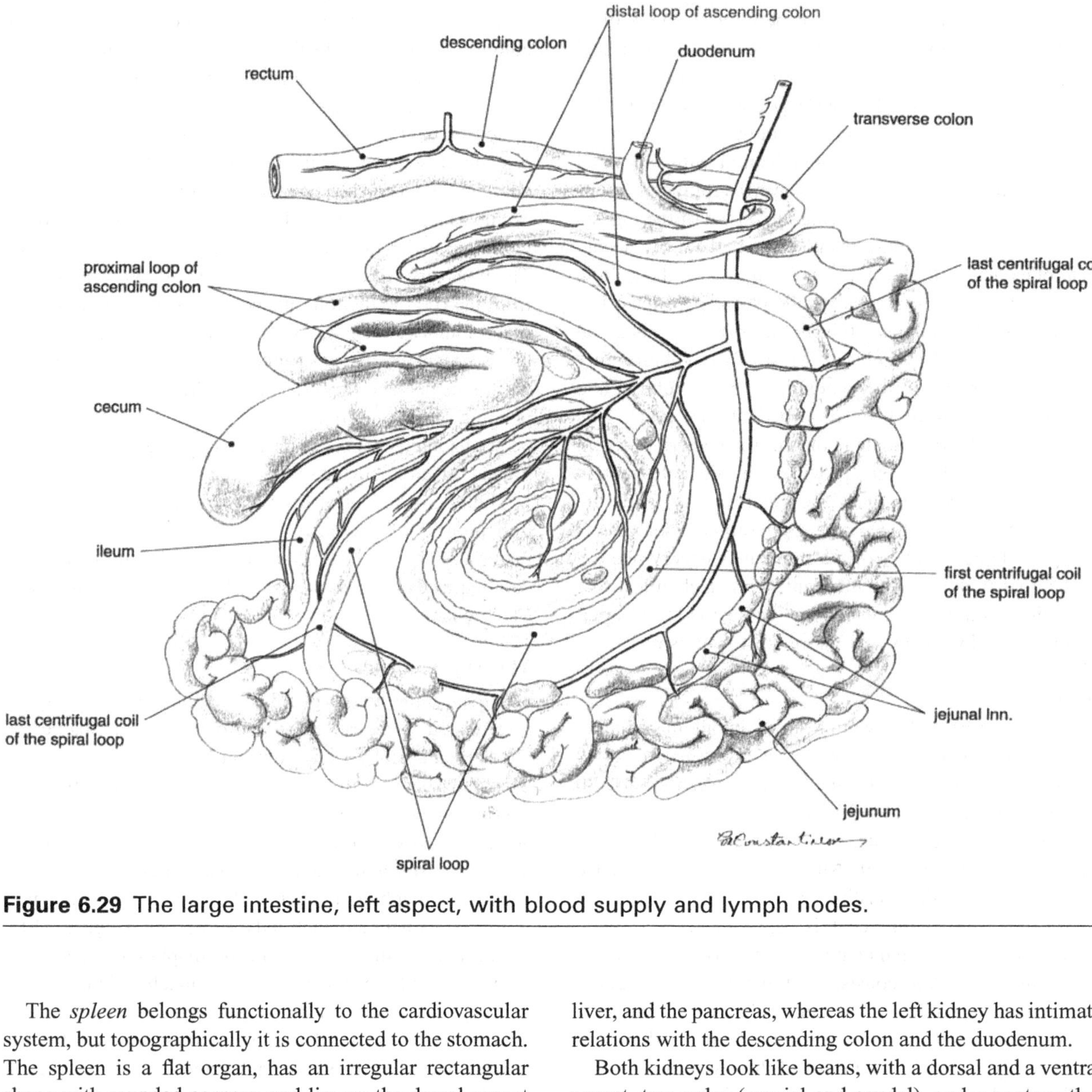


**Figure 6.29** The large intestine, left aspect, with blood supply and lymph nodes.

The *spleen* belongs functionally to the cardiovascular system, but topographically it is connected to the stomach. The spleen is a flat organ, has an irregular rectangular shape with rounded corners, and lies on the dorsal aspect of the rumen. The topography of the viscera on the right and left abdominal walls is illustrated in Figure 6.26. The nerve supply for the liver, pancreas, and spleen are identical with that of the stomach.

## THE URINARY SYSTEM

### The Kidneys

The kidneys lie against the roof of the abdominal cavity. The right kidney comes in contact with the duodenum, the liver, and the pancreas, whereas the left kidney has intimate relations with the descending colon and the duodenum.

Both kidneys look like beans, with a dorsal and a ventral aspect, two poles (cranial and caudal), and a contour that is provided on the medial side with an indentation called hilus. From the hilus, the ureter and the renal vein exit, and the renal artery enters. On the surface, the kidneys are intimately covered by a fibrous capsule and are surrounded by fat and the renal fascia.

A longitudinal section through the kidneys reveals the internal conformation. Two distinct and differently colored zones are shown: a peripheral lightly colored cortex and a central brown-reddish medulla surrounding a space called renal sinus. The medulla consists of several renal pyra-

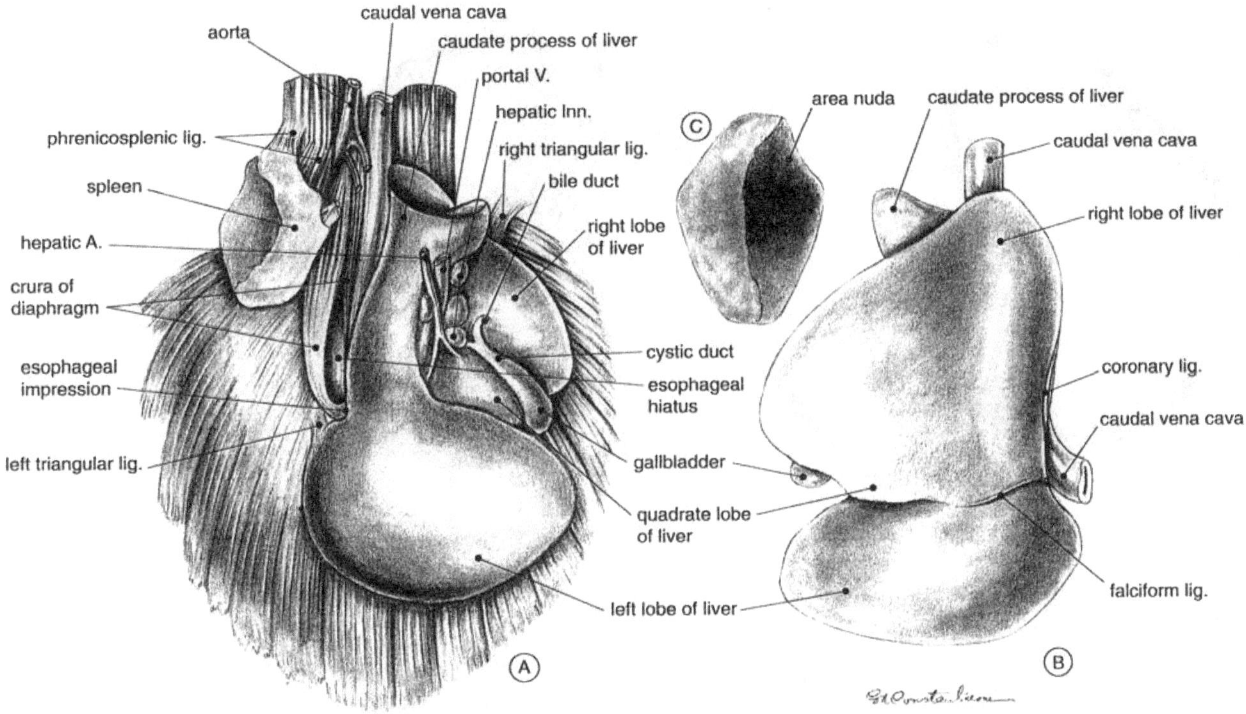


**Figure 6.30** The liver and spleen: A. Visceral aspect of the liver, in situ; B. Parietal aspect of the liver; C. Visceral aspect of the spleen.

mids, whose tips oriented toward the renal sinus open on a common renal crest. The latter empties the urine into the renal pelvis, which is the mucosal coat lining the renal sinus. The renal pelvis continues as the ureter. Both external aspect and internal conformation are shown in Figure 6.31. As to function, the kidneys process the blood by a complex mechanism of filtration, reabsorption, and secretion, and expel the result of this activity, the urine, through the urethra.

### The Ureter

The ureter is a tubular organ, with a peripheral fibrous coat (the continuation of the renal capsule), a muscular coat, and a mucosal membrane (the continuation of the renal pelvis). The muscular coat pushes the urine in peristaltic waves into the urinary bladder. The ureters run on the roof of the abdominal cavity and enter the pelvic cavity to reach the neck of the urinary bladder.

### The Urinary Bladder

The urinary bladder has the function of a temporary receptacle of urine, which is emptied through the urethra. As an organ with lumen, the urinary bladder has three coats: peritoneum on the outside, a muscular coat in the middle, and a mucosal coat inside. Pear-shaped, it lies on the floor of the pelvic cavity and slips more or less into the abdominal cavity when too full of urine. It continues with the urethra through a narrow neck. The ureters open through the dorsal wall of the urinary bladder close to the neck. They perforate obliquely the muscular coat of the urinary bladder, then travel for a short distance between the muscular and the mucosal coats before opening close to each other.

The female urethra is very short and opens on the floor of the vagina, between the vagina and vestibulum. The male urethra will be described with the reproductive organs.

## THE REPRODUCTIVE SYSTEM

### The Male Reproductive System

This consists of three categories of organs: the essential organs (the testicles) with the investing layers (the testicular tunics), the passages leading the sperm through the

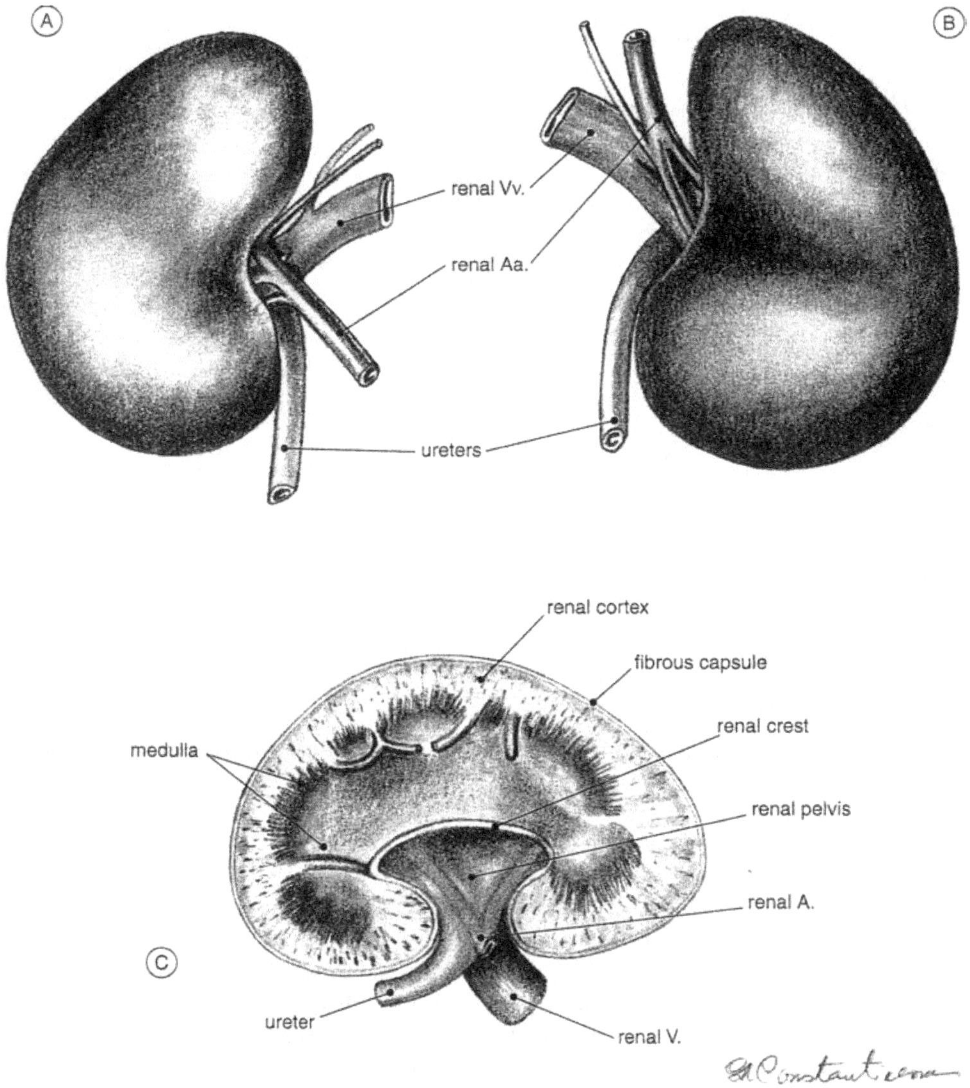


**Figure 6.31** The kidney: A. Left kidney, ventral aspect; B. Right kidney, ventral aspect; C. Longitudinal section.

penis, and the copulatory organ (the penis) within the prepuce. There are also annex glands to the urethra.

### The Testicles

As the essential male reproductive glands, the testicles have two functions: to produce spermatozoa able to fertilize the ovules, and to secrete testosterone, the male hormone. (As a curiosity, one cubic millimeter of semen of the goat contains 2,500,000 spermatozoa.)

Ovoid shaped, the testicles are located in a vertical position within the testicular tunics on the ventral and caudal aspects of the abdomen, in the inguinal region (see Figure 6.3). Each testicle weighs 130–160 g, has a head oriented dorsally, a tail oriented ventrally, two aspects (lateral and medial), and two borders (cranial free and caudal where the epididymis is connected to the testicle) (Figure 6.32). Each testicle is surrounded by a fibrous tunic called albuginea, is inextensible, and continues inside the organ as interlobular septa. The stroma (the main substance) of the testicle is divided by the septa into lobules. The septa connect with each other and form the mediastinum testis. Numerous seminiferous tubules fill the lobules and interconnect inside the mediastinum testis in a network called rete testis.

The testicles are protected by multiple layers of tunics called testicular tunics. Some of these layers are brought by the testicle during its descent (in the developmental intrauterine life), and others are modified skin and subcutaneous connective tissue, all functionally and clinically important. The tunics brought by the testicle during its descent are the internal spermatic fascia and the vaginal tunic. The former is the continuation of the endoabdominal fascia, whereas the latter is the continuation of the peritoneum (consisting of a parietal and a visceral lamina). The tunics outside of the body wall are the external spermatic fascia (a connective tissue layer) and the modified skin (the scrotal skin outside and the tunica dartos inside). While the tunica dartos makes two dartoic sacs (one for each testicle), the scrotal skin surrounds both testicles and their spermatic cords, all together. (See the spermatic cord.)

### The Reproductive Excretory Passages

There are three reproductive excretory passages in the male goat: the epididymis, the ductus deferens, and the urethra with the annex glands. The function of these passages is to allow the sperm to exit and be able to fertilize the ovules.

The *epididymis* is attached to the caudal border of the testicle and covered by an expansion of the albuginea. It starts with a head, which covers the head of the testicle, continues with a body, and ends by the tail, ventrally. From this point, it continues vertically in a proximal direction with the ductus deferens. The head contains many canaliculi (ductuli efferentes), in continuation of the rete testis. The canaliculi connect with each other and form the ductus epididymidis, which runs inside of the body of epididymis. This duct becomes very flexuous and may measure as long as 60 meters. There is a niche, the so-called testicular bursa on the lateral side, between the testicle and the body of epididymis.

The *ductus deferens* is the continuation of ductus epididymidis starting from the tail of the epididymis. It is a straight duct, included within the spermatic cord, enters the abdominal cavity, and reaches the pelvic urethra just caudal to the neck of the urinary bladder. It empties on the roof of the urethra apparently enlarged (ampulla) and bearing small glands, lateral to the colliculus seminalis. The latter is an elevation of the urethral mucosa, at the end of the urethral crest (Figure 6.33A).

The *spermatic cord* consists of the ductus deferens, the visceral lamina of the vaginal tunic, and the vessels and nerves supplying the testicle and the epididymis. This complex structure is subject to various surgical interventions and techniques used for castration, criptorchidism, inguinal hernia, etc. In the goat, the spermatic cords are long, making the testicles pendulous during the walk.

The *male urethra* is a common excretory duct for the urine and semen. It is divided in two segments: pelvic and penile urethra. The pelvic urethra is a continuation of the urinary bladder lying on the floor of the pelvic cavity, is surrounded by a spongy coat and the urethralis muscle, and suddenly narrows at the level of the ischiatic arch. Here the pelvic urethra continues as the penile urethra, following the urethral isthmus (the narrow part).

The *annex glands* are the prostate, the vesicular glands, and the bulbo-urethral glands. Their function is to produce seminal fluids for nourishing the sperm and facilitate the gliding of sperm through the excretory passages.

In the goat, the *prostate* is a disseminated gland completely surrounding the walls of the pelvic urethra. The *vesicular glands* are paired, 3–4 cm long, with lobulated surfaces and located on both sides of the neck of the urinary bladder. Their ducts open on the colliculus seminalis. The *bulbo-urethral glands* are located far caudally, on both sides of the urethral isthmus, and are rounded and small, 1 cm in diameter. Their excretory canals pass through the urethra and open directly into the urethral recess, which is illustrated in Figure 6.33B (see Figure 6.33).

### The Penis

The male copulatory organ, the penis, consists of the root, the body, and the apex. The goat has a fibroelastic penis, still firm when not erect. Its function is to allow the urine and also the sperm to be expelled. This is done by reflexes and by contraction of some of the penile muscles. The average length of the penis is 40 cm long. Starting from the ischiatic arch, the penis extends cranially and passes between the thighs ventral to the pelvis, and between the two spermatic cords, surrounded by skin and fasciae. On the ventral aspect of the abdominal cavity, the free part of the penis is protected inside of the prepuce. (The free part of the penis belongs to the body of the penis.)

The *root of the penis* is composed of the two crura and between them, the bulbus penis. The crura (sing. crus) are attached to the ischiatic arch, the very end of the coxal bones. The crura are covered by two symmetrical muscles that act in the erection of the penis. The bulbus penis is the caudal expansion of the corpus spongiosum penis, which is the erectile tissue of the penis. It is covered by a muscle.

The *body of the penis* consists of the paired corpus cavernosum penis (each of them the continuation of the corresponding crus penis), and the penile urethra surrounded by the corpus spongiosum penis. The expression "corpus cavernosum" is used for the unique structure as a result of fusion of the paired corpus cavernosum. The corpus cavernosum and the penile urethra with the corpus

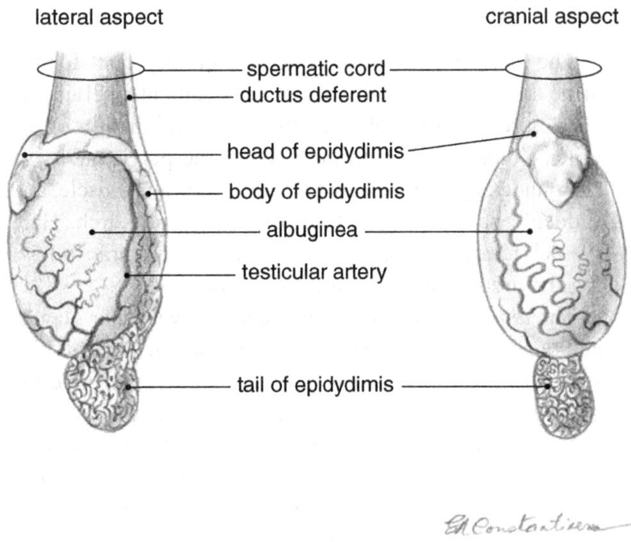


**Figure 6.32** The testicle.

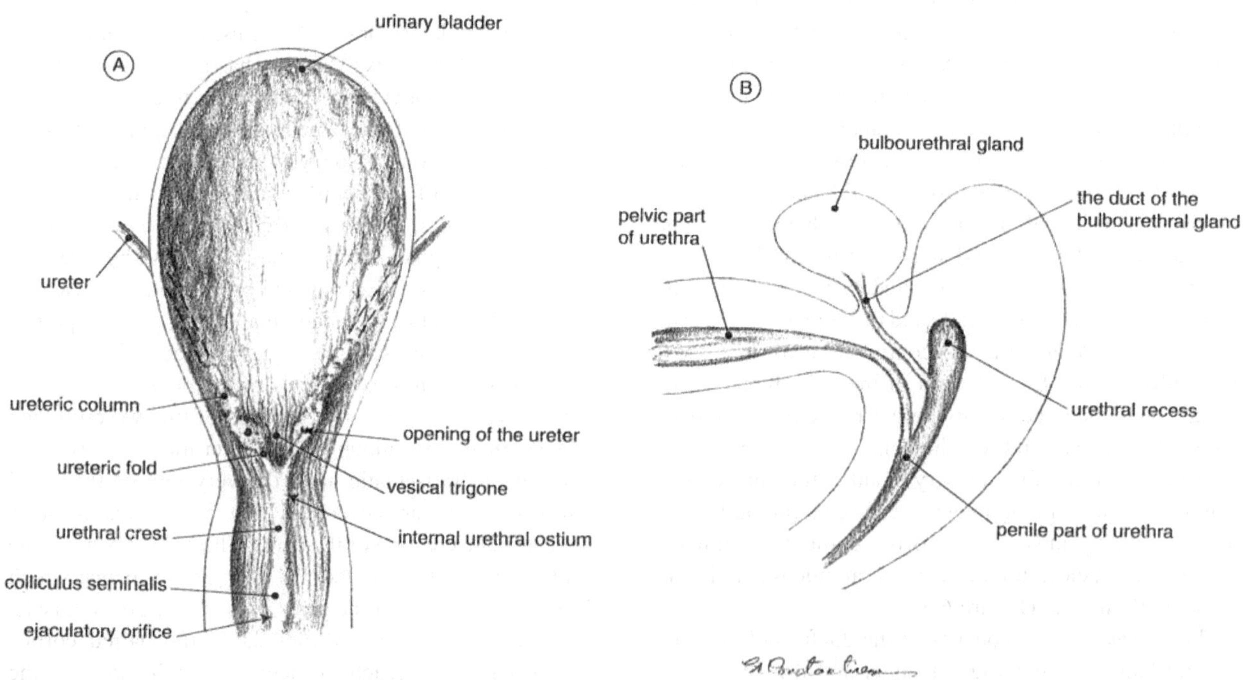


**Figure 6.33** The urinary bladder and the urethra in the male: A. The roof of the urinary bladder and of the male urethra; B. The urethral recess in ruminants.

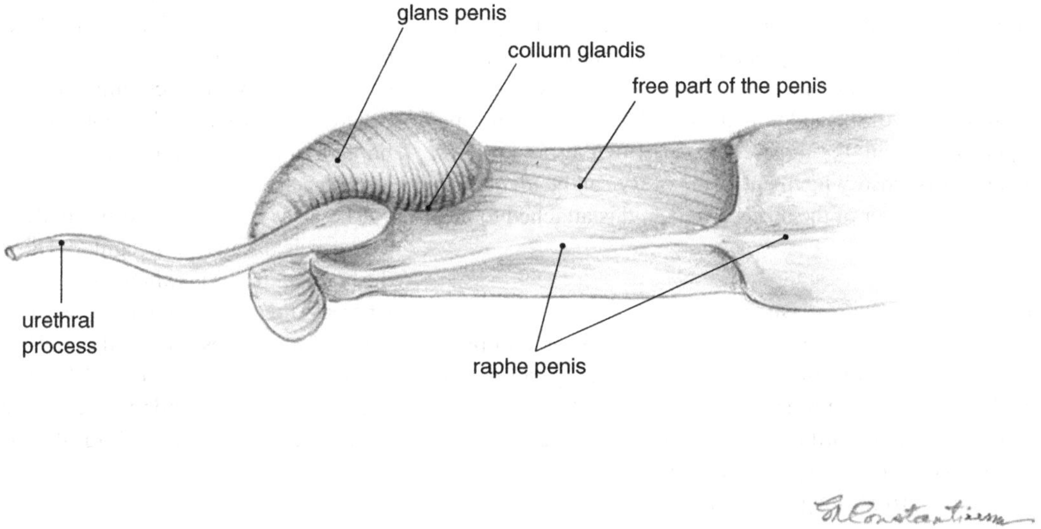


**Figure 6.34** Penis, left lateral aspect.

spongiosum penis are intimately surrounded by two separate tunica albuginea. The albuginea sends inside the corpus cavernosum several septa and trabeculae that form a characteristic framework for the blood vessels. The spaces outlined by the framework are called cavernae. There is a groove on the ventral aspect of the corpus cavernosum, the urethral groove, which protects the passage of the penile urethra. The body of the penis is double (S-shaped) flexed before it reaches the level of the spermatic cords. The first flexure is dorsal to the second. There is a paired muscle associated with the ventral aspect of the body of the penis, which originates from the caudal vertebrae. It attaches to the ventral flexure and continues cranially diminishing in size, and disappearing before reaching the glans penis. The penile urethra exceeds the glans penis by a urethral process, 2.5 cm long (Figure 6.34).

The *glans penis* is the voluminous end of the free part of the penis. It is the head of the penis, contains the corpus spongiosum glandis, and is sharply separated from the free part of the penis by a neck.

### The Prepuce

The skin of the ventral aspect of the abdomen surrounds and protects the free part of the penis under the name of prepuce. This part of the prepuce is called the external preputial lamina. The skin reflects itself inside of the preputial cavity at the preputial orifice, and continues with the internal preputial lamina. The latter attaches to the free part of the penis and continues cranially with the skin of the penis (including a seamlike band called "raphe penis") (see Figure 6.34) and glans penis. The preputial orifice is provided with an abundance of hairs. There are cranial and caudal preputial muscles that move the prepuce during micturition.

## The Female Reproductive System

Similar to the male, the female reproductive system consists of three categories of organs: the reproductive glands (ovaries), the tubular genital organs (uterine tubes, uterus, vagina, and the vestibule), and external genital organs (vulva, clitoris, and urethra). In addition, during pregnancy, the placenta develops and it will be described separately.

### The Ovaries

The function of the ovaries consists of producing ovules, and secreting different hormones depending on the reproductive cycle.

Paired structures, the ovaries are located within and on the floor of the pelvic cavity. Ovoidal shaped and flattened, their surface is irregular because of the follicles and the corpora lutea (sing. corpus luteum). Each ovary consists of a superficial epithelium, an albuginea (a dense capsule), and a cortical and a medullary zone. The cortex (the cortical zone) contains the follicles. The medulla (the medullary zone) is vascular. Both follicles and vessels are surrounded by the stroma, the connective tissue framework of the ovary.

Each ovary has a tubal end, close to the infundibulum of the uterine tube, and a uterine end, opposite to the previous end, where the ovary is attached to the uterus by the proper ligament of the ovary. Each ovary is suspended from the roof of the abdominal cavity by the mesovarium, a peritoneal fold, the most cranial part of the broad ligament. A suspensory ligament of the ovary can be identified at the free border of the mesovarium, and is attached to the diaphragm.

### The Tubular Genital Organs

The uterine tubes, salpinges (sing. salpinx), or fallopian tubes have the function of protecting the spermatozoa while "swimming" toward the ovaries, and the ova (sing. ovum) moving toward the uterine horns after the fertilization of the ovules. They are two very small and flexuous, mobile, 10–16 cm long tubes located between the ovaries and the uterine horns. Each uterine tube has an abdominal opening, which is provided with an infundibulum (funnel-shaped end) oriented toward the ovary. It is followed by an ampulla (the wide part) and the isthmus (the narrow part), which opens into the corresponding uterine horn. In the goat, the salpinx continues gradually with the uterine horn. The infundibulum is provided with a fringe of processes called fimbriae, some of them attached to the ovary. Each uterine tube is suspended by the second segment of the broad ligament, called mesosalpinx. The ovary, mesovarium, and mesosalpinx outline a cavity called ovarian bursa. See Figure 6.35.

### The Uterus

The uterus is a unique structure, consisting of two symmetrical horns, one body, and the neck called the cervix. The whole uterus is a so-called uterus bicornis. Its function is to shelter and nourish the embryo/fetus until birth.

The *uterine horns* are coiled, each of them between 12 and 15 cm long. Before touching the body of the uterus, the horns are connected by the intercornual ligament. Inside of the uterus, the horns are separated by the uterine velum. The horns have a free border, convex, and a mesometrial border, where the mesometrium (as part of the broad ligament) is attached.

The *body of uterus* is as short as 2–3 cm long. It is suspended by the mesometrium, a peritoneal fold, and part of the broad ligament. The musculature of the whole uterus is called myometrium, whereas the mucosal layer is called endometrium. The endometrium of the uterine horns and body is provided with small prominences called caruncles. There are more than 100 caruncles disposed on four rows. They are pediculated, with a flat or slightly concave surface. During the gestation (pregnancy), the caruncles are surrounded by the placenta. The areas of the placenta that come in contact with the caruncles are called cotyledons, and the combination caruncle-cotyledon is called placentome. The cotyledons send villi, which implant into the spongy surface of the caruncles (see The Placenta).

The *cervix* continues to the body of uterus for 3–4 cm. The cervical canal is irregular. The communication with the body of uterus is called internal uterine ostium (orifice), whereas the communication with the vagina is called external uterine ostium (orifice). The latter protrudes into the vagina and is surrounded by a circular recess called vaginal fornix. The endometrium of the cervix is provided with three to four circular folds that interdigitate with each other.

### The Vagina

The vagina is a tubular organ, with a lumen much wider than the uterus, and extends up to the external urethral ostium (on the floor). The mucosa is disposed in longitudinal folds. It facilitates the lubrication and smooth gliding of the penis during copulation, and of the fetus during birth. The vestibule is the continuation of the vagina, caudal to the external urethral ostium. See Figure 6.35.

### The External Genital Organs

The *vulva* is located within the area ventral to the anus and known as the perineum. Two labia (sing. labium) and two commissures (dorsal and ventral, the points where the two labia meet) outline the pudendal fissure. The labia are thick, yet soft structures. The dorsal commissure is rounded, whereas the ventral commissure is pointed toward the ground. See Figure 6.35.

The *clitoris* is the rudimentary homologue of the penis located in the ventral wall of the vestibule. Two crura and the body of the clitoris are embedded into the mucosa of the vestibule. Only the glans clitoridis can be seen. The whole clitoris measures 2–2.5 cm long.

The *female urethra* lies on the floor of the pelvic cavity, covered by the floor of the vagina. It opens by the external urethral ostium on the floor of the vagina, between it and the vestibule. At this level, the opening of the urethra is associated with the suburethral diverticulum, a small pouch caudal to the opening.

### The Placenta

The connection between the embryo/fetus and the maternal uterus starting at a certain phase of the development is

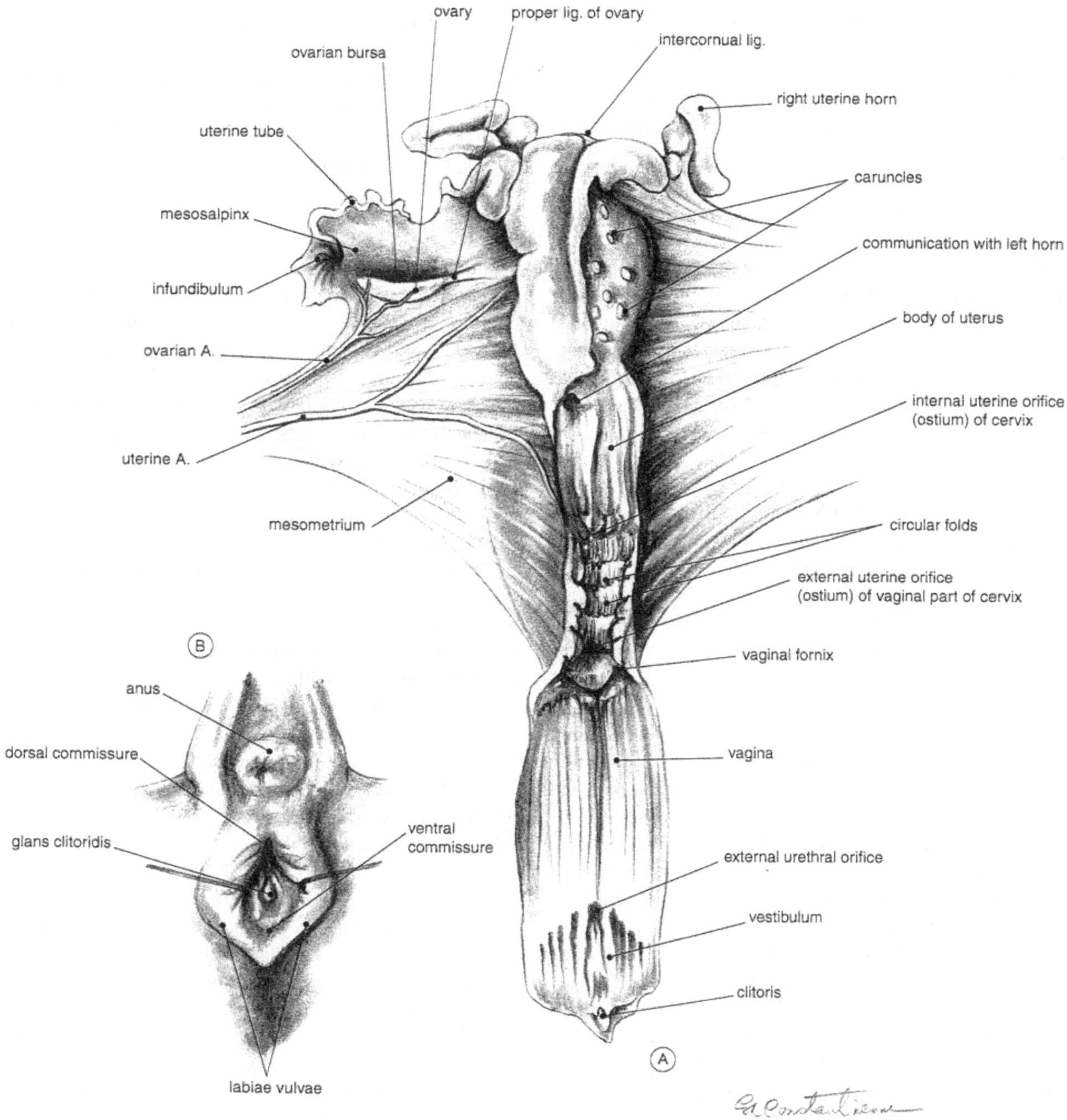


**Figure 6.35** The genital apparatus in the female: A. Dorsal view, dorsal wall partially removed; B. Vulva.

called placenta. The placental phase starts during the transformation of the embryo into the fetus. The placentation is a term that defines the formation and the attachment of the placenta to the uterus. The placenta is an organ responsible for the nutrition of the fetus, the exchanges with the mother, and the protection of the fetus. It consists of a maternal part and a fetal part. The combination between the caruncles (of the uterus) and the cotyledons (of the placenta) was briefly described with the uterus. This type of placenta is called "cotyledonary or multiplex placenta."

As a whole, the placenta has three membranes: the chorion, the allantois, and the amnion. The chorion is the most external membrane of the placenta, in contact with the maternal uterus. The allantois joins the chorion and contributes to the so-called chorio-allantoic placenta. The amnion is the most internal membrane, which surrounds the fetus and allows it to accommodate movements of the head, body and limbs in the amniotic fluid.

*The umbilical cord* extends between the embryonic umbilicus and the placenta, and makes the connection between the embryo/fetus and the placenta. A gelatinous substance in which one vein carrying oxygenated maternal blood and two arteries returning the venous blood (all of them called umbilical) are embedded, forms the umbilical cord.

## THE MAMMARY GLANDS

The mammary gland is a modified cutaneous (sweat) gland. Its function is to produce milk. By definition, one mammary complex consists of one body and one papilla (teat). The usual number of mammary glands in the goat is two, under the collective name of the udder. There may be variations in number. The two mammary glands are separated outside by an intermammary sulcus.

The udder is attached to and suspended from the ventral body wall in the inguinal region (see Figures 6.1, 6.4, and 6.5). The udder of the goat is relatively large in comparison with that of the ewe.

The entire mammary gland is intimately surrounded and protected by a capsule. A suspensory apparatus supports the full heavy udder. There are lateral and medial laminae, connective tissue sheets originating from the tunica flava abdominis, from the symphyseal tendon, and from the aponeurosis of the external abdominal oblique muscle (see The Muscular System). From both lateral and medial laminae, suspensory lamellae enter the mammary glands at different levels and are able to support the weight of the udder. The medial laminae are elastic, and this shows when the udder is full of milk. In a full udder, the two teats look abducted. The intermammary sulcus drops because of the elastic tissue of the medial laminae, while the lateral laminae are inextensible.

The body of each mammary gland is cone-shaped and ends with a voluminous teat. There is a circular constriction at the transition between the body and the teat. Each gland consists of glandular tissue and a duct system (Figure 6.36a).

### The Glandular Tissue

The glandular tissue, whose function is to produce milk, is separated in lobes by connective tissue septa. Each lobe is divided in lobules, clusters of up to 200 alveoli that secrete into a central ductule (the lactiferous alveolar ductule). The lobules are separated from each other by a thin layer of connective tissue.

### The Duct System

The duct system starts with the alveolar ductules, which continue with intralobular and interlobular ductules. All ductules are located within a lobe. From the end of each lobe, the lactiferous ducts convey milk to the lactiferous sinus. The lactiferous ducts are visible without magnification, and they become larger and larger as they approach the lactiferous sinus. There are 6–9 large lactiferous ducts according to Nickel et al. (1981), and 12–15 according to Barone (1978). They are located caudolateral (right and left) to the glandular tissue (which is perforated by small- and medium-sized ducts).

The lactiferous sinus, or cistern, is the dilated distal part of the duct system formed by the confluence of the lactiferous ducts, and consists of two parts. The glandular part, or gland cistern, is located in the ventral end of the mammary gland, whereas the papillary part, papillary or teat cistern, is located within the papilla (the teat). In full lactation, the teat cistern is up to 7 cm long and 2.5 cm wide. The papillary cistern continues in the tip of the papilla, connecting the papillary part of the lactiferous sinus with the outside by the papillary orifice or ostium. The papillary duct, also called teat or streak canal, is conically shaped and has a sphincter (Figure 6.36b). The rosette of Fürstenberg can be noticed at the junction of the papillary cistern and the papillary canal (Pugh, 2002) and consists of mucosal folds.

In older goats, the lactiferous sinus extends under the skin and can be palpated. There is a small accessory teat 0.5–1 cm long at the base of each mammary gland. Occasionally, caudal to the main teats, supernumerary teats can be observed. Also, abnormal teats may be found, such as double teats, forked teats, etc. Rudimentary developed and even functional mammary glands also exist in some male individuals, in front of the scrotum.

The blood supply to the udder is very important for assessing the milk production. It is provided by the external pudendal artery and by the mammary veins. The external pudendal artery branches in the lateral and the largest medial mammary arteries. The teat is abundantly supplied by arterial papillary retia (sing. rete). The veins originate from venous papillary retia (networks), which form at the

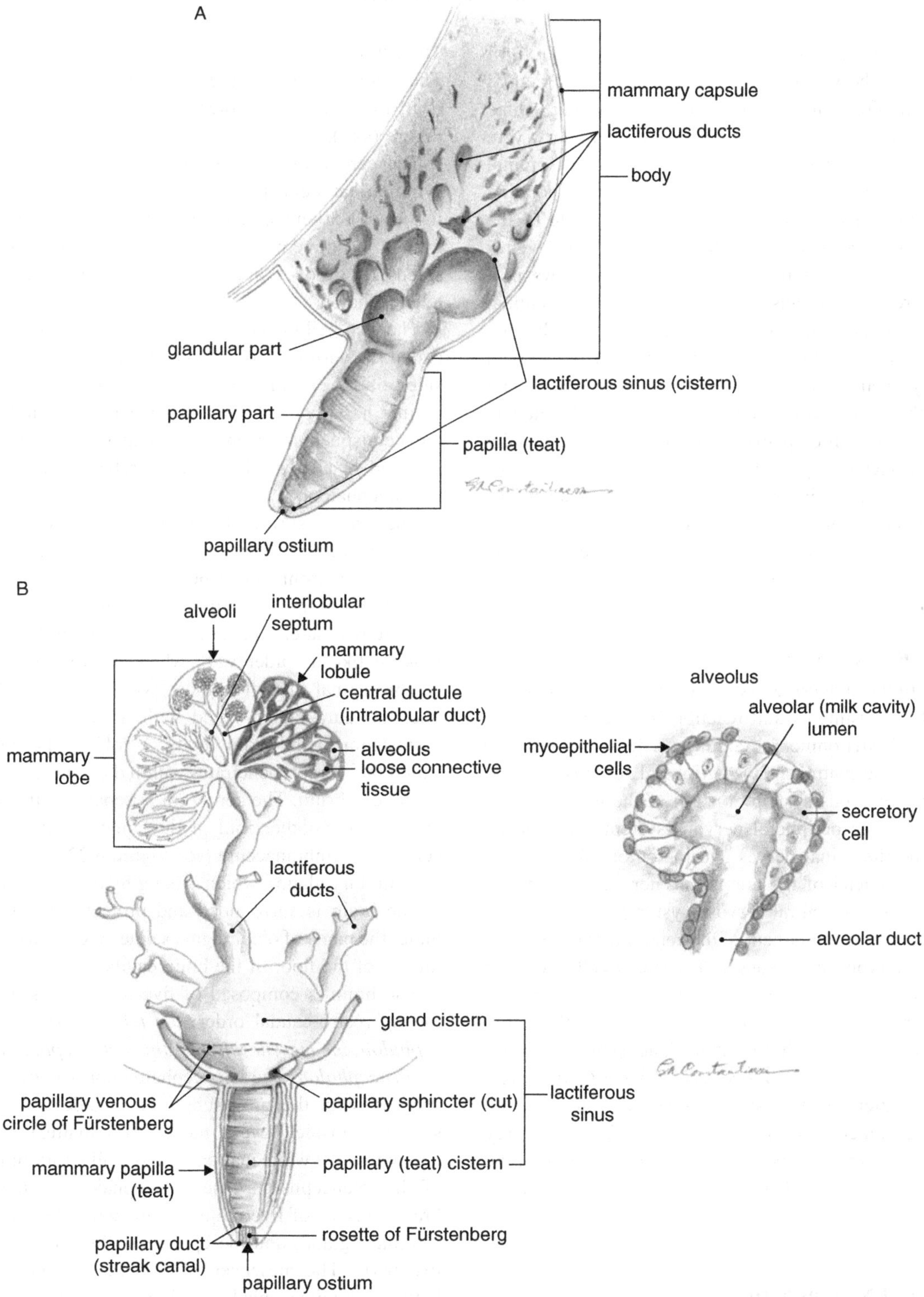


**Figure 6.36 A.** The udder, median section **B.** The internal organization of a mammary gland.

base of the teat the so-called papillary venous circle of Fürstenberg. The venous blood is drained in both cranial and caudal directions; the former passes through three subcutaneous veins: one middle cranial mammary vein (the main vein), and two lateral, right and left, cranial mammary veins in continuation to the corresponding external pudendal veins. The latter take the route of the external pudendal veins through the ventral labial veins. When the goat is not in lactation, the blood flows in a caudal direction, because the valvules of the cranial mammary veins do not allow the blood to flow through them. In lactation, the blood flows through the cranial mammary veins, whose valvules become inefficient because the veins become considerably larger. The median mammary vein can be palpated under the skin, especially during the lactation period. The lymph drainage is provided by the superficial inguinal lymph nodes, known also as mammary lymph nodes. They are located in a caudo-dorsal position to each mammary gland, and are accessible under the skin.

## THE NERVOUS SYSTEM

Along with the endocrine system, the nervous system correlates the adjustments and reactions of the organism to internal and environmental conditions. It is anatomically divided into the central and the peripheral nervous systems. The central nervous system consists of the brain and the spinal cord. The peripheral nervous system includes the cranial and the spinal nerves and the autonomic nervous system (the nuclei of the autonomic nervous system are located within the central nervous system).

The functional component of the central nervous system is the perikaryon, or nervous cell. It has a cell body and two types of processes: dendrites and axons. The dendrites bring the nervous influx to the neuron cell, and the influx leaves the cell through the axon. The central nervous system is organized into two categories of structures: the grey matter and the white matter. Neuron cells and nuclei (agglomeration of neuron cells) form the grey matter, whereas the axons of the neuron cells represent the white matter, organized in fascicles and tracts (of nerve fibers).

### The Central Nervous System

As function, the central nervous system receives stimuli from the environment, and the whole body processes them and sends motor commands to the effectors (muscles, glands, etc.) via the peripheral nervous system.

#### The Brain

The brain consists of the cerebrum and the cerebellum. Both are located and protected within the brain cavity (see Figure 6.22).

The cerebrum shows on the dorso-lateral aspects two symmetrical hemispheres separated by the longitudinal fissure. The hemispheres are provided with gyri (sing. gyrus), prominent elevations separated by sulci (sing. sulcus = groove) and fissures, and are parts of the telencephalon.

The cerebellum consists of a median prominent part called the vermis, and two symmetrical structures on both sides of the vermis, called cerebellar hemispheres. The vermis has gyri and sulci transversely oriented, whereas the hemispheres are provided with small gyri and sulci oriented in all directions. The cerebellum is part of the metencephalon.

The brain is surrounded by meninges, two soft and one tough connective tissue layers, which protect the brain from the environment (especially shocks). All irregularities of the brain are intimately covered by the soft and delicate pia mater. The latter is followed by the arachnoid, looking like a spider web, which allows spaces for the cerebro-spinal fluid. The most external layer is the tough one, the dura mater, fused with, or even replacing the periosteum of the brain cavity. A fold of the dura mater called falx cerebri penetrates within the longitudinal fissure of the cerebrum. The cerebrum is separated from the cerebellum by another fold of dura mater called tentorium cerebelli membranaceum (see Figure 6.22). The hypophysis (pituitary gland), which is hanging on the ventral aspect of the brain is surrounded and protected by dura mater under the name of diaphragma sellae, anchored on the sella turcica of the floor of the brain cavity.

The brain is composed of five major parts, in the following rostro-caudal order: *the telencephalon*, *the diencephalon*, *the mesencephalon*, *the metencephalon*, and *the myelencephalon*. The telencephalon consists of two hemispheres, with the grey substance outside and the white substance inside. Several nuclei (grey matter) are embedded within the white matter. Also the olfactory brain is part of the telencephalon. The diencephalon or intermediate brain consists of five major components. The hypophysis (pituitary gland) and the epiphysis (the pineal gland) are two parts. The mesencephalon is the midbrain, situated between the diencephalon and the pons. The metencephalon consists of a ventral structure, the pons and a dorsal structure, the cerebellum. The myelencephalon is also called medulla oblongata, or simply medulla, which connects the brain to the spinal cord.

There are several spaces inside of the brain, filled with cerebrospinal fluid. There are four ventricles and one duct: two lateral ventricles (first and second) within the cerebral hemispheres; the third ventricle in the diencephalon, the mesencephalic aqueduct (within the mesencephalon); and the fourth ventricle located between the pons and medulla ventrally, and the cerebellum dorsally. The fourth ventricle communicates with the central canal of the spinal cord.

All major parts of the cerebrum (and cerebellum) are shown in Figure 6.37 and Figure 6.38.

#### The Spinal Cord

The second major component of the central nervous system, the spinal cord, extends from the first pair of cervical nerves to the end of the filum terminale, and is protected within the vertebral canal. As the vertebral column is divided into cervical, thoracic, lumbar, sacral, and caudal vertebrae, there are five parts of the spinal cord, corresponding to the five regions of vertebrae. There are two enlargements, one in the cervical region and the other between the lumbar and sacral regions. They are called intumescences, and they represent concentrations of neuron cells of the nerves supplying the thoracic and pelvic limbs.

The spinal cord ends as the medullary cone, located in the sacral vertebral canal, and continues as the filum terminale far beyond the sacral canal, within the caudal canal. The filum terminale consists of supporting nervous cells called glial cells.

The spinal cord is surrounded and protected by meninges with the same names as those for the brain (pia mater, arachnoid, and dura mater).

On a transverse section (Figure 6.39), the spinal cord shows an H- or butterfly-shaped grey matter surrounded by white matter, and the central canal. The grey matter is organized into three symmetrical horns (dorsal, middle, and ventral) and the white matter into three symmetrical columns (dorsal, middle, and ventral). The dorsal horns and columns are sensory, the ventral horns and columns are motor, the lateral horns are related to the autonomic nervous system, and the lateral columns are both some sensory and some motor.

### The Peripheral Nervous System

The peripheral nervous system is represented by the cranial nerves, the spinal nerves, and the autonomic nervous system. Its function is to send stimuli to the central nervous system, and to send orders from the central nervous system to the effectors.

#### The Cranial Nerves

Some of the cranial nerves (CN) are sensory, some of them motor, and the others are mixed. The sensory nerves bring information from the sense organs to the brain to be processed into sensations (CN I, II, and VIII). Five cranial nerves are purely motor (CN III, IV, VI, XI, and XII). The remaining four cranial nerves are mixed (CN V, VII, IX, and X). Four pairs of cranial nerves have, in addition, parasympathetic fibers (CN III, VII, IX, and X). The origin of the sensory nerves is inside of specific organs (olfactory mucosa of the nasal cavity, the retina, and the vestibulocochlear organ within the inner ear). The origins of the other cranial nerves are in the brain. The cranial nerves supply all of the structures of the head including most of the head skin.

#### The Spinal Nerves

The nerves originating from the spinal cord are called cervical, thoracic, etc., spinal nerves. They supply muscles, joints, bones, skin, and other structures related to the body wall and the limbs, except viscera and sense organs.

A spinal nerve originates by two roots, dorsal sensory, and ventral motor. Each dorsal root is provided with a spinal ganglion. The two roots join within the vertebral canal and form the spinal nerve. Exiting the vertebral canal, the spinal nerve sends four categories of branches: dorsal, ventral, communicating, and meningeal. The dorsal and ventral branches supply structures located dorsal to, and ventral to, the vertebral column, respectively. The communicating branch is part of the autonomic nervous system, whereas the meningeal branch reenters the vertebral canal to supply the meninges. Both dorsal and ventral branches are mixed, with sensory and motor fibers. One of the most important spinal nerves is the phrenic nerve, which supplies the diaphragm. It originates from some of the last cervical spinal nerves.

#### The Autonomic Nervous System

The autonomic nervous system consists of the sympathetic and the parasympathetic nervous systems. They are provided with nerve fibers and ganglia (sing. ganglion). From the origin to the ganglia, the fibers are called preganglionic, and from the ganglia to the organ supplied they are called postganglionic. As a rule, the sympathetic preganglionic fibers are short, and the postganglionic are long. The parasympathetic preganglionic fibers are long, and the postganglionic are short. Two categories of ganglia are associated with the sympathetic system: prevertebral and paravertebral. The former are located ventral to the thoracolumbar vertebrae, whereas the latter are two symmetrical

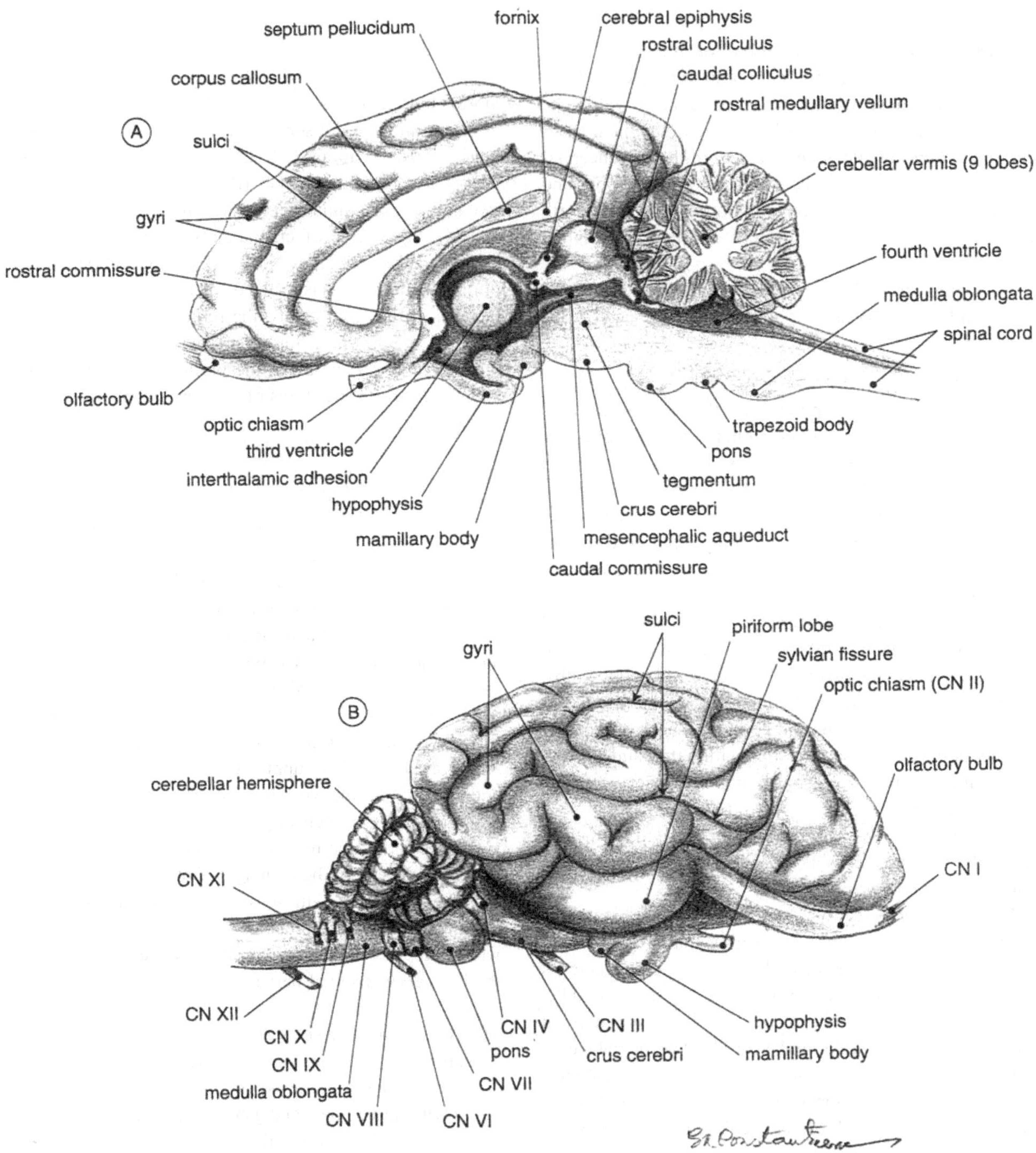


**Figure 6.37** The encephalon: A. Median aspect; B. Lateral aspect.

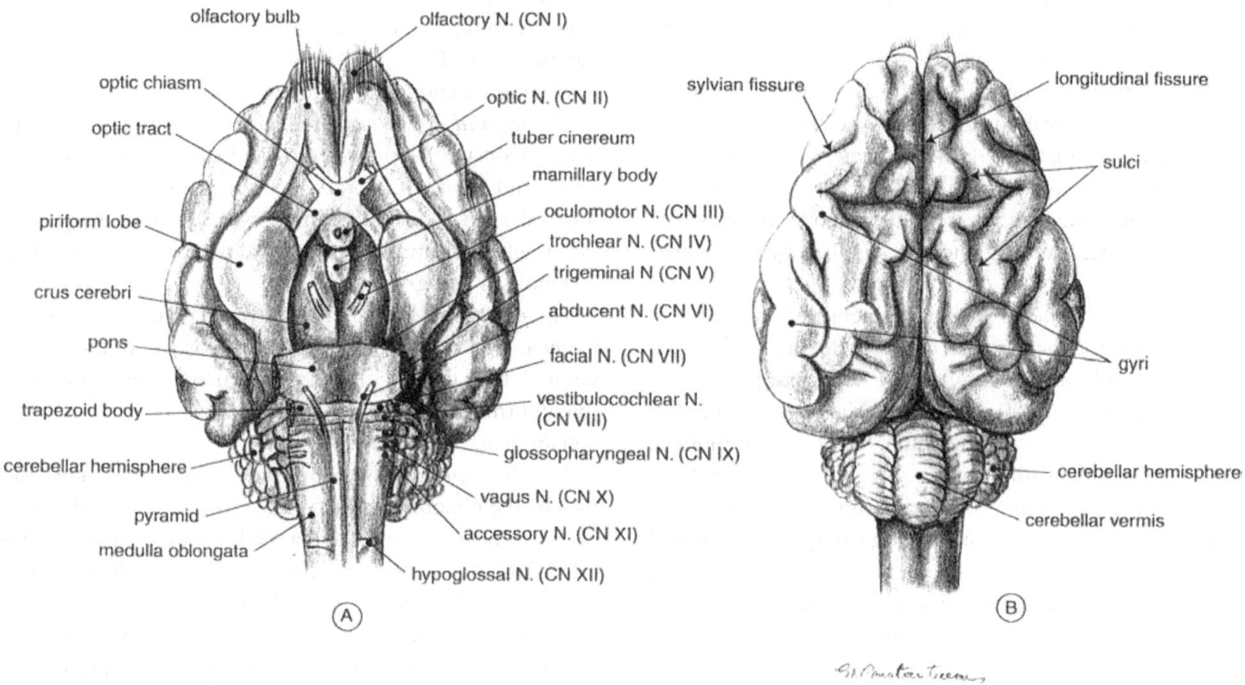


**Figure 6.38** The encephalon: A. Ventral aspect; B. Dorsal aspect.

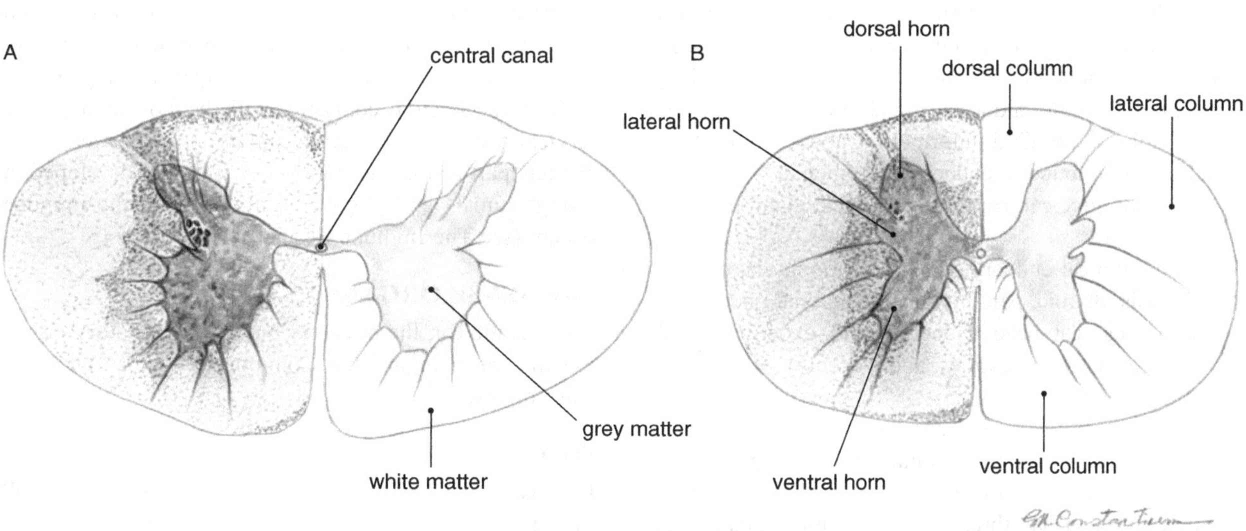


**Figure 6.39** Transverse section through the spinal cord A. Cervical region; B. Thoracic region.

ganglionic chains lateral to the thoracolumbar vertebrae. Two categories of ganglia are associated with the parasympathetic system: juxtavisceral and intramural. The juxtavisceral ganglia are those located in the vicinity of structures that are supplied, whereas the intramural are located within the walls of the organs supplied.

The autonomic nervous system is that part of the nervous system that regulates and controls the visceral activity and the activity of other structures that do not fall under the voluntary control. In general, the sympathetic and the parasympathetic systems are antagonistic.

The sympathetic nervous system originates from the thoracolumbar spinal cord, and is associated with sympathetic ganglia. Viscera of the thoracic, abdominal, and pelvic cavity, as well as the iris of the eye, the muscles of hair follicles, the arterial tree, all smooth muscles, the cardiac muscle, and the glands are supplied by the sympathetic system.

The parasympathetic nervous system is polarized into the brain and the sacral spinal cord. Nuclei in the brain and ganglia outside of the brain cavity supply organs of the head via the CN III, VII, IX, and X. The vagus nerve (CN X) travels up the descending colon, supplying all thoracic viscera (including the lungs and the heart) and all the rest of the gastrointestinal tract (up to the descending colon). The pelvic nerves originating from the sacral spinal cord supply the descending colon, viscera within the pelvic cavity, and some of the external genital organs. The nerves form the pelvic plexus, which contains pelvic ganglia.

## THE ENDOCRINE GLANDS

Ductless glands, the endocrine glands secrete hormones into the bloodstream. Due to the participation in maintaining the balance of the internal environment of the body, it is sometimes described with the nervous system in the so-called neuroendocrine system. The anatomy, physiology, and pathology of the endocrine glands are altogether called endocrinology. Detailed information about all of these aspects can be found in endocrinology books.

There are three categories of endocrine glands/organs:

- the five typical endocrine glands (hypophysis, cerebral epiphysis, thyroid, parathyroid, and adrenal glands)
- the organs that have endocrine and exocrine glands (pancreas, testicles, ovaries, and placenta)
- the thymus

The *hypophysis* (or the pituitary gland) is part of the diencephalon (see Figure 6.37), lies on the floor of the brain cavity, and has three lobes: glandular, neural, and intermediate. It hangs from the ventral aspect of the brain by a funnel-shaped stalk called infundibulum. The hypophysis is also called the endocrine brain, because it functions by governing the activity of the other endocrine glands and many more structures by specific hormones. Some of its hormones will be described in Chapter 7, Applied Reproductive Physiology.

The *cerebral epiphysis* (or the pineal gland) is located opposite to the hypophysis and is anchored above the third ventricle (see Figure 6.37). The epiphysis is known for producing melatonin, with an antigonadotrophic effect.

The *thyroid gland* is paired, consisting of two separate lobes inconstantly connected by an isthmus. The gland lies on the right and left lateral aspects of the first five to seven tracheal rings, and the isthmus, when present, surrounds the trachea ventrally. The thyroid hormones regulate metabolism and growth.

The *parathyroid glands*, usually two on each side of the trachea, are small structures located close to, or embedded within, the thyroid lobes. The parathyroid hormone (parathormone) regulates various aspects of calcium metabolism.

The *adrenal glands*, paired and small, lie against the roof of the abdominal cavity, and close to the cranial pole of the kidneys. A cortex and a medulla are evidently separated from one another. The cortex produces the corticoid hormones and some sex steroids. The medulla is developmentally associated with the sympathetic nervous system and produces the sympathetic transmitters epinephrine and norepinephrine.

The *thymus* is a paired lymphoid organ extending from the larynx to the caudal third of the neck. It continues as an asymmetrical structure and enters the thoracic cavity in the cranial mediastinum. The thymus has a lobulated appearance. A cortex and a medulla can be isolated from each other. The cervical part of the thymus is unpaired in the lower neck and is connected with the thoracic part. The cervical part regresses rapidly and disappears, whereas the thoracic part can last up to 5 years. It is well developed in young animals and plays a distinct role in the immune system (see The Immune System).

## THE SENSE ORGANS

The eye, the ear, the olfactory organ, the gustatory organ, and the touch organ (the Common Integument) are the sense organs.

### The Eye

The eye is the organ of vision, consisting of the eyeball and the adnexa. Paired organs, the eyes are located and protected within the bony orbits (Figure 6.40).

*The eyeball* (Figure 6.41) consists of three tunics and the lens. From the external to the internal tunics, they are the fibrous, the vascular, and the nervous tunics. Only a small area of the major part of the fibrous tunics, the sclera, can be seen from outside. It is an opaque structure, pierced on the posterior end by numerous holes allowing the optic

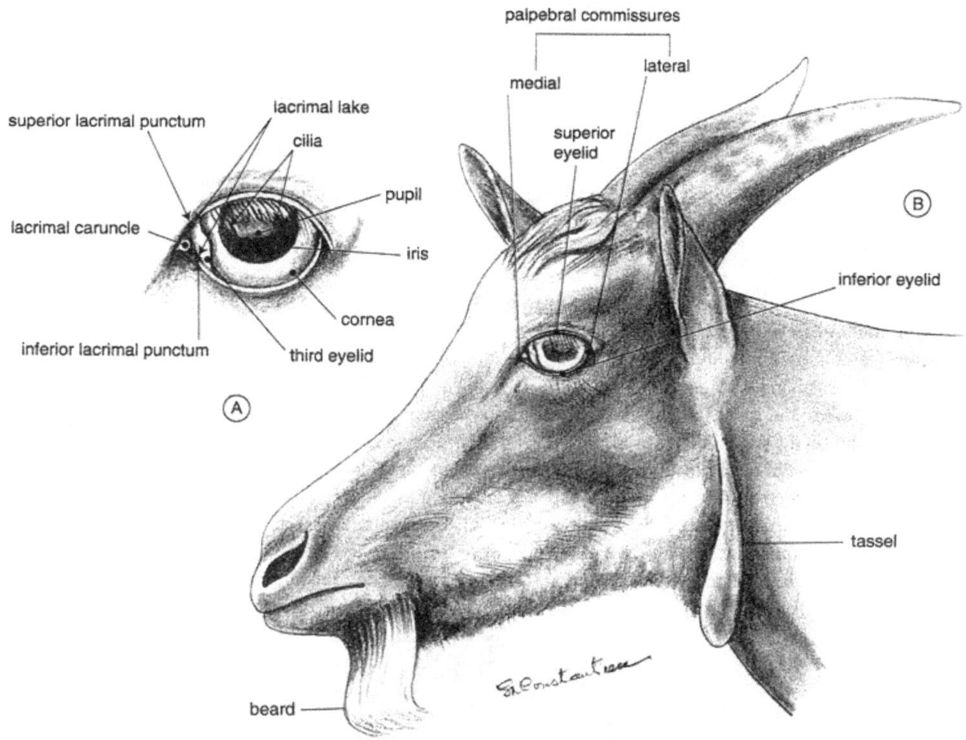


**Figure 6.40** The external structures of the eye: A. Frontal aspect of the eye and lacrimal apparatus; B. Eyelids and the eye.

nerve to exit the eyeball. The sclera continues anteriorly with the transparent cornea, the first refractive media of the eye for the light beams.

The vascular tunic (the uvea) consists of the choroid, the ciliary body, and the iris. The choroid lines intimately the sclera. At the sclero-corneal junction, the choroid continues with the ciliary body (ciliary muscle and processes). The ciliary muscle is responsible for the accommodation of the lens, whereas the ciliary processes produce the aqueous humor. The iris is an incomplete vertical wall between the cornea and the lens and is perforated by the pupil. Thus, the iris separates the space between the cornea and the lens into two chambers: the anterior and the posterior chambers, which communicate with each other through the pupil and are filled with the aqueous humor.

The nervous tunic is represented by the retina. It lines the choroid intimately and continues anteriorly with the nonpigmented epithelium of the ciliary body. The retina contains sensory cells (receptors) that send the visual excitation via the optic nerve to the brain to be processed into visual sensations. The beginning of the optic nerve is the optic disc, which lacks receptors. In 15–20% of goats only, minute blood vessels covered by pigmented epithelium of the optic disc can be seen, an important characteristic while looking at the fundus of the eye with the ophthalmoscope.

The lens is an elastic round and biconvex transparent structure, held in place by a ligament from the ciliary body. The space between the lens and the retina is called the vitreous chamber, filled with the vitreous body. The latter is a transparent refractive medium, with the aspect of a fresh egg albumen. Thus, the refractive media of the eye are the cornea, the aqueous humor, the lens, and the vitreous body.

The *adnexa* (the accessory organs) of the eye consist of fasciae, muscles, vessels and nerves, eyelids, and the lacrimal apparatus.

The fasciae are tough connective tissue structures enveloping the eyeball, muscles, vessels, and nerves, and sending septa between muscles. The muscles move the eye in all directions. There are straight and oblique muscles of the eyeball and the levator of the upper lid. The vessels are mainly provided by the external ophthalmic artery, whereas the nerves belong to the optic nerve (CN II), to the oculomotor, trochlear, and abducent motor nerves (CN III, IV,

1–anterior surface of lower eyelid
2–palpebral conjunctiva
3–palpebral fissure
4–bulbar conjuctiva
5–lacrimal sac
6–superior fornix
7–inferior fornix
8–orbicularis oculi M.
9–superior tarsus (continuation of the orbital septum)
10–tarsal gland
11–levator palpebrae superioris M.
12–lacrimal gland
13–cilia (eyelashes)
14–third eyelid
15–cartilage of the 3rd eyelid
16–gland of the 3rd eyelid
17–sclera
18–cribriform area of sclera
19–cornea
20–corneal limbus
21–choroid
22–tapetum lucidum
23–ciliary body
24–iris
25–pupillary margin
26–pupil
27–iridocorneal angle (with lig. pectinatum and iridocorneal spaces
28–retina
29–ora serrata
30–optic disc
31–optic N.
32–branches of central A. of the retina
33–anterior chamber of the eyeball
34–posterior chamber of the eyeball
35–vitreous chamber of the eyeball
36–lens
37–suspensory apparatus of the lens (ciliary zonule)

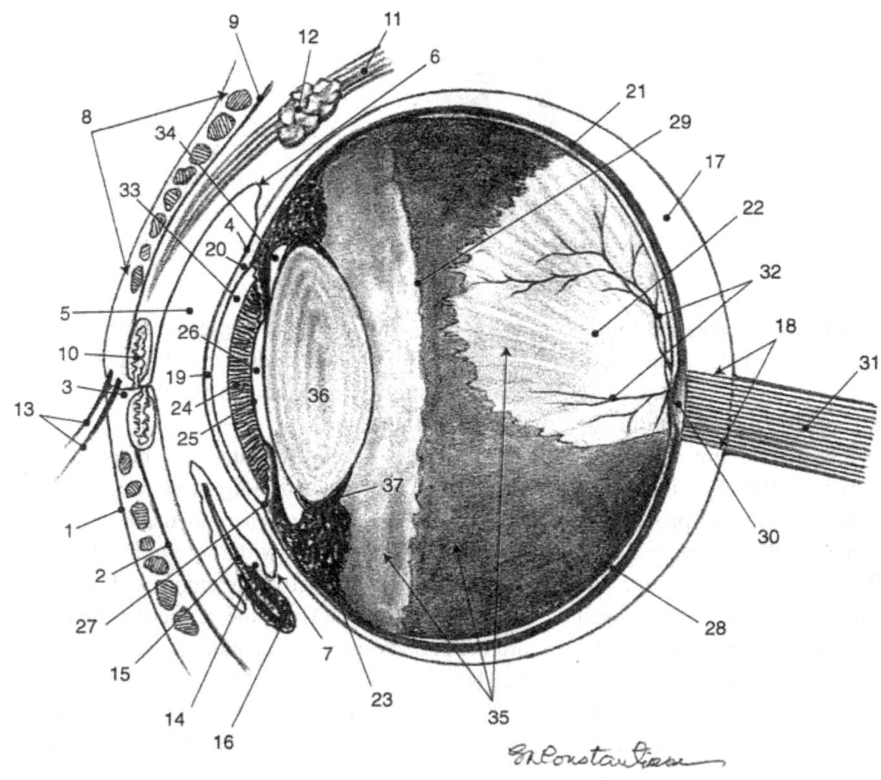


**Figure 6.41** Median section through the eye, magnification ×3.8.

and VI), and to the olfactory and maxillary sensory nerves of the trigeminal nerve (CN V). There are upper and lower eyelids, and a third eyelid, the latter being a fold of the conjunctiva. The conjunctiva lines the upper and lower eyelids and covers part of the eyeball, up to the cornea. The upper and lower eyelids are provided with a plate of dense connective tissue, glands, and cilia (eyelashes). The third eyelid has a T-shaped cartilaginous skeleton and an accessory lacrimal gland. The lacrimal apparatus consists of the lacrimal gland, the lacrimal caruncle, the lacrimal canals, the lacrimal sac, and the nasolacrimal duct. The lacrimal gland is located dorsolateral to the eyeball, the tears run toward the medial angle of the eye, where a little elevation, the lacrimal caruncle, is located. Through two lacrimal canaliculi, the tears are accumulated into the lacrimal sac; and from there through the nasolacrimal duct, they exit on the floor of the nasal vestibule.

### The Ear

The ear is the organ of hearing and equilibrium and consists of three parts: the external, the middle, and the internal ear.

The external ear is represented by several structures, but the only one that is externally visible is the auricle. The position of the auricle in the goat is vertical. The auricles are big (see Figure 6.40) and located under the horns. The internal surface of the auricle is concave, called scapha, and is provided with several irregularities.

The middle ear consists of the tympanic cavity, the tympanic membrane, the auditory ossicles, and the auditory tube. The tympanic cavity is sculpted within the temporal bone. The tympanic membrane (the drum) separates the external ear from the middle ear. It vibrates under the various intensities and qualities of sounds, which are transmitted to the internal ear via the auditory ossicles (malleus, incus, stapes, and the little lenticular bone). The stapes transmits the sound vibrations to the internal ear, whereas the malleus is in contact with the drum. The middle ear is connected with the nasopharynx by the pharyngotympanic (auditory, eustachian) tube.

The internal ear is located in the temporal bone, in the so-called osseous labyrinth. It converts the mechanical stimuli of sounds, and the gravity and positional changes of the body into nerve impulses. The impulses are transmitted to the brain by means of the CN VIII (the vestibulocochlear nerve). Within the osseous labyrinth, the membranous labyrinth is represented by ducts and cavities filled with a fluid called endolymph. The space between the osseous and the membranous labyrinth is filled by a shock-absorbing fluid called perilymph. The saccule, the utricle, the three semicircular ducts, and the cochlear duct complete the membranous labyrinth.

### The Olfactory Organ

The olfactory organ or the sense of smell is not very well developed in the goat, a reason that this species is among the large group of microsmatic animals (versus the dog, a macrosmatic animal with an abnormally keen olfactory sense). The olfactory organ has a chemical sense starting from chemoreceptors located within the olfactory mucosa of the dorso-caudal part of the nasal cavity. Neurosensory olfactory cells (the chemoreceptors), sustentacular cells, and basal cells are included in the olfactory epithelium. For detailed information, please consult histology books (for example, Eurell and Frappier, 2008).

### The Gustatory Organ

The gustatory organ consists of taste buds located on the surface of the tongue and pharynx. These buds send the information to the brain where it is processed in sensations of taste or gustation. The taste buds consist of taste cells, which are the receptors. It is mandatory that in order to be tasted, a substance must be dissoluble. The fungiform and vallate papillae are gustatory. The taste buds on the surface of fungiform papillae in the goat are abundant in comparison with the other ruminants. The taste buds of the vallate papillae are located in the epithelium surrounding the papillae. All taste buds have the role of perception. It is known that the goat is able to discriminate bitter, sweet, salty, and sour tastes (Smith and Sherman, 1994). Three cranial nerves send the information to the brain, where it is processed as the sense of taste. For details please consult histology books as previously advised.

### The Touch Organ (Common Integument)

The common integument including the skin and cutaneous organs is very complex. It consists of epidermis, dermis, hair, horns, pads, hooves, cutaneous glands, and mammary glands. Some of these were already described. The cutaneous sense will be briefly described along with particular hairs, cutaneous appendages, and cutaneous glands.

The sensory organs of the skin are receptors for temperature, pressure, tension and pain, and they allow the central nervous system to keep contact with the environment. The skin, therefore, is the organ of touch (Nickel et al., 1981), but primarily, it functions as a protective barrier against the environment.

The skin consists of three major layers: epidermis, dermis (corium), and hypodermis. The epidermis is the external layer of the skin, which becomes keratinized in the horn of the hoof (see The Digital Organ), and the external layer of the horn. The corium is thicker than the epidermis. Skin glands (sweat and sebaceous) are located in, and hairs originate from, the corium. The hypodermis is the connective tissue layer connecting the skin to the underlying structures.

The sensory organ of the skin (the cutaneous sense) consists of the specific sensory cells, which are the receptors, and the sensory nerves whose influx reaches the central nervous system. There the influx is transformed in sensations. There are highly specialized pressure, touch, pain, cold, and heat receptors.

The hair presents the bulb and the root implanted into the dermis and the shaft that extends from the surface of the skin. The invagination of the epidermis and connective tissue enclosing the root of the hair is called the hair follicle. Some hairs are associated with thin bundles of muscles, which raise the hair (arrectores pillorum muscles), and with sebaceous glands. The hair muscles are supplied by the sympathetic nervous system. The sebaceous glands located between the base of the tail and the anus produce a typical waxy secretion. There are different and various types of hairs, among them, the beard (see Figure 6.40), which is specific to the goat. Accumulation of hair follicles and sebaceous glands located in skinfolds caudomedial to the base of the horns are called horn (scent) glands. They are partially responsible for the characteristic odor of buck goats in the breeding season. Nevertheless, much smaller glands are found in female and in castrated male goats. Interestingly enough, the goat does not possess tactile hairs on the chin and cheek. The long fine wool of the Angora goat is called mohair, whereas that of the Cashmere goat is processed into cashmere wool (Frandson et al., 2003).

The tassel (wattle, or cervical appendix) is a symmetrical cutaneous appendage on the latero-ventral surfaces of the neck, close to the caudal border of the masseter muscles, and contains a bar of cartilage (see Figure 6.40). The presence of wattles is controlled by an autosomal-dominant gene (Pugh, 2002).

## THE IMMUNE SYSTEM

The immune system is an intricate complex of cellular, molecular, and genetic components, strongly interrelated for providing a defense against foreign organisms or substances and aberrant native cells (Nickel et al., 1981). The defense is the so-called "immune response." The foreign organisms or substances constitute the "antigen," and the result of the reaction of the immune system is the production of "antibodies." A very specific reaction occurs between the antigen and the antibodies, at the end of which an "active immunity" is established. Also the active immunity can be obtained by inoculation in a healthy organism with an attenuated or killed antigen (vaccine). In an individual already attacked by an antigen, specific antibodies obtained from a donor (called antiserum) can be inoculated and can act immediately against the antigen. This is the so-called "passive immunity." Comparable to passive immunization is the use of colostrum in the first several days of the neonates. The colostrum contains antibodies from the mother's immune system. In all mammals, the lymphatic system (which is part of the cardiovascular system) is deeply involved in immune reactions. The lymph nodes, the lymphatic structures in the mucous membranes, thymus, spleen, and bone marrow are the structures producing antibodies and immune reactions resisting the action of the antigens. They all contribute to the defense mechanism of the organism. Those lymphatic structures in the mucous membranes are located in the pharynx, larynx, intestine, prepuce, vagina, etc.

The immune system consists of diverse organs producing different types of antibodies and maintaining the integrity of the body. The lymphocytes, macrophages, and other "cells" possess immune properties. The thymus and the bone marrow produce T- and B-lymphocytes, respectively, which are later produced in the lymph nodes, spleen, and tonsils. The macrophages belong to the reticulo-endothelial system (in the reticular connective tissue) and have the property of phagocytosis (inclusion of the antigen into their cells and breaking it down) and storage (of the phagocytized material). For details please consult books describing the immunity and the immune system, such as "Veterinary Immunology" by Tizard (2000) and "Veterinary Immunology" by Outteridge (1985). Even in the current texts, the caprine immune system is not addressed directly.

## SUMMARY

This chapter, Functional Anatomy of the Goat, was written as basic knowledge. It covers the essential information to start with, for any direction the reader will take to pursue a professional education, to be enrolled in a graduate school, or to start a research program. The anatomical structures are described in a systemic and systematic order, using the international anatomical nomenclature. For details, different sources of information are suggested.

The 15 sections of this chapter cover the anatomy of the goat in different proportions, based on the role (functionality) of the anatomical structures. Thus, larger spaces are attributed to the horn and digital organ; to the cardiovascular, respiratory, digestive, and reproductive systems; to the sense organs; and to the mammary gland.

The chapter is illustrated with a minimum of original drawings. Additional figures may be seen in the cited literature (Constantinescu, 2001). The additional suggested readings are provided for the reader to look for more details.

## REFERENCES

Barone, R. 1978. Anatomie Comparée des Mammifères Domestiques, Laboratoire d'Anatomie École Nationale Vétérinaire, tome 3-ème, fascicule second, Lyon.

Eurell, J.A. and B.L. Frappier. 2008. Dellmann's Textbook of Veterinary Histology, 6th ed., Blackwell Publishing, Ames, Iowa.

Frandson, R.D., W.L. Wilke, and A.D. Tails. 2003. Anatomy and Physiology of Farm Animals, 6th ed., Lippincott Williams & Wilkins, Philadelphia, PA.

Nickel, R., A. Schummer, and E. Seiferle. 1981. The Anatomy of the Domestic Animals, vol. 3, Verlag Paul Parey, Berlin.

Outteridge, P.M. 1985. Veterinary Immunology, Academic Press. Orlando, FL.

Pugh, D.G. 2002. Sheep & Goat Medicine, Saunders, Philadelphia, PA.

Smith, M.C. and D.M. Sherman. 1994. Goat Medicine, Lea & Febiger, Philadelphia, PA.

Tizard, I.R. 2000. Veterinary Immunology. An Introduction, 6th ed., Saunders, Philadelphia, PA.

## ADDITIONAL SUGGESTED READINGS

Barone, R. 1976. Anatomie Comparée des Mammifères Domestiques, Laboratoire d'Anatomie École Nationale Vétérinaire, tome 3-éme fascicule premier, Lyon.

Constantinescu, G.M. 2001. Guide to Regional Ruminant Anatomy Based on the Dissection of the Goat, Iowa State University Press, Ames, Iowa.

Constantinescu, G.M. and R. Palicica. 2006. Comparative Anatomy (domestic mammals, birds, humans) and notions of Animal Physiology, Editura Orizonturi Universitare (Rom.), Timişoara.

Dyce, K.M., W.O. Sack, and C.J.G. Wensing. 2002. Textbook of Veterinary Anatomy, 3rd ed., Saunders, Philadelphia, PA.

Nickel, R., A. Schummer, and E. Seiferle. 1979. The Anatomy of the Domestic Animals, vol. 2, Verlag Paul Parey, Berlin.

Nomina Anatomica Veterinaria. 2005. 5th ed. electronic version.

Schaller, O., G.M. Constantinescu, R.E. Habel, W.O. Sack, P. Simoens, and N.R. de Vos. 2007. Illustrated Veterinary Anatomical Nomenclature, 2nd ed. Enke Verlag, Stuttgart.

Schatten, H. and G.M. Constantinescu. 2007. Comparative Reproductive Biology, Blackwell Publishing, Ames, Iowa.

Stedman's Medical Dictionary. 2000. 27th ed. Lippincot, Williams & Wilkins, Baltimore, MD.

# 7
# Applied Reproductive Physiology

*J. Greyling, PhD*

## KEY TERMS

Puberty—age at which the female starts cycling and male produces sperm.
Seasonality—trait where animals show sexual activity during certain seasons of the year regulated by photoperiod.
Breeding season—when does and bucks are sexually active and bred to produce offspring.
Estrous cycle—the sexual cycle of the doe, from one estrous period to the next, and the period is on average 21 days.
Ovulation—the shedding of an ovum by a mature ovarian follicle.
Postpartum anestrous period—interval following kidding during which the doe shows no signs of estrus.
Synchronization—hormonal induction of estrus to occur within a limited period of time.
Multiple ovulation and embryo transfer—the induction of superovulation, flushing evaluation, and transfer of embryos to recipients.
Accelerated kidding—when does are managed to breed three times in 2 years.
Dystocia—difficult births where assistance must be given.

## OBJECTIVES

By completing this chapter, the reader will acquire knowledge on:

- The reproductive potential of the goat
- The principles and reproductive physiology involved in applied reproduction of goats
- The different physiological stages in the doe and factors that play a role
- The seasonal sexual activity in the doe and buck
- The insight into the reproduction physiology of the male
- The principles and the application of certain reproductive technologies in the doe
- The pregnancy diagnosis in the female
- The most important aspects to be considered in a goat reproduction program

## INTRODUCTION

Roughly 95% of the 850 million goats in the world are found in the developing countries of Asia, Africa, and South America. These goats are multipurpose animals that provide a vast range of products, including meat, milk, skins (leather), and fiber; and for too long, the role of goats and their contribution to the livelihoods of people have been ignored (Peacock et al., 2005). In recent years, the demand for goat products has also increased in the developed countries, accompanied by the number of commercial goat farmers. As reproduction can be seen as the cornerstone in any animal production chain, its importance cannot be overemphasized. This is also relevant to the goat. Thus, to help exploit the restricted goat resources and help alleviate the demand for animal protein via human consumption, it is essential to increase the reproduction efficiency (conception, fecundity, and kidding rate) also in goats (Greyling, 1988).

The level of reproductive performance depends on the interaction between genetic and environmental factors, but it is particularly susceptible to the influence of the latter. These factors include nutrition, ambient temperature, stocking density, age groups, etc. Although goat breeds have an excellent ability to accommodate and adapt to fluctuations in environment, this often involves some degree of reproductive failure (Riera, 1982; Zarazaga et al., 2005).

Reproductive efficiency in does is determined by many different processes. These processes include the length of the breeding season, cyclic activity, ovulation rate, fertilization rate, the postpartum anestrous period, growth, and viability of the offspring. The reproductive efficiency as such is then measured and expressed in terms of kidding rate, kidding interval, weight of kids born or weaned, or length of the reproductive life of the dam (Greyling, 1988). With the goat's general hardiness, high fecundity, good mothering ability, and extended breeding season, this specie is an ideal candidate for applied or manipulated reproduction physiology to, among others, increase the reproduction efficiency (Devendra and Burns, 1983).

However, to manipulate the reproductive efficiency of the female goat using accelerated breeding techniques such as artificial insemination (AI), synchronization, and superovulation or multiple ovulation and embryo transfer (MOET), a thorough knowledge of the hormonal status and physiology of this animal is essential. The reproductive physiology must be seen in conjunction with the functional anatomy for reproduction addressed in Chapter 6.

## PUBERTY

### Female

One of the first factors important in determining lifetime production performance in the goat is puberty. Puberty occurs as a consequence of the activation of the gonadotrophin surge mechanism by the positive (stimulatory) feedback action of oestradiol on the hypothalamus. In the female, puberty is defined as the stage when the doe starts and exhibits regular cyclic estrous activity. So for example, some of the information on puberty relates to body weight and age of the doe at first standing estrus. It must be kept in mind that many of the endocrine mechanisms leading to ovulation and first estrus are capable of operating long before they are called upon to function (Greyling, 2000; Zarazaga et al., 2005).

Generally, breeding in does should be delayed until the animal has attained 60–70% of its mature body weight. Angora goats should weigh a minimum of 27 kg, while the larger dairy goat breeds should weigh between 35 and 45 kg before breeding. These weights must however not be confused with the weight at puberty. There is good evidence of inadequate nutrition (besides retarding growth rate) having adverse effects on pituitary function and hence gonadotrophin (sex hormone) secretion in the immature animal. Once a critical body weight is reached, differences in live weight have little influence on the time of onset at puberty. In most goat breeds, the does are pubertal between 5 and 7 months of age (Smith, 1980; Jainudeen et al., 2000).

The potential for improved reproductive efficiency increases as the age at first kidding decreases, but management practices are often introduced to delay mating for full development of the female, to increase the conception rate, the frequency of multiple births, and to assure the survivability of the offspring. It is important to realize that climate, nutrition, and the presence of the male are factors that could modify the age at puberty in the doe. There is also ample evidence to indicate that seasonality is an important factor in the attainment of puberty in goat kids (male and female). The onset of cyclic activity is usually stimulated during the season of shortening days and its termination occurs during the lengthening (photoperiod) days. The presence of the buck, besides modifying the age of puberty in the goat, also influences estrous behavior, with some degree of synchronization of estrus. There may be two distinct physiological actions involved in the presence of males, namely a neurohormonal action, which advances the preovulatory discharge of LH, and a solely neural action, which gives rise to the "male effect" (Riera, 1982; Chemineau et al., 1992; Gordon, 1997; Bukar et al., 2006).

The mean body weight at first estrus or puberty for Boer goat female kids born in January (summer) and August (late winter) is 31.3 kg and 27.4 kg, respectively, while that in the Saanen is 30.0 kg for animals born in March/April (spring). The mean age at puberty in Boer goat does born during late winter and midsummer was recorded as 191.1 and 157.2 days, respectively. Kids weaned in autumn (during the natural breeding season) generally exhibited estrus significantly earlier than animals weaned outside the natural breeding season. It would thus seem as if season is one of the main cues for initiating the onset of puberty. The age at the onset of puberty in the Boer goat compares well and in fact is sooner than in most other goat breeds, for example the Saanen (217.9 days), Angora (240 days), Black Bengal (196.5 days), and Barbari Nannies (213 days). From the estrous response (percentage does exhibiting estrus) obtained, it is evident that the permanent presence of the male has a marked beneficial effect on animals exhibiting estrus or puberty in- and outside the natural breeding season. Research regarding puberty in the buck however is generally very sparse (Amoah and Bryant, 1984; Greyling and Van Niekerk, 1990b; Papachristoforou et al., 2000).

### Male

In the buck, the onset of puberty is associated with a significant increase in testosterone production and subsequent spermatogenesis. This phenomenon occurs at 4–6 months of age. Factors that affect the onset of puberty include photoperiod and nutrition. Sexual maturity (when bucks can be used for breeding) is generally at approximately 18 months of age (60–80 kg, depending on the breed).

Both internal and external cues affect time of puberty, and it is difficult to isolate a single factor involved in the attainment of puberty. Generally a sequence of events is involved in the process of sexual maturation and an interaction possibly exists between the male effect, seasonality, and nutrition (Greyling, 1996; Nishimura et al., 2000; Todini et al., 2007).

## SEASONALITY AND THE BREEDING SEASON

Regarding the breeding season, different climatic factors (such as temperature and photoperiod) regulate the physiological response. The length of the breeding season as such is primarily the result of genetic and environmental interactions. The breeding season commences as the daylight length becomes shorter. There is evidence to indicate that in some species the pineal gland, through its secretion of melatonin, is involved in mediating the effects of photoperiod on gonadal function. Melatonin levels are high during the dark periods and low during light periods. There is also evidence to suggest that the premaxillary area of the hypothalamus is an important target for melatonin to regulate the reproductive activity (Hunter, 1980; Zarazaga et al., 2005).

Breeds differ in their length of the breeding season. For example, the Angora doe has a short breeding season, varying between 94 days (or 4–7 estrous cycles) and $117 \pm 3.5$ days. Proper management of the introduction of the buck can be used to slightly hasten the onset of the breeding season and also synchronize estrus among the does involved. It would appear as if the Boer goat is seasonally polyestrous, though periods of complete anestrus are not observed. The pattern of the seasonal occurrence of the animal sexual activity for the adult Boer goat doe and the female Boer goat kid (aged 5–17 months) is very similar. The peak of sexual activity occurs during autumn or the period of short daylight length, while the period of lowest sexual activity occurs from late spring to midsummer (see Figure 7.1). The mean number of estrous periods per annum per doe recorded for the Boer goat is quoted as being $11.7 \pm 4.4$ (approximately 280 days). One buck per 35–40 does is recommended when flock mating is implemented, and it is important to endeavor to mate the young maiden does separately from the mature does (Devendra and Burns, 1983; Restall, 1992; Ahmed et al., 1997; Rivera et al., 2003).

Generally the male does not exhibit such a pronounced mating season, although there is a tendency for decreasing photoperiod (daylight length) to stimulate follicle stimulating hormone (FSH), luteinizing hormone (LH) secretion, and testosterone production. Certain buck breeds tend to be more sensitive to daylight length and thus be more seasonal, for example, the Angora (Greyling and Grobbelaar, 1983; Barkawi et al., 2006; Todini et al., 2007).

## PHYSIOLOGY OF THE FEMALE REPRODUCTION

### The Estrous Cycle

The normal cyclic doe undergoes a repeated secretion of steroid sex hormones that influences both the reproductive tract and the animals' sexual behavior. The period of estrous behavior, when the female shows characteristic sexual behavior in the presence of the male, is the only period when physical mating is allowed. Ovulation or the shedding of the ovum occurs at the end or shortly prior to the end of the period of estrus and is usually taken as a reference point for the description of the estrous cycle. In any event, it is the period of estrus itself that is of greatest concern to the farmer (especially when implementing artificial insemination [AI]). Recognition of this receptive period relies on overt signs in the presence of other animals (Hunter, 1980; Greyling and Van Niekerk, 1987).

Currently little is known of the sequence of endocrine events during follicular and luteal development in the goat during the estrous cycle. The mean plasma progesterone concentration on the day of estrus is extremely low (0.2 nanograms per milliliter [ng/ml]), after which the concentration increases to a maximum of approximately 4 ng/ml at about day 10 of the 21-day estrous cycle, and decreases

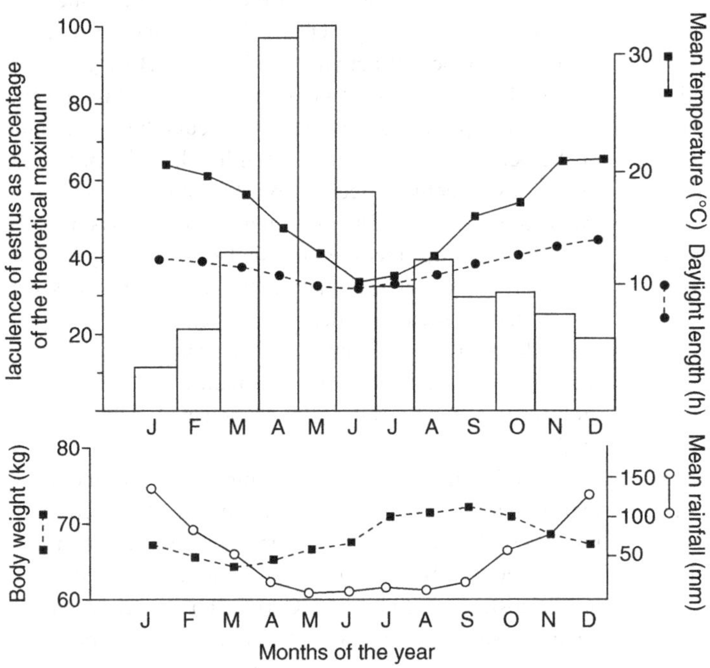


**Figure 7.1** The mean annual sexual activity and body weight of mature Boer goat does (Greyling, 1988).

rapidly during the last 3 days of the cycle. A consistency in the interval between peak LH release (preovulatory surge) and ovulation has been recorded to be approximately 24 hours confirming the importance of the role of LH in the ovulatory mechanism.

The duration of the estrous cycle in goats has been reviewed. It is evident that extreme variation exists, from cycles as short as 3 days to cycles as long as 62 days. The majority of estrous cycles, however, are 19–21 days in length. Data indicate that short cycles observed early in the goat-breeding season are predominantly anovulatory and associated with preovulatory LH peaks of markedly reduced magnitude. The duration of the estrous cycle in the mature doe has been set at $20.7 \pm 0.7$ days, with a high incidence of short (13 days) and long (25 days) estrous cycles. A certain proportion of the short cycles can be related to the season of the year and time interval postpartum. Oestrous cycles in goats have been categorized into short, medium, and long cycles, at a frequency of 19.7%, 68.8%, and 11.5% and a mean duration of 6.4, 19.8, and 37.5 days for each category, respectively. The phenomenon of short cycles observed is of interest, and these short cycles are associated with the formation of short-lived corpora lutea.

The duration of the estrous period also appears variable (24–48 hours) in length. Great care should be exercised when evaluating data reporting on the length of the estrous period because of the techniques used to detect estrus and the frequency at which it is measured. Literature is abundant on information regarding the duration of estrus in different goat breeds, different countries, seasons of the year, months, etc. For example, estrous cycles and periods were found to be significantly shorter during periods of the year with extreme cold-dry and hot-wet periods. The duration of the estrous period of the Angora doe was found to be shorter at the onset and end of the breeding season, compared to the peak months of sexual activity (autumn, May to July). The most common duration of estrous periods in goats is reported to be 36 hours, with a variation ranging from 22–60 hours. The natural duration of estrus in the Angora doe is quoted as 33.4 hours, compared to that of the Boer goat doe of 37.4 hours (Riera, 1982; Greyling, 1988; Jainudeen et al., 2000).

The "male effect" is prominent in does that have been isolated from males, for example with the introduction of bucks during the transitional period (time from the anestrous to the breeding season). The ovarian response of anovular does is stimulated by an androgen-dependent pheromone secretion of the buck. It is generally accepted that the "male effect" is responsible for the acceleration of the occurrence of estrus and could also have an influence in shortening the duration of the estrous period in the doe

**Table 7.1** Ovarian activity and ovulation rate 26, 32, and 38 hours following the onset of estrus in does during the breeding season.

| | Time following the onset of estrus | | |
|---|---|---|---|
| | 26 h | 32 h | 38 h |
| No. of does | 15 | 10 | 15 |
| Mean (±SD) body weight of does (kg) | 49.5 ± 12.8 | 47.2 ± 14.6 | 49.2 ± 14.3 |
| Mean (±SD) weight (total) of ovaries (g) | 4.17 ± 2.0 | 3.82 ± 1.49 | 3.9 ± 1.47 |
| Mean number of follicles/doe > 0 (±SD) | 8.15 ± 3.5 | 9.5 ± 4.1 | 13.1 ± 5.0 |
| Mean (±SD) size of follicles/group (mm) | 4.2 ± 2.3 | 4.2 ± 2.0 | 3.7 ± 1.5 |
| Mean (±SD) number of ovulations/doe ovulated | 1.0 ± 0.0 | 1.0* | 1.87 ± 0.92 |
| % does ovulated/does in estrus | 13.3 | 10.0 | 86.7 |

*one observation only
Source: Greyling, 1988.

(Chemineau, 1983; Kassem et al., 1989; Chemineau et al., 1992).

### Time of Ovulation and Ovulation Rate

The doe is a spontaneous ovulator, ovulating near the end of estrus. Ovulation rate as such refers to the number of eggs or ova liberated from the ovary at a given estrous period, with the time of ovulation being very important especially in AI programs and even with natural mating. Ovulation in does is generally reported as occurring a few hours after the termination of standing estrus or toward the end of estrus (24–36 hours after the onset of estrus) so that the processes of mating and ovulation are synchronized (Table 7.1). For successful fertilization, the timing of these two processes is critical. It has been shown that the maximum number of follicles ovulating is at approximately 48 hours after the onset of estrus in Black Bengal nanny goats. The mean ovulation rate recorded in Boer goat does was 1.72 ± 0.9 ovulations per doe (ranging between 1 and 4 ovulations per doe). This is higher than the ovulation rate of 1.43 quoted for Barbari nannies, the 1.2 for Angora does, and the 1.28 in Norwegian goats, while being substantially lower than the 4.0 quoted for Black Bengal nanny goats. Ovulation rate is normally reflected by the kidding rate (prolificacy) and as can be expected, there is considerable variation in the kidding rates among the different goat breeds. Among the environmental factors influencing ovulation rate, season and the level of nutrition are the most important (Table 7.2). Generally ovulation rates are higher early in the natural breeding season than later, but factors such as body size, body weight and condition, and genotype may also contribute to an increase in ovulation rate (Greyling, 1988; Gordon, 1997; Jainudeen et al., 2000).

**Table 7.2** Mean (±SE) reproductive traits in Damascus does under semiarid conditions.

| Parameter | |
|---|---|
| Estrous cycle length (d) | 19.5 ± 0.58 |
| Gestation length (d) | 149.4 ± 0.35 |
| Litter size | 1.74 ± 0.1 |
| Birth weight (kg) | 2.16 ± 0.1 |
| First service conception (%) | 64.2 |
| Fertility (%) | 70.0 |
| Kid mortalities to weaning (%) | 23.4 |

Source: Shalaby et al., 2000.

### The Gestation Period

The gestation period in farm animals is usually defined as the period of time from conception to parturition. The gestation period in goats is normally 149 days, varying between 144 and 150.8 days. There are a variety of factors, which could influence the gestation period. These include season of the year, parity, age of the dam, sex, and number of offspring at birth. The mean gestation length in the Boer goat is set at 148.2 ± 3.7 days, consistent with other reports on goats. Heredity plays an important role in determining gestation length, and the genotype of the fetus accounts for almost two-thirds of the variation in gestation length. For example, male kids are carried longer than female kids, spring-born kids longer than fall-born kids, and singles longer than twins. The effect of nutrition on fetal development during certain months of pregnancy does tend to shorten or lengthen the gestation period, but the variation due to this factor is only 1.5 days. If periods as short as this are considered to represent real differences in gestation length, then the discrepancy that exists in determining

**Table 7.3** The effect of season on semen production and motility for the different collection techniques in Boer and Angora goat bucks.

| | | Artificial vagina Boer goat | | Electrical stimulation | | | |
|---|---|---|---|---|---|---|---|
| | | | | Boer goat | | Angora goat | |
| Month | Mean monthly Temp. °C | Semen volume (ml) | Motility (1–5) | Semen volume (ml) | Motility (1–5) | Semen volume (ml) | Motility (1–5) |
| Jan | 20.8 | 1.54 ± 0.78 | 2.77 ± 0.93 | 3.56 ± 2.32 | 2.29 ± 0.99 | 1.86 ± 0.64 | 3.19 ± 1.19 |
| Feb | 19.0 | 1.52 ±± 0.73 | 3.73 ± 0.52 | 2.89 ± 2.28 | 2.13 ± 1.36 | 1.68 ± 1.37 | 3.63 ± 0.88 |
| Mar | 17.4 | 1.31 ± 0.46 | 3.46 ± 0.78 | 1.78 ± 0.30 | 2.80 ± 1.01 | 1.33 ± 0.43 | 3.20 ± 0.71 |
| Apr | 16.6 | 1.73 ± 0.75 | 3.42 ± 0.51 | 1.89 ± 0.97 | 2.36 ± 0.75 | 1.54 ± 0.78 | 3.13 ± 0.95 |
| May | 12.4 | 1.50 ± 0.68 | 3.65 ± 0.34 | 1.62 ± 1.05 | 2.17 ± 0.82 | 1.83 ± 0.85 | 3.38 ± 0.48 |
| Jun | 9.4 | 1.53 ± 0.54 | 2.83 ± 0.88 | 1.46 ± 0.59 | 1.65 ± 0.71 | 1.46 ± 0.46 | 3.00 ± 0.78 |
| Jul | 10.4 | 2.03 ± 1.21 | 1.96 ± 0.72 | 2.78 ± 1.60 | 1.75 ± 0.85 | 1.87 ± 1.00 | 2.94 ± 0.78 |
| Aug | 18.7 | 1.60 ± 0.30 | 2.69 ± 0.46 | 1.50 ± 0.67 | 1.75 ± 1.08 | 1.75 ± 0.79 | 2.42 ± 1.66 |
| Sept | 15.3 | 1.20 ± 0.30 | 2.75 ± 1.06 | 2.40 ±1.02 | 2.31 ± 0.37 | 1.40 ± 0.82 | 2.06 ± 1.24 |
| Oct | 19.3 | 1.56 ± 0.84 | 2.04 ± 0.99 | 3.05 ± 1.17 | 1.50 ± 0.89 | 1.25 ± 0.54 | 3.69 ± 0.46 |
| Nov | 18.3 | 1.52 ± 0.57 | 3.38 ± 0.83 | 3.05 ± 1.35 | 2.00 ± 0.93 | 1.88 ± 1.06 | 3.13 ± 1.30 |
| Dec | 19.9 | 1.73 ± 0.51 | 3.14 ± 0.82 | 3.82 ± 2.01 | 2.05 ± 0.98 | 1.69 ± 0.54 | 3.45 ± 0.86 |
| Mean | 16.46 | 1.56 | 2.99 | 2.48 | 2.06 | 1.63 | 3.10 |

Source: Greyling and Grobbelaar, 1983.

the onset of gestation becomes important (Table 7.3). Nutritional management is important during gestation for the maintenance of the nutritional status of the dam (with body condition score being an indicator) and also for meeting the needs of the developing fetus(es). It is known that approximately 70% of fetal growth occurs during the last trimester of pregnancy (Greyling, 1988; Jainudeen et al., 2000).

In the goat, it would seem as if there is a slight time lapse prior to stimulation of the corpus luteum to maintain pregnancy. A response of major biological significance to the presence of a developing embryo in the uterus is the prolongation of the lifespan of the corpus luteum with a continuing secretion of progesterone. This is referred to (and somewhat misleading) as the conversion of the corpus luteum of the estrous cycle into one of pregnancy, but no dramatic form of conversion occurs. The physiological events are more concerned with the prevention of regression of the corpus luteum, although the synthesis of progesterone by the luteal structures may increase after this critical stage is passed because of the luteotrophic influence of the embryo and its membranes and also trophic hormone secretion by the anterior pituitary gland. It is a known fact that in the goat, the corpus luteum is essential for the maintenance of pregnancy throughout gestation, contrary to the ewe where the corpus luteum is essential only during the first 3 months of pregnancy. It does seem as if the caprine placental production of progesterone is small, and it would be unlikely to influence the level of this hormone in the maternal circulation. In does bearing twins, a substantial rise in the serum progesterone is noted during the last weeks of pregnancy, which could suggest the caprine placenta produces small but detectable amounts of progesterone. Insufficient evidence of any dramatic increase in serum progesterone concentration above the levels attained during the luteal phase of the estrous cycle suggests the corpus luteum as being the main source of progesterone (Hunter, 1980). A positive relationship between multiple pregnancies and maternal serum progesterone levels has not been observed. Serum LH data demonstrate that luteinizing hormone (LH) levels in the Boer goat during pregnancy are not static but fluctuate apparently at random, relatively low levels. This is consistent with evidence that the physiological levels of progesterone (which is high during gestation) decrease the frequency of the LH peaks. The mean serum LH concentration was found to be 1.3 ± 0.6 ng/ml during the luteal phase of the estrous cycle, compared to 0.4 ± 1.2 ng/ml throughout pregnancy in the goat. Among Boer goats under extensive production systems in South Africa, a kidding percentage of 180% is common, with triplets being common. The percentage of singletons, twins, triplets, and quadruplets in goats are quoted as being 24.5, 59.2, 15.3, and 1%, respectively (Gall, 1981; Greyling, 1988; Hafez and Hafez, 2000).

### Postpartum Anestrous Period

An important aspect in the reproduction chain is the postpartum anestrus period. This interval between parturition and the first postpartum estrus determines how soon after parturition conception can possibly occur. The mean post partum anestrous period is quoted as being 56.0 ± 5.1 days and 45.2 ± 5.8 days in primiparous and biparous Barbari does, compared to duration of 55.5 ± 24.9 days in the Boer goat. This interval varies between breeds, nutritional regimes, and lactational status of the female. Reducing the suckling intensity may result in an earlier return to estrus and a greater percentage of does rebreeding within 60 days postpartum. A significant lower conception rate (19.1%) at an early postpartum estrus (15–30 days) was observed in goats, compared to does exhibiting estrus later (66.7–87.5% conception). No significant differences in the postpartum anestrous period between does bearing singletons, twin, triplets, or quadruplets have been recorded, although suckling as such exerts an inhibitory effect upon the synthesis and/or release of LH during the postpartum period. Season of kidding could also affect the postpartum interval to first estrus. The postpartum anestrous periods for does kidding in autumn (breeding season) is quoted as being significantly shorter (37.3 ± 12.5 versus 59.9 ± 18.0 days, respectively), when compared to does kidding in spring (nonbreeding season). The stimulatory effect of the presence of the male (pheromones) in the postpartum doe is a factor not to be ignored when determining or manipulating the interval from parturition to first estrus (Riera, 1982; Chemineau, 1983; Torres-Acosta et al., 1996; Hafez and Hafez, 2000).

## PHYSIOLOGY OF THE MALE REPRODUCTION

The neuro-endocrine control of testicular function is similar to that observed in the female. Briefly it can be said that FSH (spermatogenesis stimulating hormone [SSH]) and LH (interstitial cell stimulating hormone [ICSH]) in the male are the two gonadotrophic hormones mainly responsible for spermatogenesis in the seminiferous tubules and testosterone production in the cells of Leydig, respectively. Generally the photoperiodic control of sexual activity in the buck is less pronounced than in the doe. Some researchers allege that semen production (and not so much semen quality) is affected by seasonal changes. This, however, is not always the case. It was found that semen motility tended to be lower from June to October (onset winter to midspring) for the Boer goat and Angora goat in South Africa. Spermatogenesis or the formation of sperm is generally a lengthy process, taking approximately 50 to 60 days (from spermatogonia to sperm formation) and involving different stages (spermatocytogenesis and spermiogenesis) in the seminiferous tubules. When spermatogenesis is completed, the sperm move to the epididymis where they mature for 2–3 weeks. Thus, for mature, viable sperm to be produced, it takes from 60–80 days. In the event of temporary sterility (for example, due to fever following blue tongue vaccination), the time required for complete spermatogenesis must be kept in mind if optimal fertilization is to be obtained (Greyling and Grobbelaar, 1983, 1988; Jainudeen et al., 2000; Salamon and Maxwell, 2000).

The technique of semen collection is also an aspect to be noted when determining semen quality (Table 7.3 and Table 7.4). It is accepted that sperm motility and the percentage of live sperm determine the quality of the semen. The density or concentration of an ejaculate and the percentage of live sperm were found to be higher in semen collected by means of the artificial vagina. However, the final and most important criterion in the evaluation of semen quality lies in the fertilizing capacity of the sperm (Greyling and Grobbelaar, 1983; Karagiannidis et al., 1999; Barkawi et al., 2006).

## CARE AND MANAGEMENT OF THE BUCK

The importance of the buck in the whole breeding program is not to be underestimated. Although only actively used for approximately 6 weeks in a flock mating system or intensively for a week or two when used in an artificial insemination program, the buck has to be well managed.

From weaning at 3 months of age, following the necessary dosing for tapeworm and vaccinations as recommended by the veterinarian, males should be maintained on an above-maintenance energy diet. The average body weight of Boer goat bucks at 100 days of age (weaning) is 27 kg and the body weight at 12, 18, 24, and 36 months of age is 55 kg, 75 kg, 88 kg, and 115 kg, respectively. Acceptable body conformation, size, and appearance are prerequisites when used for breeding. Because goats are predominantly browsers (70%), they can efficiently use bushes and shrubs and in time of shortages or droughts be nutritionally supplemented to maintain an acceptable body condition (even if not used for breeding).

**Table 7.4** Least squares means (±SE) of physical semen characteristics in Zaraibi bucks as affected by season of the year.

| Traits | Season | | | | |
|---|---|---|---|---|---|
| Number of animals | Autumn 10 | Winter 9 | Spring 8 | Summer 8 | Mean |
| SC (cm)[1] | 25.4 ±± 0.22$^{c}$ | 25 ± 0.25$^{c}$ | 26.2 ± 0.34$^{b}$ | 27.3 ± 0.27$^{a}$ | 25.9 ± 014 |
| Volume (ml) | 0.98 ±± 0.03$^{a}$ | 0.56 ± 0.03$^{c}$ | 0.35 ± 0.05$^{d}$ | 0.91 ± 0.03$^{b}$ | 0.75 ± 0.02 |
| Concentration ($\times10^6$/ml) | 4,700 ±± 10.1$^{b}$ | 5,288 ± 12.7$^{a}$ | 4,903 ± 18.4$^{a}$ | 4,783 ± 9.9$^{b}$ | 5,072 ± 5.48 |
| Total output ($\times10^6$) | 4,565 ± 15.6$^{a}$ | 3,013 ± 19.1$^{b}$ | 1,791 ± 28.6$^{c}$ | 4,429 ± 15.4$^{a}$ | 3,782 ± 9.17 |
| pH | 6.69 ± 0.03$^{b}$ | 6.68 ± 0.03$^{b}$ | 6.79 ± 0.04$^{b}$ | 6.79 ± 0.03$^{a}$ | 6.72 ± 0.01 |
| Gross motility | 4.3 ± 0.08$^{a}$ | 4.3 ± 0.09$^{a}$ | 3.7 ± 0.13$^{b}$ | 4.3 ± 0.08 | 4.2 ± 0.04 |
| Progressive motility (%) | 79.5 ± 1.37$^{a}$ | 80.3 ± 1.68$^{a}$ | 73.1 ± 2.5$^{b}$ | 81.2 ± 1.35$^{a}$ | 78.9 ± 0.69 |
| Live (%) | 82.5 ± 0.83$^{b}$ | 77.2 ± 1.02$^{c}$ | 76.4 ± 1.5$^{c}$ | 87.2 ± 0.82$^{a}$ | 81.8 ± 0.45 |
| Abnormality (%) | 8.8 ± 0.46$^{c}$ | 18.3 ± 0.56$^{a}$ | 17.9 ± .84$^{a}$ | 11.4 ± 0.46$^{b}$ | 13.5 ± 0.29 |
| Semen index ($\times10^6$)[2] | 3,152 ± 13.2$^{a}$ | 1,982 ± 16.2$^{b}$ | 1,139 ± 24.1$^{c}$ | 3,354 ± 13.0$^{a}$ | 2,630 ± 7.63 |

$^{a,b,c,d}$Within columns, different superscript letters (a–d) are different ($P < 0.01$).

[1]SC, scrotal circumference

[2]Semen index = semen volume × concentration × live (%) × progressive motility (%)

Source: Barkawi et al., 2006.

**Table 7.5** Correlation coefficients of semen characteristics with testicular measurements and live weight in Rayini goats.

| | Scrotal circumference | Testis width | Testis length | Live weight |
|---|---|---|---|---|
| Semen volume | 0.85*** | 0.79** | 0.51 | 0.73** |
| Sperm concentration | 0.32 | 0.36 | 0.54* | 0.35 |
| Percent live sperm | 0.13 | 0.10 | −0.04 | 0.01 |
| Percent abnormal sperm | −0.16 | −0.16 | −0.15 | 0.01 |
| Total sperm number | 0.86*** | 0.83*** | 0.63* | 0.76** |
| Number of live sperm | 0.71** | 0.66** | 0.45 | 0.61* |
| Number of live and normal sperm | 0.75*** | 0.72** | 0.52 | 0.63* |

*$P < 0.05$

**$P < 0.01$

***$P < 0.001$

Source: Zamiri et al., 2006.

In general, bucks should be maintained in a good body condition according to body condition scores (BCS) throughout the year (3 out of a score of 5). From 2–3 weeks prior to the mating season, bucks will benefit from the supplementary feeding of concentrates (energy) also known as flush feeding (300 g/goat/day). All required annual vaccinations should be administered at least 6 weeks before the intended breeding season and the hooves trimmed to facilitate easy walking and mating ability. During the rainy season, foot dipping to prevent foot rot is recommended every 3–4 weeks. Dosing for internal parasites should also be implemented 1–2 weeks prior to mating. Undoubtedly the most important routine evaluation in bucks to be performed prior to mating is testing for fertility (semen and scrotal consistency) and sexually transmitted diseases (brucellosis). These tests are performed preferably within 30 days of the breeding season (Table 7.5) (Smith, 1980; Gall, 1981; Devendra and Burns, 1983; Bath and De Wet, 2000).

## APPLIED REPRODUCTIVE TECHNOLOGIES

### Introduction

The active interest in the control of ovulation, AI, and embryo transfer in farm animals, and in this case, the goat, revolves around the improvement of reproduction efficiency, genetic improvement, and the output potential of these animals. Thus, for the last 4 decades, researchers have examined the possibility of employing hormones and different techniques in the control of estrus and the improved reproductive performance in farm animals. Controlled breeding in goats may imply breeding the animals after the end of the normal anestrous period. It may mean breeding females to permit a compact kidding season. It may mean breeding does with top quality bucks by AI. It may also mean a rapid buildup of a superior flock by embryo transfer or the early diagnosis of pregnancy to increase reproduction efficiency. Reproduction efficiency as such is one of the main factors that determines the overall productivity of an animal. It also determines the rate of expansion of the flock (for selection purposes), the number of excess stock for sale, and the eventual availability of meat and milk for human consumption (Holtz, 2005).

The manipulation of the estrous cycle in the doe involves two alternative approaches, either (1) removing or reducing the effect of the corpus luteum (CL) on the estrous cycle or (2) suppressing follicular development by an artificially extended luteal phase. Suppression of follicular development can be achieved by the use of progesterone or its analogues, while luteolysis of the CL can be achieved by implementing prostaglandin $F_{2\alpha}$. If the time of estrus or ovulation is controlled, it would obviate the need for frequent detection of estrus and permit fixed-time artificial insemination and a compact kidding season (Robinson, 1976; Boshoff, 1980).

One of the most exciting developments in controlled breeding of farm animals is the use of embryo transfer. Superovulation of does and the subsequent recovery and transfer of embryos to appropriately synchronized recipients has been proven to be an effective means of increasing the contribution of outstanding females to the gene pool of a specific specie and decreasing the generation interval. This technique is now used in practice commercially, with millions of embryos being transported worldwide. The number of superior offspring that a doe can produce can also be greatly increased by this technique (Van der Nest, 1997; Greyling et al., 2002; Holtz, 2005).

Early pregnancy diagnosis (PD) in goats is a management tool that has the distinct advantages of identifying infertile animals and can be used to adjust a fodder flow program (for pregnant animals with multiple fetuses), while also preventing the slaughter of pregnant animals and facilitating the management program. The main prerequisites in using the technique of pregnancy diagnosis in goats are that it must be accurate, relatively inexpensive, repeatable, and practical (Wildeus, 1999).

### Synchronization of Estrus in Does

Seasonality in the reproduction activity limits the reproductive rate of the doe, and manipulation of reproduction could increase the frequency of breeding per year and the litter size in the specie. The effective synchronization of estrus and control of ovulation by means of intravaginal progestagen and/or injections of prostaglandin has been employed in goats. This modification of the ovarian cycle has been based on the hypothesis that maximum fertility potential depends on a finely balanced endocrine relationship (Greyling and Van Niekerk, 1986, 1990b; Greyling and Van der Nest, 2000).

Although sheep have shorter estrous cycles (16–17 days) than goats (21 days), an increase in the treatment period of intravaginal progestagen sponges from 12 to 18 days has no significant effect on the estrous response, follicular activity, and ovulation rate. Fertility levels following intravaginal progestagen sponges (fluorogestone acetate [FGA]) in goats favor a longer rather than a shorter treatment period (16 days and longer). The shorter treatment with progestagen generally leads to a delayed induction of the onset of estrus. However, when using a prostaglandin $F_{2\alpha}$ treatment at this intravaginal progestagen treatment withdrawal, complete luteolysis can be achieved with acceptable synchronization efficiency. Furthermore, administration of pregnant mare serum gonadotrophin (PMSG) or equine chorionic gonadotrophin (eCG) at intravaginal sponge withdrawal is necessary to enhance fertility following AI at a fixed time, by inducing a more compact synchronization of estrus. Ovulation in the goat following synchronization generally occurs between 27 and 31 hours following the onset of estrus, or 6.5 hours prior to the end of the estrous period. The ovulation rate per doe following synchronization quoted in the literature varies between 4.0 in the Black Bengal nanny goat, 1.72 in the Boer goat, 1.2 in the Angora, and 1.4 in Barbari goats. The mean interval from intravaginal sponge withdrawal to the onset of estrus in Boer goat does is $58.4 \pm 22.7$ hours, compared to $62.6 \pm 19.1$ hours for Boer goats synchronized with prostaglandin. The mean duration of the induced estrous period ($30.6 \pm 6.8$ hours) was also found to be shorter than that of the natural estrous period

**Table 7.6** The time of ovulation, ovulation rate, follicular activity, and mean serum progesterone concentrations at the onset of estrus in Boer goat does, following treatment with progestagen.

| | Duration of progestagen treatment (days) | | | |
|---|---|---|---|---|
| Item | 12 | 14 | 16 | 18 |
| Time of ovulation relative to onset of estrus (h) | 30.8 ± 2.8 | 29.6 ±7.2 | 33.6 ± 10.4 | 29.8 ± 2.4 |
| Range (h) | 26.0 – 32.8 | 26.0 – 42.5 | 26.0 – 48.5 | 25.5 – 32.5 |
| Mean ovulation rate (No. of ovulations/doe ovulating) | 1.4 ± 0.9 | 1.5 ± 0.6 | 1.4 ± 0.6 | 1.25 ± 0.5 |
| No. follicles >0.2 cm at time of ovulation | 5.7 ± 1.4 | 5.1 ± 2.0 | 4.9 ± 1.7 | 5.8 ± 1.6 |
| Mean serum progesterone concentration at onset of estrus (ng/ml) | 0.38 ± 0.14 | 0.39 ± 0.13 | 0.36 ± 0.06 | 0.37 ± 0.07 |
| Does ovulating per does showing estrus (%) | 71.4 | 62.5 | 57.1 | 80.0 |

Note: Values in the body of the table are mean ± S.D.
Source: Greyling, 1988.

(37.4 ± 8.6h) of the doe. The serum progesterone levels throughout the induced estrous period have been shown to vary between 0.31 and 1.4ng/ml, and the plasma LH surges occurring after the administration of exogenous progesterone have been shown to be lower than those occurring during unsynchronized LH surges. Subnormal circulating LH levels could be associated with lower fertility. The position of the preovulatory LH surge (indication of ovulation) and amplitude relative to the onset of the induced estrus was found not to be significantly affected by the synchronization of estrus. These are all aspects to keep in mind when using controlled breeding in goats (Table 7.6) (Corteel et al., 1988; Gordon, 1983, 1997; Greyling and Van der Nest, 2000).

The administration or route of administration of eCG (FSH) has no significant effect on the estrous response of goats, during the breeding season. A low dose of eCG (375IU), however, does result in a more predictable and precise synchronization of estrus. Outside the natural breeding season, the use of eCG is recommended. The route of PMSG administration seemingly has a minor impact on the time to estrus, although there is a tendency for goats treated intramuscularly to have a more compact and faster estrous response, compared to those injected eCG subcutaneously. The duration of the induced estrous period tends to be significantly longer following the exogenous administration of FSH due to a possible elevated plasma oestrogen level following follicular stimulation. The effect of this prolonged estrous period on fertility is unsure. eCG treatment in some cases may cause problems in sperm transport and sperm survival and hence possible reduced fertility (Table 7.7) (Greyling and Van Niekerk, 1990a).

The proof of successful synchronization in goats lies in the conception rate obtained. Factors affecting the efficiency of synchronization, and hence the fertility results obtained, include the dose of intravaginal progestagen (perhaps the most important), time relationship between the induced estrous and the preovulatory LH surge, body condition of the does, season, age of the doe, and parity. There is speculation that the amount of steroid actually released from the intravaginal sponge and the uniformity of the release over the treatment period is more important than the actual dose impregnated in the sponge (Table 7.8). Conception rates (AI 48 and 60 hours following sponge withdrawal) of 70 and 80% have been reported in indigenous SA feral and Boer goats, respectively, when using different progestagen doses. In both the smaller indigenous (mean body weight 24.7 ± 4.8kg) and larger (mean body weight 54.7 ± 6.8kg) Boer goat does, halved medroxy acetate progesterone (MAP) sponges were as efficient as whole 60mg MAP intravaginal sponges in the synchronization of estrus (Gordon, 1997; Greyling and Van der Nest, 2000; Motlomelo, 2000).

Estrus in goats can also be efficiently synchronized when using two injections of prostaglandin (inside the natural breeding season) 11–14 days apart. Prostaglandin causes luteal regression between days 5 and 16 of the estrous cycle in the doe. CLs outside this stage of the cycle are not always sensitive to prostaglandin, the so-called refractory period. Synchronization with two injections of prostaglandin outside the normal breeding season leads to poor estrous responses and very low conception rates indicating that prostaglandin is only effective in the active breeding season and with an active corpus luteum present. The use of a short progestagen treatment (8 days) followed

**Table 7.7** Effect of different synchronization techniques on the reproductive performance of Boer goat does outside the natural breeding season.

| Item | 500 IU eCG | | | No eCG | | |
|---|---|---|---|---|---|---|
| | Sponges | Sponges plus PGF | PGF plus PGF | Sponges | Sponges plus PGF | PGF plus PGF |
| Number of does | 15 | 15 | 10 | 15 | 15 | 15 |
| Conception rate % | 73.3[a] | 66.7[a] | 0 | 53.3[a] | 60.0[a] | 6.7[b] |
| Kidding rate % | 146.7[a] | 166.7[a] | 0 | 106.7[a] | 126.7[a] | 13.3[b] |
| Fecundity | 2.0 | 2.5[a] | 0 | 2.0[a] | 2.11[a] | 2.0[b] |

[a,b]Means in the same row, with the same superscript indicate no significant difference.
Source: Greyling, 1988.

**Table 7.8** The mean (±SE) estrous response of does following MGA or FGA treatment to synchronize estrus.

| Variables | Treatment (X ± SEM) | |
|---|---|---|
| | MGA (n = 12) | FGA (n = 12) |
| Time to estrus (h) | 86.7 ± 3.9[a] | 44.4 ± 1.5[b] |
| Time to LH peak (h) | 100.5 ± 2.4[c] | 59.7 ± 2.4[d] |
| Withdrawal to ovulation (h) | 126.8 ± 3.2[c] | 82.1 ± 1.6[d] |
| Estrus to LH peak (h) | 14.9 ± 1.8[a] | 15.3 ± 0.9[a] |
| Estrus to ovulation (h) | 40.1 ± 2.3[a] | 37.6 ± 0.5[a] |
| LH peak to ovulation (h) | 26.2 ± 1.1[c] | 22.4 ± 0.8[d] |
| Peak LH concentration (ng/ml) | 19.8 ± 2.9[a] | 22.3 ± 3.2[a] |

Note: Values in a row with different superscript letters (a–d) are significantly different ($P < 0.05$).
MGA = Melengestrol acetate; FGA = Fluorogestone acetate.
Source: Martinez Alvarez et al., 2007.

by a single prostaglandin injection is a technique that gives acceptable conception rates (66.7%) in goats, inside and outside the natural breeding season (Table 7.8) (Greyling and Van Niekerk, 1986; Haresign, 1978).

It would thus seem as if especially intravaginal progestagen sponge synchronization (with eCG) programs (as used in sheep) achieve acceptable high estrous responses and conception rates in goats and can be used as tools in the reproductive management of goat does, especially in combination with AI (Greyling and Van der Nest, 2000; Motlomelo et al., 2002).

## AI in the Doe

The primary purpose of artificial insemination (AI) in goats is to make optimum use of bucks of exceptional quality, thereby accelerating genetic progress. By using this method of breeding, it is possible to impregnate 400 does with semen from a single buck within 3 weeks, compared with 30–50 does per buck in a conventional mating system.

The technique of artificial insemination involves obtaining semen from the male artificially and the application of live part volumes of the semen to a number of does.

The collection and handling of semen with AI must be seen as an interference with the normal, natural process, and it can be deleterious to the semen. The aim must therefore always be to simulate natural conditions as closely as possible. However, it must be emphasized that AI is not a cure for infertility.

Following collection of the semen (preferably with the aid of an artificial vagina), AI in brief, involves the deposition of the evaluated semen (0.05 ml undiluted or 0.1 ml diluted) in the mouth of the cervix at a fixed time (minimum of 50–60 × $10^6$ sperm/insemination dose) 48 and 60 hours following sponge withdrawal or a single insemination 54 hours following sponge withdrawal. Only semen with a motility score of 3 (out of a scale of 5) and above is suitable for use in an AI program.

In conclusion, it can thus be said that the success of an AI program depends exclusively on the care and accuracy with which it is applied. Moreover, fertility results that vary between 60 and 90% are common when fresh semen is used. Correctly applied, AI can only be beneficial to the goat industry worldwide (Evans and Maxwell, 1987; Greyling and Grobbelaar, 1988; Wildeus, 1999; Hafez and Hafez, 2000).

### Multiple Ovulation and Embryo Transfer (MOET) in Goats

One of the most exciting developments in the reproduction of farm animals is embryo transfer, the so-called second generation reproductive biotechnology, after artificial insemination (AI). The successful transfer of embryos is largely dependent on the selection of acceptable donor and recipient does. Generally more inferior does in terms of genetic traits are used as surrogate females. However, these recipient animals should still be fertile, show good mothering ability, and produce adequate milk for sustaining their young. Survival of the transferred embryos is maximized if the recipients are synchronized to show estrus on the same day or one day after the donor.

The major problem in the use of MOET is the variable ovulatory response to superovulation in the donor animals. The superovulation rates in goats can vary from 1–18 ovulations, depending on the dosage and type of exogenous hormones used. These ranges may, however, also be influenced by several factors, including season of treatment, body condition of the doe or plane of nutrition, and method of breeding. Genetic effects, age, and type of gonadotrophin agents used may also contribute to this variation in ovarian response (Warnes et al., 1982). Embryo recovery rates have improved over the years, and with the development of laparoscopy, it has now become possible to collect embryos from the same female donor repeatedly, with the minimum complication of adhesions. Repeated hormonal superstimulation on the same individual has, however, been reported to lead to a decreased ovarian response due to an immune response by the donor (Lehloenya, 2008).

The superovulatory treatment of does aims at producing the maximum number of viable oocytes within one estrous period. The most commonly used gonadotrophins in embryo transfer programs are eCG (PMSG) or FSH. Experiments using eCG (long half-life) have generally resulted in low ovulation rates and were accompanied by the occurrence of a high number of large unovulated follicles. Later studies revealed an increased ovulation rate, ova recovery rate, and number of transferable embryos that could be obtained using pure FSH. Trials on the superovulatory response in goats to these exogenous gonadotrophins revealed an ovulatory response per doe of 16.6 for pFSH and 11.7 for PMSG or eCG. Factors affecting the superovulatory response also include individual variation in terms of response within a specie and seasonality (Table 7.9 and Table 7.10) (Van der Nest, 1997; Hafez and Hafez, 2000).

Fertilization in superovulated does (donors) can be achieved by using three methods, namely natural mating, cervical insemination, or intrauterine (laparoscopic) insemination. In goats, the technique of laparoscopy is usually implemented for intrauterine insemination because it yields a very high fertilization rate with lower insemination doses (36 to 48 hours following intravaginal progestagen removal) being required.

Generally surgical flushing of embryos in goats is performed 6–7 days following AI. The flushings are collected, the embryos recovered and examined, and then evaluated. Prior to flushing, the number of CL on each ovary serves as an indication of the ovulation rate and the possible number of embryos that could possibly be recovered. A recovery rate of 80% and more is acceptable in the retrieval of embryos (Warnes et al., 1982; Greyling et al., 2002). Only high quality embryos (Grades or Classes I and II) are generally used for fresh embryo transfer to synchronized recipients within 5 hours of flushing (Table 7.11). Embryos at this stage of development postestrus include the following developmental stages: (1) compact morula, (2) early blastocyst, or (3) expanded blastocyst with no signs of degeneration, protruding cells, or vacuolization of the blastomeres. Recipients responding to synchronization (showing estrus) receive two embryos by laparoscopy (one per uterine horn). The kidding performance of recipients and the survival of embryos has a major effect on the profitability and success of any embryo transfer program. A pregnancy rate of 60% and higher is acceptable in a MOET program with goats. The overall ovulation rate, embryo recovery rate, and pregnancy rates obtained in goats demonstrates MOET as an effective and acceptable accelerated breeding technique (Table 7.12) (Holtz, 2005; Melican and Gavin, 2008).

### Accelerated Kidding

With most goats (with maybe the exception of the Angora doe), being polyestrous and having an extended breeding

**Table 7.9** The superovulation response, embryo recovery rate, and pregnancy rate in Boer and indigenous feral goats following superovulation and embryo transfer.

| | Boer goats | | Indigenous goats | |
|---|---|---|---|---|
| | Donor | Recipient | Donor | Recipient |
| Number of animals | 8 | 20 | 16 | 40 |
| Total number of CLs (ovulations) | 70[a] | 60[a] | 205[b] | 67[a] |
| CL/doe in estrus | 18[a] | | 15[a] | |
| Embryos recovered (%) | 94[a] | | 80[a] | |
| Classes I and II embryos (%) | 97[a] | | 64[a] | |
| Number of unfertilized ova (%) | 2[a] (2.9) | | 2[a] (1.0) | |
| Does pregnant (%) | | | | |

[a,b]Within rows, means with same superscripts are not significantly different at 0.05 level.
Source: Greyling et al., 2002.

**Table 7.10** Mean (±SD) estrous response, onset and duration of the induced estrous period in Boer goat does superovulated with pFSH (control) or pFSH/GnRHa protocols.

| Treatment* | N | Estrous response (%) | Time to estrus (h) | Duration of estrus (h) |
|---|---|---|---|---|
| pFSH (control) | 11 | 100 | 30.6 ± 9.1 | 18.2 ± 3.7 |
| pFSH/GnRHa | 11 | 100 | 31.1 ± 8.8 | 18.9 ± 4.0 |

*No significant differences
pFSH, porcine FSH; GnRHa, GnRH analog.
Source: Lehloenya, 2008.

**Table 7.11** Embryo grading for goats prior to transfer or cryopreservation.

| | |
|---|---|
| Grade/class 1 | Embryos morphologically intact and compact with morulae or blastocyst |
| Grade/class 2 | Embryos (morulae to blastocyst) with minor morphological deviations (few extruded blastomeres) |
| Grade/class 3 | Embryos (morulae to blastocyst) with an uneven cell organization, loose structures, and numerous free blastomeres |
| Degenerate | Embryos at 8-cell stage or embryos earlier (little or no cleavage) |

Source: Lehloenya, 2008.

season, the proposition of accelerated kidding (three times in 2 years) is a definite option. The kidding season must be selected to fit in with the period when feed is most plentiful, under intensive conditions with above average management, then kidding can occur every 7–8 months. However, in this practice, there are certain prerequisites like high level of nutrition throughout, early weaning (±2 months of age), and early reconception (within 90 days of kidding). With accelerated breeding, the kidding rate (number of kids born per does mated) can be increased to well over 200% per year. Factors to be kept in mind when implementing such a practice are that during certain seasons the nutrition of the goats needs to be supplemented, and artificial breeding (synchronization and AI) during at least one of the cycles may be necessary. The Boer goat among others is one of the leading meat goats worldwide that lends itself excellently to such an accelerated kidding program (Erasmus, 2000; Malan, 2000).

**Pregnancy Diagnosis in the Doe**

Early detection of pregnancy in the doe has not only the advantage of identifying nonpregnant females, but money can be saved on feed costs (supplementary feed), and nonpregnant does can immediately be remated or culled. By the identification of multiple births, feed and grazing can also be used more efficiently.

Ultrasonic scanning or sonography by means of the sonar is a practical, inexpensive, and accurate means of diagnosing pregnancy in goats. In practice, it is recommended that does be scanned at approximately 6 weeks following mating. Scanning in small stock is the most accurate when performed between 42 and 100 days of gestation. Besides the use of the abdominal probe where does can be scanned in a standing position, a rectal probe is currently also available. This probe provides a better

**Table 7.12** Mean (±SE) ovulation rate (OR), total (TEMB) and freezable (FEMB) embryos, fertilization rate (FR), freezability rate (FRR), ova recovery rate (RR) and embryo recovery rate (ERR), in ewes treated with two FSH[1] preparations.

| | Total (n = 34) | oFSH (n = 16) | pFSH (n = 18) |
|---|---|---|---|
| OR | 8.0 ± 6.2 | 9.3 ± 1.6[a] | 6.8 ± 1.5[a] |
| TEMB | 5.0 ± 5.0 | 7.1 ± 1.3[a] | 3.1 ± 1.2[b] |
| FEMN | 4.2 ± 5.1 | 5.9 ± 1.3[a] | 2.6 ± 1.2[a] |
| | Total (n = 26) | oFSH (n = 16) | pFSH (n = 10) |
| FR (%) | 92.5 ± 22.0 | 94.6 ± 5.5[a] | 89.2 ± 6.7[a] |
| | Total (n = 26) | oFSH (n = 16) | pFSH (n = 9) |
| FRR (%) | 68.2 ± 40.0 | 66.4 ± 10.2[a] | 71.6 ± 13.5[a] |
| | Total (n = 34) | oFSH (n = 16) | pFSH (n = 18) |
| RR (%) | 60 ± 30 | 80 ± 10[a,***] | 30 ± 10[b,***] |
| ERR (%) | 50 ± 30 | 80 ± 10[a,***] | 30 ± 10[b,***] |

Means in the same rows with different superscripts (a–b) differ significantly (*$P < 0.05$ or ***$P < 0.001$).

[1]oFSH, ovine FSH; pFSH, porcine FSH.

Source: Bettencourt et al., 2008.

image, and pregnancy can be diagnosed from 20–30 days postmating. An aspect to keep in mind is that the larger the fetus, the more difficult it is to determine the presence of multiple births. The fact that 60–70% of fetal growth takes place during the last trimester of gestation makes the strategic nutritional supplementation during this period in pregnant animals very important. By identification and the strategic feeding of the pregnant does, a higher birth weight of the kids (and higher survival rate) can be practically assured, as well as increased milk production by the dam (higher ADG by the kids) (Schutte et al., 1986).

Other techniques (although less practical and sometimes less accurate) for the diagnosis of pregnancy are also available. These include radiographic techniques (X-rays), ultrasonic techniques (fetal pulse detection, rectal Doppler probe), rectal-abdominal palpation, progesterone and hormonal tests, vaginal biopsies, and immunological tests. Even laparoscopy can be used as a technique to determine pregnancy in goats. The prerequisite of the technique used is the practical ease of application and cost involved.

Pregnancy diagnosis as a tool and aid in the efficient reproduction management program of goats is an aspect not to be underestimated—not only can time be saved, but also money and the fodder flow program can be managed accordingly (Holtz, 2005).

## APPLIED REPRODUCTION MANAGEMENT IN GOATS

Good, efficient reproduction management is a prerequisite for any successful and profitable small stock enterprise. By the implementation of a few basic practices, the overall efficiency of the goat herd can be increased notably. By the implementation of an effective health and nutritional program through breeding, gestation, lactation, and the postpartum period of the doe, the production of viable offspring can be practically guaranteed. Generally this reproduction performance is measured in terms of the kidding rate or weaning rate. There are, however, certain critical management areas that warrant highlighting. These briefly include aspects covered in the following sections (Devendra and Burns, 1983; Schutte et al., 1986; Bath and De Wet, 2000).

### Recordkeeping

The keeping of reliable records is an integral part of an effective reproduction management program. It is the only means of recording the performance of the herd, while monitoring progress. Certain parameters warrant recording, and these include mating dates, kidding dates, birth weights of the kids (and sex of the kids) and where natural mating is used, buck fertility, mating ability, and libido of the bucks.

### Herd Health Management

A healthy animal is generally productive and one that normally performs well in terms of growth, reproduction, and lactation. The aim of any health program must be to be proactive, and any effective health program starts with the goat kids. If suppressed in growth at an early stage, it is reflected later in the animal's productive life.

### Nutritional Management

The efficiency of nutritional management (regarding reproduction) is normally reflected by the body condition of the animal, an indication of the nutritional status of the animal. Nutrient intake influences the body reserves and is related to the reproductive performance of the animal. A balance should be maintained between the overfeeding of energy, which is related to a lower conception rate, abortions, and retained placentas, and underfeeding of energy, which in turn could be related to problems of delayed puberty, lower semen production, the occurrence of silent heats, decreased libido, and irregular estrous cycles. The strategic feeding of breeding animals is important in that nutritional flushing of energy 3 weeks prior to mating could have a dramatic effect on ovulation and conception rate, while nutrition during pregnancy is crucial, because 70% of fetal growth occurs during the last trimester of pregnancy.

### Pregnancy Diagnosis

The diagnosis of pregnancy is not only a managerial tool in terms of determining the reproduction efficiency of the herd, but also in determining the fodder flow program in terms of does with singletons, twins, triplets, or those that are not pregnant. Much time and money can be saved by the timely use of this practice. Currently sonography (abdominal 6–7 weeks or rectal 4–5 weeks after breeding) is the standard procedure for determining pregnancy and multiple births in goats.

### Mating

Timing of a restricted breeding season (preferably in the natural breeding season) of approximately 6 weeks is crucial, with breeding delayed in maiden does until the animal has attained 60–75% of its mature body weight. The availability of adequate nutrition is essential during and prior to the breeding season and also during the kidding season.

### Dystocia

Difficult births are generally not a problem experienced in goats and can be controlled and manipulated by a managerial program. The main factors generally contributing to this phenomenon are kid size, body condition, frame size, and litter size.

### Buck Soundness

An effective managerial program would not be complete without the evaluation of the male. Reproductive aspects such as libido, semen quality, body condition, and health need to be constantly monitored. In the case of temporary sterility (irrespective of the cause), a period of 60–70 days must be allowed for the buck to once again produce mature, viable, and fertile sperm. Make sure that the buck suites your breeding goals. Look at the performance records and breeding history of the animal.

## SUMMARY

The goat is a hardy (adapted to arid conditions), fertile, and versatile (milk, fiber, meat, skins) animal with a generally extended breeding season. By using controlled breeding practices, goat's reproductive and genetic potential can be exploited to the fullest. The goat with its excellent reproductive qualities (high fecundity) and low maintenance requirements also makes it an ideal animal for the rural small-scale farmer in serving as a source of animal protein. Under intensive feeding conditions and accelerated breeding practices, very high kidding rates are feasible. With an average daily gain of 160–220 g reported under intensive production systems, goat meat production as such is also a viable proposition.

## REFERENCES

Ahmed, M.M.M., S.A. Makawi, and A.A. Gadir. 1997. Reproductive performance of Saanen bucks under tropical climate. Small Rumin. Res. 26:151–155.

Amoah, E.A. and M.J. Bryant. 1984. A note on the effect of contact with male goats on the occurrence of puberty in female goat kids. Anim. Prod. 38:141.

Barkawi, A.H., E.H. Elsayed, G. Ashour, and E. Shehata. 2006. Seasonal changes in semen characteristics, hormonal profiles and testicular activity in Saraibi goats. Small Rumin. Res. 66:209–213.

Bath, G. and J. De Wet. 2000. Sheep and goat diseases. Tafelberg Publishers, Cape Town.

Bettencourt, E.M., C.M. Bettencourt, J. Chagas e Silva, P. Ferreira, C.I. Manito, C.M. Matos, R.J. Ramao, and A. Rocha. 2008. Effect of season and gonadotrophin preparation on superovulatory response and embryo quality in Portuguese Black Merinos. Small Rumin. Res. 74: 134–139.

Boshoff, D.A. 1980. The effect of lactation and season on sexual activity and synchronization of oestrus in lactating Karakul ewes. Ph D Thesis, University of Stellenbosch, Stellenbosch, South Africa.

Bukar, M.M., J.D. Amin, M.N. Sivachelvan, and A.Y. Ribadu. 2006. Postnatal histological development of the ovaries and uterus and the attainment of puberty in female kid goat. Small Rumin. Res. 65:200–208.

Chemineau, P. 1983. Effect on oestrus and ovulation of exposing Creole goats to the male three times of the year. J. Reprod. Fertil. 67:65–72.

Chemineau, P., A. Daveau, F. Maurice, and J.A. Delgadillo. 1992. Seasonality of estrus and ovulation is not modified by subject female Alpine goats to a tropical photo period. Small Rumin. Res. 8:299–312.

Corteel, J.M., B. Leboeuf, and G. Baril. 1988. Artificial breeding of adult goats and kids induced with hormones to ovulate outside the breeding season. Small Rumin. Res. 1:19–33.

Devendra, C. and M. Burns. 1983. Goat production in the tropics. Commonwealth Agricultural Bureaux, Farnham Royal, Bucks, England.

Erasmus, J.A. 2000. Adaptation to various environments and resistance to disease of the improved Boer goat. Small Rumin. Res. 36:179–187.

Evans, G. and W.M.C. Maxwell. 1987. Salamon's artificial insemination of sheep and goats. Butterworths, London.

Gall, C.L. 1981. Goat production. Academic Press, London, New York, Toronto, San Francisco.

Gordon, I. 1983. Controlled breeding in farm animals. Pergamon Press, Oxford, New York.

Gordon, I. 1997. Controlled reproduction in sheep and goats. CAB International, Wallingford, U.K.

Greyling, J.P.C. 1988. Certain aspects of reproductive physiology in the Boer goat doe. PhD Thesis, University of Stellenbosch, South Africa.

Greyling, J.P.C. 1996. Induction of puberty in female Boer goat kids. Proc. 3rd Biennial conference of the African SRR network, UICC.

Greyling, J.P.C. 2000. Reproduction traits in the Boer goat doe. Small Rumin. Res. 36:171–177.

Greyling, J.P.C. and C.H. Van Niekerk. 1986. Synchronization of oestrus in the Boer goat does: Dose effect of prostaglandin in the double injection regime. S. Afr. J. Anim. Sci. 16:146–150.

Greyling, J.P.C. and C.H. Van Niekerk. 1987. Occurrence of oestrus in the Boer goat doe. S. Afr. J. Anim. Sci. 17:147–149.

Greyling, J.P.C. and C.H. Van Niekerk. 1990a. Effect of pregnant mare serum gonadotrophin (PMSG) and route of administration after progestagen treatment on oestrus and LH secretion in the Boer goat. Small Rumin. Res. 3:511–516.

Greyling, J.P.C. and C.H. Van Niekerk. 1990b. Puberty and the induction of puberty in female Boer goat kids. S.A. J. Anim. Sci. 20, 193–200.

Greyling, J.P.C. and J.A.N. Grobbelaar. 1983. Seasonal variation in semen quality of Boer and Angora goat rams using different collection techniques. S. Afr. J. Anim. Sci. 13(4):250–252.

Greyling, J.P.C. and J.A.N. Grobbelaar. 1988. Artificial insemination in sheep. Farming in South Africa. Mutton Sheep B.7/1988.

Greyling, J.P.C. and M. Van der Nest. 2000. Synchronisation of oestrus in goats: dose effect of progestagen. Small Rumin. Res. 36:201–207.

Greyling, J.P.C., M. Van der Nest, L.M.J. Schwalbach, and T. Muller. 2002. Superovulation and embryo transfer in South African Boer and indigenous feral goats. Small Rumin. Res. 43:45–51.

Hafez, E.S.E. and B. Hafez. 2000. Reproduction in farm animals. 7th Edition Lippincott, Williams & Wilkens.

Haresign, N. 1978. Ovulation control in sheep. In: Control of ovulation. Chrighton, D.B., N.B. Haynes, G.R. Foxcroft, and G.E. Lamming (Eds.). Butterworths, London-Boston.

Holtz, W. 2005. Recent developments in assisted reproduction in goats. Small Rumin Res. 60:95–110.

Hunter, R.H.F. 1980. Physiology and technology of reproduction in female domestic animals. Academic Press, London, New York, Toronto.

Jainudeen, M.R., H. Wahid, and E.S.E. Hafez. 2000. Sheep and goats. In: Reproduction in farm animals, 7th edition. Hafez, B. and E.S.E. Hafez (Eds.). Lippincott Williams & Wilkins, Philadelphia, Baltimore.

Karagiannidis, A., S. Varsakeli, and G. Karatzas. 1999. Characteristics and seasonal variations in the semen of Alpine, Saanen and Damascus goat bucks born and raised in Greece. Theriogenology 53:1285–1293.

Kassem, R., J.B. Owen, and I. Fadel. 1989. The effect of premating nutrition and exposure to the presence of rams on the onset of puberty in Awassi ewe lambs under semi-arid conditions. Anim. Prod. 48:393–397.

Lehloenya, K.C. 2008. MOET in Boer goats. Ph D Thesis, University of the Free State, South Africa.

Malan, S.W. 2000. The improved Boer goat. Small Rumin. Res. 36:165–170.

Martinez-Alvarez, L.E., J. Hernandez-Ceron, E. Gonzalez-Padilla, G. Perera-Marin, and J. Valencia. 2007. Serum LH peak and ovulation following synchronized estrus in goats. Small Rumin. Res. 69:124–128.

Melican, D. and W. Gavin. 2008. Repeat superovulation, non-surgical embryo recovery and surgical embryo transfer in transgenic dairy goats. Theriogenology 69:197–203.

Motlomelo, K.C. 2000. Synchronization of oestrus in indigenous goats: The use of different progestagen treatments. M Sc dissertation, University of the Free State, Bloemfontein, South Africa.

Motlomelo, K.C., J.P.C. Greyling, and L.M.J. Schwalbach. 2002. Synchronisation of oestrus in goats: The use of different progestagen treatments. Small Rumin. Res. 45:45–49.

Nishimura, S., K. Okano, K. Yasukouchi, T. Gotoh, S. Tabata, and H. Iwamoto. 2000. Testis developments and puberty in the male Tokara (Japanese native) goat. Anim. Reprod. Sci. 64:127–131.

Papachristoforou, C., A. Koumas, and C. Photiou. 2000. Seasonal effect on puberty and reproductive characteristics of female Chios sheep and Damascus goats born in autumn or in February. Small Rumin. Res. 38:9–15.

Peacock, C., C. Devendra, C. Ahuya, M. Roets, M. Hossain, and E. Osafo. 2005. Goats. In: Livestock and Wealth Creation. Owen, E., Kitalyi, A., Jayasuriya, N., Smith, T. (Eds.), pp 361–385. Nottingham University Press.

Restall, B.J. 1992. Seasonal variation in reproductive activity in Australian goats. Anim. Reprod. Sci. 27:305–318.

Riera, S., 1982. Reproductive efficiency and management in goats. Proc. 3rd Intern. Conf. on Goat Prod. and Disease, Tucson, Arizona, USA.

Rivera, G.M., G.A. Alanis, M.A. Chaves, S.B. Ferrero, and H.H. Morello. 2003. Seasonality of estrus and ovulation in Creole goats of Argentina. Small Rumin. Res. 48: 109–117.

Robinson, T.J. 1976. Controlled breeding in sheep and goats. Tomes, G.L., D.E. Robertson, and R.J. Lightfoot (Eds.). Institute of Technology, W. Australia 423–437.

Salamon, S. and W.M. Maxwell. 2000. Storage of ram semen. Anim. Reprod. Sci. 62:77–111.

Schutte, A.P., E.M. Van Tonder, J.M. Van der Westhuizen, I.A. Herbst, and J.J. Steyn. 1986. Reproduction, artificial insemination and applied guidance in small stock. Taurus Co-op. Perskor Printers, Johannesburg.

Shalaby, A.S., S.M. Sharawy, N.H. Saleh, and M.S. Medan. 2000. Reproductive performance of Damascus goats in semi-arid areas of Egypt. Proc. 7th Intern. Conf. on Goats, France, 15–21 May, 424–425.

Smith, M.C. 1980. Caprine reproduction. In: Current therapy in theriogenology diagnosis, treatment and prevention of reproductive diseases in animals. Morrow, D.A. (Ed.). W.B. Saunders Company, Philadelphia, London.

Todini, L., A. Malfatti, G.M. Terzano, A. Borghese, M. Pizzillo, and A. Debenedetti. 2007. Seasonality of plasma testosterone in males of four Mediterranean goat breeds and three different climatic conditions. Theriogenology 67:627–631.

Torres-Acosta, J.F., A. Ortega-Pacheco, R.C. Montes-Perez, and J.M. Blanco-Molina. 1996. First oestrus post-partum in Criollo goats under sub-humid tropical conditions in Mexico. Proc. 6th Intern. Conf. on Goats, Beijing, China, 6–11 May.

Van der Nest, M. 1997. Accelerated breeding techniques in Boer and indigenous goats: The evaluation of synchronization and MOET. M Sc dissertation, University of the Free State, South Africa.

Warnes, G.M., A.P. Pfitzner, and D.T. Armstrong. 1982. Embryo transfer procedures in the goat: Factors which have a major influence on success rate. Proc. sympos. on embryo transfer in cattle, sheep and goats. Canberra, Australia. Aust. Soc. Reprod. Biol 44–46.

Wildeus, S. 1999. Current concepts in synchronization of estrus: Sheep and goats. Proc. Am. Soc. Anim. Sci. pp 1–14.

Zamiri, M.J. and A.H. Heidari. 2006. Reproductive characteristics of Rayini male goats of Kerman province in Iran. Anim. Reprod. Sci. 96:176–185.

Zarazaga, L.A., J.L. Guzman, C. Dominguez, M.C. Perez, and R. Prieto. 2005. Effect of plane of nutrition on seasonality of reproduction in Spanish Payoya goats. Anim. Reprod. Sci. 87:253–267.

# 8
# Digestive Physiology and Nutrient Metabolism

*S.G. Solaiman, PhD, PAS and F.N. Owens, PhD, PAS*

**KEY TERMS**

Fermentation—microbial conversion of substrates into acids and gases under anaerobic (oxygen-free) conditions.

Rumination—the process of regurgitation, re-mastication, re-insalivation, and re-swallowing of feeds, specific to and characteristic of ruminants.

Digestion—mechanical, chemical, and enzymatic breakdown of food nutrients within the gastrointestinal (GI) tract.

Absorption—passage of nutrients from intestinal lumen into the intestinal wall (epithelium) usually followed by passage into the blood.

Required nutrients—nutrients needed to sustain life.

Digestive disorders—diseases resulting from dysfunction of the GI tract.

Prehension—capacity to grasp food for consumption.

Mastication—chewing food to reduce particle size.

Salivation—production of saliva that lubricates ingested foods for swallowing and assists in food digestion.

Ionophores—polyether antibiotics that modify microbial fermentation in the rumen.

**OBJECTIVES**

By completing this chapter, the reader will acquire knowledge on:

- Animal classification based on feed choice
- Work of digestion by goats
- The role of microorganisms in ruminant digestion
- Digestion by preruminant (young) kids
- Ruminal dysfunction
- Nutrients required within a goat's diet

## INTRODUCTION

The gastrointestinal (GI) tract of domestic animals differs, and this difference is related to their preference for and use of dietary components. Animals are classified into three groups based on their diet type. Herbivores are animals that have a digestive system adapted to eat and digest fibrous plant materials; carnivores are animals with a digestive system adapted to digest meat and meat components; and omnivores are those with a digestive system adapted to consume a mixture of plant and animal matter.

As a ruminant, the adult goat can consume and digest fibrous plant materials and therefore is classified as an herbivore. Other ruminants include cattle, sheep, deer, elk, and many other wild ruminant species. These all are even-toed ungulates from the suborder *Ruminantia* of the mammal order *Artiodactyla* (meaning an even number of toes). Among the ungulates, ruminants are the most numerous, widespread, and diverse grass-eating mammals. Of about 250 known genera, 68 still exist, of which 4 (cattle, sheep, goat, and bison) are widely domesticated agricultural species (Leek, 1993b). The ecological success of this group can be attributed to predigestive fermentation. This microbial fermentation process results in catabolism of fibrous plant cell walls (including cellulose) into energy-rich fermentation acids, release of nutrients enclosed within plant cells, and synthesis of microbial protein from nutrients recycled to the rumen such as nonprotein nitrogen (urea) and phosphorus. In addition, the fermentation process yields vitamin K and B complex vitamins provided there are adequate precursors (cobalt for $B_{12}$; sulfur for thiamin and biotin) for synthesis.

The term "ruminant" reflects the capacity of an animal to ruminate (chew their cud). Anatomically, ruminants possess a unique multicompartment (typically subdivided histologically into three or four sections) fermentation vat generally called the rumen (or the reticulo-rumen) anterior to their gastric stomach. This nonsecretory forestomach and associated organs are designed to support microbial fermentation. Following fermentation, the digesta flows to the ruminant's secretory stomach and intestines for digestion. Subsequent to the rumen, digestion generally parallels that of nonruminant animals. The capacity for this microbially assisted digestion within the rumen permits herbivores like goats to consume diets very rich in fibrous cell walls. Generally, over half of the dry matter consumed will be fermented to microbial products within the rumen reducing the quantity of material that passes to the gastric stomach (abomasum) for postruminal digestion. Anatomical and physiological adaptations of ruminants are geared to providing a favorable ruminal environment for maintaining a high degree of symbiosis between ruminal microorganisms and the host ruminant. These alterations of their GI tract permit ruminants to consume and digest feed resources of much wider variety than nonruminants (Hofmann, 1988).

## DIVERSITY OF RUMINANTS IN FEED TYPE

Ruminant species can be subdivided into classes based on their ecological spectrum and feeding type (Hofmann, 1989). At one end of the spectrum are the very selective feeders or concentrate selectors (CS) such as antelopes and giraffes. This class consumes frequent meals that consist of highly digestible nutrients, they ruminate frequently but for short periods of time, and relative to other ruminants, they have a small forestomach compartment. At the other end of the spectrum are the coarse grazers or grass/roughage eaters (GR) that include cattle and sheep. This class can consume very large quantities of food because of their large forestomach capacity that allows slow but efficient fermentation of high fiber feeds. They eat intensively but only a few times of the day, and they ruminate less frequently but for longer periods than CS. Most agriculturally important pasture species are readily consumed by the GR group. Between these two classes are the intermediate group/type eaters (IM) including goats. Within the IM class, the goat is most similar to the CS class having a relative small forestomach when compared to cattle, with less capacity for fermentation of roughage. However, they are extremely adaptable to different environments. When given a choice, goats select and consume the less fibrous portions of plants as well as shrubs and browse. But when fed fibrous feeds and pasture, the volume of the forestomach of goats will increase.

## PREHENSION, MASTICATION, SALIVATION

### Prehension

Grazing ruminants use both visual cues and taste when selecting a diet. Ruminants will exhibit preferences for certain plant parts and specific forage when grazing and also may select and sort feed components when fed dried forage (hay), but preferences can differ with feeding conditions. Smell and taste appear responsible for selectivity or "palatability" of a diet. Palatability is measured by allowing animals to choose among various diets. Palatability should not be confused with total feed consumption. When animals are not given a choice among a more and a less preferred feed, total dry matter consumption typically is no greater for a feed that is more preferred in a "palatability" study. Much remains to be learned about physical and chemical factors that are responsible for differences in food preference, diet sorting, and total dry matter intake. Some results from preference trials are noted below. Cattle and goats appear to recognize and have preference for a sweet taste. Given a choice, cattle dislike a salty taste, but goats prefer it mildly with pygmy goats having a higher preference for salt (Goatcher and Church, 1970). Goats can distinguish between bitter, sweet, salty, and sour tastes, and are more tolerant than sheep and cattle to a bitter taste (Bell, 1959). Goats enjoy a wide variety of plants that are distasteful to other ruminants presumably because of their greater tolerance to a bitter taste. Grazing goats prefer grasses to legumes and clover over alfalfa (De Rosa et al., 1997, 2002). Goats also show a preference for forage with higher available carbohydrate content (Burns et al., 2001). Why goats prefer some feeds and avoid others is not known, but they appear to have a preprogrammed capacity to recognize nutrients and toxins (Euphagia) and to consume certain plants because of their pleasant smell and taste (Hedyphagia). In most cases, body morphology and size may be determinant factors for food preference. Goats have a smaller size, and thus prefer more digestible and nutritious feeds. When grazing a tall-standing forage, ruminants do not graze to ground level. Presumably, this allows the grazer to keep their eyes above the grazing horizon to be on the outlook for predators. Consequently, grazing animals with a longer snout or with eyes atop their head would have a selective advantage.

### Mastication

The mouth is designed to harvest food, mechanically reduce particle size, and mix it with saliva. Saliva is essential as a lubricant to facilitate swallowing. Goats have a pointed tongue and jaws, mobile thin lips, and a deep mouth. Grazed forage is gathered between the labial surface of the lower incisor teeth and the upper dental pad (ruminants lack upper incisors), and forward movement of the muzzle cuts through the forage. Chewing of food is irregular with variable amplitude. In contrast, during rumination, the cud is chewed more slowly and regularly. Premolars and molar teeth aid in mastication. The upper jaw is wider than the lower jaw, so lateral (circular) jaw movement and the shape and spacing of the molar teeth result in shredding of tough plant fibers.

Mastication results in particle size reduction to increase in ports of entry for fermenting microorganisms. In addition, movements of teeth excite mechanoreceptors in the mouth that provide stimuli for production of saliva by salivary glands. Mouth movement also increases the rate and amplitude of primary and secondary cycle contractions of the reticulo-rumen. Feedstuffs that fail to promote chewing (low particle size) result in reduced rumination, saliva secretion, and forestomach motility.

### Salivation

Although salivary production is relatively continuous, volumes are greater when ruminants are eating and ruminating. Daily saliva production averages 6–16 liters by sheep and may be greater by goats. Saliva produced by ruminants contains no enzymes, but it is particularly rich in buffers ($HCO_3$ and $HPO_4$); these account for its alkalinity (pH = 8.1). Being swallowed, saliva aids in preventing ruminal pH from becoming too low for microbial growth. In the rumen, pH is maintained within a range (5.5–7.0) ideal for microbial growth. The $HPO_3$ in saliva allows phosphate to be recycled for rumen microorganisms to synthesize nucleoproteins, phospholipids, and nucleotides. Recycled urea, being up to 77% of total salivary nitrogen, also provides ammonia for formation of microbial protein. Urea, recycled in saliva, together with efficient kidney renal tubular urea resorption, appears critical for survival of ruminants consuming forage or feeds with very low protein content (NRC, 2007; Leek, 1993a). In addition, recycling of urea reduces water excretion; combined with very efficient resorption of water by the large intestine, which helps ruminants survive when the quantity or quality of water is limited.

As browsing ruminants, goats have relatively large salivary glands and higher rates of secretion than sheep. The primary functions of saliva are to provide a copious and continuous supply of alkaline buffers to counterbalance the volatile fatty acids (VFA; primarily acetate, propionate, and butyrate) produced during fermentation and to provide an aqueous suspension for rumen solids. Secondary functions of saliva include urea recycling as a source of non-protein nitrogen (NPN) for microbial protein synthesis and phosphate for synthesis of microbial nucleic acid and membrane phospholipids. Saliva also acts as a wetting agent for ingesta, provides an antifoaming agent to prevent frothy bloat, and supplies proline-rich proteins that bind and deactivate dietary tannins.

Saliva provides a medium for short-term adaptation to changes in diet composition, namely, the presence of plant secondary metabolites such as tannins. Salivary proteins influence taste and digestive function. Proline-rich proteins in saliva that bind tannins are present in saliva of browsers but not grazers. Lamy et al. (2008) showed that protein profiles of saliva in the 25–35 kiloDalton (kDa) range differ between goat and sheep saliva (25–35 kDa range). Austin et al. (1989) reported that tannin-binding proteins were present in the saliva of deer but not of sheep and cattle. Hofmann et al. (2008) indicated that in ruminants, the mass of salivary glands is correlated positively with body mass but negatively correlated with the ratio of grass to browse in the diet, perhaps reflecting the need for complex salivary compounds, such as tannin-binding proteins, by browsing ruminants. Silanikove (2000) confirmed that the ratio of salivary gland mass to body size was high for goats indicating their capacity to consume browse plants.

## FUNCTIONS OF RUMEN AND RETICULUM

The major physiological activities related to rumen and reticulum will be addressed here. Readers are referred to

Chapter 6 for more detailed descriptions of the functional anatomy of these organs.

The reticulo-rumen is the primary site for fermentation by diverse but specialized microbes. Rumen motility originates in the reticulo-rumen for the processes of rumination and eructation. Environmental features unique for the reticulo-rumen that prove useful for action of ruminal microbes include (1) a reasonably isothermal environment regulated by the homeothermic metabolism of the animal, (2) constant influx of water and feeds, (3) a relatively constant pH achieved through absorption of fermentation acids and salivary buffers, and (4) removal of fermentation end-products by absorption or passage.

### Rumen Motility, Rumination, and Eructation

Powerful contractions by the reticulo-ruminal wall include a primary cycle (a mixing cycle or "A" sequence) as well as a secondary cycle (eructation cycle or "B" sequence). Both are initiated by excitation of vagal nerve fiber receptors distributed throughout the reticulum and rumen regions. The primary cycle consists of a double contraction of the reticulum followed by caudally moving contractions of the dorsal ruminal sac and ventral ruminal sac. This serves to pump ruminal fluids atop the floating raft in the rumen allowing fluid to percolate through the raft removing small particles for removal from the rumen. The secondary cycle usually occurs at the end of alternate primary cycles and consists of contractions of the caudoventral ruminal blind sac, a cranially moving contraction of the caudodorsal ruminal blind sac followed by the middorsal ruminal sac and the ventral sac. At the end of this cycle, the point where the esophagus enters the rumen is cleared of liquid so that the headspace gas can escape. Compared to motility before a meal, rate of motility often is doubled during and after feeding. The motor activity responsible for these contractions originates in the bilaterally paired gastric centers in the medulla oblongata of the hindbrain. Reticulo-ruminal motility is important for mixing of digesta, rumination, particle size reduction through attrition, eructation, and VFA absorption.

Two types of sensory receptor mechanisms are responsible for the vagal nerve input: tension receptors and epithelial/mucosal receptors. Tension receptors are located in the muscle layers of all parts of the GI tract. They monitor the tension present in the muscular wall imposed by passive distension and thus may be responsible for limiting intake of low quality feeds based on bulk or mass. The epithelial receptors are located close to the luminal epithelium of the forestomach whereas mucosal receptors are located close to the luminal mucosa of the abomasum and small intestine. These receptors are excited by both mechanical and certain chemical stimuli and may be responsible for chemostatic regulation of intake.

Ingested feed enters the reticulum through the cardia. Heavy materials (stones, sand) may immediately drop to the bottom of the reticulum and remain there. Less dense and fibrous materials float high in the reticulum and cranial sac of the rumen forming a raft of entangled particles. Reticular and rumen contractions move ingested particles caudally to join the fibrous raft in the rumen. Ruminal contents, particularly the raft, are meshed, kneaded, and slowly rotated with powerful rumen contractions. Soupy material common to the reticulum, cranial sac, and ventral sac is randomly pushed out of the reticulum to the omasum and postruminal tract by reticulo-ruminal contractions when the reticulo-omasal orifice is open.

The process of rumination includes four steps: regurgitation, re-insalivation, re-mastication, and re-swallowing. During regurgitation, a bolus of feed gathered near the cardia area of the rumen is pulled back to the mouth, an extrareticular contraction that precedes the usual biphasic contraction and opening of the cardia with antiperistaltic movement of the esophagus bringing the bolus to the mouth. There, the bolus is re-chewed, mixed with additional saliva, and re-swallowed.

Much of the fermentation in the rumen occurs within the fibrous raft. Rumen contractions persistently attempt to break the raft and release the gases for removal via eructation (belching). During the secondary cycle, aided by contraction of the dorsal sac, gases are moved cranially into the reticulum while the raft and fluids are pushed ventrally. If and when the gas layer clears the cardia of its fluid, eructation is evoked.

### Fermentation

The primary sites of microbial fermentation within the digestive tract of goats are the rumen and reticulum. The host animal and the diverse population of microbes in a symbiotic relationship ferment feeds yielding products that are useful nutritionally. Microbes that inhabit the rumen and reticulum, being provided with the unique anaerobic environment by the host ruminant discussed above, perform several functions: (1) fermentation of structural (cellulose and hemicellulose) and nonstructural (sugars and starch) carbohydrates into readily metabolized energy sources such as volatile fatty acids, (2) conversion of non-protein nitrogen from plants and metabolically recycled urea to a high biological value microbial protein, (3) synthesis of vitamin K and most B complex vitamins, and (4) conversion of metabolically recycled phosphate to microbial nucleic acid and membrane phospholipids.

#### Effect of Diet on Microbial Fermentation

The amount and composition of feed consumed can alter the microbial population and distribution of types found in the reticulo-rumen. Thus, diet can impact rate of digestion, rate of passage, and retention time of digesta in the rumen. Feeds rich in protein can alter the microbial populations to predispose an animal to fermentation disturbances such as ammonia toxicity or other toxins in the feeds. Diets rich in protein stimulate growth of protein-digesting (proteolytic) microorganisms. High fiber diets will increase the proportion of cellulose digesting (cellulolytic) microorganisms within the rumen, and high starch diets promote growth of starch-digesting (amylolytic) microorganisms. Diets rich in protein and starch (up to 80% concentrate diets) are associated with higher fermentation rates, higher digestibility of dry matter, and possibly improved intake. However, diets that contain less than 15% roughage tend to have a negative impact on fermentation and digestion (Vieira et al., 2008) by goats. High concentrate diets lead to greater production of fermentation acids in the reticulo-rumen and place a greater load on rumen buffering system. Goats like other ruminants have more tolerance to downward than to upward shifts in rumen pH (Silanikove, 2000). Generally, methane-producing (methanogenic) and cellulolytic microorganisms in the rumen are more sensitive to lower rumen pH and have a reduced prevalence under such conditions.

Nitrogen, sulfur, and essential minerals must be supplied for optimum microbial fermentation in the reticulo-rumen both for microbial growth and protein synthesis and for fiber digestion. Saliva provides buffering via bicarbonates of sodium and potassium. Through hydrolysis to ammonia, salivary urea also provides buffering for rumen contents. These components must be at optimum levels for optimum rates of digestion and feed intake.

Passage rate of feed residues from the rumen is associated with the plant cell wall content of the diet, action of microorganisms, and rate of particle size reduction. When passage rate is high, more slowly available substrates such as cellulose will escape ruminal digestion leading to a lower extent of fiber digestion. Grinding or pelleting feeds will increase passage rates of particles and thereby will negatively impact fiber digestion relative to feeding long particle forage. Because performance generally is proportional to intake of digested dry matter, a compensatory increase in feed intake may counterbalance this negative impact on digestibility and lead to improvements in both rate and efficiency of gain.

#### Types of Microorganisms in the Rumen

Microorganisms in the rumen are either facultative or strict anaerobes. A limited supply of oxygen can be tolerated so long as it is actively removed by facultative anaerobes. Because ruminal microbes are obligate anaerobes, inoculation or cross-inoculation from one animal to another via feed, saliva, or air is limited. Inoculation of protozoa can be prevented by maintaining an animal in isolation, but inoculation by bacteria is impossible to prevent. Fecal contamination appears to be the major source of anaerobic fungi in the reticulo-rumen.

Active fermentation in the reticulo-rumen requires semi-continuous influx of substrates and a specialized group of bacteria. Although the majority of rumen bacteria are obligate anaerobes, some facultative anaerobes exist in the rumen and may play roles in rumen dysfunction. Although bacteria at some $10^{10}$ to $10^{11}$ bacteria per ml, can account for about half of the rumen biomass, they are responsible for most of the fermentation within the rumen. Larger organisms (protozoa at $10^4$ to $10^6$/ml) can account for the other half of the biomass in the rumen, but they play a smaller but yet significant role. In addition, fungi can be found within the rumen (Van Soest, 1994).

Bacteria within the reticulo-rumen are diverse and specialized according to the substrates used, the products formed, or their nutrient requirements. The primary groups of rumen bacteria are those that degrade diet components while a secondary group will use end-products of the primary groups as substrates. Functions within these primary groups overlap considerably. The secondary groups (more than 60% of the total) are very important for adjusting fermentation output, providing growth factors for the primary groups, and maximizing efficiency of fermentation. Major primary bacterial and protozoan species found in the rumen are shown in Figure 8.1.

#### Fermentation of Fiber

The primary cellulolytic bacteria have the unusual capacity of being able to hydrolyze the β1,4 linkages of plant polysaccharides and convert released cellobiose (2 glucose units with a β1,4 linkage) and glucose into volatile fatty acids (VFA) and a transient intermediate formate. Cellulose and starch are converted to glucose, fructosans to fructose, and hemicellulose and pectin to xylose, as transient intermediates. These intermediates are converted

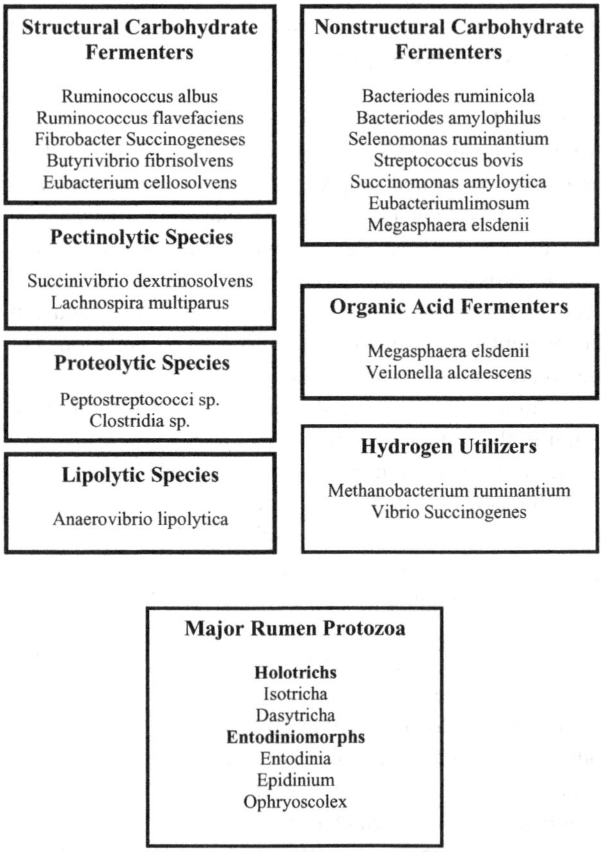


**Figure 8.1** Major rumen microorganisms, bacteria and protozoa (adapted from Van Soest, 1994).

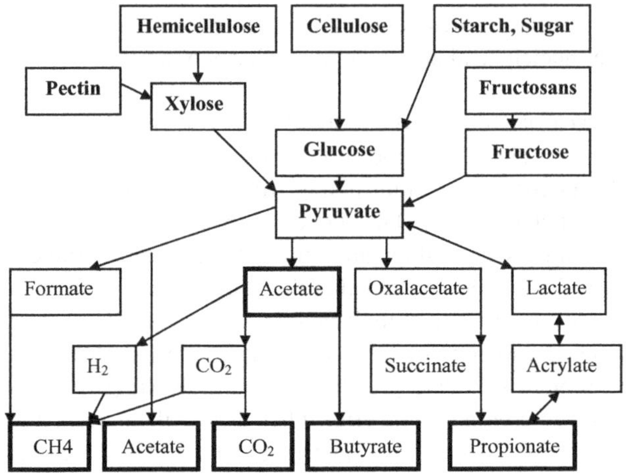


**Figure 8.2** Carbohydrate metabolism in the rumen.

further anaerobically to phosphoenolpyruvate (PEP) and pyruvate. Acetate and to a lesser extent butyrate and formate (as a transient intermediate) originate from pyruvate. Via β-OH butyrate, pyruvate yields butyrate. Oxaloacetate and succinate serve as transient intermediates leading to propionate while lactate and acrylate also yield propionate. Propionate production via oxaloacetate to succinate is the most common pathway used by rumen bacteria for producing propionate although production via acrylate is a highly efficient pathway favored when animals consume a high grain diet. *Megasphaera elsdinii*, a secondary fermenter, may be the main organism responsible for this pathway. Methane, generated when acetate and butyrate are formed, and propionate are products of secondary fermenters, and their production is important in optimizing fermentation. Secondary methanogenic bacteria convert formate to methane (Figure 8.2).

Cellulolytic bacteria generally have low metabolic rates and higher generation intervals (18 hours). Fermentation is slow and may require B vitamins, minerals, $NH_3$, $CO_2$, branch chain fatty acids, and a proper pH (6.2–6.8). This pH requirement matches the pH in the rumen of ruminants fed forage-based diets. Small amounts of iso-acids are required as growth factors for cellulolytic bacteria when grown in culture. These iso-acids are isobutyrate, isovalarate, and 2-methylbutyrate that arise from deamination of branched chain amino acids of valine, leucine, and isoleucine, respectively. However, diet supplementation with iso-acids is seldom needed because these acids are available due to crossfeeding from other bacteria and regularly available from protein catabolism. The typical acetate: propionate: butyrate ratio generated from fermentation of cellulose fermentation is 70:15:10.

#### Fermentation of Starch

Primary amylolytic bacteria hydrolyze α1,4 and α1,6 linkages in starch (amylose and amylopectin) to form maltose, isomaltose, and glucose and converting them to VFA, metabolic acids (lactic acid), and formate as a transient intermediate. Although substrates and bacterial species differ, the fermentation pathways generally are similar to those described for fiber fermentation. Amylolytic bacteria have higher metabolic rates and lower generation intervals (15 minutes to 4 hours). Fermentation is rapid and requires $NH_3$, amino acids, and a proper pH (5.5–6.6). This pH optimum matches the ruminal pH of ruminants fed diets rich in grain. The typical acetate:

propionate : butyrate ratio from starch fermentation is 55 : 25 : 15.

Secondary groups of rumen bacteria convert formate to methane and lactic acid or other acids to propionate. Secondary fermenters have lower metabolic rates, higher generation intervals (16 hours), and a higher optimum pH (6.2–6.8) than amylolytic bacteria. Therefore, when an animal is abruptly transferred to high concentrate diet, rapid fermentation often leads to high acid (lactic acid) production and a rumen pH lower than optimum for amylolytic bacteria but too low for secondary fermenters. Hydrogen disposal is hindered, acids accumulate, and clinical or subclinical rumen acidosis can occur.

### Fermentation of Proteins

Proteolytic bacteria represent a small portion (12–38%) of total reticulo-rumen bacterial mass and are responsible for ammonia production. Ammonia is required for growth of many cellulolytic and amylolytic bacteria. Peptides, amino acids, and isoacids may be stimulants for their growth and activity. The main dietary substrate for this group of bacteria is dietary protein that can be divided into two classes: ruminally degraded protein (RDP) or degraded intake protein (DIP) and ruminally undegraded protein (RUP) or undegraded intake protein (UIP). The later generally passes to the lower tract undegraded by rumen microorganism. Generally about half of the dietary protein is degraded in the rumen. Extracellular proteases from bacteria degrade dietary protein to yield peptides, amino acids, and ammonia. Amino acids can be used by some ruminal microbes, but most can synthesize their own amino acids from ammonia. When protein supply is high, bacteria readily degrade it yielding ammonia and various metabolic acids. Amino acids are deaminated to keto acids that are fermented to VFA and small amounts of the branched chain VFA—isobutyrate, iso valerate, and 2-methylbutyrate—that arise from valine, leucine, and isoleucine, respectively. Dietary plant amides, nitrites and nitrates, and endogenous salivary urea (nonprotein nitrogen compounds) when fermented also yield ammonia. Bacterial urease activity is high in the reticulo-rumen and readily converts dietary or recycled urea to ammonia. Urea is an important substrate for microbial protein synthesis. The capacity of rumen microbes to use ammonia and convert it to microbial protein depends on the availability of energy, largely from carbohydrate fermentation. The greater the supply of fermented carbohydrate, the greater the ammonia uptake and synthesis of bacterial mass.

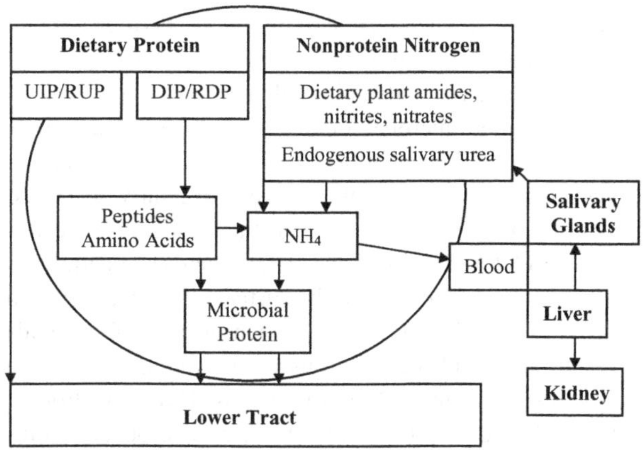


**Figure 8.3** Protein and nonprotein nitrogen metabolism in the rumen.
UIP = undegraded intake protein;
RUP = ruminally undegradable protein.
DIP = degradable intake protein;
RDP = ruminally degradable protein.

Conversion of peptides via amino acids to ammonia is a slow process of oxidative deamination of amino acids. This process requires the transfer of hydrogen via NADH. Ionophores inhibit proteolytic activity in the rumen either through inhibiting transfer of hydrogen needed for deamination of amino acid and ammonia formation, or by alteration of rumen microflora. Because peptides and amino acids are degraded in the rumen, flow of nitrogen to the abomasum and intestines consists largely of undegraded dietary protein and microbial protein (Figure 8.3).

Diet formulators seek to balance the supply of readily available carbohydrates and proteins in the diet to ensure that an adequate supply of ammonia (for protein synthesis) and carbohydrate (for energy) are available for microbial protein synthesis but to avoid excessive protein that will be wasted. Excessive production of ammonia, particularly from nonprotein nitrogen (esp., urea) will increase the risk of ammonia toxicity. Energetically, protein degradation in the rumen is wasteful, and for the animal, energy is required to convert ammonia to urea within the liver and for the kidneys to excrete urea.

### Fermentation of Lipids

Ruminants fed forage-based diets typically consume only a small amount (3–7% dry matter [DM]) of structural lipids in the plant leaves. The majority of these lipids are phospholipids and galactolipids with less than 50% being triglycerides that in turn are composed of free fatty acids, mainly palmitic, linoleic, and linolenic. Diets often contain oilseeds that consist primarily (65–80%) of triglycerides that contain the free fatty acids palmitic, oleic, and linoleic. Reticulo-rumen microbes rapidly hydrolyze dietary lipids and synthesize their own body lipids. They hydrolyze triglycerides to form glycerol and respective fatty acids and galactolipids to galactose. Glycerol and galactose are fermented to VFA. The majority of fatty acids is neutralized with calcium and adheres to the surface of bacteria or feed particles (McAllen et al., 1983). Unsaturated fatty acids are saturated, and the positions of the remaining double bonds are distributed along the chains. Hydroxylation of fatty acids and production of keto acids also occurs. Trans fatty acids are more stable, have higher melting points, and are difficult to hydrogenate, resulting in higher concentrations than their counterpart cis acids. Saturated fatty acids are not degraded further but pass to the abomasum as calcium salts. Because unsaturated fatty acids are largely hydrogenated within the rumen, the supply of fatty acids available for absorption from the small intestine (SI), are largely saturated. This contributes to the more saturated nature and higher melting point of tallow (ruminant fat). Biosynthesis of microbial fats in the reticulo-rumen can involve formation of odd and branched chain fatty acids. Fatty acids with 15 carbons and branches are found in microbial lipids.

Metabolism of lipids in the anaerobic environment of the reticulo-rumen is quite limited. Diets with more than 7% of DM as fat may reduce methanogenic and cellulolytic fermentations. Although unsaturated fatty acids do not compete directly with methane production, they serve as a separate sink for excess hydrogen and indirectly reduce methane production. Ionophores added to the diet similarly inhibit hydrogen transport, increase the prevalence of propionate, and may inhibit lipid hydrolysis in the rumen.

#### Other Fermentation Products

Microbes in the reticulo-rumen synthesize the B-vitamins provided an adequate amount of cobalt for $B_{12}$ synthesis, and an adequate supply of sulfur is available for synthesis of biotin and thiamin. For synthesis of sulfur-containing amino acids (methionine, cystine), sulfur also is required. This can become critical for fiber production because wool and mohair are rich in cystine and cysteine, amino acids that contain sulfur. By complexing with sulfur and reducing its availability, molybdenum and copper can reduce sulfur availability and induce a deficiency. Conversely, excess sulfur complexing with copper can reduce copper availability. A cofactor for tyrosinase, copper is needed for melanin (pigment) formation and the normal crimp in wool, so a deficiency can reduce hair coloration and cause wool to become straight and steely. Rumen microbes are beneficial through partially hydrolyzing oxalates and phytates to increase mineral bioavailability from plants and in addition can detoxify many plant metabolites that are toxic for nonruminants.

#### Rumen Protozoa and Fungi

Two main groups of ciliate protozoa are found in the rumen. The entodiniomorphid protozoa engulf particles and together with attached or internal bacteria can hydrolyze some structural carbohydrates (cellulose and hemicellulose). Bacteria associated with protozoa may contribute as much as one-third of the total rumen cellulolytic activity. The holotrichs prefer nonstructural carbohydrate such as starches and sugars. Entodiniomorph protozoa also consume and digest bacteria as substrate and use bacterial amino acids for producing protozoal protein. The end-products of protozoal fermentation in the reticulo-rumen include organic acids, $CO_2$, ammonia, and hydrogen. The extent to which protozoa can digest fiber and synthesize amino acids is unclear because ruminal bacteria always are so closely associated with protozoa so that cultures of ruminal protozoa free of bacteria have never been grown.

Although protozoa are not necessary for rumen fermentation, their role in balancing the fermentation process in the ruminants must not be underestimated. Protozoa engulf starches and sugars to retard rapid fermentation and thereby reduce the prevalence of lactic acidosis. Engulfed starches and sugars can pass to the abomasum and intestines digestion by the host. These are favorable functions of protozoa. Major rumen protozoa are listed in Figure 8.1.

Anaerobic fungi also are present in the rumen, but their contribution to the ruminal microbial mass is quite small, and their turnover rate is slow and similar to that of protozoa. Fungi usually are associated with slowly passed forage particles. Fungi prefer lignified coarse cell walls as a substrate, and contribute to VFA, gases, traces of ethanol, and lactate in the rumen. Their contribution to the rumen digestion is not yet fully understood.

#### The Fate of Fermentation End-Products

The main end-products of carbohydrate fermentation are acetic, propionic, and butyric acids. Fermentation of protein also yields these VFA plus valeric acid and branched chain VFA and ammonia. Branched chain VFA deficiencies for microbial protein synthesis would be more likely when dietary crude protein comes largely from nonprotein nitrogen (NPN).

Volatile fatty acids, being weak acids (pK = 4.75 – 4.87), they almost entirely dissociated in the reticulo-rumen at pH 6.6. About one-half of the VFA are absorbed through reticulo-rumen epithelium in their nondissociated form by passive diffusion with absorption rates being correlated positively with length of the VFA and negatively correlated with the rumen pH. The remaining VFA are absorbed as anions by facilitated diffusion in exchange for bicarbonate ions. Epithelial cells of the rumen contain carbonic anhydrase to promote production of carbonic acid from $CO_2$ and water. Carbonic acid in turn is dissociated into bicarbonate and H ions. These hydrogen ions convert VFA anions into acids for passive diffusion through the rumen epithelium. Transfer of hydrogen from carbonic acid (a weaker acid) to form VFA (stronger acids) lowers the pH and promotes further VFA absorption. Bicarbonate ions released into the rumen during absorption will neutralize more than half of the VFA in the rumen. The remaining VFA are neutralized largely by salivary alkali.

During absorption, some VFA are metabolized. Most of the butyrate is converted to β-hydroxybutyrate (β-OH-BUT) with the remainder being similarly metabolized by the liver. Thus, absorbed butyric acid enters the circulation as β-OH-BUT for metabolism by most ruminant tissues as a source of energy as well as fat, being the four-carbon primer used for synthesis of short and medium chain fatty acids unique to the ruminant's milk fat.

Almost one-third of the propionic acid absorbed is metabolized by the rumen wall to lactic acid. Propionate is completely converted by the liver to oxalacetate and that enters the Kreb's cycle. Lactate is converted to glucose for storage as glycogen by the liver or released as glucose into the circulating portal blood. Propionate and valerate are the only VFA used for gluconeogenesis.

The majority of acetate enters circulating portal blood unchanged except for a small amount that is metabolized to $CO_2$ by the rumen epithelium. The most abundant VFA in the blood, acetate, is readily used by tissues to form Acetyl-Co-A that yields energy via the Kreb's cycle or forms fatty acids. Mammary tissues use acetate and β-OH-BUT to form the short and medium chain length fatty acids found in milk.

Lactic acid produced by amylolytic bacteria is present as a transient acid in low concentration, and can be used by secondary rumen bacteria to produce propionate. However, the low rumen pH associated with high concentrate diets may inhibit propionate producing bacteria more than amylolytic bacteria; this can result in an accumulation of both isomers of lactic acid (+) and (–) that accentuates the rumen pH decline resulting in ruminal acidosis. Lactic acid absorption rate from the rumen is only one-tenth that of VFA. The liver metabolizes the L (+) lactic acid isomer more rapidly than the D (–) isomer. L lactic acid also is produced by muscle tissue during anaerobic exercise, but D lactic acid is primarily a product of bacterial metabolism found in fermented feeds including silages at up to 10% of dry matter.

Ammonia is produced in the rumen from deamination of dietary proteins, or hydrolysis of dietary NPN or urea from saliva, or diffusing into the rumen from blood. Given an adequate supply of energy from fermented carbohydrates, ammonia is used for synthesis of microbial protein. Excesses of ammonia are absorbed through the rumen wall, readily removed from portal blood, and detoxified by the liver by conversion to urea. Ammonia that escapes liver metabolism (for example, absorbed by the lymphatic system) can prove toxic for ruminants.

Gas production in the reticulo-rumen peaks some 2–4 hours after feeding. The major rumen gases include carbon dioxide (60%), methane (30–40%), and nitrogen, with small amounts of hydrogen sulfide, hydrogen, and oxygen. Carbon dioxide is produced either by decarboxylation during fermentation, or by neutralization by acids of bicarbonate ions that enter the rumen via saliva or exchange across the rumen wall during VFA absorption. Methane is produced from reduction of $CO_2$ and formate. Loss of methane from the rumen accounts for about 6% of ingested energy. Methane is a contributor to global warming, so reducing methane production by ruminants and by anaerobic bacteria in wetlands, and by release from petroleum are of worldwide concern. Hydrogen sulfide, being toxic to rumen microbes and animals, is produced from ruminal reduction of sulfates in the diet or derived from amino acids that contain sulfur. Hydrogen, present in only small amounts in the rumen, can increase when animals are abruptly switched to a high concentrate diet and fermentation is abnormal. Oxygen enters the rumen together with ingested feed and water and by diffusion into the rumen from blood; it is quickly removed by facultative ruminal bacteria.

Practically no peptides or amino acids are flushed onto the abomasum or small intestine because they are rapidly catabolized by ruminal bacteria. However, the feed and microbial proteins that pass out of the reticulo-rumen undergo lysis by abomasal lysozyme. Microbial proteins, lipids, vitamins, and starch are digested in the intestines and make a substantial contribution to the nutrient supply of the host.

## FUNCTION OF OMASUM

The role of omasum in digestive physiology of ruminants is briefly summarized in this section, but further information is presented in Chapter 6. Size of the omasum is considerably larger for cattle than for goats. It is an additional site for fermentation and absorption, but its primary function appears to be regulation of the flow of digesta from the reticulo-rumen to the abomasum. Ingesta from the reticulum flows to the omasum through the reticulo-omasal orifice that is open during the secondary phase of the primary cycle contractions of the reticulum. Prolonged and powerful contractions of the omasal body tend to empty the materials trapped between leaves into the abomasum. Particle size of digesta in the omasum (about 1 mm) is similar to that found in the reticular area adjacent to the orifice, and its physiochemical conditions resemble those of the cranial and ventral regions of the reticulo-rumen. The omasum has a large surface area (leaves) relative to its volume, a factor that gives this organ a large capacity for absorption of VFA, electrolytes, and water. In this organ, chloride (instead of bicarbonate) plays a major role in VFA absorption.

## FUNCTION OF ABOMASUM

The abomasum or "true stomach" receives more or less continuous (but at variable rates) input of ingesta containing partially fermented materials, fluids, or particle clumps (of variable composition) from the omasum and, following acidification, passes digesta in a reasonably constant flow to the duodenum. The abomasum plays two important roles in ruminant digestion: (1) transfer of partially digested feed and (2) chemical and enzymatic breakdown of ingesta. The cardiac and fundus regions of the abomasum are responsible for nonacid secretion while the antrum/pyloric region secretes acid.

Distension of the pyloric region, a rise in abomasal pH, and presence of VFA and lactic acid all serve to stimulate gastric secretions and contractions. The presence of acidic conditions and fat in the duodenum inhibit gastric motility and gastric emptying. The G cells in the pyloric gland area release gastrin hormone into the blood that stimulates parietal cells to release hydrochloric acid. Feedback control of acid release comes from a low pH (approaching pH = 2) of gastric contents that stimulates release of somatostatin. A low duodenal pH also can inhibit acid release, likely through inhibiting gastric emptying.

Pepsinogen, an inactive proteolytic enzyme, is released by chief cells of the gastric mucosa. Pepsinogen is autolytically activated and converted to pepsin in the presence of hydrochloric acid. In addition, lysozymes catalyze the hydrolysis of specific glycosidic bonds in mucopolysaccharides that constitute some bacterial cell walls. Ruminant animals secrete large amounts of specially adapted lysozymes into the abomasal lumen. Active at low pH, lysozymes resist pepsin digestion. Acid, pepsin, and lysozymes chemically and enzymatically digest microbial protein and other digesta, preparing the chyme to enter the duodenum for further digestion.

## INTESTINAL DIGESTION

Acidic chyme (a mixture of partially digested feeds and digestive juices) enters the duodenum for further enzymatic digestion in and absorption of monomers from the SI. Since the SI is a primary site for both digestion and absorption, luminal flow is regulated so as to provide mixing of luminal contents with digestive juices, time for luminal digestion of nutrients, and exposure of digested nutrients to the intestinal wall for absorption.

Presence of acid in the duodenum provides the stimulus for the intestinal wall to release the hormone secretin in the portal blood that in turn stimulates the pancreas and gallbladder to release bicarbonate-rich fluid into the duodenum to partially neutralize acidic chyme. Cholecystokinin (CCK) from the SI wall is released in the presence of fats or proteins in the duodenum to stimulate the pancreas to release enzyme-rich digestive juices (and some bicarbonate) and the gallbladder to release bicarbonate as well as bile acids and salts. The pancreas secretes all of the enzymes necessary to digest lipids, proteins, and carbohydrates; however, in ruminants most (50–90% of starch) of the readily available carbohydrates and lipids (small amounts of triglycerides and galactolipids in the diet) are cleaved to free fatty acids in the rumen so supply of these nutrients is limited. This may explain why pancreatic juice of ruminants is not rich in lipolytic and amylolytic enzymes.

Pancreatic proteolytic enzymes are secreted in proenzyme form, being exopeptidases such as carboxypeptidase A and B or endopeptidases such as trypsinogen, chymotrypsinogen, and elastase. Trypsinogen is activated by intestinal enterokinase to form trypsin that in turn activates remaining trypsin and other proenzymes. The end-products of pancreatic proteolytic digestion are amino acids and oligopeptides (up to 6–10 amino acid chain). Amino acids and some oligopeptides are actively transported into the epithelial cells where more than 90% of the oligopeptides are hydrolyzed to amino acids and actively transported into the blood. The remaining 10% of dipeptides and tripeptides may diffuse directly into the bloodstream.

Sucrose is not digested in the small intestine, but most will be fermented before reaching the small intestine.

Starch is hydrolyzed by pancreatic amylase to maltose while maltase and isomaltase are hydrolyzed to glucose that is actively absorbed. This process is facilitated by the intestinal enzyme maltase present at the brush border of the epithelial cell.

Digestion of lipids (rumen protected) begins by emulsification of fats as they combine with bile acids/salts and form micelles (hydrophilic outer side). Pancreatic lipase then hydrolyzes emulsified triglycerides to form β-monoglycerides and free fatty acids that can diffuse into epithelial cells. Ruminants may absorb triglycerides without further digestion (Wrenn et al., 1978). In the intestinal epithelial cells, fatty acids and monoglycerides are re-esterified to form triglycerides. Combined with cholesterol, cholesterol esters, phospholipids, and a small amount of proteins (lipoproteins), they form chylomicrons for uptake into portal blood and transport. Under most conditions supply of triglycerides is limited because most lipids have been hydrolyzed to glycerol and fatty acids in the rumen, and unsaturated fatty acids also have been saturated therein.

Soluble vitamins and minerals are primarily absorbed from the SI though some are absorbed from the large intestine. Within the colon of ruminants, fermentation may begin again (postgastric fermentation), but the extent of fermentation is limited because chyme already has been fermented in the rumen and enzymatically digested in the SI. However, water and some electrolytes are absorbed in the small and particularly in the large intestine of animals selected under desert conditions with limited access to water.

## DIGESTION IN YOUNG RUMINANTS

### Newborn Phase

When born, ruminants have a small, nonfunctional forestomach with no microorganisms and no acid or pepsinogen secretion. Colostrum directly enters the abomasum and flows to the duodenum where immunoglobulins (IgM antibodies, γ-globulin) are absorbed intact through phagocytosis by the intestinal mucosa. Colostrum contains an antitrypsin factor that prevents digestion of antibodies within the duodenum. Through this process, the newborn kid develops immunity to most diseases to which the adult doe had been exposed. This absorption proceeds for only some 24–48 hours after birth. Colostrum also provides nutrients including lactose and some microbes (lactobacilli). Limited energy reserves due to limited glycogen storage by the liver and inefficient use of lactose as a source of energy are the main causes of death in newborn kids. Exposure to the dam, to fecal matter, and to the environment readily exposes newborn kids to aerobic and anaerobic microbes, the latter eventually colonizing the rumen.

### Preruminant Phase

During this period (up to 3 weeks of age), the newborn is nursing. Most milk bypasses the rumen and flows directly through the reticular groove of the reticulo-rumen to the omasum and abomasum. Sucking stimulates salivary and abomasal secretion more than drinking from a bucket. Saliva of the newborn contains esterase, and the abomasum secretes rennin (chymosin) and hydrochloric acid at this stage of life. Exposed to rennin, milk clots and separates into a hard curd containing butterfat and protein curd (casinogen) that remains in the abomasum for digestion. The remaining whey fraction (albumins, globulins, lactose) leaves the abomasum in bursts. Butterfat is hydrolyzed to glycerol and fatty acids by lipase originating either from milk (mammary glands) or from estrase in saliva. Curd proteins are exposed to further hydrolysis by rennin and acid. In the intestine, the curd, whey proteins, and lactose are completely digested. The intestine of newborns has low maltase activity and therefore cannot fully utilize starch. Intermediary metabolism is driven by glucose, and blood glucose concentrations are sensitive to insulin.

### Transitional Phase

In this phase (3–8 weeks of age), young ruminants will ingest progressively larger amounts of roughages and dry food. These stimulate development of salivary glands and the reticulo-rumen. A population of ruminal microbes becomes established due to ingestion of food, water, cud, fecal matter, and other environmental contaminants. Fermentation of feeds produces VFA that stimulate the growth of reticulo-rumen papillae and omasal leaves. To handle the gases produced and larger particles, the muscle wall has the rumen development and rumen motility initiating the processes of eructation and rumination. As this development progresses, intermediary metabolism shifts from a glucose driven- to a VFA-driven system that is less insulin-sensitive.

### Pre-weaning and Post-weaning Phase

During this period (8 weeks to adulthood), reliance on milk reduces because milk production by the doe decreases. Reticular groove closure becomes erratic and usually is absent in adults. The forestomach proportions and motility change to attain those of an adult, salivary esterase diminishes and salivary urea is present, and abomasal rennin is replaced by pepsinogen, all of which reflect inborn physiological changes allowing the newborn and preruminant to becomes a functional ruminant.

## RUMEN DYSFUNCTION

The major digestive disorders common to ruminants will be discussed briefly. Readers are referred to a more detailed discussion of these problems in textbooks on veterinary physiology (Reece, 2004). Because most dysfunctions reflect specific aberrations in normal rumen function, such discussion helps to reinforce the reader's knowledge about normal rumen function.

### Bloat

Bloat is defined as the distension of the reticulo-rumen associated with accumulation of gases produced by fermentation of certain feeds with failure in gas removal by eructation. Bloat is not common in free ranging ruminants consuming grasses but is manifested by mismanagement of the animal and the diet. Van Soest (1994) divided bloat into two types: legume or frothy bloat, and grain bloat that may be either acute or chronic. Legume or frothy bloat usually is associated with animals grazing rapidly growing legume (alfalfa or white clover) pastures grown in temperate climates. Proteins involved with carbon dioxide fixation apparently uncoil and float to the top of the rumen, and this leads to a foam or froth that traps gas in a form that cannot be eructed. Tropical legumes, temperate trefoil, vetch, or sanfoin contain higher amounts of tannins that precipitate proteins. This prevents formation of stable foam. Grain bloat usually is the result of feeding high grain diets (typically wheat and barley) and pelleted feeds. High concentrate diets that have a very small particle size from grinding and pelleting are often implicated. Such feeds are associated with less saliva production per unit of feed and less salivary mucin that helps protect animals from bloat. High amylolytic activity that produces rumen acids (lactic acid) also may reduce rumen motility to promote bloat. Grain bloat is commonly chronic while legume bloat usually is acute. Acute bloat causes death by placing pressure on the heart or blood vessels preventing flow that results in cardiovascular collapse. Oils or detergents that reduce surface tension help to suppress foam formation. In severe cases the rumen can be punctured to release gases as a last resort. Goats, being intermediate food selectors, have relatively larger salivary glands and their habit of browsing helps makes them more resistant to bloat.

### Acidosis

Acidosis is defined as an imbalance in rumen fermentation elicited by abrupt introduction of rapidly fermented carbohydrates (starches and sugars) to the diet. When an animal is fed a high concentrate feed without being adapted, rapid fermentation by facultative bacteria produces lactic acid and reduces the rumen pH into a range favorable for amylolytic bacteria. Under normal circumstances, secondary rumen bacteria convert lactic acid or other acids to propionate. But secondary fermenters have lower metabolic rates, higher generation intervals (16 hours), and a higher optimum pH (6.2–6.8) than amylolytic bacteria. Therefore, hydrogen disposal is hindered, acids accumulate, and rumen acids accumulate. Rumen pH may drop as low as 4 and cause rumenitis and rumen parakeratosis, a condition where the rumen epithelium is sloughed from the basal membrane. Lactic acid may be absorbed through the rumen wall into the blood and carried to the liver where the natural L (+) lactate form can be metabolized, but the bacterial form of lactate (the D [−] isomer) may accumulate in the blood causing systemic acidosis and death from a reduced oxygen carrying capacity of hemoglobin. Treatment often involved dosing with bicarbonate buffers while prevention involved feeding more roughage, particularly long hay that induces rumination and saliva production. In severe cases, the rumen may be partially or fully evacuated and replaced with fresh rumen fluid from a healthy animal.

### Displaced Abomasum

Ruminants fed finely ground, very high concentrate diets tend to have highly acidic and fluid ruminal digesta with little fibrous material. High acidity and low fiber content of digesta reaching the abomasums is associated with reduced motility and altered gastric function. Impaired motility causes abomasal distension. The abomasum may become engorged with fluids and gas and become displaced laterally. Left displacement is associated with a chronic condition, whereas right displacement is associated with blockage of digesta flow and may prove fatal. The actual cause of this symptom is not clearly understood, but its association with high starch, fine particle diets has been documented.

### Urea/Ammonia Toxicity

Urea or ammonia toxicity usually occurs when an excessive amounts of urea is consumed by an animal. Animals grazing lush forage pastures with highly soluble proteins also may be at risk. Bacterial urease rapidly hydrolyzes urea-forming ammonia. In the presence of an adequate supply of readily available carbohydrates such as starch and sugars that are rapidly fermented (amylolytic fermentation), rumen microbes will use available ammonia to produce microbial protein. However, when supply of energy is low, microbial use of ammonia for protein synthesis will be reduced. When ammonia accumulates in the rumen, this increases ruminal pH. An elevated pH increases the amount of ammonia in the un-ionized, absorbable form exacerbating ammonia absorption. The liver can detoxify ammonia by converting it to urea, but some ammonia may bypass the liver through uptake into the lymph system and enter the blood directly. Ammonia toxicity is associated with systemic alkalosis that, probably through altering calcium and magnesium status, intoxicates the central nervous system. Preventive measures include avoiding high urea diets or feeding readily fermented carbohydrates when feeding urea or when animals graze lush, rapidly growing, high protein pastures. Ruminal bacteria adapted to urea form ammonia less rapidly, so adaptation is recommended. Finally, chemical inhibitors of urease or slow release urea complexes can slow the rate of ammonia production and will help avoid toxicity. Treatments include administering VFA (vinegar) to lower rumen pH and reduce the rate of absorption of ammonia. In severe cases, rumen evacuation may be necessary.

### Nitrite-Nitrate Toxicity

Nitrite/nitrate toxicity occurs when ruminants consume forage from stressed, often overfertilized grass pastures and crops. Such plants accumulate nitrate as its potassium salt. Accumulation by plants is enhanced by drought and, with cool-season grasses, by low temperature and cloudiness (low light). Nitrate itself is not toxic, but during normal rumen fermentation, nitrate is reduced to nitrite (a toxic intermediate). Nitrite can be converted further into ammonia that in turn can be used to produce microbial protein. If the nitrite intermediate accumulates in the rumen, it is absorbed into the blood and there it unites with hemoglobin and produces methemoglobin. Compared to hemoglobin, methemoglobin has a reduced ability to transport oxygen, and impaired oxygen delivery to tissues will cause death by asphyxiation. Gradually increasing the nitrate concentration of the diet will permit microbes or animals to adapt to nitrate and increase their tolerance to it.

## SECONDARY PLANT METABOLITES

Certain compounds found in feeds, forbs, and browse are produced as defenses against invasion by pathogens and, in some cases, against consumption by herbivores. Compared with cattle and sheep, goats have a very high tolerance to such compounds.

### Organic Nitro Compounds

Glycoside compounds found in various *Astragalus* species such as crown vetch (Coronilla varia L.) and timber milkvetch (Astragalus miser varia T.) are rapidly hydrolyzed in the rumen producing toxic nitro compounds, that is, 3-nitropropionic acid (NPA) and 3-nitroproponal (NPOH). These compounds can be metabolized in the rumen or absorbed. NPA is toxic while NPOH is converted to NPA by the liver. Rumen microbes may partially metabolize these compounds by reduction of aliphatic nitro groups to their corresponding amines, 3-aminopropanol, and β-alanine, respectively. This explains why ruminants have greater tolerance to these compounds than nonruminants do.

### Mimosine Toxicity

Mimosine is an alkaloid, β-3-hydroxy-4 pyridone amino acid, found in genus Leucaena and few other Mimosa species. Leucaena is a legume shrub/tree that provides a high quality feed in certain areas of the tropical and subtropical regions of the world. However, when these feeds comprise more than 30% of the diet, they may cause severe toxicity. Goats may lose hair when fed more than 50% leucaena in the diet. Rumen microbes convert mimosine to 3,4 dihydroxy pyridone (3,4 DHP), a toxic goitrogenic intermediate. However, certain strains of bacteria found in tropical ruminants can degrade mimosine to nontoxic products. Toxicosis has been reported in Australia, Papua New Guinea, Africa, and Florida. Pure colonies of bacteria that can degrade 3,4 DHP were isolated (*Synergestes jonesii*) and have been inoculated successfully into the rumen, increasing tolerance to mimosine (Hammond, 1995).

### Tannins

Condensed tannins (CT) are polyphenolic compounds present in plants that when eaten may have either positive or negative effects on animals depending on the concentration in the forage and ability of the rumen environment to degrade tannins. Proline rich proteins present in the saliva of certain herbivores including goats bind tannins and reduce their adverse effect on rumen microorganisms. However, the complexes formed are indigestible and are excreted in the feces (Hagerman et al., 1992). Condensed tannins at low levels, 2–5% of dry matter (DM), will bind with ruminally degraded proteins to increase the supply of protein reaching the SI and, when digested, these proteins can improve the amino acid balance of animals. However, higher levels of CT (above 5% of DM) generally have negative effects on ruminal digestion and decrease digestibility. Tannins have shown promise as an anthelmintic for combating nematode infestation of small ruminants in humid and warm environments (see Chapter 11 for more details). This topic deserves further research attention.

## METABOLIC DISORDERS

Most prominent disorders are caused by improper feeding management and nutrient imbalances in the feeds offered to goats. Animals selected for rapid growth or high rates of milk production may have a genetic predisposition to these disorders.

### Grass Tetany or Hypomagnesemia

Grass tetany is a metabolic disorder associated with hypomagnesemia (low magnesium concentrations of blood). This condition is most prevalent when animals are milking heavily and have a higher requirement for magnesium or early in spring when animals are grazing rapidly growing pastures that are heavily fertilized. Grass tetany occurs when pastures are low in magnesium but rich in nitrogen and potassium that result in a low magnesium : potassium ratio. In addition, plants involved may have high concentrations of tricarbalyllic acid (propane-1,2,3-tricarboxylic acid) that binds magnesium. Clinical signs depend on the severity of the magnesium deficiency and whether animals also are hypocalcemic (low blood calcium concentrations). Affected animals have low feed intake and low milk production. The condition may be chronic and if undetected will predispose an animal to milk fever. Feeding magnesium supplements as magnesium oxide is advised. Magnesium can be mixed with grain for feeding to pregnant and lactating does. Feeding ionophores like monensin to growing animals may increase activity of the sodium-linked magnesium transport system in the rumen and increase efficiency of magnesium absorption.

### Milk Fever

Milk fever is a metabolic disorder apparent as hypocalcemia (low blood calcium concentration) of milking does that occurs just after kidding. This disorder usually is related closely to hypophosphatemia (low blood phosphorus concentration) and hypomagnesemia (low blood magnesium concentration). During milk fever, the calcium homeostatic mechanism of goats fails to maintain blood calcium at a normal level. Because calcium is essential for muscle contractions and nerve function, hypocalcemic animals cannot rise or eat. At parturition with the onset of milk production, the drain on blood calcium for milk must be replenished either by additional absorption of dietary calcium or mobilization from bone reserves. Under normal conditions, hypocalcemia triggers parathyroid hormone (PTH) release reducing urinary calcium loss, stimulating calcium resorption, and promoting synthesis of 1,25-dihydroxyvitamin D to enhance intestinal transport of calcium. The lack of a timely response of any of these three mechanisms provokes milk fever. An acid-base imbalance of the body at parturition is a predisposing factor. Metabolic alkalosis impairs the ability of PTH to function normally. Injected calcium gluconate can keep the animal alive until calcium homeostasis is restored. Preventive measures include feeding an acidic (low calcium) diet during late pregnancy to provoke PTH release and prepare the metabolic system to mobilize bone reserves and increase the efficiency of urinary and intestinal calcium transport. Preventive practices include feeding less salt and potassium and improving phosphorous and magnesium intake; adding anions such as ammonium, calcium chloride, and magnesium chloride and sulfate to induce mild metabolic acidosis; and feeding low calcium diets to stimulate PHT release prior to parturition. Oral calcium drenching and vitamin D supplementation at kidding also can help.

### Ketosis or Pregnancy Toxemia

Ketosis is a metabolic disorder caused by a negative glucose balance that, combined with an energy drain, provokes fat mobilization. Ketosis usually occurs during late gestation (pregnancy toxemia) or 2–4 weeks after parturition. It commonly is associated with accumulation of triglycerides in the liver combined with depressed glycogen levels. A high amount of glucose is required either for development of multiple fetuses at late gestation or for milk production; this is responsible for the hypoglycemic condition of the animal. As a result of the energy drain, body fat is mobilized leading to an increase in nonesterified fatty acid (NEFA) concentrations in blood that flood the liver with lipid. Esterified fatty acids normally would be exported from the liver or stored. But the mechanism to export triglycerides in ruminants is slow, leading to fat accumulation in the liver. With incomplete oxidation of fat, NEFA are converted to ketone bodies, primarily acetoacetate and β-hydroxybutyrate that are released into blood. Ketone bodies often increase in blood when the amount of energy needed for milk production exceeds energy intake. Feeding management that reduces the severity and length of the negative energy balance can help prevent ketosis. Extra supplemental concentrate fed 1 week before and after kidding reduces the incidence of fatty liver and ketosis. Preventive measures include increasing the feed supply the 2 weeks around kidding; avoiding overconditioning animals during pregnancy;

changing diets gradually; and avoiding environmental stress. Feeding glucogenic compounds like sodium propionate or oral doses of propylene glycol will reduce NEFA formation at kidding and ketone body formation after kidding and may alleviate ketosis. Injection of glucose or glucose-forming compounds will temporarily elevate insulin to suppress fat mobilization and ketone body formation.

## REQUIRED NUTRIENTS

Many nutrients are required in the goat's diet for metabolism, both for maintenance of body functions, and for production that includes tissue accretion (growth), reproduction, and production of meat, milk, and fiber. The National Research Council (2007) published an extensive review of nutrient requirements of goats based on current scientific information on goats. Where information about goats was lacking, information from cattle and sheep was used. The nutrient requirements for different classes of goats are presented in Appendix B. For more detailed information on nutrient requirements of goats, the reader should refer to NRC (2007) and Sahlu et al. (2004).

Specific classes of nutrients include carbohydrates and lipids that provide energy; protein or nonprotein nitrogen that provide amino acids and energy; vitamins; minerals; and water. Though often ignored, water is classified as a nutrient that is necessary for digestion, metabolism, and products.

### Energy

Energy, as fuel for the body, is defined as the potential to do work. The international unit of energy is the joule (J); however, the calorie (cal) is used most often in the United States. One calorie, equal to 4.184 J, is defined as the amount of energy required to raise the temperature of one gram of water from 15.5–16.5°C at one atmospheric pressure. This is the calorie, sometimes called a small calorie. One calorie or joule is a small amount of energy; therefore, feed or body energy utilization typically is expressed in terms of kilo (1,000; kcal or kJ) or mega (1,000,000; Mcal or MJ) calories or joules. All functions of the body including prehension, digestion, and metabolism require energy. Energy in the body is in constant flux. Amounts of energy required will vary with breed, sex, age, climatic conditions, and activity. Energy is expended for maintenance, growth, reproduction, and formation of products.

#### Flow of Energy in the Body

Ingested energy (gross energy [GE]) represents input while its destination is expressed as net energy (NE) for body maintenance, growth, and production. Energy loss occurs at each stage of digestion and metabolism. Readers are referred to NRC (1981a) for more detailed information on nutritional energetics. Gross energy is measured as the amount of heat released when 1 gram of feed is oxidized to carbon dioxide and water in bomb calorimeter. The GE value of feeds is proportional to the carbon and hydrogen contents of the feed's organic matter, carbohydrates, proteins, and lipids. Because all GE is not available to animals, GE is not precisely related to usefulness of a feed to an animal.

Digestible energy (DE) is an index of the amount of feed energy value presumably available for meeting an animal's requirement for energy. Within the GI tract, available GE is digested while waste is removed from the system as feces (fecal energy [FE]). Apparent digestible energy is the difference in energy between GE and FE. The term "apparent" reflects the fact that some energy in feces does not come from feed but instead represents inherent loss of enzymes, sloughing tissues, and microbial cells that collectively are called "metabolic fecal matter." Depending on the nature of feeds, DE can be as high as 80% for concentrate diets and as low as 50% for forage diets. Straw may be even lower (45% DE).

Total digestible nutrients (TDN) is another index of available feed energy that accounts for the higher energy content of lipids. TDN is the sum of digestible nitrogen free extract (NFE) (carbohydrates), digestible crude protein (CP), digestible crude fiber (CF), and 2.25 times the digestible ether extract (lipids) for animals fed at an energy intake equal to maintenance. TDN can be approximated from DE, with 1 kg TDN being about 4.4 Mcal DE, or by empirical equations using feed composition data such as CP, neutral detergent fiber (NDF), and acid detergent fiber (ADF) together with their predicted digestibility. The TDN value of a feed may be lower when feed intake is above maintenance due to reduced time for digestion of nutrients by the GI tract. Total digestible nutrient values also appear to overestimate the energy availability from diets rich in fiber relative to high concentrate diets (Moore et al., 1953).

Metabolizable energy (ME) deducts energy lost in urine (UE) and energy lost in rumen gases (GasE) or methane from DE. Compared with DE or TDN, ME more precisely estimates the usable energy available to support tissue accretion, milk, and conceptus. The energy loss in urine (4–5% of GE) and heat of gases loss in fermentation (4–5% of GE) can be considerable. The UE losses are higher for ruminants than nonruminants because microbial nucleic acid by-products as well as urea are excreted in urine. Gaseous energy loss increases as the level of dietary fiber increases, but it is reduced as a percent of GE when energy intake increases. Consequently, methane losses are less when animals have high feed intakes or consume either high-concentrate diets (Johnson and Johnson, 1995) or when browse comprises an important dietary component (Woodward et al., 2001). Components of browse including essential oils or unsaturated fatty acids might inhibit protozoa or methanogenic bacteria to reduce GasE. Loss of energy in UE and GasE are quite predictable and result in a high correlation between DE and ME. The ME for ruminants generally is calculated as ME = DE × 0.82 for forage-based diets; however, this may be an underestimate with diets rich in concentrate or browse that tend to have lower gaseous losses.

Because sufficient data are not available to calculate the net energy (NE) of feeds or the net energy requirements of goats, NE is used less commonly than DE, TDN, and ME for formulating diets. The NE considers an additional loss of energy, heat increment (HiE). Heat increment is subtracted from ME as NE = ME − HiE. Heat increment is defined as the increase in heat production following feed consumption in a thermoneutral environment (NRC, 1981a). It includes both the heat of fermentation in the GI tract and the heat of metabolism (i.e., heat released when nutrients are metabolized). Heat increment is useful to keep animals warm when they are exposed to low environmental temperatures but presents a burden otherwise, and must be dissipated. Heat increment losses account for 25–40% of GE, increasing as fiber content of the diet increases, and as feed intake and tissue gain increase. Net energy relates more closely to animal performance than DE, TDN and ME, each of which overestimate energy value of feeds.

#### Net Energy Requirements

Net energy required for maintenance is that portion of energy used for basal metabolism, muscular activity, tissue repair, involuntary metabolic processes, and voluntary activities that are necessary to sustain life. Life-sustaining activities account for walking to seek and obtain food, browse, shade, feeders and waterers, or other related activities such as social activities specific to goats like jumping, playing or fighting. Maintaining body temperature under extreme environmental conditions also will impact the maintenance energy requirement. Energy available in excess of maintenance is available for a wide variety of activities such as tissue gain, reproduction, lactation, hair/wool production, or physical work.

Composition of tissue gain varies depending on the age of the animal. Younger animals tend to gain more water and protein, but as animals age, more water and fat are deposited. Gender of the animal also can influence the pattern that animals deposit protein or lipid. Deposited water plus protein (lean) requires less energy (1.2 Mcal/kg) than deposited fat plus water (8 Mcal/kg). Other factors such as genotype, rate of gain, or energy density of diet may affect energy requirements for gain.

Energy required for reproduction depends on the stage of pregnancy, number of fetuses carried, and development of mammary tissues. The additional energy required for pregnancy typically is so small that they can be ignored until about 100 days in gestation. Most of the fetal and mammary growth occur during the last 50–60 days of gestation for goats. Data are insufficient to determine requirements for pregnancy of goats accurately. Estimating or measuring the energy requirements for and efficiency of lactation also are complicated because animals mobilize body energy reserves to produce milk while, later in lactation, body energy reserves are replenished simultaneous with lactation. Efficiency of converting energy from feed to energy in milk for ruminants averages 0.62; however, efficiency can be higher (0.84; ARC, 1980) when mobilized energy is used for milk production.

Energy requirements for fiber growth in Angora or other fiber-producing goats depend on rate of fiber growth, the amount of energy in the fiber, and energetic efficiency of fiber growth. Energy also is expended for exercise or work. According to the NRC (1981b), additional energy needs for low, medium, and high activities are 25, 50, and 75% of maintenance energy requirements, respectively.

#### Energy Sources

Energy is released during oxidation of nutrients in the body, being produced with oxidation of carbohydrates, fats, proteins, and other organic compounds by the body. Any organic compound capable of being fermented and converted to VFA, or of being digested and absorbed as a monosaccharide or fat from the digestive tract can enter the Kreb's cycle and be converted to energy. One gram of carbohydrate or protein upon oxidation will produce 4–5 calories of energy while a gram of fat, with a higher energy density, can produce 9 calories. With most practical diets, carbohydrates provide the majority of energy for ruminants with only a small portion coming from fats. Proteins generally are the most expensive part of the diet and are fed to meet protein (amino acid) requirements, but excesses will be oxidized yielding energy.

##### *Carbohydrates*

Carbohydrates, also called polysaccharides, usually comprising 60–70% of ruminant diets, are the main source of energy for ruminants. They include diverse compounds classified as monosaccharides (single sugar unit), such as glucose, fructose, or polymers of monosaccharides, that is, oligosaccharides (2–10 sugar units), such as maltose and lactose, and polysaccharides, such as cellulose, hemicellulose, and starch. Carbohydrates also are divided into two main groups based on their availability for fermentation or digestion: (1) nonstructural carbohydrates (NSC) being found inside the cells of plants that are readily available and fermented such as sugars and starch, and (2) structural carbohydrates (SC) being present as plant cell walls that resist digestion but are partially fermented in the rumen and in the cecum/colon.

Nonstructural carbohydrates include sugars, starches, organic acids, and other reserve carbohydrates such as fructans. Starches generally comprise 50–100% of NSC in most plants. NSCs are well digested and provide the main source of energy for both ruminal microbes and the host ruminant. Nonstructural carbohydrates typically are measured by enzymatic methods and differ slightly from nonfibrous carbohydrates (NFC) that are calculated by subtraction as NFC = 100 – (%NDF + % CP + % Fat + % Ash). The difference between NSC and NFC includes pectin and organic acids. Pectin is not included in NSC. The levels of NSC in the diets of dairy cattle should not exceed 30–40% of the ration dry matter while the maximum level for NFC can be slightly (2–3%) greater. The optimal concentration of NSC or NFC in the diet of dairy goats has not been determined.

Structural carbohydrates separated from cell contents by solubilizing away materials solvent in acid and base yield crude fiber (CF), or in a pH neutral detergent solution yield neutral detergent fiber (NDF), or an acid detergent solution yielding acid detergent fiber (ADF). Neutral detergent fiber represents much of the fiber in the plant cell walls and includes cellulose, hemicellulose, and lignin. Acid detergent fiber is equal to NDF minus hemicellulose. Generally, ADF is less digestible than NDF, and ADF concentration in a feed is negatively related to energy digestibility of that feed. However, digestibility of NDF is affected by its source (from forage or grain), and the proportions of its components: cellulose, hemicellulose, and lignin. NDF from nonforage and present in small particles is less effective for maintaining rumen pH than NDF from coarser forage particles. The recommended dietary requirements for dairy cattle for NDF to support optimum milk production with no depression of milk fat is set at 25% of diet dry matter with not less than 19% of NDF from forage (NRC, 2001). There is no recommendation stated for dairy goats.

*Lipids*

Lipids are organic compounds defined by being soluble in a nonpolar solvent like ether or chloroform. Lipids will include long chain fatty acids (FA), triglycerides, phospholipids, and other substances such as sterols and cholesterol. Lipids are generally classified as (1) simple lipids, mainly neural fats and waxes; (2) compound lipids, including phospholipids, such as lecithin, cephalins, and nonphosphorylated lipids, such as glycolipids and lipoproteins; and (3) derived lipids, such as fatty acids and sterols. Fats are lipids, but not all lipids are fats. For example, petroleum products are lipids but are not fats. Fat generally refers to stored, energy-rich compounds that have high concentrations of long-chain fatty acids including triglycerides, phospholipids, nonesterified fatty acids, and salts of long-chain fatty acids. Fats are dense sources of energy providing more than twice the energy per unit of weight of carbohydrates and proteins with more than 85–90% of this additional energy being available for metabolism. Fats also provide fat-soluble vitamins such as A, D, and E that rumen microbes are unable to synthesize in the rumen, and the essential fatty acids such as linoleic acid and linolenic acid. Fat is present in grains and forage at levels of 2–5% of dry weight. In addition, fats often are added to diets of ruminants in the form of oilseeds, animal or animal-vegetable blends (with a saturated : unsaturated FA ratio of 1 : 1), dry granular fat, or rumen-protected fat. Feeding more than 6–7% of diet DM in the form of fat will decrease intake and has been associated with depression in cellulose digestion in the rumen.

## Proteins

Proteins are large molecular weight nitrogenous compounds composed of amino acids. Proteins are vital to living cells and play important roles (1) as enzymes, hormones, and structural components of the cells, (2) for immunity and heredity, and (3) for oxygen transport, muscle contraction, acid-base balance, osmotic pressure, and blood clotting. Proteins generally are classified as either (1) simple proteins that upon hydrolysis yield mainly amino acids and their derivatives, (2) conjugated proteins, or (3) derived proteins. Simple proteins are classified based on their solubility as globular proteins (albumin, globulin, glutelins, and prolamines) or fibrous, less soluble proteins (collagen, elastin, and keratin). Conjugated proteins are simple proteins with an additional nonprotein prosthetic group, such as nucleoproteins, glycoproteins, metaloproteins, etc. Derived proteins include peptones and metalloproteins. Globular proteins are present in all feeds, whereas fibrous proteins are more abundant in feeds of animal origin. Seed proteins are rich in glutelins and prolamines, whereas leaf protein is primarily albumin. Feedstuffs also contain NPN compounds, peptides, free amino acids, nucleic acids, amides, amines, and ammonia. Grasses and legumes have the highest and most variable content of NPN (nonprotein nitrogen, typically being any protein that is not precipitated by a protein precipitant like tungstic or trichloroacetic acid). The NPN content of fresh forage is mainly short peptides, free AA, and nitrates, whereas fermented forage is rich in free AA, ammonia, and amines but lower in peptides and amines.

#### Metabolizable Protein

Metabolizable protein is used for maintaining and repairing body tissues, tissue gain, conceptus gain, and milk or wool production. The goal of feeding protein to ruminants is to provide amino acids to complement the microbial protein and supply dietary protein that escapes destruction within the rumen and to reach the small intestine to be digested therein and provide AA for absorption. The crude protein and digestible protein systems commonly were used in the past as indices of ruminants' protein needs. These have been displaced through the metabolizable protein (MP) concept that should define requirements more accurately. True MP supply is the total of microbial protein synthesized in the rumen plus the amount of dietary protein that escapes fermentation (UIP) that in turn is supplied by the diet and by recycled nitrogen. Microbial protein synthesis in the rumen requires both a nitrogen source and available energy. Most cellulolytic bacteria require ammonia for growth while amylolytic bacteria may require amino acids. Protozoa meet their nitrogen need from digestion of bacteria, feed, or fungal protein. Outflow of microbial protein from the rumen is associated with both liquids and solids. Faster rates of rumen outflow will increase microbial yield by reducing the time that microbes spend in the rumen and thereby the amount of energy bacteria used for maintenance. Protozoa have a slower generation rate and limited outflow from the rumen and thereby have a very low efficiency for production of microbial protein. Passage rate also can affect how much of the intake protein is degraded in the rumen. Although dietary protein is divided into two fractions, degradable intake protein (DIP) and undegradable intake protein (UIP), extent of degradation of DIP will vary with time available for degradation that in turn is altered by level of feed intake. This in turn influences the amount of MP available. In the small intestine, amino acid absorption rate appears more variable for UIP than that for microbial protein. Small intestinal protein digestion is assumed to be 80–85% and MP is estimated to be between 60–64% of CP.

#### Nitrogen Recycling

The liver synthesizes urea from ammonia in blood. Ammonia in turn is absorbed through the rumen or intestinal wall or is produced during nitrogen metabolism by tissues. In goats, some 18–85% of blood urea is recycled to the rumen either via saliva or directly via diffusion through the rumen wall (NRC, 2007). The intakes of ruminally degradable intake protein and available fermentable carbohydrates regulate the degree to which intake or recycled urea is used whereas salivary flow and rumen pH influence the extent to which urea is recycled and to which ammonia is retained within the rumen. Recycled nitrogen (urea) contributes to rumen microbial protein, fecal nitrogen, and urine nitrogen. The metabolizable protein represents the proportion of protein that is digested (DP) and absorbed and not eliminated by the kidneys. Nitrogen recycling by ruminants is a significant survival tool that helps conserve protein when quality of feed is marginal. However, urea recycling may be simply a fortunate side benefit of adaptation to a desert environment. Through recycling N, excretion of urine is reduced and this in turn reduces the need for water, a factor important for survival under desert conditions.

### Minerals

Minerals are essential components of a diet and play multiple major roles in the body. Collectively they are assayed as and thereby called "ash" and comprise the inorganic portion of the diet. Minerals do not yield energy or produce protein for the animal, but their presence is crucial for nutrient metabolism. Minerals also provide structure with bone and teeth formation; play significant roles as electrolytes in acid-base balance and body fluid volume regulation; maintain osmotic pressure, membrane permeability, and nervous transmission; regulate cell replication and differentiation; and act as coenzymes or cofactors in metabolic activities and body immune function. Fourteen different mineral elements have been identified as required in the diet of goats.

Minerals generally classified based on the quantities required as either (1) macrominerals or major minerals that are needed in "gram per day" quantities and include calcium, phosphorous, sodium, chloride, potassium, magnesium, and sulfur or (2) microminerals or trace minerals that are needed in very small amounts as "mg per kg" and include cobalt, copper, iodine, iron, manganese, selenium, and zinc. Additional trace elements that may be needed under certain conditions include aluminum, arsenic, boron, chromium, fluorine, silicon, tin, and vanadium (Underwood and Suttle, 1999). Most minerals in excess can prove to be toxic. Minerals whose excesses are less tolerated by animals include fluorine, molybdenum, lead, arsenic, aluminum, cadmium, and mercury.

The main source of minerals is various feeds and occasionally water and air. Soil also can provide minerals directly through consumption during grazing and feeding. Plants generally have a mineral composition characteristic of the soil though soil pH can alter the availability of soil minerals for plants. Acid soils lead to decreased availability of most macrominerals but will increase the availability of many trace minerals. Alkaline soils may increase molybdenum or selenium availability but will decrease the availability of most trace minerals.

Most forage is a rich source of potassium and iron but often is deficient in sodium, copper, selenium, and possibly iodine. Cereal grains are deficient in most minerals. High protein feeds generally are richer in mineral content than forage and cereal grains. Legume forage usually is richer in minerals than grasses, and leaves are richer than stems. As forage matures, protein content decreases while phytates and oxalates will increase; these changes reduce mineral content and mineral bioavailability.

Nutrient balance among minerals will alter mineral absorption. Specific imbalances, induced by man through mismanagement or natural conditions, can result in mineral interactions and reduced mineral absorption. For example, acid soils promote molybdenum absorption by plants, and high molybdenum intakes may reduce copper uptake and induce copper deficiencies in animals. Goats given the opportunity for browsing tend to balance their mineral needs; however, in confinement, deficiencies or toxicities can occur.

One general symptom of mineral deficiency is "pica." Goats exhibit pica as a peculiar craving that results in eating or chewing on wood or digging or licking soil. Assessment of mineral status of animals can help to identify mineral excesses or deficiencies and diagnose the source of a pica problem. This assessment includes collective knowledge of mineral content of the feeds, water, soil, animal fluids and tissues, as well as clinical signs and symptoms. Diagnosis is confirmed when the problem is corrected and the animal recovers.

Physiological stages of production may impact mineral requirements and if not corrected, induce deficiencies. Minerals are needed for body metabolism and for all phases of growth and production, particularly for skeletal growth of young animals, conceptus growth of pregnant does, and milk and fiber production. Mineral requirements of goats are not fully understood or investigated, and established requirements are extrapolated largely from information from cattle and sheep. Recent research findings on mineral requirements of goats confirm major differences between species and have been documented for copper (Solaiman et al., 2001).

### Vitamins

Vitamins are complex organic compounds that are needed only in small amounts, that perform multiple physiological functions, and that are involved in many metabolic processes. Vitamins are generally classified by their solubility as either (1) water soluble vitamins, that include the B complex vitamins and vitamin C, and (2) fat soluble vitamins, A, D, E, and K. Although fat-soluble vitamins only have oxygen, hydrogen, and carbon in their structure, water-soluble vitamins are more variable in composition with some containing nitrogen, sulfur, and cobalt. Water-soluble vitamins are synthesized in the rumen by fermentation in the presence of adequate cobalt for vitamin $B_{12}$ and sulfur for thiamin and biotin synthesis. Although ruminants may be self-sufficient for synthesizing their own B vitamins, supply of certain B vitamins may limit production under some conditions. Ruminants can synthesize vitamin C (ascorbic acid) because they have the enzyme L-gulanolactone oxidase to convert gulanic acid to gulanolactone during ascorbic acid synthesis. Fat-soluble vitamin A is present in forage as its precursor $\beta$-carotene, but vitamin E often is added to diets. Vitamin D can be obtained through exposure of skin to sunlight but must be supplied when animals are indoors or have limited exposure to ultraviolet arrays. Vitamin K can be produced by rumen fermentation when the rumen is free from antivitamin K activities (coumarin) produced by molds that grow on white clover and produce the antivitamin dicoumarol. Although some fat-soluble vitamins are required in the goat's diet, research data to establish requirements are lacking; therefore, requirements have been extrapolated from data for cattle and some from data from sheep.

### Water

Water is the largest single component of the animal's body making up between 50 and 81% of total weight at various stages of development. Water has unique properties and functions in metabolism and physically. Water facilitates cellular reactions with its high dielectric constant and hydrogen-binding property that promotes ionization of electrolytes and allows oppositely charged ions to move independently. Water helps in the transportation of metabolites, nutrients, and hormones throughout the body. In the gastrointestinal tract, water provides moisture for rumen fermentation and aides in excretion of waste. Certain characteristics of water, including its high specific heat, high thermal conductivity, and latent heat of vaporization, assist in body temperature regulation. Water requires heat to raise its temperature, has a high heat transfer capability, and requires heat to change from liquid phase to vapor or to solid phase. Vaporization of water is the main route by which the body will lose heat. Other functions include lubrication of joints and conduction of sound through the body.

Body water is gradually displaced by body fat as animals grow and mature. Milk production increases the water requirement and body water content. Males tend to have more body water than females. Turnover rate of water in the body under normal conditions ranges from 2–8 days but may increase with potassium and salt intake, temperature, and humidity.

An animal obtains its water from drinking, from water in feeds (called bound water or preformed water), and from metabolic water. Metabolic water is the water released during oxidation of the carbohydrates, fats, and proteins. Metabolic water of nutrients relates to their oxidation state. Fats are the least oxidized compounds and have a high ratio of carbon and hydrogen to oxygen. One hundred grams of fat when metabolized will release 109 grams of water, carbohydrates yield only 60 grams, and proteins yield only 42 grams. Disposal of end-products of protein metabolism (urea) requires additional water. Preformed and metabolic water can meet the water needs of many grazing animals.

The body loses water through four main channels: (1) kidneys, (2) skin, (3) lungs, and (4) intestines. Kidney water excretion through urine is under hormonal adrenocorticotropic hormone (ACTH) control. Water loss through skin is either insensible (radiation) or sensible (perspiration) and is affected directly by solar radiation input, temperature, humidity, and wind velocity. Lungs lose water through vaporization; this loss is affected directly by temperature and animals' activity. Water loss through the lungs and skin, accounts for one-third of total water loss. Intestinal water loss through feces is affected by animal species, being lower for sheep and goats than cattle. Water lost by any channel except lungs will also include loss of electrolytes.

Water requirements are not fixed. Under normal conditions, voluntary water intake is proportional to feed intake. However, high ambient temperature will increase water consumption. An animal may drink only 3 liters of water per kg of feed at 5–6°C versus about 15 liters at 40°C. Season of the year also can change the water requirement. Daily water intake generally is estimated at 5–6% of body weight but may be 10% of body weight in extreme cases. Dietary factors as well as environmental factors can influence water intake. Dietary factors that increase water intake include intake of dry matter, protein, fat, and salt. Feeds with higher water content such as pastures and fresh cuts may displace intake of imbibed water. However, water intake often increases with intake of ensiled feeds because of their higher osmolarity. The primary environmental factors that affect water intake are high temperature and high humidity; both increase water intake. Design, accessibility, and cleanness of water also may alter water consumption by goats.

#### Water Distribution

Total body water (TBW) is distributed into two major compartments: (1) extracellular fluid (ECF) that comprises some 31–38% of TBW, and (2) intracellular fluid (ICF) that comprises the remaining 62–69% of TBW. The ECF includes blood plasma (25%) and intestinal fluids (75%) with volume being regulated by maintaining a constant Na concentration. The ICF includes water plus other regulatory electrolytes, potassium, other inorganic ions, and different proteins within cells. The volume of ICF is closely regulated by volume sensors, hormones, and water transport mechanism supported by the liver, heart, and kidneys. Fluid exchange between the two pools is important for the survival of animals under desert conditions and the Mediterranean environment. Water content of the forestomach will comprise from 10–30% of TBW and is a major temporary water reservoir whose volume will change over time with meals and drinking bouts. Withholding food and water for 20–24 hours is a routine procedure that will minimize the variation in body weight measurement or "shrunk weight." Alternatively, body weight measured without food and water restriction is called "unshrunk weight."

#### Water Requirements

Requirements for water are not determined directly but instead the requirement typically is calculated as voluntary intake of water for animals given free choice access to water. Water intake usually is calculated from dry matter intake according to the following equation (NRC, 1985):

$$\text{Total Water Intake (liter/day)} = 3.86 \times \text{Dry Matter Intake} - 0.99$$

Adjustments are needed to consider effects of season, environmental temperature, breed of animal, and physiological stages of growth. Water intake usually is higher in summer and with hot temperatures. Breeds of animals adapted through natural selection to a shortage of water may consume less water and lose less water with feces. Gestation may increase water intake an average of 126%, being even greater for animals bearing twins and triplets. Water needed for lactation is estimated at 3.5 liter of water/kg of milk production. The water content of milk is usually sufficient for growing the newborn up to 2–3 weeks of age; however, providing free access to extra water is recommended. The water need for nursing offspring is set at 8–13 mL/g body weight gain for growth plus another 120 – 140 mL/kg of metabolic body weight ($BW^{0.75}$) for maintenance. Water needs for weaned animals for growth is set at 7–8 mL/g of body weight gain with another 143 mL/kg $BW^{0.75}$ required for maintenance (NRC, 2007).

## SUMMARY

Goats are one of four existing true ruminant domesticated agricultural species among ungulates that continue to be very successful forage-consuming mammals. Goats benefit from a predigestive fermentation GI tract that allows for the breakdown of fiber to yield energy-rich fermentation products and the synthesis of microbial protein from nonprotein nitrogen sources or recycled metabolic nitrogen. The other benefits include production of vitamin B complex. Goats use visual cues and taste to select for diets based on the concentrate part of the plants and for shrubs and browse. They have larger salivary glands that secrete tannin-binding protein in the saliva to detoxify tannin associated with browse species. The newborn kid and up to 3 weeks of age is preruminant, its intermediary metabolism is glucose driven, and blood glucose is insulin sensitive. By 8 weeks as a young ruminant, the intermediary metabolism shifts to VFA driven and is less insulin sensitive.

Rumen dysfunction is not common in free-ranging ruminants, and it appears to be manifested by mismanagement of normal balances between animal and the diet. They may include bloat, rumen acidosis, displaced abomasums, urea toxicity, and nitrate toxicity. Goats require nutrients for maintenance, growth, gestation, and production of milk, meat, and fiber. Carbohydrates and lipids provide required energy; proteins are provided through true protein or nonprotein nitrogen; vitamins and minerals are provided through feeds or supplements; and water is provided free choice. Knowledge of work of digestion and nutrient metabolism in goats will help in proper management and decision making when raising goats. Some of the information presented in this chapter has been extrapolated from sheep and cattle data. Information regarding digestive physiology and nutrient requirements for goats is very limited and needs further investigation.

## REFERENCES

ARC. 1980. The nutrient requirements of ruminant livestock. Technical Review by an Agricultural Research Council Working Party. Sci. 80:2429–2441.

Austin, P.J., L.A. Suchar, C.T. Robbins, and A.E. Hagerman. 1989. Tannin-binding proteins in saliva of deer and their absence in saliva of sheep and cattle. J. Chem. Ecol. 15 (4):1335–1347.

Bell, F.R. 1959. Preference thresholds for taste determination in goats. J. Agric. Sci. 52:125–128.

Burns, J.C., D.S. Fisher, and H.F. Mayland. 2001. Preference by sheep and goats among hay of eight tall fescue cultivars. J. Anim. Sci. 79:213–224.

De Rosa, G., L. Moio, F. Napolitano, F. Grasso, L. Gubitosi, and A. Bordi. 2002. J. Chem. Ecol. 28:269–281.

De Rosa, G., V. Fedele, F. Napolitano, L. Gubitosi, A. Bordi, and R. Rubino. 1997. Dietary preferences in adult and juvenile goats. Animal Science 65 (3):457–463.

Goatcher, W.D. and C.D. Church. 1970. Taste responses in ruminants. III. Reactions of pygmy goats, normal goats, sheep and cattle to sucrose and sodium chloride. J. Anim. Sci. 31:364–372.

Hagerman, A.E., C.T. Robbins, Y. Weerasuriya, T.C. Wilson, and C. McArthur. 1992. Tannin chemistry in relation to digestion. J. Range Manage. 45:57–62.

Hammond, A.C. 1995. Leucaena toxicosis and its control in ruminants. J. Anim. Sci. 73:1487–1492.

Hofmann, R.R. 1988. Anatomy of the gastrointestinal tract. In: The Ruminant Animal. D.C. Church, ed. Prentice-Hall, Englewood Cliffs, N. J.

Hofmann, R.R. 1989. Evolutionary steps of oecophysiological adaptation and diversification of ruminants: comparative view of their digestive system. Oecologia 78:443–457.

Hofmann, R.R., W.J. Streich, J. Fickel, J. Hummel, and M. Clauss. 2008. Convergent evolution in feeding types: Salivary gland mass differences in wild ruminant species. J. Morphol. 269 (2):240–257.

Johnson K.A. and D.E. Johnson. 1995. Methane emissions in cattle. J. Anim. Sci. 73:2483–2492.

Lamy, E.G. Da Costa, F.C.E Silva, J. Potes, A.V. Coelho, and E.S. Baptista. 2008. Comparison of electrophoretic protein profiles from sheep and goatparotid saliva. J. Chem. Ecol. 34 (3):388–397.

Leek, B.F. 1993a. Clinical diseases of the rumen: a Physiologist's view. Vet. Rec. 113:10–14.

Leek, B.F. 1993b. Digestion in ruminant stomach. In: Duke's Physiology of Domestic Animals. Eleventh edition, M. Swenson and W. Reece, eds. Comstock Publishing Associates: Ithaca, New York.

McAllen, A.B., R. Knight, and J.D. Sutton. 1983. The effect of free and protected oils on the digestion of dietary carbohydrates between the mouth and duodenum of sheep. Br. J. Nutr. 49:433–440.

Moore, L.A., H.A. Irvin, and J.C. Shaw. 1953. Relationships between TDN and energy value of feeds. J. Dairy Sci. 36:93–99.

NRC. 1981a. Effect of environment on nutrient requirements of domestic animals. Washington, DC: The National Academies Press.

NRC. 1981b. Nutrient requirements of goats: Angora, Dairy, and Meat Goats in Temperate and Tropical Countries. Washington, DC: The National Academies Press.

NRC. 1985. Ruminant Nitrogen Usage. The National Academies Press, Washington, DC.

NRC. 2001. Nutrient requirements of dairy cattle, 7th rev. ed. Washington, DC: The National Academies Press.

NRC. 2007. Nutrient Requirements of Small Ruminants, Sheep, Goats, Cervids, and New World Camelids, The National Academies Press.

Reece, W.O. 2004. Dukes' Physiology of Domestic Animals. 12th ed. Comstock Publishing Associates, Ithaca, New York, U.S.A.

Sahlu, T., A.L. Goetch, J. Lou, I.V. Nsahlai, J.E. Moore, M.L. Galyean, F.N. Owens, C.L. Ferrell, and Z.B. Johnson. 2004. Energy and protein requirements of goats: Developed equations, other considerations and future research to improve them. Small Ruminant Res. 53:191–220.

Silanikove, N. 2000. The physiological basis of adaptation in goats to harsh environments. Small Ruminant Res. 35:181–193.

Solaiman, S.G., M.A. Maloney, M.A. Qureshi, G. Davis, and G. D'Andrea. 2001. Effect of high copper supplements on performance, health, plasma copper and enzymes in goats. Small Ruminant Res. 41:127–139.

Underwood, E.J. and N. Suttle. 1999. The Mineral Nutrition of Livestock. 3rd ed. New York: CAB International.

Van Soest, P.J. 1994. Nutritional Ecology of the Ruminants, 2nd ed. Ithaca, NY: Comstock.

Vieira, E.L., A.M.V. Batista, A. Guim, F.F. Carvalho, A.C. Nascimento, R.F.S. Araujo, and A.F. Mustafa. 2008. Effects of hay inclusion on intake, in vivo nutrient utilization and ruminal fermentation of goats fed spineless cactus(Opuntia ficus-indica Mill) based diets. Anim. Feed Sci. and Tech. 141:199–208.

Woodward, S.L., G.C. Waghorn, M. Ulyatt, and K.R. Lassey. 2001. Early indication that Lotus will reduce methane emissions from ruminants. Proc. N.Z. Soc. Anim. Prod. 61: 23–26.

Wrenn, T.R., J. Bitman, R.A. Waterman, J.R. Weyant, D.L. Wood, L.L. Strozinski, and N.W. Hooven, Jr. 1978. Feeding protected and unprotected tallow to lactating cows. J. Dairy Sci. 61:49–58.

# 9
# Ingestive Behavior, Diet Selection, and Feed Intake

*H. Dove, PhD*

## KEY TERMS

Grazing—consumption of herbaceous material from the plant biomass.
Browsing—consumption of more lignified material from the plant biomass, often above head height.
Ingestive behavior—the behavioral components that result in the animal selecting and consuming forage.
Diet selection—the process of choosing which plant parts and species are consumed.
Diet composition—the consequence of diet selection; the composition of the diet in terms of plant parts/species.
Intake—the amount of forage consumed; "voluntary intake" is the amount of a given diet consumed voluntarily by the animal when the amount available is nonlimiting.
Bipedal foraging—foraging above the normal reach by browsing while on the hind legs.
Buccal architecture/dentition—the morphology of the jaw, mouth, and teeth/the condition of the teeth, especially incisor teeth.
Condensed tannins—a major class of plant secondary compounds, capable of complexing dietary and gut proteins.
Tannin-binding salivary proteins (TBSP)—salivary proteins that complex with dietary tannins, thus reducing their effect on dietary proteins; proline-rich salivary proteins (PRP) are a specialized subset of TBSP.

## OBJECTIVES

By completing this chapter, the reader will acquire knowledge on:

- The balance of browsing versus grazing is different in goats compared with other domestic ruminants.
- The browsing process in goats is only partly "driven" by bipedal grazing.
- That as a consequence, goats will select different plant species than sheep or cattle, when confronted with the same biomass.
- That within goats, diet selection will be affected by age, sex, buccal architecture/dentition, and physiological state.
- Analyzing the validity of statements about dietary preferences and diet selection in goats.
- That the evidence suggesting that goats digest forage better than sheep, or that they eat more of a given forage than sheep, is not conclusive.
- The importance of dietary condensed tannins as one component of the diet selection process.
- How to synthesize this knowledge into a comparative understanding of the processes of diet selection and intake in goats, compared with other domestic herbivores.

## INTRODUCTION

Goats have the ability to use browse as well as the herbaceous components of the forage resource, and can provide animal products from rangeland or semiarid environments that would not suit sheep and cattle. Goats thus assume major economic importance in less-favored environments, often in the world's less developed economies.

The interaction of foraging goats with the plant biomass involves components relating to the animals themselves, the plant resource available to them, and the extent to which the system of livestock management permits the processes of diet selection and intake to proceed. Ingestive (eating) behavior, diet selection, forage intake, and animal production are components of a continuum of mechanisms linking plant resource and animal product.

## EATING BEHAVIOR AND DIET SELECTION

Free-ranging livestock, including the goat, usually select a diet that differs in botanical composition and in quality from the average of the available feed. However, the demonstration that the diet composition differs from the plant biomass does nothing to elucidate why this is so or why certain plants were selected or rejected. The fact that goats often select a diet higher in browse species than would be selected by sheep or especially cattle tells us little about why more herbaceous components of the plant biomass were not selected. Therefore, caution is required in interpreting diet selection data.

Some of the factors influencing feeding behavior and diet selection in grazing ruminants are listed in Table 9.1, and in part have also been discussed elsewhere (AFRC, 1998; NRC, 2007; Animut and Goetsch, 2008). Examination of the information in Table 9.1 leads to two closely related conclusions. First, the process of diet selection is exceedingly "multifactorial." Second, a number of the listed factors are closely related (for example, season and the species composition of the plant biomass, or age of goat and the state of their dentition). A degree of confounding between the factors is thus not only possible but highly likely. Unfortunately, studies of diet selection in goats have not always paid sufficient attention to this.

**Table 9.1** Indicative summary of some of the factors shown to influence ingestive behavior and diet selection in goats.

| Factor |
|---|
| 1. Species composition of the biomass/ season of year |
| 2. Clover content of and clover position in the "sward" |
| 3. "Presentation" of the biomass—zero grazing versus grazing; species in rows cf. mixed sward; species offered separately or together |
| 4. Sward height, biomass; stocking rate |
| 5. Condensed tannin content of species on offer |
| 6. Dry matter content of the forage |
| 7. Odor/flavor of the forage |
| 8. Previous experience and adaptation of grazing animals; species of animal that grazed biomass previously |
| 9. Differences in possible instantaneous intake rate between biomass components |
| 10. Breed, gender, age, and physiological state of goats |
| 11. Buccal architecture and integrity of teeth |
| 12. Goats grazing separately or co-grazing with other species |
| 13. Grazing height, bipedal versus quadripedal foraging |

### General Observations on Diet Selection by Goats

Frequently, ruminant livestock have been classified as being grazers (grass/forb consumers), as concentrate selectors/browsers, or as intermediate or mixed feeders. Sheep and especially cattle are regarded as grazers, with relatively broad muzzles and a cornified tongue tip designed for grass consumption at low biomass. Goats, on the other hand, have a narrow mouth and mobile lips and tongue that should allow selective browsing, and are regarded as "intermediate feeders" (see NRC, 2007; Animut and Goetsch, 2008). The evidence from many published studies indicates conclusively that when faced with a mixture of browse, forbs, and grasses, goats will select a diet containing much more browse than would be selected by sheep or especially cattle. In a Mediterranean shrubland, Papachristou (1994) reported that goats took 40–60% of their bites from browse species. However, when the availability of the herbaceous component was high, it could make up >50% of the diet (Papachristou, 1994). By contrast, in an oak forest environment, cattle co-grazing with goats took more than 97% of their bites from the herbaceous component of the biomass. In a more recent comparison in the Sahelian region of Africa, Sanon et al. (2007) reported that cattle spent 4.5–6.6% of their time browsing, depending on season. By contrast, goats browsed for 43–52% of their time; sheep were intermediate at 4.8–28% of their time. A feature of many published studies is that diet selection is measured indirectly in terms of ingestive behaviors or features of the plant biomass. It can be difficult to convert plant biomass data, or data about percentages of times spent grazing different plant components, to actual intakes of those components (Dove and

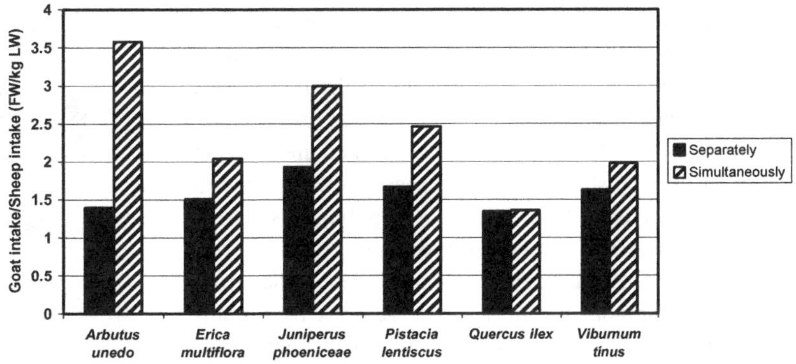


**Figure 9.1** Effect of feeding a range of browse species, either separately or simultaneously, on the relative intake of goats versus sheep (recalculated from the data of Rogosic et al., 2006a).

Mayes, 2005). A recent study in Spain (Celaya et al., 2007) avoided such problems in co-grazing cows, ewes, and does in heathland in Spain, by estimating diet selection directly, using plant wax alkanes as markers (see Dove and Mayes, 2005). Their data confirm behavioral observations. The herbaceous component made up 75–90% of the intake of cows whereas in ewes, shrubs made up 36–59% of the diet, and in does, 49–85% of the diet. There is thus little doubt that, where diet choice is possible, goats will consume more browse than sheep and particularly, more than cattle.

## Factors Influencing Diet Selection

### Season of Year and Species Composition of the Plant Biomass

In many studies, selection of particular plant species in the diet has been, in part at least, a function of the seasonal availability of different species. As would be expected, the species composition of the biomass, both in its own right and as influenced by season, has a marked effect on diet selection by the goat. The preference of goats for particular plant species also depends on whether different plants are fed together or separately. For example, Rogosic et al. (2006a) offered six different browse species to either housed goats or housed sheep. When the shrubs were offered individually, goats always ate more than sheep. Recalculation of their data indicates that the ratio of goat:sheep intakes (both expressed as g DM/kg LW) ranged from 1.35–1.93 across the different plant species (Figure 9.1). When all six plant species were offered simultaneously to the two species of animals, goats again always ate more than sheep, but the intake ratios ranged from 1.36–3.58 between plant species. In other words, the preference for different plant species was more extreme when all plant species were offered together. This means that caution is required when extending indoor data obtained by feeding individual plant species to field situations where a range of plant species is on offer at the same time.

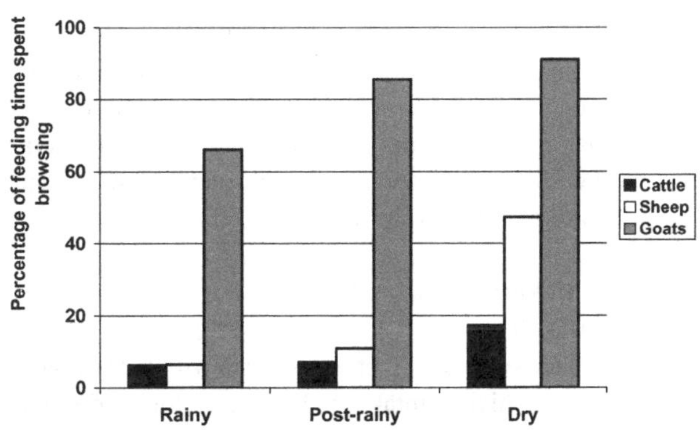


**Figure 9.2** Comparison of the time spent browsing by cattle, sheep, and goats in a natural pasture in the Sahelian region of Africa (recalculated from the data of Sanon et al., 2007).

Marked seasonal effects on diet composition have been reported in many studies. For example, Sanon et al. (2007) reported data for goats, sheep, and cattle grazing natural pasture in the Sahelian region of Africa. Their data for the time spent browsing have been recalculated as a percentage of the total feeding time; differences between animal species and seasons are shown in Figure 9.2. As the seasons progressed, there was little change in the time cattle spent browsing, despite the progressive loss of the herbaceous component of the plant biomass. The apparent increase in browsing by cattle in the dry season was due more to a reduction in their time spent feeding (from 72–38% of the day), rather than an actual increase in the time spent browsing. By contrast, goats browsed for more than 60% of their feeding time during the rainy season, and as the season progressed and the herbaceous component declined, the time spent browsing rose to more than 90% of the feeding time. The marked increase in browsing time by sheep in the dry season was not due to a reduction in their time spent feeding but was a real increase in the actual time spent browsing.

Although goats often consume most of their diet from the browse component, during some seasons or circumstances the grass/herbaceous component can be a major contributor to the diet. Published data indicate that when this occurs, herbaceous species can comprise more than half of the diet of goats. For example, Papachristou (1994) found that the herbaceous component could make up >50% of the diet of goats in a Mediterranean shrubland when herbage availability was high. This serves as a reminder that although goats have a much stronger preference for browse than sheep or cattle, goats are intermediate feeders, and the herbaceous component of the plant biomass can still contribute an important fraction of their diet.

**Figure 9.3** Browsing goat showing typical bipedal stance (photo courtesy Dr. S.G. Solaiman).

#### Foraging Height—Bipedal Versus Quadripedal Foraging

It is frequently suggested that a major contributor to goats' achieving their higher intake of browse is their adoption of a bipedal stance, as shown in Figure 9.3. Indeed, some reports almost suggest that the adoption of a bipedal stance is what confers on goats their browsing success. Given the assumed importance of this, it is remarkable that there is so little quantitative evidence to support the notion. One must also remember that in a ruminant, the adoption of a bipedal stance comes at an energy cost, which might be expected to influence the time spent in bipedal browsing. How do goats browse, relative to other ruminants?

Sanon et al. (2007) recently reported foraging heights for co-grazing goats, sheep, and cattle in the Sahel. The mean and the maximum foraging heights recorded for the species follow:

- goats—1.65 m and 2.10 m
- cattle—1.47 m and 1.90 m
- sheep—0.87 m and 1.17 m

However, these heights are obviously influenced by the size of the species themselves. Differences in foraging heights become even more marked if they are recalculated in terms of the shoulder heights of the three animal species. In these terms, sheep and cattle foraged to similar mean heights (sheep 1.18, and cattle 1.21 times their respective shoulder heights) and maximum heights (sheep 1.58 and cattle 1.57 times shoulder height). By contrast, goats foraged to a mean of 2.42 times their shoulder height and a maximum of 3.08 times their shoulder height. Clearly, this could not have been attained without the goats adopting a bipedal stance for some of their foraging.

There are few estimates of the proportion of their feeding time that goats actually do spend in bipedal foraging. Pfister et al. (1988) measured the percentage of their foraging time that sheep and goats spent foraging within a series of strata defined as "ground level" (0–5 cm), "animal height" (6–75 cm), "overhead quadripedal foraging" (76–115 cm), or bipedal foraging (116–160 cm) in the wet, dry, and late-dry seasons in the Brazilian caatinga. Their recalculated data (Figure 9.4) confirm that goats spent much more time than sheep browsing above animal height, though it should be noted that in all seasons, both animal species spent a large proportion of their grazing time foraging at animal height or at ground level. This study also showed that most of the browsing above animal height occurred using quadripedal browsing; goats spent much longer than sheep browsing in this stratum. Bipedal browsing essentially only occurred with the goats, but even in this species, it accounted for only

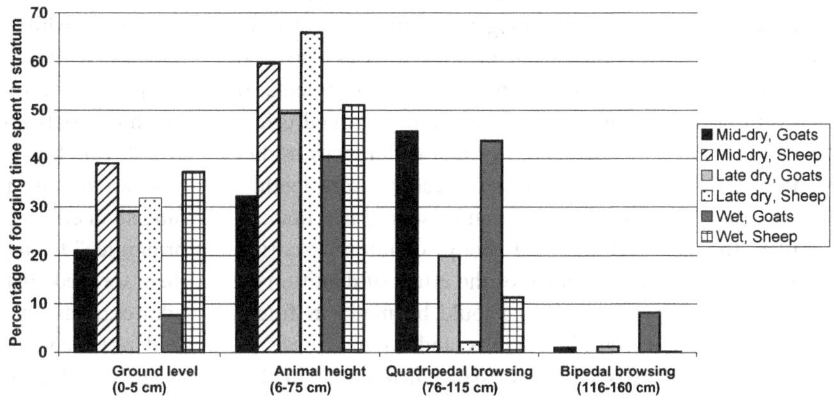


**Figure 9.4** Comparison of the proportion of their foraging time spent by goats and sheep in four defined strata, during three seasons of the year (mid-dry season, late-dry season, wet season) in the Brazilian caatinga (recalculated from the data of Pfister et al., 1988).

about 8% of foraging time in the wet season. In the other two seasons, it accounted for much less.

In the data reported to date, goats have spent 10% or less of their foraging time in bipedal mode. Depending on the relative availabilities of forage in the different strata, goats may on occasions be forced to spend more of their foraging time in the bipedal stance. They are also known to climb into trees to obtain forage. To confirm the quantitative importance of these behaviors, we need more measurements of ingestive behavior under field conditions.

### Breed, Gender, Age, and Physiological State of Goats

All of these factors can influence ingestive behavior and diet selection. There is evidence that Saanen goats browse for more of their total foraging time than Angoras, which graze for a much greater proportion of their foraging time. A practical consequence of this difference in feeding behavior is that helminth burdens may be more of a problem in Angoras, presumably because they forage in the layer occupied by nematode larvae. The interaction between feeding behavior, nutritional status, and helminth infestation is an area that would be worthy of detailed study in goats.

In general, juvenile goats spend 20–90% more of their day foraging than do adult goats. Differences in diet composition have also been reported for bucks versus nonpregnant, nonlactating does. Bucks consume a higher proportion of browse and tend to avoid grasses more than does. It has been suggested that sex dimorphism in buccal architecture may have contributed to differences in diet composition. Superimposed on the above effects can be an effect of physiological state. Depending on availability, forbs and grasses tend to make up a larger proportion of the diet of pregnant or lactating goats, compared with nonpregnant, nonlactating goats, which consumed more browse in their diet. Hence, there are effects of breed, gender, age, and physiological state on ingestive behavior and diet composition. Given the magnitude of the effects, future studies need to pay much more attention to controlling or at least reporting these factors.

### Buccal Architecture, Dentition

The oral and dental features of goats are often mentioned when discussing their ingestive behavior, but the actual quantitative contribution of these mouth features has not been well defined. Mellado et al. (2007) recently used multivariate statistical procedures to relate diet choice by crossbred goats in Mexico to a range of oral measurements. Overall, there was a good relationship between the measured oral characteristics and the main plant species ingested (canonical correlation = 0.81). Muzzle width, incisor breadth, and jaw length were the main oral measurements associated with the extent of ingestion of spiny shrubs. Incisor length and the distance between the first molars had the dominant effect on the ingestion of grasses and coarse shrubs. The importance of incisor length is consistent with the known importance of incisors in the prehension and ingestion of grass.

A related effect is the state of the incisor teeth themselves (for example, missing, loose, or broken incisors), as influenced by such things as age. In ewe/lamb systems, it has been amply demonstrated that, independently of the effect of age, bad incisor dentition can reduce feed intake and animal performance. This is especially important in lactating ewes. In goats, Mellado et al. (2005)

estimated diet composition of dairy × criollo bucks and does and also recorded the integrity of their incisors. In general, goats with worn teeth avoided grasses and also had a higher preference for saltbush and for succulents. It is worth noting in passing that despite the marked effect of worn teeth on diet selection, gender differences in diet composition were greater than the effects of worn teeth.

The state of incisor dentition can be scored in a manner similar to the well-accepted body condition scoring systems. Given the magnitude of the effect of dentition on diet composition, more effort should be made in future studies to obtain dentition scores and to use these as explanatory variables when assessing differences in diet composition.

#### Clover Content and Clover Position Within the Sward

One of the most common assertions in popular writings about goat husbandry is that goats have a strong preference against clover and would rather not eat it. To take an extreme example, Hetherington (1977) suggested that "... (goats) will only eat ryegrass under sufferance and entirely ignore white clover." The evidence does not support either of these assertions.

In addition, one should also bear in mind that in such discussions, what is actually meant by "clover" is "white clover" (*Trifolium repens*). The preference of goats for other *Trifolium* species has not been reported, though goats are known to eat other legumes (e.g., Papachristou, 1994).

In a study with esophageal-fistulated Cashmere × feral goats grazing a perennial ryegrass/white clover pasture, Fraser and Gordon (1997) found that perennial ryegrass leaf laminae made up 66% of the diet of goats. Their diet also contained 9–18% white clover. The goats appeared to select positively for green clover leaf, relative to its availability in the total sward, but the clover proportion in the diet was similar to that at the sward surface, where the goats appeared to concentrate their grazing.

Penning et al. (1997) also examined the dietary preferences for perennial ryegrass and white clover in sheep or goats. Sheep grazed for longer than goats, and their extra grazing time was spent grazing clover. Dietary preference for clover, defined as the percentage of the total grazing time spent grazing clover, was 70% for sheep. The equivalent preference for goats was 52%, which certainly does *not* imply rejection of clover, but merely that goats have a lower preference for white clover than do sheep.

As a result of the somewhat lower preference for white clover demonstrated by goats, it has often been observed that whereas sward clover content falls under grazing by sheep, is relatively constant or rises when goats are grazing a ryegrass/white clover sward. This has practical consequences within mixed farming systems, because it raises the possibility of using a leader-follower grazing system in which goats are used to increase the clover content of a sward, which is then grazed by sheep.

In an attempt to understand the mechanisms underpinning such effects, De Rosa et al. (2002) extracted the "odor principles" from perennial ryegrass and white clover. They then sprayed distilled water or aqueous solutions of the odor extracts onto straw pellets. Goats had an almost two-fold preference for straw pellets sprayed with perennial ryegrass odor over those sprayed with distilled water. Pellets sprayed with white clover odor were less preferred than those sprayed with ryegrass extract, but were still preferred to those sprayed with distilled water. Once again, these data provide no support for the idea that goats have a low preference for perennial ryegrass or will "not eat" white clover.

Hence, taken together, these data provide little support for the notion that goats will reject or select strongly against white clover. Rather, they support the suggestion that goats will eat white clover but prefer it less than sheep, or that no selection against clover exists when it is part of the grazing horizon of the goat. Goats can and do consume perennial ryegrass as a major portion of their diet, when grazing sown swards of ryegrass/clover.

#### Presentation of the Biomass: Sward Height, Biomass, and Stocking Rate

In studies of ingestive behavior and diet selection, goats have been presented with different plant species in many ways: as individual plant species indoors; as mixtures of plants outdoors; or as the same material both indoors and outdoors. The method of presentation can itself alter the observed diet composition, because compared with housed goats, grazing goats are better able to select as they forage in a sward.

The published data for white clover selection by goats indicate that they are "top-down" grazers with a shallow biting depth, whereas sheep are more inclined to graze deep into the sward. In turn, this should render goats more susceptible to the effects of sward height, which does seem to be the case. Merchant and Riach (1994), for example, reported that goats found it more difficult to maintain intake and live weight gain as sward height decreased.

Based on this and on studies with sheep, a reduction in the amount of biomass on offer, or of sward height, due to increased stocking rate will influence both diet selection and intake. In general, stocking rate effects in goats have been similar to those seen in sheep. In co-grazing sheep and goats, the time spent eating and the grazing time as a proportion of total time both usually increase as stocking rate increases, while ruminating time and idle time tend to decrease.

#### Separate Versus Co-grazing: Previous Experience

The results of a number of studies indicate that the species of animal that previously grazed an area can potentially influence the diet choice of later animals by, for example, altering the legume content of the biomass as described above. Similarly, previous experience, preconditioning, or adaptation of animals to particular plant species can alter their future diet choice (e.g., Distel and Provenza, 1991). This should be taken into account when designing or interpreting studies of ingestive behavior and diet selection.

Ingestive behavior and diet selection can differ depending on whether the animal species involved are co-grazing or grazing as separate species. A detailed discussion of the merits or otherwise of co-grazing is outside the scope of this chapter. For a recent discussion of the major issues as they relate to goat production, the reader is referred to the excellent review by Animut and Goetsch (2008). One major advantage of co-grazing in relation to investigations of ingestive behavior is that it removes the confounding, which can occur between grazing site and animal species, when separate grazing occurs. For example, the use of co-grazing of the same area by ewes, does, and cows (Celaya et al., 2007) allowed a direct assessment of the degree to which their respective diet compositions overlapped.

#### Forage Odor/Flavor and Dry Matter Content

The diet choice of goats can certainly be influenced by flavor and by odor, as demonstrated by De Rosa et al. (2002). The DM content of forage can also influence ingestive behavior, though the response is not always clear-cut. Bateman et al. (2004) offered goat kids fresh forage from cereals and brassica crops and found that there was a 3-fold range in intake, which was strongly correlated with forage DM content ($r = 0.958$; $P < 0.01$). Their results also provided some evidence that goats prefer those feeds that allow a higher short-term intake rate.

#### Condensed Tannin Content of the Forage

Condensed tannins are a major class of plant secondary compounds that are often found in the browse species consumed by goats. It is beyond the scope of this chapter to present a detailed treatment of the effects of condensed tannins or other secondary compounds on animal nutrition and health. For a detailed discussion of these aspects, the reader is referred to excellent reviews by Foley et al. (1999), Mueller-Harvey (2006), and Shimada (2006).

Reference is often made to goats being able to consume more tannin-rich browse than sheep under similar conditions. Similarly, goats are said to somehow exhibit resistance to tannins or at least are less affected by plant secondary compounds than other animals. At times, this ability to cope with tannins has been linked to goats having proteins in their saliva that can bind to the tannins, though this remains a controversial area. For example, Snyder et al. (2007) asserted that goats are more tolerant of tannins "… because they have been reported to have proline-rich proteins … in their parotid salivary glands. …" They cited Mehansho et al. (1987) in support of this assertion, though there is no mention of goats in the latter paper.

There are actually two components embedded within comments like that made by Snyder et al. (2007). The first is that goats do have tannin-binding salivary proteins (TBSPs) and the second is that these TBSPs are in fact the proline-rich salivary proteins (PRP) that function as specialist tannin-binding proteins in some animals. It is true that TBSPs occur in the saliva of true browsers like deer and moose. It is perhaps because of this fact and the fact that goats often browse, that the idea has arisen that goats also cope with dietary condensed tannins through tannin-binding proteins in saliva. This idea merits closer examination.

There is no doubt that, through their greater intake of browse that may contain condensed tannins, goats may encounter a greater dietary intake of plant secondary compounds than sheep or especially cattle. Moreover, goats have a higher tolerance of bitter flavor than did sheep or cattle.

What is not clear from the research to date is whether TBSPs are present in goats, whether they are proline-rich proteins (PRPs), and whether they are functionally significant. As Foley et al. (1999) emphasized, the word "functionally" is most important. Even if TBSPs are detected, if they are present in very small amounts or have very low affinities for condensed tannins, then their functional significance is questionable.

Austin et al. (1989) found that there were TBSPs in the saliva of deer but not in the saliva of sheep and cattle (goats were not tested). Mole et al. (1990) subsequently tested for the presence of PRP in the saliva of lagomorphs, rodents, marsupials, and sheep and cattle. While salivary PRPs were regarded as "detectable" in the saliva of sheep and cattle, when scaled to body weight, these species had only 0.01–0.02% of the amount in the saliva of rodents, marsupials, or hares and rabbits. The PRP detected in sheep and cattle saliva also had a much lower affinity for tannins. These studies thus provide no evidence of there being functional PRP in the saliva of sheep and cattle.

More recent studies reviewed by Mueller-Harvey (2006) and by Shimada (2006) indicate that sheep and goat saliva contained (unidentified) TBSP with a reasonable affinity for condensed tannin. However, these TBSPs do not appear to be the PRPs usually associated with tannin-binding ability. Lamy et al. (2008) recently reported a comprehensive electrophoretic comparison and characterization of the salivary protein profiles of both sheep and goats eating a low-tannin diet. They also specifically tested for the presence of salivary PRP in both animal species and could *not* detect them. They suggested there was a need to repeat such measurements in animals fed high levels of tannin, but other evidence indicates that TBSP cannot be induced in the saliva of ruminant livestock by feeding high-tannin diets.

In summary, the search for TBSP in the saliva of ruminant livestock in general and goats in particular suggests that some TBSP may be present in saliva. Equally, it indicates that they are not the PRPs associated with tannin binding in true browsers. Whether the unidentified TBSP would be of physiological importance as a defense against high-tannin diets in goats is not yet established (Shimada, 2006).

Goats can nevertheless consume high-tannin diets without ill effect and because they excrete only a small proportion of dietary condensed tannin in feces, some other defense mechanism against condensed tannins must be operating. Two other possible mechanisms could allow the goat to tolerate high-tannin diets:

1. Quantitative studies of the fate of condensed tannins during gut transit have shown substantial losses between intake and feces (e.g., Perez-Maldonado and Norton, 1996), suggesting the possibility of microbial metabolism of tannins in the gut.
2. One might expect that a herbivore's intake of a feedstuff containing a secondary compound would be limited by its ability to detoxify the compound. The predicted consequence of this is that herbivores would choose a mixed diet to minimize the effects of the consumption of any particular secondary compound (see Foley et al., 1999). According to this hypothesis, the goat would consume its typically mixed diet not *because* of its overall capacity to detoxify secondary compounds, but in order to *avoid* a high intake of any one such compound. It should also follow that provision of a mixed diet of tannin-containing feeds would allow a greater total intake than could be achieved by eating any one feed. There is evidence to support this suggestion in goats. Rogosic et al. (2006b) found that feeding a mixture of three tannin-rich shrubs (*Quercus ilex, Arbutus unedo* and *Pistacia lentiscus*) to goats resulted in more than twice the intake achieved when the shrubs were fed separately.

The quantitative significance of these two mechanisms and the interactions between them have yet to be determined. This would be a fruitful area for future research.

Although goats can clearly tolerate high dietary intakes of condensed tannins, diet selection and intake can nonetheless be altered by supplementing foraging goats with polyethylene glycol (PEG), a polymer that binds to dietary condensed tannins. For example, Landau et al. (2002) supplemented Damascus goats with PEG (MW 4,000) in a rangeland dominated by lentisk (*Pistacia lentiscus*). Supplementation had no effect on the time spent foraging (Table 9.2) and in goats, which were not also offered alfalfa hay, resulted in a large increase in the proportion of their foraging time spent eating the tannin-rich lentisk. The time spent eating dry grasses decreased. When the goats were also offered alfalfa hay while on the range, the PEG supplement had no effect on the amount of time spent

**Table 9.2** Effect of polyethylene glycol (PEG, MW 4000) on the feeding behavior of Damascus goats in a rangeland dominated by tannin-rich lentisk (*Pistacia lentiscus*).

| Behavior | Control (−PEG) | +PEG |
|---|---|---|
| % of time spent foraging | 44 | 39 |
| *Without hay supplement* | | |
| % of foraging time eating: | | |
| Lentisk | 41 | 73* |
| Dry grasses | 28 | 12* |
| *With hay supplement* | | |
| % of foraging time eating: | | |
| Hay | 30 | 26 |
| Lentisk | 46 | 62* |
| Dry grasses | 13 | 10 |

Source: Adapted from Table 2 of Landau et al., 2002. Lentisk condensed tannin content = 22% DM.

*Control and +PEG values differ significantly ($P < 0.05$).

eating hay but still significantly increased the time spent browsing lentisk (Table 9.2). Goats offered PEG thus changed their patterns of diet selection. They also had a higher intake and maintained weight, whereas those not given PEG lost weight.

## DIGESTIBILITY AND FEED INTAKE

There are many factors that influence the voluntary feed intake of grazing or browsing ruminants, including many of the factors that influence diet selection. These factors have been reviewed extensively recently (see for example, Dove and Mayes, 2005; NRC, 2007). The AFRC (1998) presents equations for the prediction of intake in housed goats. More recently, Luo et al. (2004) conducted an extensive review of the feed intake literature for "... lactating, Angora, growing and mature goats ...," but once again, the emphasis was on the development of equations for predicting intake in housed animals. There is much less reviewed information on intake in the free-ranging goat, in part because of the difficulty in estimating digestibility and intake in free-ranging animals (Dove and Mayes, 2005). Of necessity, much of the information discussed below has thus been derived from housed goats. A degree of caution is required in extending this to the pasture or range situation.

An indicative, but by no means complete, summary of the factors influencing diet digestion and voluntary intake is given in Table 9.3. Clearly, the amount of plant material available, its spatial distribution, and its content of nutrients and plant secondary metabolites are major factors influencing intake. However, it is the animal itself that can be regarded as setting the upper limit to intake, through its need for and capacity to use nutrients at a particular age, body condition, and physiological state. This "potential intake" (CSIRO, 2007; NRC, 2007) is then constrained by features of the plant biomass on offer, by the animal's previous experience, by transactions within the gut, and by the intake of milk or supplement. A cornerstone of the approach is the notion of a "standard reference weight" (SRW) for a given breed, that is, the weight of the mature female with a condition score in the middle of the range used. Potential intake is scaled to this weight (see CSIRO, 2007; NRC, 2007) but will be different if the animal's current weight is below the SRW either because it is younger or if its condition differs, or both.

Both CSIRO (2007) and NRC (2007) took the view that intake equations derived for sheep could be applied equally to goats, arguing that published data did not warrant a separate treatment for goats. This assumption is examined further later in this chapter.

**Table 9.3** Indicative summary of some of the factors shown to influence herbage intake in the free-ranging ruminant.

| Plant factors | Animal factors |
|---|---|
| Species composition of plant biomass | Capacity to use energy/ physiological state |
| Mass, height, proportions of leaf, stem, or thorns in the plant biomass | Standard reference weight ("mature live weight") |
| Spatial heterogeneity—patchiness, height distribution | Current age, weight and body condition |
| Stage of plant development—vegetative versus reproductive | Previous experience with plant species on offer |
| Digestibility of the species on offer | Social interactions with other animals |
| Concentrations of plant secondary compounds in the plant species | Digesta load |
| Protein content of the species on offer | Intake of milk or of supplementary feed |

### Forage Digestion and Intake in Goats

A selection of the published comparisons of forage digestion and of intake by goats and sheep is given in Table 9.4. Almost all of the data are derived from studies with housed animals. Although details of the breeds and live weights of the animals are often provided, it is usually very difficult to assess age, body weight, and body condition in terms that would allow the use of the prediction equations in the published nutrient requirements. Moreover, the diets consumed by the two species are not always identical so that diet and animal species are confounded. These points should be kept in mind when assessing whether goats and sheep really differ in their digestion and intake of forage.

#### Forage Digestibility

Diets consumed in the reports summarized in Table 9.4 have ranged from high-quality (e.g., Doyle et al., 1984, diet A) to high-fiber, low-quality materials such as wheat straw (Reid et al., 1990). In general, published results do suggest that higher-fiber diets are digested better by goats than sheep, but overall, the data in Table 9.4 do not provide convincing evidence that diet digestibility is predictably and substantially higher in goats than in sheep. Tolkamp and Brouwer (1993), for example, surveyed published data

**Table 9.4** Indicative summary of comparisons of digestibility and intake in goats and sheep.

| Reference | Animals | Animal species with higher digestibility | Animal species with higher intake |
|---|---|---|---|
| Doyle et al., 1984 | Angora goats; Merino sheep | Diet A,C—Neither<br>Diet B—Goats | Diet A; Goats (g/d, not g/kg $W^{0.75}$) |
| Brown and Johnson, 1985 | Toggenburg × Saanen castrated kids; Suffolk cross wether lambs | Overall, neither; sheep digested 65% wheat straw diet better | Goats (g/d); Neither (g/kg $W^{0.75}$) |
| Pfister and Malechek, 1986 | Free-ranging castrated "hair sheep"; undefined goats | Overall, neither; sheep ate higher digestibility diet, wet season | Neither |
| Domingue et al., 1990 | Angora × feral castrated goats; Border Leicester × Romney castrated sheep | Goats (winter only) | Summer—Goats<br>Winter—Neither |
| Reid et al., 1990 | Toggenburg × Alpine goats; Suffolk sheep | Neither | Neither |
| Tolkamp and Brouwer, 1993 | Literature survey | Goats, but by mean of only 0.8 percentage units | ND |
| Papachristou, 1994 | Free-ranging—breeds not defined | Cleared pasture—neither overall, sheep October only slashed/ native pastures—goats | ND |
| Perez-Maldonado and Norton, 1996 | Castrated Cashmere goats; castrated Border Leicester × Merino sheep | Neither (OM, NDF, ADF, N) | Sheep (g/d)<br>Neither (g/kg $W^{0.9}$) |
| Hadjigeorgiou et al., 2001 | Castrated Blackface sheep; castrated Cashmere goats | Sheep (for DM, OM, NDF, ADF and N) | Goats (g/kg $W^{0.75}$) |

ND = not determined.

on digestibility and concluded that while goats did tend to show higher digestibility capability than sheep, the mean difference between species was only 0.8 percentage units of digestibility, an amount which can be regarded as too small to be of any practical significance.

#### Forage Intake

On sown pastures dominated by perennial ryegrass, Merchant and Riach (1994) reported daily DM intakes of 26–88 g/kg$^{0.75}$ (1.1–3.7% of live weight) for castrated Cashmere goats. Intakes in lactating mixed-genotype goats were 57–140 g/kg$^{0.75}$, similar to ranges reported elsewhere for lactating sheep. Intakes by unweaned kids of 25–30 g/kg$^{0.75}$ are lower than published values for lambs and were also markedly lower than daily DM intakes in weaned kids, as might be expected. In general, all the above values are similar to those cited in the extensive review by Luo et al. (2004), which were as follows (minimum–maximum): growing goats 42.6–97.2 g/kg$^{0.75}$ (1.8–4.5% of live weight); mature, nonlactating goats 27.6–92.0 g/kg$^{0.75}$ (1.6–3.2% live weight); and lactating goats 31.7–151.3 g/kg$^{0.75}$ (1.0–6.3% of live weight).

In comparisons of feed intake by goats and sheep, it has sometimes been asserted that goats have the higher intake and that this is because of their higher ME requirement for maintenance, compared with sheep. There is strong evidence that goats do have a higher ME requirement for maintenance (AFRC, 1998; Sahlu et al., 2004) and that their requirement/kg$^{0.75}$ is more like that of cattle than of sheep. From this, it would follow logically that to maintain weight, goats would have to eat more of a given diet than sheep. However, this does not necessarily mean that *voluntary* feed intakes will be higher in the goat. This distinction between "the intake required to maintain weight" and "voluntary feed intake" needs to be more clearly acknowledged in future comparisons of the voluntary feed intake of goats with other species.

A selection of direct comparisons of voluntary intakes by goats and sheep is given in Table 9.4 (column 4). There are more instances where the intake by goats is higher, but given the number of instances where there is no difference, taken at face value, these empirical comparisons do not provide a convincing case for regarding daily intakes by goats as being higher than those of sheep. However, these data should be treated with some caution, for a number of reasons. First, there is a wide range of goat and sheep genotypes involved in the comparisons, and even when live weights are reported, it is very difficult to assess the extent to which the animals have reached mature weight, or whether there are differences in body condition between the goats and sheep involved in the comparison. Both relative size (current weight as a proportion of SRW) and body condition will affect intake (NRC, 2007; CSIRO, 2007). Moreover, for the field studies (Pfister and Malechek, 1986; Papachristou, 1994), the diet composition differed between goats and sheep and was thus confounded with animal species.

Hence without further corroborating evidence, one cannot yet conclude that goats and sheep differ in voluntary feed intake. Although there seems little doubt that the feed intakes required to maintain live weight are higher in goats than sheep, the issue of whether goats have higher voluntary feed intakes is still unresolved.

## FEEDING HOUSED GOATS, SUPPLEMENTARY FEEDING, DROUGHT FEEDING

Goat production systems can often rely on the natural supply of pasture and/or browse, but there will be seasons, years, or periods of high production when extra feed has to be supplied by the manager (supplementary feeding) or when animals have to be fed during drought. Under these circumstances, questions of preference and feeding behavior become even more important. In addition, dominance hierarchies within the goat herd, which may not usually influence daily management, become a major consideration. The management of feeding under all of these circumstances has a strong economic dimension, for as Solaiman (2006) points out, from one-half to two-thirds of the variable costs of production are those related to feeding. Unless supplementary or drought feeding is managed well, enterprise profit will rapidly be eroded by the costs of feeding. For useful and detailed discussions of feeding management issues, the reader is referred to Bateman et al. (2004), McGregor (2005), and Solaiman (2006).

### Feeding of Housed Goats

The highly selective feeding of goats applies equally indoors, though goats that are too hungry or too full will discriminate less between feeds (Morand-Fehr, 2003). Goats in general have a strong preference for dry feed to be in the pelleted form rather than flour or meal form (Morand-Fehr, 2003; Bateman et al., 2004). The main reason for this seems to be the irritation caused to the upper respiratory tract by the consumption and inadvertent inhalation of fine, dry feeds. This preference appears to apply equally to young goats (Bateman et al., 2004) as well as to adults (Morand-Fehr, 2003) and in part also relates to the fact that pelleted diets allow a more rapid eating rate. As might be expected, pelleted rations that consist of 5- to 10-mm pellets, as might be fed to sheep, are preferred by goats to larger pellets (15–20-mm) typical of the kind fed to cattle (Pinkerton and Pinkerton, 2008). Goats prefer dry diets fed in meal/flour form to have slightly higher moisture content because this allows a degree of agglomeration and, again, a faster intake rate with less irritation (Morand-Fehr, 2003). However, liquid diets such as those based on molasses do not appear to be preferred by young goats (Bateman et al., 2004).

When trough-fed, housed goats consume both "main meals," which account for the bulk of their intake, and smaller "secondary meals." Main meals usually occur immediately after feeding, and the amount consumed will be a function of the rate of intake which the ration permits. The manager must therefore ensure that the physical form of the diet allows the animals to consume the desired quantities. Housed goats will also consume a feed more readily if they have had some previous experience of it (Distel and Provenza, 1991).

An aspect that assumes much more importance in housed animals than in free-ranging goats is the effect of dominance hierarchies on eating behavior. When it comes to consumption of feed from a trough with limited available space, the comment by Pinkerton and Pinkerton (2008) is relevant:

> Gracious, equitable sharing is simply not a caprine characteristic; greed, aggression and sheer size conspire to "cheat" the more civil and/or smaller goats.

The manager must therefore ensure that this "greed and aggression" is minimized, mainly by ensuring that trough space is adequate for all goats to be able to feed. As a generalization, 25–30 cm of trough space/kid should be allowed, or 45–50 cm/adult. A smaller allowance of trough space will almost certainly result in aggressive behavior and in the more submissive goats consuming less.

### Supplementary Feeding under Field Conditions

A number of the above comments about the feeding of housed goats apply equally well under field conditions, especially if goats are grazing sown pasture. When supplements are offered to animals in the field, true supplementation (i.e., the supplement is offered and there is no reduction in forage intake) is actually a rare event. The most common outcome is for the consumption of the supplement to result in a reduction of the forage intake. The extent of the reduction, per kg supplement fed, is referred to as the "substitution rate." If supplementary feeding is managed properly, substitution rate will usually be <1 so that total intake will in fact be increased, as intended. Moreover, since the quality of the supplement (for example, digestibility) is often better than the average of the forage on offer, the digestible DM intake may be increased by an even greater proportion.

A range of factors influences the degree of substitution between supplement and forage. Generally speaking, substitution will be higher under the following circumstances:

1. Substitution rate increases when more forage is available or when forage is of higher quality, or if higher quality supplements are fed, because of associative effects between the digestion of supplement and the digestion of the fiber of the forage.
2. It is usually greater (though not always) if more supplement is fed or if the supplement is fed infrequently.
3. For a given quantity and quality of forage on offer, the degree of substitution can be altered by the animals' physiological state. For a given forage base and supplement, animals with a higher nutrient demand (such as lactation) will usually exhibit a lower substitution rate.

Under field conditions, supplements may be offered by trailing them along the ground in a thin stream. This can ensure that all animals, including the so-called "shy feeders," will get a share, but there is also an interaction with feeding frequency. If supplements are trailed out infrequently, then dominant animals can overeat on the day of feeding, with resultant digestive disturbances if, for example, whole grain is being fed. If supplements are fed in troughs, then the above comments about allowing sufficient trough space apply equally in the field, if not more so. For example, allowing too much trough space can also result in aggressive behavior by dominant goats, by allowing them the space to claim and defend a "territory" (Solaiman, 2006).

The economics of supplementary feeding are much more favorable if animals take only a short period to begin eating the supplement. As discussed above, this is in part a function of their previous experience. Goats will more readily eat a supplement if they have experienced it before, even pre-weaning. For supplementary feeds that the herd has not previously experienced, a useful management tool is to include within the herd a few "experienced" (usually older) animals that have had some experience of the supplement. This has been shown to increase the supplement intake of the inexperienced animals.

In low-cost, more extensive production systems, often characterized by a generally lower-quality forage base, it may be appropriate to supplement animals using feed blocks, constituted to provide nutrients in poor supply in the diet (for example, minerals, rumen-degradable N). In general, the evidence that the provision of feed blocks results in an increase in forage intake by goats is inconclusive. However, where the forage contains plant species with condensed tannins, feed blocks may be a cost-effective way of supplementing with PEG, to overcome the negative effects of condensed tannins and to increase intake.

### Drought Feeding of Goats

Drought feeding differs from other forms of supplementary feeding in that the emphasis is on survival of the herd and not on high levels of animal production. This alters what can and ought to be done. For a detailed and very practical treatment of the issues involved in drought feeding of the goat herd, the reader is referred to McGregor (2005). In drought feeding, the issues of prior experience of the feed, or of a training or introductory period for feeds such as whole grain, are equally relevant. Substitution may be less of an issue, because by definition, drought will involve a shortage of forage. However, for exactly the same reason, animals in drought will often be hungry, so careful management of feeding is very important to avoid the dominance effects described above. If the drought feed is offered as a feed trail, it may be necessary to offer a space of somewhat more than 0.5 m/animal.

Drought conditions may also imply hot conditions, and under these circumstances, dominant goats will not only fight for space at the feed trough, but also for access to water and shade (McGregor, 2005). They may even exclude submissive animals from these areas. This means that the location of drought feeding areas and the frequency of feeding must be very carefully considered in relation to the location and nature of shade areas and watering points.

## SUMMARY

The capacity of the goat to select and consume substantial quantities of browse places it in a unique position, compared with other domesticated ruminants, to contribute to animal production and thus human well-being, especially in regions where sown pastures cannot be grown. It does this in part by being able to actively browse in both the quadripedal and the bipedal modes. Popular writings on goat feeding behavior claim that bipedal grazing is the key to browsing success, but when measurements have been made, a surprisingly small proportion of time is spent browsing in this mode. Nevertheless, goats undoubtedly browse much more than do sheep or cattle.

However, as this chapter has also demonstrated, the goat can also use sown pastures well and is thus a successful intermediate feeder. Claims that goats will reject legumes like white clover are not supported by experimental results, which show that goats can produce well on perennial ryegrass/white clover pastures.

Published evidence suggests that goats do tend to digest forage slightly better than do sheep, but by an amount which would be of little practical significance. Similarly, the evidence that goats eat more of a given diet than sheep is equivocal. Goats do appear to have a higher ME requirement for maintenance, so to maintain weight, they would need to eat more of a given diet than sheep. However, it does not necessarily follow that goats will have higher *voluntary* intakes than sheep.

A fruitful area for future work on ingestive behavior and intake by the goat would be to resolve the issue of how it deals with the high condensed tannin loads with which it is occasionally confronted. There is no doubt that it does cope, but the mechanisms remain unclear. Recent experimental results suggest that the role of tannin-binding proteins in saliva may have been overstated, but further work is required to clarify this.

## REFERENCES


AFRC. 1998. The Nutrition of Goats. AFRC Technical Committee on Responses to Nutrients. CAB International: Wallingford.

Animut, G. and A.L. Goetsch. 2008. Co-grazing of sheep and goats: Benefits and constraints. Small Ruminant Research 77:127–145.

Austin, P.J., L.A. Suchar, C.T. Robbins, and A.E. Hagerman. 1989. Tannin-binding proteins in saliva of deer and their absence in saliva of sheep and cattle. Journal of Chemical Ecology 15:1335–1347.

Bateman, H.G., T.W. White, C.C. Williams, and S. Alford. 2004. Goat preference for concentrates or forages is influenced by physical and chemical characteristics of the feed. The Professional Animal Scientist 20:198–204.

Brown, L.E. and W.L. Johnson. 1985. Intake and digestibility of wheat straw diets by goats and sheep. Journal of Animal Science 60:1318–1323.

Celaya, R., M. Oliván, L.M.M. Ferreira, A. Martínez, U. García, and K. Osoro. 2007. Comparison of grazing behaviour, dietary overlap and performance in non-lactating domestic ruminants grazing on marginal heathland areas. Livestock Science 106:271–281.

CSIRO. 2007. Nutrient Requirements of Domesticated Ruminants. CSIRO Publishing: Melbourne.

De Rosa, G., L. Moio, F. Napolitano, F. Grasso, L. Gubitosi, and A. Bordi. 2002. Influence of flavour on goat feeding preferences. Journal of Chemical Ecology 28:269–281.

Distel, R.A. and F.D. Provenza. 1991. Experience early in life affects voluntary intake of blackbrush by goats. Journal of Chemical Ecology 17:431–449.

Domingue, B.M.F., D.W. Dellow, P.R. Wilson, and T.N. Barry. 1990. Comparative nutrition of deer and goats, goats and sheep. Proceedings of the New Zealand Society of Animal Production 50:39–42.

Dove, H. and R.W. Mayes. 2005. Using n-alkanes and other plant wax components to estimate intake, digestibility and diet composition of grazing/browsing sheep and goats. Small Ruminant Research 59:123–139.

Doyle, P.T., J.K. Egan, and A.J. Thalen. 1984. Intake, digestion, and nitrogen and sulphur retention in Angora goats and Merino sheep fed herbage diets. Australian Journal of Experimental Agriculture and Animal Husbandry 24:165–169.

Foley, W.J., G.R. Iason, and C. McArthur. 1999. Role of Plant Secondary Metabolites in the Nutritional Ecology of Mammalian Herbivores: How Far Have We Come in 25 Years? Pp. 130–209 in Nutritional Ecology of Herbivores, edited by H-J.G. Jung and G.C. Fahey. American Society of Animal Science: Savoy.

Fraser, M.D. and I.J. Gordon. 1997. The diet of goats, red deer and South American camelids feeding on three contrasting Scottish upland vegetation communities. Journal of Applied Ecology 34:668–686.

Hadjigeorgiou, I.E., I.J. Gordon, and J.A. Milne. 2001. The intake and digestion of a range of temperate forages by sheep and fibre-producing goats. Small Ruminant Research 39:167–179.

Hetherington, L. 1977. All About Goats, pp. 37–38. Farming Press: Ipswich.

Lamy, E., G. da Costa, A. Capela e Silva, J. Potes, A.V. Coelho, and E.S. Baptista. 2008. Comparison of electrophoretic protein profiles from sheep and goat parotid saliva. Journal of Chemical Ecology 34:388–397.

Landau, S., A. Perevolotsky, D. Kababya, N. Silanikove, R. Nitzan, H. Baram, and F.D. Provenza. 2002. Polyethylene glycol affects goats' feeding behaviour in a tannin-rich environment. Journal of Range Management 55:598–603.

Luo, J., A.L. Goetsch, I.V. Nsahlai, J.E. Moore, M.L. Galyean, Z.B. Johnson, T. Sahlu, C.L. Ferrell, and F.N. Owens. 2004. Voluntary feed intake by lactating, Angora, growing and mature goats. Small Ruminant Research 53:357–378.

McGregor, B.A. 2005. Nutrition and management of goats in drought. A report for the Rural Industries Research and Development Corporation (RIRDC Publication 05/188), at http://www.rirdc.gov.au/reports/NAP/05–188.pdf (verified 13 August 2008).

Mehansho, H., L.G. Butler, and D.M. Carlson. 1987. Dietary tannins and salivary proline-rich proteins: interactions, induction and defense mechanisms. Annual Review of Nutrition 7:423–440.

Mellado, M., A. Rodríguez, J.A. Villareal, R. Rodríguez, J. Salnas, and R. López. 2005. Gender and tooth wear effects on diets of grazing goats. Small Ruminant Research 57:105–114.

Mellado, M., L. Olivares, W. Pitroff, H. Díaz, R. López, and J.A. Villareal. 2007. Oral morphology and dietary choices of goats on rangeland. Small Ruminant Research 71: 194–199.

Merchant, M. and D.J. Riach. 1994. The intake and performance of Cashmere goats grazing sown swards. Grass and Forage Science 49:429–437.

Mole, S., L.G. Butler, and G. Iason, 1990. Defense against dietary tannin in herbivores: a survey for proline-rich salivary proteins in mammals. Biochemical Systematics and Ecology 18:287–293.

Morand-Fehr, P. 2003. Dietary choices of goats at the trough. Small Ruminant Research 49:231–239.

Mueller-Harvey, I. 2006. Unravelling the conundrum of tannins in animal nutrition and health. Journal of the Science of Food and Agriculture 86:2010–2037.

NRC (National Research Council). 2007. Nutrient Requirements of Small Ruminants. National Academies Press, Washington D.C.

Papachristou, T.G. 1994. "Foraging behaviour and nutrition of goats grazing on shrublands in Greece." In: Grazing Behaviour of Sheep and Goats, edited by I.J. Gordon and R. Rubino. Cahiers Options Mediterraneens 5:83–90.

Penning, P.D., J.A. Newman, A.J. Parsons, A. Harvey, and R.J. Orr. 1997. Diet preferences of adult sheep and goats grazing ryegrass and white clover. Small Ruminant Research 24:175–184.

Perez-Maldonado, R.A. and B.W. Norton. 1996. The effects of condensed tannins from *Desmodium intortum* and *Calliandra calothyrsus* on protein and carbohydrate digestion in sheep and goats. British Journal of Nutrition 76:515–533.

Pfister, J.A. and J.C. Malechek. 1986. The voluntary forage intake and nutrition of goats and sheep in the semi-arid tropics of northeastern Brazil. Journal of Animal Science 63:1078–1086.

Pfister, J.A., J.C. Malechek, and D.F. Balph. 1988. Foraging behaviour of goats and sheep in the *caatinga* of Brazil. Journal of Applied Ecology 25:379–388.

Pinkerton, F. and B. Pinkerton. 2008. Supplemental winter feeding of goats. At http://www.saboergoats.com.asp/other/suppl-winter-feeding.asp (Verified 15 August 2008).

Reid, R.L., G.A. Jung, J.M. Cox-Gasner, B.F. Rybeck, and E.C. Townsend. 1990. Comparative utilization of warm- and cool-season forages by cattle, sheep and goats. Journal of Animal Science 68:2986–2994.

Rogosic, J., J.A. Pfister, F.D. Provenza, and D. Grbesa. 2006a. Sheep and goat preference for and nutritional value of Mediterranean maquis shrubs. Small Ruminant Research 64:169–179.

Rogosic, J., R.E. Estell, D. Skobic, A. Martinovic, and S. Maric. 2006b. Role of species diversity and secondary compound complementarity on diet selection of Mediterranean shrubs by goats. Journal of Chemical Ecology 32: 1279–1287.

Sahlu, T., A.L. Goetsch, J. Luo, I.V. Nsahlai, J.E. Moore, M.L. Galyean, F.N. Owens, C.L. Ferrell, and Z.B. Johnson. 2004. Nutrient requirements of goats: developed equations, other considerations and future research to improve them. Small Ruminant Research 53:191–219.

Sanon, H.O., C. Kaboré-Zoungrana, and I. Ledin. 2007. Behaviour of goats, sheep and cattle and their selection of browse species on a natural pasture in a Sahelian area. Small Ruminant Research 67:64–74.

Shimada, T. 2006. Salivary proteins as a defense against dietary tannins. 2006. Journal of Chemical Ecology 32: 1149–1163.

Snyder, L.J.U., J.-M. Luginbuhl, J.P. Mueller, A.P. Conrad, and K.E. Turner. 2007. Intake, digestibility and nitrogen utilization of *Robinia pseudoacacia* foliage fed to growing goat wethers. Small Ruminant Research 71:179–193.

Solaiman, S.G. 2006. "Feeding Management of a Meat Goat Herd." In: Notes on Goats—Technical Paper No. 06–11, 11 pp. Tuskegee University: Tuskegee.

Tolkamp, B.J. and B.O. Brouwer. 1993. Statistical review of digestion in goats compared with other ruminants. Small Ruminant Research 11:107–123.

# 10
# Feeds and Feeding Management

*S.G. Solaiman, PhD, PAS*

**KEY TERMS**

Feed ingredients—components of the ration that when eaten will provide nutrients or energy for animals.
Forage—plant materials rich in fiber that are included in diets.
Concentrates—ingredients rich in energy or protein included in diets.
Manufactured feeds—mixtures of ingredients that often include agro-industrial by-products, urea, and binders.
Polyethylene glycol (PEG)—a tannin-neutralizing compound.
Diet—mixture of feed ingredients consumed by an animal.
Ration—amount of a diet consumed by an animal.
Feeding Management—practice instituted by man to provide feeds to animals.
Feeding System—interactions among animals, plants, and the environment designed to provide feed to sustain life and productivity.
Intensive Feeding System—feeding system for animals fed in confinement.
Extensive Feeding System—feeding system where animals browse or have free range with little or no supplemental feeds.

**OBJECTIVES**

By completing this chapter, the reader will acquire knowledge on:

- Different feeds
- Feed composition
- Management practices involved in feeding goats
- Differences in nutrient needs of animals based on their physiological status including growth, pregnancy, and milk production
- Feeding management for maintenance and growth
- Feeding management for meat production
- Feeding management for milk production
- Feeding management for fiber production
- Feeding system

## INTRODUCTION

The single greatest cost in any livestock enterprise is the cost of supplying feeding. With intensive feeding systems, feed cost averages 64% of the variable cost of an operation (excluding labor) or some 45% of the variable cost of an operation including labor. Management practices that reduce the cost of feed and feeding will improve profitability of a livestock enterprise. The major points in the feeding management of goats will be discussed to help readers understand differences among feeds and feeding practices that are humane, promote good health, and encourage strong immune function of goats with emphasis on efficient and economical production of meat, milk, and fiber.

## FEEDS

Specific feeds are not required. Instead, the nutrients from feeds, for example, protein, energy, minerals, and vitamins, are required and must be provided by feed ingredients provided in the diet. Feed ingredients include those provided by humans as well as feed components gleaned by animals from available forage or browse. First, protein is needed. Protein may come partly from nonprotein nitrogen sources like urea. Next, energy is required; that typically is obtained from dietary carbohydrates (sugars, starches, and fiber), fats, and to a lesser degree, excess protein. Accurate feed management is a three-step process. First, nutrient requirements of the animal must be estimated; second, the composition of available feeds must be assessed; and third, one seeks to provide appropriate amounts of nutrients from specific feeds to meet the nutrient requirements. Nutrient requirements of goats for maintenance, growth, pregnancy, and production of meat, milk, and fiber are published in the NRC (2007). For practical purposes, selected nutrient requirement tables are provided in Appendix B.

Feed analysis is usually based on chemical analysis of a representative sample of forage, supplements, or concentrate feeds. Preferably, feeds should be analyzed individually. Feed analysis laboratories use specific chemical procedures to assay feeds. Typical measurements include dry matter (100 minus moisture), nitrogen, ether extract, and ash, all of which are determined by "proximate analysis." The carbohydrate portion of feeds often is assayed by detergent fiber analysis (Van Soest et al., 1991) where cell contents are separated from cell wall constituents. Cell wall constituents, also called neutral detergent fiber (NDF), include cellulose, hemicellulose, and lignin. Because of their bulkiness and slow rate of digestion, feed intake will be reduced when diets rich in NDF are fed. Acid detergent fiber (ADF), cellulose plus lignin, usually is proportional to digestibility of a feed because lignin is indigestible and cellulose is fermented slowly. A general scheme outlining chemical composition for typical feedstuffs is illustrated in Figure 10.1.

Feeds for goats can be divided into two groups: forage/roughage and concentrates. The ideal proportion of forage to concentrate in the diet depends on an animal's physiological stage and the desired level of performance. Goats managed for milk production have higher nutritional requirements and usually are fed higher amounts of grain to maintain milk production for 9–10 months when compared to goats managed for meat production (only 3–4 months of milk production for the kids).

Goats consume a wider spectrum of plants than other ruminants. As selective browsers, this grazing behavior facilitates the ability of goats to survive under harsher, semiarid conditions than either sheep or cattle. Goats selectively consume and digest a wide variety of shrubs, woody plants, weeds, and briers.

### Forage/Roughage

Forage/roughage generally is rich in fiber (18% crude fiber or more) and low in energy, adds bulk to the goat's diet and thus enhances passage, and increases the butterfat content of the goat's milk. Most forage/roughage consists of the green vegetative parts of the plants though some are by-products (straw, hulls). Forage feeding often is economical and in such cases should be maximized in practical goat-feeding operations. Forage/roughage includes both wet forage/roughage (pastures, green chop, trees, shrubs, and silages) and dry forage/roughage (hay, pelleted forage, and some by-products).

Fresh forage may be grazed or fed to goats in stalls in a zero grazing system. Goats consume large amounts of forage when grazing, but intake may be even greater when fed forage in a cut-and-carry feeding system. Goats select long hay over chopped hay and leaves over stem portions of forage. Fresh forage often is preferred, followed next by hay, and finally by silages. When goats graze pasture or browse, they typically choose a higher quality diet than that offered in the trough. When given a choice, per unit body weight (BW), goats consume more legumes than grasses, more alfalfa than clover, and more Italian ryegrass or corn and sorghum than orchardgrass and fescue. Within each plant species, voluntary intake will vary with stage of plant growth. Intake is maximum 1–2 weeks before grain production with grasses and 1 week before budding with legumes. The relationship between the nutritive value of forage and its voluntary intake is well

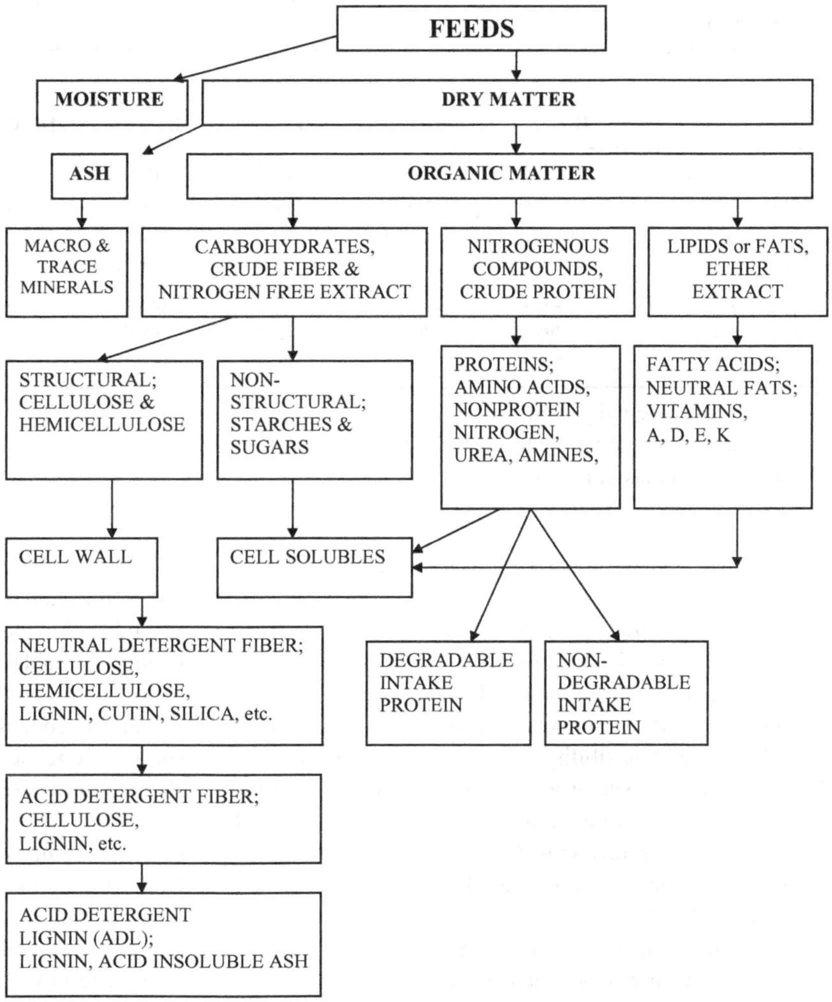


**Figure 10.1** Analytical scheme for determining nutrient content of feeds.

established. Forage preference is positively correlated with organic matter digestibility of the individual forage whereas it is negatively related to the fiber characteristics (NDF, ADF, ADL). When dry matter (DM) is less than 10–12%, low forage DM may limit intake.

Amount of refused feed also will alter intake of forage by goats. Goats will eat more forage (depending on type of forage) when the allowed refusal is larger because they have greater selection opportunity. Mixed fresh forage along with dry forage usually is preferred over a single forage, but performance will vary with feeding value of the forage.

#### Dry Forage

Adding dry forage to high-concentrate diets will increase rumen-buffering capacity that in turn optimizes rumen fermentation and improves animal performance. Dry forage include hays, pelleted forage, and by-products like straw and seed hulls. Feeding a higher quality hay also allows diet formulators to reduce the protein content of the grain mix fed and thereby to reduce the cost of the grain mix. Factors that influence the quality of hay include date of the harvest, leafiness, lack of seed heads, absence of coarse stems, lack of foreign material, and green color. Hay intake will vary depending on the plant species and maturity stage. With roughage-based diets, hay intake is related inversely to its fiber content, and fiber content increases as plants mature. Grass hay usually is of lower quality and feeding value than legume hay. Although goats prefer long hay, feed intake is increased when hay is pelleted and chopped. Intake of pelleted and dehydrated forage will be similar to intake of good quality hay.

**Table 10.1** Diet digestibility and growth of goats with sericea lespedeza meal replacing alfalfa meal in the diet.

| | Sericea Lespedeza, % | | | | | *P*-value[a] | | |
|---|---|---|---|---|---|---|---|---|
| Item[1] | 0 | 10 | 20 | 30 | SEM | Linear | Quadratic | Cubic |
| DM Digestion % | 70.6 | 66.7 | 66.3 | 65.9 | 1.3 | 0.08 | 0.28 | 0.57 |
| ADG, g | 129.6 | 77.0 | 94.9 | 129.9 | 14.1 | 0.77 | 0.01 | 0.45 |
| ADFI, g DM | 1064.8 | 1082.4 | 1195.4 | 1295.6 | 1414.5 | 0.01 | 0.70 | 0.84 |
| Gain:Feed, kg | 0.12 | 0.07 | 0.07 | 0.09 | 0.009 | 0.09 | 0.002 | 0.32 |
| Scrotal circum., cm | 26.2 | 22.7 | 24.4 | 24.1 | 0.76 | 0.21 | 0.07 | 0.96 |
| Height, cm | 66.3 | 61.7 | 62.5 | 64.8 | 1.27 | 0.05 | 0.01 | 0.53 |
| Heart girth, cm | 98.8 | 93.2 | 98.3 | 93.2 | 2.31 | 0.68 | 0.57 | 0.11 |

[1]ADFI, average daily feed intake.
[a]Based on orthogonal contrasts for equally spaced treatments.
Source: Thomas et al., 2008.

Cassava hay and peanut hay can be successfully fed to goats.

Sericea lespedeza gained popularity as forage because it reduces the parasite load of goats (Terrill et al., 2007). When sericea lespedeza meal (SLM) was fed at 30% of the total diet DM to completely replace alfalfa meal in the diet, dry matter digestibility and gain efficiency decreased. However, when goats were fed 30% SLM alone, they gained weight at a rate similar to goats fed alfalfa meal alone, and both groups gained more than goats fed acombination of alfalfa meal and SLM (Thomas et al., 2008; Table 10.1).

Straws, most hulls, and stovers have low quality (digestibility) as a feed but do not necessarily have low value when fed in a mixture. Cottonseed hulls have higher palatability and feeding value than peanut hulls and oat hulls and add bulk factor of diet (Perry et al., 1999). Soybean hulls (an exception) in combination with hay are excellent as forage for goats.

### Pasture

Globally, mixed farming systems (crop and livestock) provide more than 54% of total meat and 92% of total milk production by cattle and goats followed by landless systems (feedlots); grazing systems (no crop) provide less than 10% of the world's production of meat and milk. However, grazing systems are important in certain areas of the world such as Central America and South America where population density is low, and within the U.S. Most of the livestock in Asia is found in mixed farming systems. In the future, more livestock likely will be produced in grazing systems and mixed farming systems.

The soil and climate in most Southeastern states, in California, and in the Midwest regions of the U.S. are favorable for production of high quality grasses and legumes suitable for grazing. Pasture management, either under continuous or rotational grazing, is discussed below. Warm or cool season grass/legume pastures are site specific, so readers wishing deeper discussion of this topic should study pasture and grazing management manuals.

#### *Continuous Grazing*

Under a continuous grazing system, animals are allowed to graze the pasture throughout the grazing season. The number of animals grazing per hectare (grazing intensity) is predetermined so that plants have sufficient time to regrow. This system of grazing over time results in an accumulation of less-desirable plant species. This is the result of selective consumption of more-desirable plant species and reduced regrowth of those species. Understocking and overstocking may alter total output in terms of animal production and pasture quality. Understocking results in more mature plant stands in the pasture with higher fiber content and lower protein and energy contents that result in low digestibility of plants and lower quality pastures. Trampling loss associated with understocked pastures also increases waste of herbage.

Overstocking, on the other hand, reduces available forage per animal and will not sustain optimum animal production. In addition, parasite management for goats must be intensified using an overstocking system of grazing. The parasite burden will increase if proper measures are not practiced. The proper stocking rate in terms of maximum animal gain usually is not the same as that to

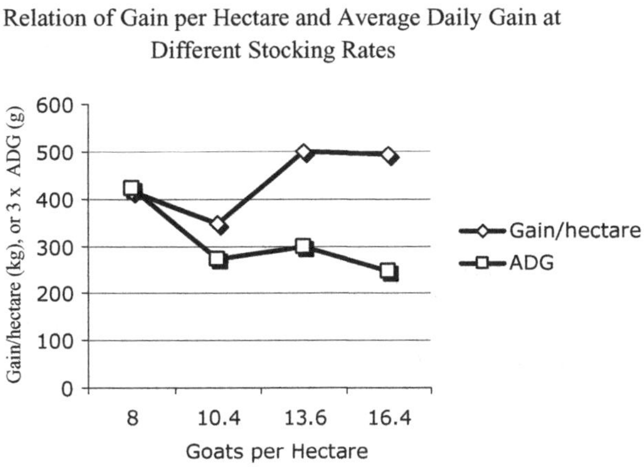


**Figure 10.2** Average daily gain and gain per hectare for goats grazing continuously annual ryegrass pastures at different stocking rates. Source: Solaiman et al., 2006.

produce the maximum gain per hectare. Higher stocking rates often will reduce average daily gain (ADG) even though gain per hectare will be increased as illustrated in Figure 10.2 (Solaiman et al., 2006).

Because investments in fencing and watering systems are small with continuous grazing, it is not capital intensive and requires few management decisions because animals are not moved from pasture to pasture. Research data on optimum stocking rates for pastures continuously grazed by goats are lacking, and most of the available information has been transferred to goats using cattle data considering that five or six goats are equal to one animal unit. However, parasite management is difficult with continuous grazing systems and warrants further investigation. Also, forage preferences and accumulation of undesirable plant species may be greater with goats than cattle due to greater plant selectivity by goats.

*Rotational Grazing*

Rotational grazing or controlled grazing is an economical way to provide forage for goats, but this system requires careful planning, fencing, and more intensive animal and pasture management. This also is called management-intensive grazing where goats are allowed to rigorously graze one pasture before being rotated or transferred to another similar pasture. Rotational grazing permits plants to regrow on the grazed pasture and is helpful for parasite management. With this system, multiple paddocks are needed so that grazing animals can be rotated among pastures. The timing of animal rotation is based on the growth characteristic of the forage in the pasture and occasionally may depend upon the stage of the parasite cycle prevailing in that environment.

**Table 10.2** Grazing efficiency or utilization rate.

| Number of Paddocks | Approximate Days on Each Paddock | Grazing Efficiency includes Maintaining Minimum Stubble |
|---|---|---|
| Continuous | — | 40% or less or (80% overgrazed, low yield) |
| 4 to 6 paddocks | 7 to 9 days | 40 to 55% |
| 8 to 10 paddocks | 4 day | 55 to 65% |
| 24 to 45 paddocks | 1 day or less | 70 to 80% |
| Hay | — | 70 to 80% |

Source: Brann, 2005.

The size of the paddocks dictates the need for fencing and the design of the water supply system. Paddock size is positively correlated with animal numbers, animal size, level of intake, and days on pasture, but negatively correlated with available forage and grazing efficiency. The equation below can be used to determine paddock size for a rotational grazing system.

$$\text{Hectare/Paddock} = \frac{(\text{No. Anim.}) \times (\text{Anim. Wt.}) \times (\text{DMI, \% BW}) \times (\text{Freq. of Rotation})}{\left[\begin{array}{l}(\text{cm of forage}) \times (\text{kg per ha, cm}) - \\ (\text{DM from grain mix})\end{array}\right] \times (\text{\% GE})}$$

Where, DMI = dry matter intake; BW = body weight; GE = grazing efficiency.

Dry matter intake for goats on pasture will range from 3.5–4% of body weight. If grain supplement is provided on pasture, dry matter contributed from grain mix can be deducted from total forage DM required. Frequency of rotation (days on paddock) is a managerial decision, but grazing for 3–7 days per pasture usually is recommended. Grazing efficiency (utilization rate) of pasture varies depending on number of paddocks used and height of grazed forage (Table 10.2). The recommended grazing height and rest days for pasture are listed in Table 10.3. For goats, it is highly recommended to move goats out of pasture when vegetation height approaches 13 cm (5 in.).

**Table 10.3** Recommended grazing heights and rest periods.

| Forage Type | Begin Grazing Height, cm | End Grazing Height, cm* | Rest Time Days |
|---|---|---|---|
| Tall fescue, orchard grass, and legume, annual ryegrass, or small grains | 20 (8″) | 10 (4″) | 14–45 |
| Bermudagrass crabgrass | 13–20 (5–8″) | 7.6 (3″) | 14–45 |
| Alfalfa | 30–38 (12–15″) | 7.6 (3″) | 24–32 |
| Sudan grass, pearl millet | 30–46 (12–18″) | 20 (8″) | 14–30 |
| Native warm season grasses | 30–46 (12–18″) | 20 (8″) | 30–50 |

*Due to parasites and desired intake it is recommended that goats not graze below a height of 13 cm (5″). Rotation timing should be based on minimum forage height, not days grazing (adjusted for goats by Ann Peischel).

Source: Brann, 2005.

Stocking rate usually is determined by the availability of forage and number of paddocks. Overstocking and understocking rules apply in this system, too, so the optimum stocking rate must be predetermined for a specific type of forage and environmental conditions that alter forage availability.

Strip grazing has gained popularity in goat production systems. In this system, goats are allowed to graze a strip of pasture a few hours each day to partially meet their daily needs. This system requires less land, but it requires more intensive management and labor. Perimeter fences are required to hold animals in pastures and to protect them from predators. Rotational and controlled grazing and strip grazing require temporary or interior fencing to subdivide the pasture into paddocks for rotation. For proper fencing and watering on pasture, the reader is referred to more specialized fencing manuals and Chapter 17, Housing Requirements.

### Browse

Browse materials, shoots, leaves, and very high quality parts of shrubs, forbs, and trees, often are the goat's staple food. Goats prefer foliage from trees and shrubs. These novel dietary ingredients are rich in nutrients even though they often contain high levels of plant toxins and antinutritive factors. The chemical composition of browse differs from that of grasses. Grasses are rich in a relatively unlignified cell wall (NDF = 50–70%) that is readily available for cellulolytic fermentation in the rumen, whereas browse materials have a relatively lower cell wall (NDF = 30–50%) that is highly lignified but is rich in readily fermented cell contents. Browse materials also contain secondary plant metabolites (SPM) that are absent in most grasses. These biologically active compounds help protect plants from herbivore and insect damage. Some SPMs of interest include tannins that limit the nutritive value of plants by forming complexes with proteins and carbohydrates, phenolic acids and terpenes that disturb rumen fermentation, and pyrrolizidine alkaloids that are hepatotoxins. One possible collateral benefit of consuming browse is that these materials contain secondary compounds that may reduce internal parasite infestations. Goats select tree foliage, shrubs, forbs, flowering parts, seeds, and nuts when they are available. Available browse can serve as a supplement in pasture-based feeding systems when forage declines in quality, and this can help sustain grazing animal performance. Conversely, pastures that are predominantly browse can be improved by supplementing animals with conventional feeds (e.g., corn, barley) to dilute secondary compounds and provide limiting nutrients.

Goats are resistant to many plant toxins and antinutritive factors. Thereby, goats can readily defoliate most plant species, many of which cattle will not use. Although the proportion will vary with availability, goats typically consume a diet that is predominantly browse (65–70%) with smaller amounts of grasses (20–25%) and forbs (5%). Goats prefer woody plants like blackberry and raspberry bushes and use their upper mobile lips to select the tender, more digestible young leaves and leave the less digestible branches and thorns. Browsing by goats with proper management can prune and stunt bushes, and this helps to encourage grass growth. However, many browse plants contain antinutritional factors like high lignin content, silica, and terpene-based organic compounds, tannins, and other alkaloids. High levels of tannin compounds reduce the availability of protein; however, when a diet is supplemented with polyethylene glycol (PEG), it reduces the antinutritional effects of tannins by improving protein availability (Decandia et al., 2000) that in turn will increase ovulation rate (Lassoued et al., 2006) in goats. Animal performance should be monitored closely. Condensed tannin-containing browse can help control anthelmintic-resistant gastrointestinal parasites. Tannins decrease fecal egg counts in sheep and goats and may decrease hatch (Min and Hart, 2003). Mimosa, fed as a cut-and-carry forage, had no anthelmintic activity when fed for 21 days to goats as compared to fresh soybean leaves. However, when goats browsed on mimosa fields, they required less parasite control than when they grazed bahaigrass pastures (Solaiman et al., 2006).

The value of browse as a partial fulfillment of a goat's diet is not fully understood. Several factors can influence its effectiveness as a part of a feeding system including selection opportunity, nutritional quality, diversity of browse to keep goats entertained, and the height where parasite larvae cannot reach. Availability of browse should be managed closely by monitoring the percentage of plant defoliation that can secure rejuvenation rather than elimination of plants; each plant can tolerate a certain percentage of defoliation before it is eliminated. Goats tend to have higher blood urea nitrogen when allowed to browse. This might reflect insufficient consumption of carbohydrates or carbohydrates and protein being bound by tannins. Also, the size of animals in comparison to height of the canopy, and number of animals per unit area must be considered when considering the availability of browse plants. Obtaining reliable browsing data requires an extensive commitment of time with very limited applicable information. Browsing studies are rare, and variation in the browse species and in their composition within season and between years complicates application to other conditions.

#### SILAGE

Silages include partially fermented grass or legume forage, and they should be free of molds. Silage or haylage produced from legumes has a higher DM and protein content and a lower cell wall content than silage produced from grasses. High dietary concentrations of silage have been associated with metabolic and other problems (acidosis, demineralization, cortical necrosis, enterotoxemia, and listeriosis) of goats. Silage should be introduced into the goat's diet gradually. However, grass silage has been used successfully in the goat diet (Trodahl et al., 1981). Also many food industry by-products such as waste from beets, apples, potatoes, banana leaves (Pieltain et al., 1998), and cassava leaves (Marjuki et al., 2008) have been successfully used as silage for goats. Ensiling cassava leaves reduced the content of cyanide (HCN), a poisonous compound for livestock normally found in cassava leaves (Tewe, 1992). When fed silage, goats tended to eat less and produced less milk than when fed either palatable green or pelleted forage. Corn silage is a good source of energy, but its intake must be limited to avoid excessive fat deposition. Supplying hay with silage helps reduce digestive and metabolic problems, and increases feed intake and milk production.

#### OTHER ROUGHAGE

Goats readily consume garden by-products. Later discussion is based on experience with limited scientific documentation but may provide some guidance for goat producers.

Rape, kale, or beets commonly are added to diets for goats. Under controlled feeding, goats adapt well to by-products and surplus feeds including discarded produce. Some surplus or damaged produce like carrots, artichokes, and turnips should be used with caution. Because crucifers like cabbage contain goitrogens that interfere with thyroid hormones, their use in diets for goats should be limited. Avoid feeding clippings from rhododendron or branches pruned from cherry, apricot, or peach trees because when wilted, they may prove toxic to goats. Sweet potato forage and its mixture with grasses provide an inexpensive source of nitrogen in the diet for growing goats (Aregheore, 2004). Little is known about the effects of pine needles in grazing species other than cattle where they may prove toxic. Sheep and goats may not abort if they consume pine needles (Short et al., 1992).

### Grains and Concentrates

Concentrate feeds are low in fiber but rich in energy, protein, or both. In addition to the roughage, grain mixes typically are added to diets for growing and nursing goats. This is the most expensive portion of the diet to purchase from outside the farm or ranch. Concentrate mixes are composed from high energy or high protein feeds. Although the majority of a goat's required nutrients often come from forage, when forages are not available or when forages are insufficient in protein and energy to support desirable performance, supplements should be provided. Supplements should be designed specifically to match dietary shortages. When protein is deficient, feeding a low-protein, high-energy supplement will not increase total nutrient intake, but a relatively small amount of a high-protein supplement may prove beneficial. Feeding grain often decreases both forage consumption and digestibility; however, with low quality forage, a grain concentrate may have a positive associative effect on forage intake (Huston, 1994). Goats may refuse to eat concentrates in an abnormal physical form, particularly if the feed is dusty. Goats prefer cubed, pelleted, or coarse feed compared to finely ground feeds. Even more than cattle and sheep, goats may refuse to eat moldy or fermented feeds. Cereals, cereal by-products, and oil seed meals are readily acceptable. However, goats may refuse to eat rapeseed meal, dehydrated alfalfa, flour meal, or animal fat.

#### Energy Concentrates

Energy concentrates are defined as feeds with less than 20% protein and less than 18% fiber. High-energy feeds include all cereal grains, some root crops, flourmill and bakery by-products, and many other food manufacturing by-products such as beet pulp and citrus pulp.

When forage or browse crops are low in digestible energy content, 0.25–0.50 kg shelled corn or whole cottonseed per goat daily often is provided to supplement the diet with digestible energy. Whole cottonseed contains high levels of CP and TDN and requires no processing, which makes it a very desirable by-product feed. The energy in WCS is primarily from fat. Whole cottonseed also contains gossypol, a polyphenolic yellow pigment that has deleterious effects on male reproduction. Adding whole cottonseed to medium- or low-quality forage diets for growing male goats has had detrimental effects on animal production; however, when limited to 15% of diet DM, whole cottonseed consumption improved dry matter digestibility and passage and had no adverse effect on growth or sperm quality in growing male goats (Solaiman et al., 2009; Table 10.4). The difference among corn, milo, barley, and wheat in energy content for goats is small, so cost per unit of energy usually determines which grain should be used. Although it is not recommended, pearl millet grain can substitute successfully for corn in complete diets for lactating dairy goats but not for growing animals according to Glaye et al. (1997). Metabolizable energy of pearl millet is only 90% of that for corn for mature goats, and its protein is not used efficiently (Terrill et al., 1998). Milo may promote urinary calculi if not balanced with proper minerals and should be used with caution for male goats. Being very rapidly fermented, wheat should not constitute more than 50% of the grain mix. Oats usually are not cost effective. Grinding promotes dustiness (undesirable for goats) and increases feed cost. Yet, cracking exposes the endosperm and improves the digestion of feed grains. Vegetable oil or animal fats also are fed to increase the energy density of the diet. However, feeding more than 7–8% fat in the diet may negatively impact rumen fermentation and depress fiber digestion.

#### Protein Concentrates

When forage is low in protein content or availability, additional protein supplements are needed for maximum production or performance. Protein concentrates usually contain at least 20% crude protein. High-protein feeds include alfalfa hay, alfalfa meal or cubes, and other high-protein concentrates such as oilseed by-product meals (cottonseed, peanut, soybean, etc.). Corn gluten meal has a

**Table 10.4** Growth performance, scrotal circumference, and semen quality of goat kids fed diets with different amounts of whole cottonseed.

| | Whole cottonseed, % DM | | | | *P*-value[a] | |
|---|---|---|---|---|---|---|
| Item | 0 | 15.7 | 32.7 | SEM | Linear | Quadratic |
| Average daily gain, g | 81.4 | 109.8 | 85.7 | 6.83 | 0.66 | 0.01 |
| Average daily feed intake, g | 948.3 | 1,295.6 | 1,084.7 | 103.0 | 0.37 | 0.05 |
| Gain:Feed (G:F) | 0.09 | 0.09 | 0.08 | 0.01 | 0.70 | 0.84 |
| Scrotal circumference, cm | 23.1 | 22.3 | 20.4 | 0.31 | 0.0001 | 0.22 |
| Sperm quality | | | | | | |
| Normal, % | 81.0 | 73.9 | 71.5 | 6.1 | 0.30 | 0.76 |
| Gross motility (scale 0–5) | 4.90 | 4.52 | 4.56 | 0.11 | 0.05 | 0.16 |
| Progressive motility, % | 72.3 | 67.2 | 59.2 | 5.89 | 0.14 | 0.84 |
| Head abnormality, % | 8.24 | 4.56 | 15.41 | 5.80 | 0.40 | 0.33 |
| Mid-piece abnormality, % | 9.49 | 10.6 | 12.1 | 2.84 | 0.54 | 0.96 |
| Tail abnormality, % | 2.52 | 2.83 | 4.62 | 1.36 | 0.32 | 0.67 |

[a]Based on orthogonal contrasts for equally spaced treatments.

Source: Solaiman et al., 2009.

poor balance of amino acids, whereas fishmeal and heat-treated soybean meal provide a desirable mix of amino acids in ruminally bypassed protein. Protein derived from ruminants, and poultry litter (potentially contaminated with ruminant derived protein) must not be fed to goats in the U.S. even though animal protein sources like blood meal, or meat and bone meal can be useful sources of high quality protein. Because these materials may transfer prions (BSE) and cause human health problems, their use is prohibited in the U.S. The most common protein sources fed to goats are the oilseed meals: cottonseed meal and soybean meal.

Urea is a nonprotein-nitrogen source hydrolyzed to ammonia by rumen microorganisms. Ammonia can be efficiently used by microorganisms of the rumen during fermentation for synthesis of microbial protein, provided sufficient energy is available for the microbes to grow. When used correctly in goat feed, urea can provide an excellent, cost-effective source of nitrogen. As a thumb rule, the amount of urea to be fed should be no more than 25% of the required protein in rations for milking does. Because urea does not provide energy, it is useful only when the supply of fermentable carbohydrates is adequate. This means that its value and usefulness with low quality forage will be nil unless sources of readily available energy, such as molasses, are provided. Urea should not be fed to young ruminants before the rumen is fully functional. Ammonia toxicity can occur when a large amount of urea is introduced abruptly into the ration or intake is excessive due to diet formulation errors or improper diet mixing. With adaptation to urea, rate of ammonia production is reduced, so ruminants should be adapted to diets containing high levels of urea over several days or weeks.

Commercial protein supplements are available in many forms: meals, pellets, liquid supplements, and liquid blocks. Typically protein is the most expensive portion of the diet purchased off the farm. Therefore, when supplementing protein, cost comparisons, the presence of other dietary components, palatability of the product, ease of handling, feeding facilities, labor cost/convenience, and the need for uniform intake should be considered carefully. Because goats are skilled at sorting feed components being fed, diets should be formulated and feeding methods should be employed to reduce or prevent sorting. Sorting typically is most prevalent when (1) feed components differ in particle size, (2) components differ in density, and (3) a surplus of feed is provided. Therefore, protein supplements selected should have a particle size and density similar to other diet ingredients, and the supply of feed or time of access to feed should be limited.

**Manufactured Feeds and By-Products**

Manufactured feeds are mixtures of feeds that often contain agroindustrial by-products, urea, binders, and preservatives that either are complete diets (providing all the nutrients required) or supplements (providing nutrients otherwise deficient within the basal diet ingredients). The products used and their proportions in manufactured feed will differ with cost and availability of by-products at the local level. Supplements that provide missing nutrients generally will increase digestibility of low quality forage and thereby will increase the rate of body weight gain. For example, olive cake–based feed blocks enriched with squeezed cactus fruits provide a cost-effective supplement to a kermes oak-based diet in olive-producing areas of the world. Such blocks substantially increase the nutritive value of kermes oak-based diets by partially neutralizing tannins (Ben Salem et al., 2003). Feed blocks are efficient carriers for macro- and micronutrients, tannin-neutralizing reagents (for example, polyethylene glycol), and anthelmintic products for small ruminants. Grain-based commercial supplements may not prove economical for growing and finishing meat goats if not properly formulated for goats. Many high-starch supplements traditionally fed to cattle when fed to goats may reduce ruminal pH and fiber digestibility. Typically, diets for growing and finishing goats and sheep contain considerably higher amounts of roughage than diets fed to cattle.

Certain by-product feeds contain fiber that is quite digestible. Such fiber sources should provide adequate rates of gain without causing the digestive problems associated with high-starch (high-grain) diets. Such feeds would include soy hulls, corn gluten feed, and wheat middlings (Moore et al., 2002). Ensiled green tea waste also has potential as a protein supplement (Kondo et al., 2004). Dehydrated citrus pulp can replace corn up to 40% in the diet and can maintain high performance of goats (Bueno et al., 2002). Mustard seeds contain about 30–35% oil and 34–39% protein. The oil-extracted mustard seed, mustard cake has a good balance of essential amino acids and a relatively high methionine content, and it can replace peanut cake completely with no adverse effect on performance (Anil Kumar et al., 2002). Olive leaves traditionally have been used in Mediterranean countries as a source of nutrients for small ruminants during periods of scarce feed supplies. Because olive leaves contain condensed tannins, they should be supplemented with barley, faba beans (Yanez et al., 2004), or PEG for optimum utilization. Condensed tannins will form complexes with dietary proteins, carbohydrates, and microbial enzymes and decrease protein availability. Good palatability and high-energy content make unripe banana fruits a suitable feed for goats that appears similar to barley but with lower protein content. Seed harvested cowpea vines are low-quality forage that can be improved by ensiling (Solaiman, 2007), and poultry waste has also been used as feed for goats in some regions of the world. Replacing alfalfa meal with broiler litter at 25% of the dietary DM in a total mixed ration reduced dry matter digestibility in goats but still yielded an acceptable diet (Bartlett, 1989; Table 10.5) while reducing diet cost. Whole cottonseed is high in energy and protein; fed at up to 15% of dietary DM, it improved average daily gain with no adverse effect on reproductive performance of goats (Solaiman et al., 2009). The distillers dried grains with solubles (DDGS), a by-product of the ethanol industry, can be successfully used in the diet of growing goats (Gurung et al., 2009; Table 10.6). This is a unique feedstuff providing protein, energy, highly digestible fiber, and minerals to goats.

**Table 10.5** Nutrient digestibility of diets with broiler litter replacing alfalfa meal in total mixed diets fed to goats.

| | Broiler Litter, % DM | | | |
|---|---|---|---|---|
| Digestibility, % | 0 | 12.5 | 25.0 | SEM |
| Dry Matter | 75.7$^{a}$ | 70.6$^{b}$ | 67.2$^{c}$ | 0.52 |
| Crude Protein | 71.6$^{a}$ | 64.1$^{b}$ | 59.1$^{c}$ | 0.92 |
| Neutral Detergent Fiber | 55.7$^{a}$ | 51.0$^{ab}$ | 48.2$^{b}$ | 1.48 |
| Acid Detergent Fiber | 48.1 | 49.8 | 45.1 | 1.65 |
| Cellulose | 54.2$^{b}$ | 53.5$^{b}$ | 60.2$^{a}$ | 1.37 |
| Hemicellulose | 65.2$^{a}$ | 52.2$^{b}$ | 52.6$^{b}$ | 4.11 |
| Ash | 46.9$^{a}$ | 22.2$^{b}$ | 18.7$^{b}$ | 4.25 |

$^{abc}$Means with different superscripts differ ($P < 0.05$).
Source: Bartlett, 1989.

### Mineral and Vitamin Supplements

Minimum requirements of minerals and vitamins must be provided for optimum goat performance, but the need for supplements will vary with composition of diet ingredients. Mineral supplements can be as simple as salt alone or may be a complex mix consisting of several macro and trace elements plus vitamins provided free choice in the form of either a loose mixture or a block. Some mineral/vitamin mixtures can be included in a total grain mix to be fed as part of a totally mixed ration (TMR). Generally calcium, phosphorus, and sodium are supplemented to diets for goats, but the need for each will differ with the diet ingredients being fed. Normally, calcium is present at adequate levels in most forage for gazing ruminants, but salt (or some source of sodium) must be supplemented. Phosphorus may be deficient in forage at certain times of the year, and a phosphorus deficiency can reduce reproductive performance. If grass tetany is a regional problem with forage, seasonal supplementation with magnesium may be necessary. In early spring or when animals are consuming fresh green and rapidly growing pastures fertilized with N, P, and K, high levels of potassium and nitrogen with low levels of magnesium and salt may lower the ratio of magnesium : potassium and provoke grass tetany. Trace mineralized salt provides a mixture of micro minerals, and vitamins A and E are supplemented in vitamin premixes.

Soils in many areas of the U.S. are deficient in copper, so feeds produced in those areas also will be deficient in copper. Adding a mineral supplement to the diet that has

**Table 10.6** Average daily gain and dry matter intake of Kiko × Spanish crossbred growing male goat kids fed different amount of distillers dried grains with solubles (DDGS).

| | Distillers dried grains with solubles, % DM | | | |
|---|---|---|---|---|
| Item[1] | 0 | 10.3 | 20.6 | 31.0 |
| Initial BW, kg | 27.9 ± 0.94 | 29.8 ±± 0.73 | 27.8 ±± 0.82 | 29.5 ±± 0.73 |
| Final BW, kg | 38.5 ±± 1.63 | 39.8 ±± 1.26 | 36.4 ± 1.42 | 38.3 ± 1.26 |
| ADG, g/d | 141.0 ± 18.4 | 134.1 ± 14.2 | 115 ± 16.0 | 117 ± 14.2 |
| Total DMI, g/d | 1,017 ± 87.3 | 1,138 ± 77.5 | 1,106 ± 93.1 | 1,003 ± 87.0 |
| Concentrate, g/d | 519 ± 41.9 | 591 ± 37.2 | 575 ± 44.7 | 520 ± 41.8 |
| Hay, g/d | 499 ± 45.8 | 547 ± 40.7 | 531 ± 48.5 | 483 ± 45.7 |
| Gain:Feed | 0.12 ± 0.01 | 0.12 ± 0.01 | 0.11 ± 0.01 | 0.12 ± 0.01 |

[1]No differences were observed ($P > 0.10$).
Source: Gurung et al., 2009.

an adequate amount of copper can eliminate most symptoms of deficiency. Copper deficiency is sporadic throughout the U.S. Soils in and feeds from the upper coastal areas on the West Coast (including California) tend to have higher copper and other trace minerals than the East Coast; the Southeast and major parts of the Midwest are deficient in copper and many other trace elements according to U.S. geochemical soil maps. However, animals raised in Western states often have lower serum copper than those raised in Southeastern and Midwestern states. In some states, copper levels in soil may be sufficient, but other minerals present in feeds (molybdenum or sulfur) may reduce copper availability. The addition of 5 or 10 mg of molybdenum to the diet can deplete copper in the liver and negatively impact immune response in goats (Solaiman et al., 2008; Table 10.7).

Soil copper concentrations may vary from location to location within a state. Therefore, it is recommended to check soil or forage mineral levels for copper, molybdenum, and sulfur. Copper deficiency symptoms vary depending on the severity of the condition. Symptoms can include frequent staphylococcus lesions on the body, a thin and faded hair coat, bald tail tips, twisting and bending of the front legs, spinal cord injuries, and anemia. Generally, the immune system loses effectiveness with a copper deficiency leaving animals more vulnerable to diseases and parasites. Although most symptoms of copper deficiency can be reversible by feeding adequate copper, other symptoms like swayback in young kids (caused by copper deficiency in pregnant does), and spinal cord injuries are not reversible. Goats are more tolerant to high levels of copper than sheep or cattle (Solaiman et al., 2001). Feeding as high as 100 mg of copper per day for 100 days improved daily weight gain and immune function of goats (Table 10.8; Solaiman et al., 2007). This confirms previous recommendations of other producers and goat experts. However, copper is accumulative, and high levels of copper are not recommended for long-term use. Dietary levels of copper above 40 ppm are not recommended for dairy goats. For more information on soil copper, the reader is directed to the USGS map for copper at http://tin.er.usgs.gov/geochem/doc/averages/cu/usa.html.

Internal parasites, either through altering GI pH and copper absorption or by increasing blood loss, can contribute to copper deficiency; conversely, parasites are more prevalent in copper-deficient animals, and parasite infestation can be managed more readily when animals are kept at optimum copper status. Other minerals can interfere with copper status. These include diets with high levels of either sulfur (>0.35% of diet dry matter), molybdenum (more than 2 ppm in feed), iron (more than 250 ppm), calcium, zinc, manganese, or cobalt. Such diets can reduce copper absorption and deplete liver copper. Soils that are heavily limed or have a high pH will render the copper unavailable. The concentration of copper in the liver is the most reliable index of copper status of the animal. Although

**Table 10.7** Mineral concentration in the liver and immune response of goats fed high amounts of molybdenum.

| | Added Molybdenum, mg | | | | *P*-value[a] | |
|---|---|---|---|---|---|---|
| Item | 0 | 5 | 10 | SEM | Linear | Quadratic |
| Liver, DM | | | | | | |
| Copper, ppm | 380.0 | 152.1 | 120.0 | 5.24 | 0.003 | 0.15 |
| Molybdenum, ppm | 5.17 | 4.93 | 6.03 | 0.32 | 0.07 | 0.11 |
| Iron, ppm | 190.0 | 136.7 | 131.7 | 10.9 | 0.002 | 0.09 |
| Zinc, ppm | 94.7 | 88.3 | 101.7 | 7.24 | 0.50 | 0.28 |
| Humoral Immune Response | | | | | | |
| Antibody Titer[b] | 86.7 | 46.7 | 33.5 | 11.5 | 0.002 | 0.35 |
| Cell-Mediated Immune Response | | | | | | |
| Skin Fold Thickness[c], cm | 4.63 | 4.06 | 3.84 | 0.14 | 0.002 | 0.30 |

[a]Based on orthogonal contrasts for equally spaced treatments.
[b]Antibody titer to ovalbumin injection; a higher titer reflects a stronger immune response.
[c]Skin fold thickness in response to PHA injection; greater thickness reflects a stronger immune response.
Source: Solaiman et al., 2008.

**Table 10.8** Effects of supplemental copper on growth and immune function of growing kids.

| | Dietary Copper, ppm | | | | *P*-value[a] | |
|---|---|---|---|---|---|---|
| Item | 0 | 100 | 200 | SEM | Linear | Quadratic |
| Average daily gain, g | 131 | 147 | 117 | 6.09 | 0.11 | 0.01 |
| Average daily feed intake, kg | 1.21 | 1.11 | 1.03 | 0.06 | 0.05 | 0.90 |
| Gain:Feed (G:F) | 0.11 | 0.13 | 0.11 | 0.01 | 0.67 | 0.02 |
| Scrotal circumference, cm | 23.1 | 22.3 | 20.4 | 0.31 | 0.0001 | 0.22 |
| Serum copper, mg/dL | 0.82 | 0.78 | 0.86 | 0.03 | 0.35 | 0.14 |
| Liver copper, mg/kg DM | 206 | 504 | 778 | 59.0 | 0.001 | 0.87 |
| Immune response[b] | | | | | | |
| Antibody titer to ovalbumin | | | | | | |
| Day 72 | 0.26 | 0.31 | 0.32 | 0.03 | 0.08 | 0.46 |
| Day 98 | 0.19 | 0.29 | 0.26 | 0.01 | 0.005 | 0.003 |

[a]Based on orthogonal contrasts for equally spaced treatments.
[b]A greater titer reflects an improved or greater immune response.
Source: Solaiman et al., 2007.

safe procedures are available to biopsy the liver of cattle, similar safety of biopsy procedures for goats is questionable. A secondary measurement for assessing copper status of goats is plasma or serum copper. Unfortunately, the liver can be markedly depleted before plasma or serum copper concentrations decline.

Copper can be supplemented through mineral mixes that are high in copper. Mineral mixes that are labeled for both sheep and goats should not be used for goats. Because sheep are much more sensitive to high copper levels than goats, mineral mixes appropriate for sheep do not have adequate copper levels for goats, or they may have high levels of molybdenum. Copper also can be supplemented through slow release copper wire needles as copper boluses (0.625–1.35 grams) given early to kids (2–4 weeks of age) or 1 gram of copper oxide in bolus form per 10 kg of body weight every 5–6 months in copper-deficient areas. The needles are deposited in the goat's stomach and release copper slowly. Some feeds (wheat, barley, and oats) are low in copper and some, like alfalfa, are rich in molybdenum. Applying 1.5–3.0 kg of copper per hectare as an organic compound such as copper EDTA, copper lignisulfonates, or copper polyflavonoids has been recommended and can increase soil copper levels for many years.

Selenium or iodine is deficient in certain regions of the U.S. Coastal areas usually have sufficient iodine and are deficient in cobalt. The Dakotas are so rich in selenium that selenium toxicity may occur; however, many other places including California are selenium deficient. Selenium injections often are used for pregnant does toward the end of the pregnancy and for young kids at birth. However, both selenium and copper can be toxic if overdosed. Feeding rather than injecting selenium is recommended unless severe deficiency is prevalent.

### Feed Additives

Additives are products of nonfeed origin that may improve feed efficiency, growth, and health of animals. Many herbal and natural remedies are used commonly by goat producers in the U.S. with little if any scientific evidence of their effectiveness. Only two additive classes are available commercially that have proven to be effective for goats.

#### Ionophores

Ionophores are antibiotics produced by varieties of actinomycetes bacteria that alter the flux of ions across gram-positive bacteria in the rumen and reduce their numbers. This fermentation shift negatively affects methane, acetate, and butyrate production, while increasing the molar proportion of propionate; ammonia production is decreased resulting in a "protein sparing" effect; protein degradation is inhibited; and lactic acid formation associated with high grain diets is reduced (Russell and Strobel, 1989). As a result, feed efficiency and the net energy available from feed increases. Ionophores also may increase milk production and milk protein synthesis and help to prevent ketosis and reduce bloat. Lasolocid and monensin are the two ionophores that are approved for prevention of coccidiosis in meat goats in the U.S., but they are not approved for dairy goats. Ionophores may improve animal health and reduce amino acid catabolism, fat mobilization, nonesterified fatty acids (NEFA), and ketone bodies formation. Readers are directed to the FDA website for more updated information on approved drugs for goats.

#### Probiotics

Probiotics are live, naturally occurring microorganisms fed directly to animals. They have been shown to relieve stress, increase dry matter intake, and improve growth. They provide a viable culture of microorganisms in the digestive tract that may decrease or prevent the establishment of pathogenic organisms. They also may alter rumen fermentation, optimize rumen pH, increase nutrient flow through the gastrointestinal tract, improve digestibility, and reduce stress by boosting immune function. Benefits from probiotics are not always apparent. A commercial probiotic supplement failed to affect growth performance, diet digestibility, and carcass traits of healthy growing meat goats (Whitley et al., 2009). Commonly used probiotics include fungal cultures (Aspergillus oryzae and Saccharomyces cerevisiae) and lactic acid bacteria (Lactobacillus and Streptococcus).

## FEEDING PRACTICES

The process of matching the nutritional requirements of animals with the nutrient content of the feed ingredients is referred to as "feeding practices." The objectives of a sound feeding practice are to (1) optimize animal performance, (2) minimize feed cost, and (3) improve sustainability of the environment. Reproductive performance of bucks and does is crucial for a successful animal production system. Reproductive efficiency of does coupled with improved kid performance will determine the economic profitability of the system. Through enhancing milk or fiber production, matching nutrient requirements with feed ingredients will improve productivity and viability of the system. Feeding practices for goat production are highly variable across the globe ranging from simply tethering animals near households in some part of the globe, extensively grazing or browsing rangelands in some areas, and intensive feeding practices in other areas. In all cases the nutritional goal is to meet fully the nutrient requirements of the animal. In the natural setting and given free choice, goats as intermediate selectors, will manage their own intake, but may or may not meet all of their nutrient and energy requirements. With intensive production systems, diets must be carefully formulated to assure that all nutrient requirements are met. In the following section, feeding management of the entire herd will be discussed starting with raising kids (either as herd replacements or for meat production), the breeding herd, and the herd producing milk or fiber.

### Feeding Kids

The first 3 days after birth are the most critical days in the life of a newborn kid. Readers are referred to Chapter 8, Digestive Physiology and Nutrient Metabolism, for more detail about digestive function of young kids. The young kid basically is a nonruminant that acquires all of its required nutrients initially from colostrum and later from milk. The kid may be separated from its doe immediately (at some dairy farms), or it may stay with its dam. If separated, the kid should receive colostrum for the first 3 days of its life. Colostrum is the secretion of the mammary glands immediately after parturition. Colostrum contains high levels of protein, milk solids, globulins, fats, and vitamin A. Most importantly, colostrum contains antibodies against the diseases to which the doe has developed immunity. When fed during the first 3 days of life, colostrum provides protection against diseases because these antibodies are absorbed intact. Absorption of large proteins will cease approximately 3 days after birth. (See Chapter 8.) Separated kids should receive milk or milk replacer until they are weaned.

Kids should nurse as long as possible or until weaned. Weaning age for meat goat operation usually is 3 months while for dairy operations it usually is 3 to 4 days but can vary with management practices. If a doe rejects its kid(s), a surrogate doe should be found to adopt the kid(s) as soon as possible. Colostrum from other animals or frozen sources should be fed to the orphan newborn three times a day, a total of 1–1.5 liters (2–3 pints) daily. After 3 days, orphan kid(s) can be fed milk or milk replacer twice daily. When using milk replacer, the manufacturer's directions for feeding should be followed. Supply of milk replacer should be gradually increased. Feeding milk or milk replacer can continue up to 8–12 weeks or until kids are weaned or consuming 0.25–0.5 kg of grain mix (kid starter) daily. Diets for kids typically include a grain mix with high fiber (11%) and protein (16%) content plus good quality hay and should be available starting the first week to encourage development of the rumen and GI tract.

### Feeding Herd Replacements

From weaning to 6 months of age, kid starter should be fed at (0.25–0.5 kg/day) together with good quality forage and pasture where available. Silage and nonprotein nitrogen at this age should not be fed because the rumen is not fully functional at this age. Fresh water should be available at all times. Macro and micro minerals should be incorporated into the grain mix or provided in loose form.

As animals grow from 6 months to breeding age, they may require 0.5 to 1 kg of grain mix daily containing at least 14–16% protein, salt-mineral mix, and vitamins A and E. Feeding more than 25% of the protein as nonprotein nitrogen is not recommended for rapidly growing replacement animals at this age. Yearling does and bucks require a proper balance of nutrients for maintenance and growth, but overfeeding so that animals gain a large amount of fat is not recommended. Replacement bucks should be kept in good body condition but not fat. Creep feeding is not recommended for replacement dairy does because fat deposited in mammary glands can negatively impact milk production in the future. The proper time of breeding depends on the age and weight of the doe. Yearlings (12 months of age and older) can be bred. The proper weight for breeding usually is 65% of adult weight for cattle, but no benchmark has been established for goats and may differ with breeds. In order to breed at specific age, doelings should be fed to achieve this weight at the needed time. To improve fertility, flushing is used. Flushing is the practice of feeding supplemental protein or energy to breeding does 30 days prior to and after the introduction of the herd sire (buck). Widely used for dairy goats, flushing may not be necessary for meat-producing animals if the quality and quantity of available forage is ample. When yearlings are bred, they can be placed with pregnant does.

### Feeding Pregnant Does

Nutrient needs are not constant for reproducing animals but vary with their reproductive status. Producers must understand these cycles to develop sound feeding practices. Feeding strategies for meat goats and dairy goats for pregnancy will differ markedly. Meat goats generally are dry (not milking) at breeding, whereas dairy goats may be still milking; and continued milk production makes rebreeding efficiency much more challenging.

#### Feeding Nondairy Does in Early Gestation

During the first 90 days in pregnancy, for meat goats, nutrients are needed for maintenance and growth, especially if the doe is still growing, but additional nutrients needed for fetal growth are miniscule because fetal growth at this stage is minimal. Goats should be kept in positive energy balance and be gaining weight at least at the later part of this period. Feed supply should match body condition of animals. A dry goat in good condition can meet her energy requirements from good quality forage (half legume, half grass) supplemented with salt, minerals, and vitamin A and E. If forage has a low quality, a small amount of protein supplement may be beneficial.

#### Feeding Nondairy Does in Late Gestation

During the final 60 days of pregnancy, does require more attention, particularly 4–6 weeks prior to kidding. A good pasture, hay, or silage as well as 0.25–0.5 kg of a 12–14% protein grain mixture should be sufficient. Use of alfalfa as a sole source of forage during this period is not recommended because alfalfa has a high calcium-to-phosphorus ratio, which is not desirable for late pregnant does. Does should be kept in good flesh but not fat during this period.

#### Feeding Early Gestation Dairy Does

Besides growth, dairy does need additional nutrients for milk production during the later part of the lactation. Little additional nutrients are needed for fetuses, but does carrying multiple fetuses may need some additional energy or protein. Starting at the fourth month of lactation, does and doelings should gain at least 1–2 kg per month to restore the tissues lost in early lactation. Providing good quality grass hay or a mixture of grass and legume hay along with fresh water, a salt-mineral mix, vitamins A and E, and a small amount of supplemental grain mix with 12–14% protein is recommended. Pregnant does should be fed grain mix so they are in good body condition but not fat. Fat pregnant does are predisposed to fatty liver (ketosis).

#### Feeding Far-Off Dry Does

The dry period for an average milking doe is the break between lactations. A dry period is needed for maximum milk production. Individual cases of high production without a dry period have been reported, but most of the studies show that a dry period will extend the productive life of the doe. The most attention should be given to the pregnant doe during the last 2 months of gestation, so the final 60 days of pregnancy is divided into this "far-off dry period" and "close-up dry period."

The far-off dry period covers the time that does have just turned dry to about 3 weeks before kidding. The dry doe should be maintained in good condition but not fat during this period. This time allows the mammary system to repair, regenerate, and acquire new cells for the next lactation. The greater the production level, the more likely that doe will have depleted her body reserves of nutrients used for milk production. The condition of the animal and the quality of the roughage become the key factors in determining grain intake. Total dry mater intake for a dry doe generally will be about 2–2.5% body weight of which only 0.5–0.6% should be grain. Good pasture, hay, or silage plus 0.5 kg of 12–14% protein grain mix, a salt-mineral mix, and vitamin A and E supplement can provide the required nutrients. The feed trough should be located some distance away to encourage exercise. Fresh water should be available at all times.

#### Feeding Close-up Dry Does or Transition Does

The last 3 weeks of gestation is referred to as the "close-up dry period" or transition period for pregnant dairy does. This is the period when nutrient requirements for fetal growth and mammary development dramatically increase while at the same time, space in the abdomen becomes limited by the developing concepta so that feed consumption can be limited. Intake often decreases by more than 35% while nutrient requirements increase. At this point, the animal needs high-energy and protein feeds. Dairy rations often use the "halfway rule of thumb" (that is, feed half the amount of feed that will be fed after does kid). Increase intake gradually. No extra buffers should be needed, but if milk fever is expected, an anionic salt program is recommended for goats as it is for dairy cattle.

Feeding anionic salts (Table 10.9) to manipulate the cation-anion balance is recommended especially for high-producing dairy cattle fed forage like alfalfa hay that is rich in calcium, phosphorus, and potassium. Feeding anionic salts lowers blood pH and urine pH, stimulates release of calcium from bones, and increases calcium absorption. The use of anionic salts in dairy goats has not been studied. Nevertheless, it seems prudent to feed low potassium feeds such as beet pulp or corn silage, reduce sodium bicarbonate and limestone in the diet, and add calcium sulfate, or calcium and ammonium chloride to reduce the cation-anion difference (CAD) and urine pH. Along with anionic salts, feeding a palatable total mixed ration properly supplemented with minerals is recommended. Avoid feeding anionic salts to young doelings because it may cause udder edema; instead, young does should be separated and monitored closely. Feeding a high-energy/protein lactation ration along with long hay, yeast, probiotics, and chelated minerals according to the body condition score is recommended (Table 10.10).

**Table 10.9** Common anionic compounds fed to dairy cattle late in pregnancy.

| Anionic Salts or Mixtures |
|---|
| • Magnesium, Calcium and Ammonium Sulfate |
| • Calcium, and Ammonium Chloride |
| • Nutri-Chlor—Canola meal treated with HCl |
| • Soy-Chlor—Heat treated soybean meal treated with HCl |

### Feeding Nursing Does

During the first few months of lactation, animals should consume enough to meet their needs for milk production and for nursing kids. High quality legume or grass hay and a grain mix containing a minimum of 16% protein balanced for vitamins and minerals is recommended. Nursing does should be fed both to produce milk and produce heavier weaned kids. Good pasture, browse, and garden products including root crops can be helpful. Grain intake should be increased if needed to optimize kid growth. The grain mix should be supplemented with an adequate salt-mineral mix and vitamins A and E.

### Feeding Herd Sires (Bucks)

To have a productive herd and make genetic progress, care must be exercised in selecting and caring for the herd sire. A herd sire should be selected from a high producing doe. Proper early nutrition is necessary to enhance the productive life of a herd sire with good performance and conformation.

#### Prebreeding Season

A young herd sire should be maintained on a balanced diet year-round for maintenance and growth. Good quality hay provided for free choice intake generally will meet the sire's nutrient requirements. The buck's activities during the off breeding season are limited, but as the breeding season approaches, necessary measurements should be

**Table 10.10** Nutrient recommendations for transition rations for lactating dairy goats.

| | Far-off | Close-up | Fresh |
|---|---|---|---|
| Net Energy Lactation (Mcal/kg) | 1.32 | 1.5–1.6 | 1.7 |
| Crude Protein, % | 12 | 15–16 | 18 |
| Neutral Detergent Fiber, % | 45–65 | 35–40 | 30 |
| Non-fiber Carbohydrates, % | 25 | 30 | 35 |
| Fat, % | 3 | 4 | 5 |

Source: Adapted from "Feeding the transition dairy cow" by Sandra Stokes, Extension Dairy Specialist, Texas A&M University System, Texas AgriLife Extension Service. http://AgriLifebooksstore.tamu.edu.

taken to prepare the buck for his active season. Some meat breeds are year-round breeders, and must be fed accordingly. Depending on the condition of the animal and his prior plane of nutrition, in addition to good quality hay provided free choice, 0.5–1 kg of 14–16% protein mix may be needed. Supplementation should start as early as 6–7 weeks or could be as late as 2 weeks prior to the breeding season depending on prior nutrition of the buck. Production of fertile sperm begins 40–60 days before it is deposited in the female reproductive tract. Grain supplements should include a sufficient amount of vitamins and minerals. Trace mineralized salt should be provided free choice and preferably in loose form. Avoid overfeeding and overconditioning.

#### Breeding Season

In the breeding season (that usually starts in August and lasts through December for seasonal breeders in the northern hemisphere), the herd sire is quite active. Males will fight with other bucks, and with their breeding activities, they may devote little time to grazing or eating. Therefore, body reserves will be depleted. Sometimes bucks completely stop eating during the mating season and lose 15–20% of their body weight. Breeding males may serve does up to 20 times a day with 350 services being possible during a limited breeding season. Therefore, bucks should be provided with good quality hay and 1–1.5 kg of concentrate mix containing 14–16% protein, adequate amounts of minerals, and vitamins. The buck should receive 2,000–6,000 retinol equivalent (RE) of vitamin A and 300–800 IU of vitamin E daily. Plenty of clean water and salt mineral in loose form should be provided at all times.

#### Post-Breeding Season

Unless year-round breeding is practiced, the herd sire should be removed from the herd at the end of the breeding season. Feed allowance should be reduced to 0.5–1 kg of grain mix (16% crude protein) per day at the end of the breeding season depending on the animal's condition. If good quality hay is provided, the animal will not need additional supplement. However, if the hay is poor in quality, additional supplement may be needed to meet maintenance energy requirements. Salt and a mineral mix preferably in loose form should be provided at all times.

### Feeding Kids to Produce Meat

After weaning at about 3 months of age, doelings may be placed in the replacement herd while bucklings (young bucks) usually are raised for meat. Creep feeding is defined as providing additional supplemental feeds to kids to increase their growth rate so that they reach their market weight (often 30–35 kg) at an early age (less than 6 months). Although it has not been a common practice in meat goat production, creep feeding should be more profitable with changes in goat marketing strategies that promote premium prices for larger, more highly finished kids. Feed is provided in a creep feeder that will allow access to feed by the kids but not by other animals. Creep feed can be simple like corn and oats, barley and oats, or a more complex commercial kid grower mix containing 14–16% protein and about 10% fiber. Adding more fibrous feeds such as beet pulp, oats, or dehydrated alfalfa to creep feed will promote rumen development. Loose mineral mix and salt should be provided at all times. The creep feeder should be conveniently located where animals spend most of their time.

Generally, most male kids and female kids that are not herd replacements are either sold as weaned kids or are fed for meat production. Weaned goats should be fed according to their nutrient requirement to reach the desired slaughter weight of 30–35 kg (in the U.S.). Some ethnic groups may desire older animals with heavier carcasses or want kids before specific holidays. Following the typical growth curve, male kids grow faster than female kids (Figure 10.3), and growth rates are most rapid between birth and 6 months of age.

For ease of handling and to improve docility, all male kids in the U.S. except those kept for breeding are castrated very early in life. However, as shown in Figure 10.4, buck kids grow faster than wethers, and castration will stunt

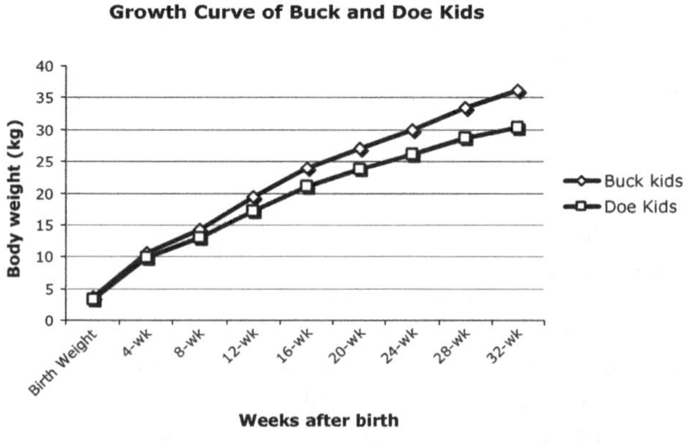


**Figure 10.3** Growth curves for bucks and does from birth to 35 weeks of age.

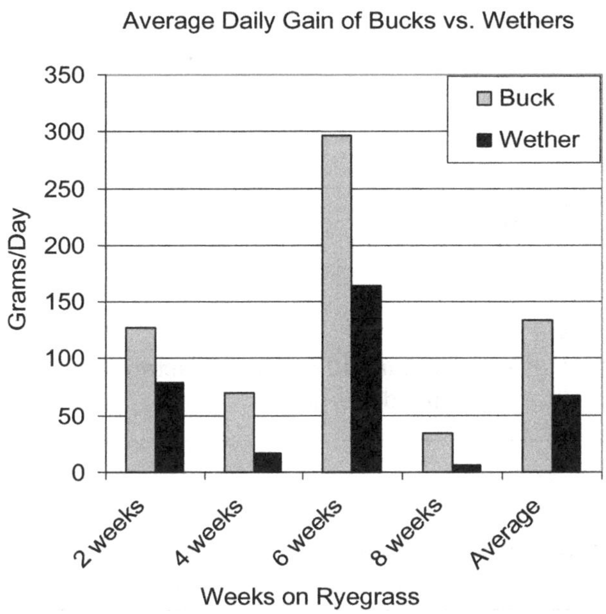


**Figure 10.4** Period and total daily gain of bucks versus wethers grazing ryegrass pastures over an 8-week period. Source: Solaiman et al., 2006.

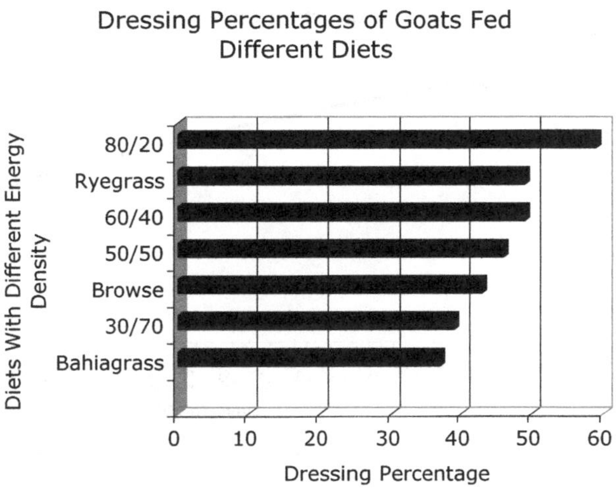


**Figure 10.5** Dressing percentages of growing goats (BW = 35–40 kg) fed different diets or various levels of concentrate: hay ratio.

growth, reduce average daily gain, and increase the time needed for goats to reach market weight. Premature castration also increases the incidence of urinary calculi by reducing the development of the urinary tract. Intact buck kids gain 20–30% more weight than castrated wether kids. If kids are slaughtered at 6 months of age or less for meat, and provided bucks are separated from does, castration can be skipped so that bucks will gain weight more rapidly. However, success in raising intact bucks for meat will depend on the management skills and practices on an individual farm.

Lightweight goats that are in poor condition or that have been transported a long distance can be reconditioned through a controlled grazing scheme without or with a limited amount of grain. Goats fed concentrate diets will have higher dressing percentages at slaughter largely because their carcasses tend to contain a higher percentage of body fat and more muscle but a smaller digestive tract with less digesta (Figure 10.5).

**Feeding for Milk Production**

Management is challenged to maintain health and maximize milk production of lactating does, particularly early in lactation when the animal is in negative energy balance. Milk production by dairy goats generally peaks at 6–8 weeks of lactation (Figure 10.6) while intake of feed and energy peaks at 3–4 months into lactation. During this period, the lactating doe relies on body reserves to produce milk, and generally is in a ketotic condition. Animals often lose 0.5–1 kg of body weight per week for the first 2 months in lactation, but starting at 4 months in lactation, they should gain weight (1–2 kg per month) to restore the tissues lost during early lactation. Lactating does should be fed according to their requirements for milk production (Appendix A). Nutrient composition of a complete mixed dairy ration is presented in Table 10.11. Adding fat at 3–4% of diet DM can increase energy density of the diet to help meet the needs for early lactation and allow feeding forage to promote optimum milk fat synthesis. However, higher levels of fat will reduce fiber digestion and calcium absorption. A higher protein level also is required in early lactation for microbial protein synthesis, and additional rumen by-pass protein will help provide the amino acids necessary for milk synthesis as well as glucogenic amino acids for lactose production. Some suggested maximum limits for ingredients in diets for dairy goats are presented in Table 10.12.

The ideal environmental temperature for milk production is between 5 and 25°C. A temperature-heat index (THI) higher than 72 reduces DM intake by 10–12% and

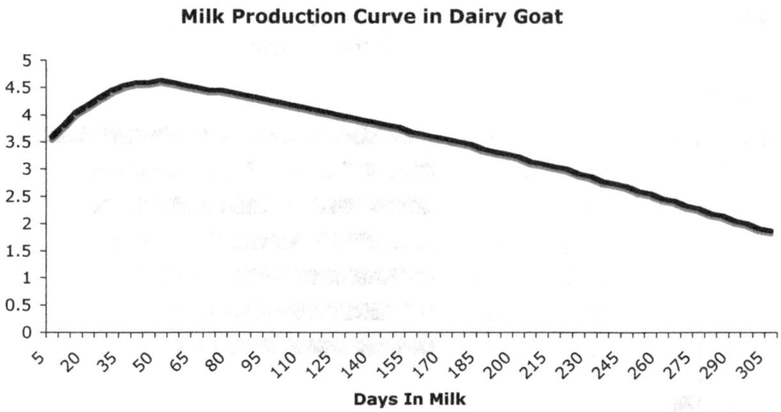


**Figure 10.6** Typical lactation curve for dairy goats.

**Table 10.11** Recommended nutrient contents for complete mixed rations for dairy goats.

| Item | Amounts |
|---|---|
| Total Digestible Nutrients | 75% for early lactation and reduced to 60% during late lactation |
| Crude Protein | 18% for early lactation and reduced to 14% during late lactation |
| Acid Detergent Fiber | 18% for early lactation and increased to 20–22% during late lactation |
| Fat | 3–4% added fat particularly for early lactation |
| Concentrate:Forage Ratio | 60 : 40 for early lactation and changing to 40 : 60 during late lactation |
| Calcium:Phosphorus Ratio | At least 1.5 : 1 |
| Phosphorus | 0.4% |
| Added Salt | 0.5% |
| Sulfur | 0.25–0.35% |
| Nitrogen:Sulfur Ratio | 10–12 : 1, with added sulfur usually needed when diets contain nonprotein nitrogen |
| Magnesium | Higher magnesium may be needed with N and K fertilized pastures to prevent hypomagnesemia |
| Copper | 15 ppm |
| Selenium | 3 ppm |

**Table 10.12** Recommended maximum limits for select feeds in diets for dairy goats.

| | % of Concentrate Mix | % of Body Weight |
|---|---|---|
| **Nitrogen Sources** | | |
| Urea | 1.5 | 0.03 |
| Fish meal | 3 | 0.08 |
| Whole soybeans | 25 | 0.5 |
| Corn gluten feeds, dry | 30 | 0.8 |
| **Energy Sources** | | |
| Tallow | 4 | 0.12 |
| Molasses | 6 | 0.15 |
| Bakery products | — | 0.3 |
| Soy hulls | 20 | 0.4 |
| Wheat middlings | 20 | 0.4 |
| Cottonseed hulls | 20 | 0.4 |
| Whole cottonseeds | 20 | 0.4 |
| Brewers grain, dry | 25 | 0.5 |
| Brewers grain, wet | — | 2.3 |
| Distillers grain | 25 | 0.5 |
| Citrus pulps, dry | — | 0.6 |

milk production by 20–30% for lactating dairy cows. Corresponding responses by goats to THI have not been determined, but heat stress definitely can reduce milk production. To reduce heat stress of dairy does, a number of management changes can prove to be useful, including (1) feeding smaller portions of the ration more frequently, (2) feeding during cooler times of the day or under shade with adequate air flow, (3) feeding only high quality forage, (4) reducing the forage: concentrate ratio (to reduce the heat increment), (5) adding dietary buffers to help maintain rumen pH if diets are high in concentrates, (6) feeding yeast to increase fiber digestion, (7) feeding fungal products and niacin to improve energy utilization, (8) supplementing fat to increase energy density for higher milk production, (9) providing cool water, and (10) providing fans, water sprinklers, or sprayers for additional cooling aid.

### Feeding for Fiber Production

Management of goats for fiber production in many ways is similar to that of meat-producing goats. But fiber-producing goats usually are smaller in adult size than meat goats and have a slower growth rate. Diets of fiber-producing animals generally will be only forage with little or no supplemental grain. Fiber proteins contain higher levels of sulfur, so when nonprotein nitrogen is fed to fiber-producing goats, supplemental sulfur should be provided to maintain a nitrogen: sulfur ratio between 10:1 and 12:1. Feeding a high quality protein diet will increase fiber diameter and result in a coarser, lower quality fiber, but it will result in greater fiber yield. Therefore, the management choices (quality versus quantity of fiber) must be considered carefully when designing the feeding regime. Fiber quality (fineness) will be at a maximum under adverse nutritional conditions. The higher the feed quality, the coarser the fiber, but the greater is the yield of fiber. Basic nutritional requirements (see Appendix A) still should be met to ensure that the herd will be healthy and profitable.

## GOAT FEEDING SYSTEMS IN THE U.S.

Meat goats in the U.S. are raised under a wide variety of environments that include the West (California), under a Mediterranean climate; the Southwest, a semiarid climate; Southeast, a temperate and subtropical climate; and the Midwest, a temperate climate. Dairy goats are raised mainly in California, Texas, and Wisconsin, and fiber goats are mainly concentrated in Texas, New Mexico, and Arizona. Feeding practices specific for goat production in these regions will be covered briefly in this section.

### Feeding Systems for the Southeastern U.S.

The Southeastern states have a mild, temperate environment with hot and humid summers, mild winters with not more than 20 freezing days, and high rainfall throughout the year. Goat production systems in this region can range from pasture and browsing woodlands to total confinement. Feeds available include grasses; legume trees (mimosa, kudzu); forbs; understory brush as forage; sweet potatoes, peanuts, and cotton by-products; and conventional supplements. The quality of forage, especially of pastures, drops drastically in the summer, therefore supplemental energy and protein are needed for optimum gain. Legume trees or small grain pastures can reduce the need for supplemental feeds. Ideal animal production systems may be localized to maximize use of available feed and land resources, and producers must develop sound feeding programs. Pasture and browse provide needed nutrients and energy for feeding goats year-round, particularly in the Southeast. A year-round foraging system for this region may include annual Marshall ryegrass (a cool season grass) for winter grazing (December–April), peas or soybean greens (legume) for late spring and early summer (May–June) grazing, mimosa browse, sorghum, or bermudagrass (warm season grasses) for summer (July–October), with hay or concentrates being used sparingly when needed (Solaiman et al., 2006). Unfortunately, this system is very management intensive and requires outside input (fertilizer, anthelmintics). Compared to goats on bahiagrass pasture and mimosa browse (gaining 46 and 80 g/d, respectively), animals in a feedlot or grazing Marshall ryegrass (winter pasture) grew faster (141 and 139 g/d, respectively) and reached their expected slaughter weight in less time. Animals on feedlot and Marshall ryegrass had superior carcass quality, but this difference was not significant. In contrast with the performance advantages for feedlot, economic returns from the Marshall ryegrass system were superior followed by the mimosa browse system.

Parasite infestation is one of the major problems in raising goats in many areas of the country, especially in the Southeast. Most common and effective anthelmintics lose their effectiveness with frequent use as parasites develop resistance after a short period of time. Bioactive forage plants containing condensed tannins may help in control of parasites. Browse plants containing tannins like mimosa can be safely incorporated in the diet of goats in this region to reduce parasite load (Solaiman et al., 2006).

### Goat Feeding Systems for the Midwest

In the Midwest region of the U.S., interest in raising dairy and meat goats has had a resurgence. Known as the agriculture heartland of the country, the Midwest has locations that provide ideal conditions for raising goats. Daily temperatures range from −10 to 8°C in the winter and 18 to 27°C in the summer. Rainfall usually is plentiful, averaging 250 to 500 cm annually. The lower Midwest region has higher temperatures and more precipitation than the colder and dryer upper Midwest. Because of the low precipitation and humidity, farmers have fewer problems with goat internal parasites than in the Southeast. The Midwest region also produces pastures and forage with excellent quality. These forage products reduce the requirement for supplemental grain. Although the lower part of this region has cool season grasses and legumes for winter grazing, the middle and upper portions of the Midwest are not as fortunate with very cold winters and grazing of cool season grasses for only a few months out of the year, usually May and June. In July and August, sudangrass or corn can be fed to goats. Harvested hay or silage can be used when the pasture is not available. Silage and haylage are produced in this region, but caution should be used when feeding silage or haylage to goats. Corn silage is an underutilized feed because of the labor involved in production and feeding. If poorly harvested and preserved (with soil contamination and presence of oxygen), it can harbor *listeria*, the bacteria that causes listeriosis (circling disease). Corn silage also is deficient in protein, calcium, and phosphorus, so, for best results, it should be supplemented with a protein source like alfalfa. Because corn silage produces more energy per acre than other crops, more goats can be maintained using corn silage than with any other harvested feed. Haylage usually is produced from alfalfa and brome forage. Though more variable than corn silage in quality, it is higher in protein and minerals. Hay or other roughage usually is fed with silage and haylage to reduce the incidence of metabolic disturbances in goats. Ryegrass, orchardgrass, bluegrass, festuloliums (cross between fescue and ryegrass), and other grasses as well as alfalfa, clover, and other legumes are common forage produced in this region. Winterkill of legumes can be a problem, so a diverse mixture of species usually is sown for pasture and hay for goats. Protein supplements such as soybean meal, linseed meal, sunflower meal and by-products such as corn stover and soy hulls are readily available. Distillers' grain and other grain by-products are plentiful in the Midwest and are excellent sources of feed energy and protein for goats. By-products of the ethanol industry of interest include wet and dried distillers' grain, wet and dry distillers' grain with solubles, modified wet cake, and condensed distillers' solubles. Although distillers' grains are good sources of energy and protein, protein availability can be reduced by high temperature drying.

### Semiarid Goat Production System

This goat production region includes rangeland, chaparral, grasslands, and woodlands where dry, sandy, rocky, and saline soils predominate and the environment (typically rainfall) will dictate both the quality and quantity of vegetation available. Parts of Texas (west), Arizona, New Mexico, California, and some other states fit under this production system. In contrast to intensive systems where many goats are produced in a confined space, most goats in this region are raised under very extensive conditions with minimal inputs of feed and labor. Low rainfall, a diversified native forage material, extensive rangeland goat production, goat herds with a very large number of head, and hundreds of acres of rangeland per ranch are characteristics unique to this system.

Rangeland vegetation, brush, browse, and other bushes and forbs are widely available. Natural grassland will provide sufficient forage for growth and meat production during the rainy seasons (spring and summer) while forbs, bushes, shrubs, and trees provide feed for maintaining these free-ranging goats during the rest of the year. The nutritive quality of vegetation diminishes drastically during the winter and dry seasons, so productivity can only be sustained with supplemental feeds. Nevertheless, feeding of supplement is seldom practiced. Due to dry environmental conditions and extensive goat production, unlike other systems, internal and external parasites pose little problem so parasite control is rarely practiced in this region. Highly variable climates and unpredictable environmental conditions from season to season and from year to year make this system quite challenging for goats and goat producers. Due to the nature, quality, and variability of vegetation, and these variable environmental conditions, it is difficult to design proper supplementation programs for optimum production. Experienced managers can assess range conditions as well as animal conditions to determine quality and quantity of supplemental feeds that are needed. Goats are usually used for range improvement and vegetation clearance in this region. Readers are directed to Chapter 16, Environmental Enhancement, for further information.

### Goat Production Practices Focused on California

California historically has led the nation in agricultural production for more than 50 years. Its moderate Mediterranean climate, fertile soils, and progressive farmers have contributed to this success. More than 250 commodities are produced in California, with this state being the national leader in production of 75 of them. Regions are ideally suited to produce fruits, vegetables, nuts, and animal products such as milk and meat. Most importantly, more than 97% of farms in California are family owned, another key factor conducive to excellent goat production. California's abundance of rangeland caters to the browsing ability of goats and mixed grazing. The mostly dry, mild Mediterranean climate in California helps reduce gastrointestinal parasitism of goats.

California has the highest product value per acre in the U.S., and a wide variety of agricultural products are grown on its fertile land. Energy-rich cereal grains such as wheat, barley, and sorghum are grown locally, and protein feeds such as cottonseed, sunflower meal, and other by-products are readily available. Roughage such as good quality hays and silages, and by-products such as rice bran, wheat bran, and sugar beet pulp are common. Garden produce, often sold in farmers' markets, provides residues that can be used as compost or as feed for goats. California's diverse topography—valleys, foothills, coastal areas, and deserts—make this diverse Western state suitable for production of a variety of agricultural products including goats.

#### North Coast and Mountains of California

This area has fewer farms of middle and large size than other areas of the U.S. and is well suited for timber and production of cattle, sheep, and some goats that can graze hills and rough terrain over wide expanses. Goats raised in this region often form part of herds used in a mixed grazing system with other livestock. Hay produced on irrigated pastures and rangeland covering private and leased public lands provides high quality forage for goat production in this region. With more than 32% of California's terrain covered by forestland, and with proper management, goats as browsers can be successfully raised for meat production in silvopasture systems using undercover brush and browse.

#### Central Valley of California

The Sacramento Valley, with its cooler winters and high rainfall, produces small grain crops and seasonal grazing on nonirrigated land. Rice is produced on irrigated lands. Row crops including tomatoes, beans, corn, milo, and sunflowers also are produced in this region.

The foothills of the Sacramento Valley support seasonal grazing. The southern portion of the great Central Valley, the San Joaquin Valley, is the most extensive and productive agricultural region in California and the nation. In this region, several favorite feeds for goats are grown including alfalfa, cotton by-products, vegetables, and field crops. Sheep, cattle, and sometimes goats, can graze on the irrigated pastures in the foothills of this region.

#### Central Coast of California

With the highest value farmland in the nation, the Central Coast is home to the country's premium agricultural products including grapes, strawberries, orchard crops, and vegetables. By-products of the fruit and vegetable industries can be used for goat feed. The Salinas Valley is the salad bowl of the nation, creating an abundance of leftover greens for goats.

#### Southern California

The South Coast region tends to have farms of smaller size and is a major producer of avocados, citrus, vegetables, and hays. Alfalfa, cotton, citrus, dates, small grains, and winter vegetables are produced in abundance in the hotter interior valley where farms are generally larger.

## GLOBAL LIVESTOCK PRODUCTION SYSTEMS

According to the FAO, livestock production systems are classified as (1) livestock production systems, where less than 10% of the total value of production comes from non-livestock farming practices, and (2) mixed farming systems, where more than 10% of the total value of production is from non-livestock farming activities (Seré and Steinfeld, 1995). A landless livestock production system is where less than 10% of animal feed is produced on-farm, and animal density is more than 10 livestock units per hectare. This system is mainly concentrated in the developed countries and is capital and feed intensive, and labor extensive. A grassland-based livestock production system is where more than 10% of the animal feed is produced on-farm, and animal density is less than 10 livestock units per hectare. This system is used in Central and South America and in developed countries. Rain-fed, mixed farming systems in developed countries and Asia produce more than 70% of the meat produced in the mixed farming system, and the remaining 30% is produced by irrigated mixed farming systems. Grazing and mixed farming systems are represented in temperate and tropical highland zones; the humid and subhumid tropical and subtropical zones; and arid, semiarid tropical and subtropical zones of the world.

Globally, mixed farming systems provide more than 54% of total meat and 92% of total milk production followed by landless systems; grazing systems contribute less than 10% of total production for meat and milk. It is expected that mixed system farming will continue to grow (Table 10.13).

**Table 10.13** Total meat production and share of meat or milk production from different livestock production systems in the world.

| Regions | Landless[1] | Grassland[2] % | Mixed Farming[3] |
|---|---|---|---|
| Developed countries | 54.6 | 37.5 | 35.8 |
| Eastern Europe | 14.9 | — | 16.1 |
| Asia | 19.3 | 8.5 | 34.9 |
| Central and South America | 8.8 | 39.0 | 7.6 |
| West Asia and North Africa | 1.9 | 2.4 | 3.4 |
| Sub-Saharan Africa | 0.5 | 12.5 | 2.1 |
| Share of meat production | 36.8 | 9.3 | 53.9 |
| Share of milk production | — | 7.9 | 92.1 |

[1]Landless Livestock System represents an intensive system.
[2]Grassland Based Livestock system represents extensive system.
[3]Mixed Farming System represents livestock and crop production system.
Source: www.fao.org.

## SUMMARY

Feeds for goats fall into two groups: forage/roughage and concentrates. The optimum proportions of forage and concentrate in the diet will vary depending on an animal's physiological stage and its level of performance. Specific nutrients (protein, energy, minerals, vitamins, and water) are required by goats and must be provided in the diet through feed. Diets must contain a protein source that for goats, like other ruminants, may be derived partially from nonprotein nitrogen sources like urea. Energy is derived primarily from dietary carbohydrates (sugars, starches, and fiber) though fats and excess protein also provide energy. The process of matching the animal's requirements with the nutrient contents of feed ingredients is referred to as "feeding practices." Sound feeding practice will (1) optimize animal performance, (2) minimize feed cost, and (3) sustain the environment. Each animal should be matched to its specific energy and nutrient needs for maintenance, growth, reproduction, and production of meat, milk, and fiber. Improper feeding practices or imbalances of nutrients will reduce production and may cause metabolic disorders. Feeding management practices specific for goat production in different regions of the United States and throughout the world will vary widely depending on the availability of feeds and natural resources.

## REFERENCES

Anil Kumar, G.K., V.S. Panwar, K.R. Yadav, and S. Sihag. 2002. Mustard cake as a source of dietary protein for growing lambs. Small Ruminant Res. 44:47–51.

Aregheore, E.M. 2004. Nutritive value of sweet potato (Ipomoea batatas [L.] Lam) forage as goat feed: voluntary intake, growth and digestibility of mixed rations of sweet potato and batiki grass (Ischaemum aristatum var. indicum). Small Ruminant Res. 51:235–241.

Bartlett, J. 1989. Passage kinetics and digestion of different broiler litter diets in Nubian dry does. M.S. Thesis. Tuskegee University, Tuskegee, AL.

Ben Salem, H., I. Ben Salem, A. Nefzaoui, and M.S. Ben Said. 2003. Effect of PEG and olive cake feed blocks supply on feed intake, digestion, and health of goats given kermes oak (Quercus coccifera L.) foliage. Anim. Feed Sci. Tech. 110:45–59.

Brann, G.L. 2005. Acres per paddocks "A must know." Agronomy Technical Notes No. TN–26. Natural Resources Conservation Service (NRCS) Nashville, Tennessee.

Bueno, M.S., E. Ferrari, Jr., D. Bianchini, F.F. Leinz, and C.F.C. Rodrigues. 2002. Effect of replacing corn with dehydrated citrus pulp in diets of growing kids. Small Ruminant Res. 46:179–185.

Decandia, M., M. Sitzia, A. Cabiddu, D. Kababya, and G. Molle. 2000. The use of polyethylene glycol to reduce the anti-nutritional effects of tannins in goats fed woody species. Small Ruminant Res. 38:157–164.

Glaye, S., T. Terril, E.A. Amoah, S. Miller, R.N. Gates, and W.W. Hanna. 1997. Nutritional value of pearl millet for lactating and growing goats. J. Anim. Sci. 75:1409–1414.

Gurung, N.K., S.G. Solaiman, D.L. Rankins, G.M. Abdrahim, and W.H. McElhenny. 2009. Effects of feeding varying levels of distillers dried grain with solubles on growth performance, blood parameters and carcass quality of meat goats. 1890 Research: Sustainable Solutions for the 21st Century, Association of Research Directors 15th Biennial Research Symposium, 188 pp.

Huston, J.E. 1994. Effects of supplemental feeding on intake by kid, yearling, and adult Angora goats on rangeland. J. Anim. Sci. 72:768–773.

Kondo, M., K. Kazumi, and H. Yokota. 2004. Feeding value to goats of whole-crop oat ensiled with green tea waste. Anim. Feed Sci. Tech. 113:71–81.

Lassoued, N., M. Rekik, H. Ben Salem, and M.A. Dargouth. 2006. Reproductive and productivity traits of goats grazing Acacia cyanophylla Lindl. with and without daily PEG supplementation. Livest. Sci. 105:129–136.

Marjuki, S.H.E., D.W. Rini, I. Artharini, Soebarinoto, and R. Howeler. 2008. The use of cassava leaf silage as a feed supplement in diets for ruminants and its introduction to smallholder farmers. Volume 20, Article #93. Retrieved April 20, 2009, from http://www.cipav.org.co/lrrd/lrrd20/6/marj20093.htm.

Min, B.R. and S.P. Hart. 2003. Tannins for suppression of internal parasites. J. Anim. Sci. 81:102–109.

Moore, J.A., M.H. Poore, and J.M. Luginbuhl. 2002. By-product feeds for meat goats: Effects on digestibility, ruminal environment and carcass characteristics. J. Anim. Sci. 82:1752–1758.

Perry, T.W., A.E. Cullison, and R.S. Lawrey. 1999. Feeds and Feeding, 5th ed. Upper Saddle River, NJ: Prentice Hall.

Pieltain, M.C., J.I.R. Castanon, M.R. Ventura, and M.P. Flores. 1998. Nutritive value of banana (Musa Acuminata L.) fruit for ruminants. Anim. Feed Sci. Tech. 73: 187–191.

Russell, J.B. and H.J. Strobel. 1989. Effect of ionophores on rumen fermentation. Appl. Environ. Microbiology, 55:1–6.

Seré, C. and H. Steinfeld. 1995. World livestock production systems: Current status, issues, and trends. FAO Animal Production and Health Paper No. 127, 51 pp.

Short, R.E., L.F. James, K.E. Panter, R.B. Staigmiller, R.A. Belows, J. Malcolm, and S.P. Ford. 1992. Effects of feeding ponderosa pine needles during pregnancy: Comparative studies with bison, cattle, goats and sheep. J. Anim. Sci. 70:3498–3504.

Solaiman, S.G. 2007. Feeding value of seed-harvested cowpea vines for goats. Technical Paper No. 07–09. Tuskegee University.

Solaiman, S.G., D. Bransby, C. Kerth, B. Blagburn, and N. Noble. 2006. A sustainable year-round forage system for goat production in the Southern U.S. Final Report, Project # LS02–141. Southern SARE.

Solaiman, S.G., M.A. Maloney, M.A. Qureshi, G. Davis and G. D'Andrea. 2001. Effect of high copper supplements on performance, health, plasma copper, and enzymes in goats. Small Ruminant Res. 41:127–139.

Solaiman, S.G., N.K. Gurung, Q. McCrary, H. Goyal, and W.H. McElhenney. 2009. Feeding performance and blood parameters of male goat kids fed EasiFlo® cottonseed. Small Ruminant Res. 81:137–145.

Solaiman, S., S. Roper, K. Beguesse, G. Reddy, N. Gurung, and K. Copedge. 2008. Effects of induced copper deficiency with added molybdenum on health and immune responses of male goat kids. J. Anim. Sci. 86(Suppl. 3): 30.

Solaiman, S.G., T.J. Craig Jr., G. Reddy, and C.E. Shoemaker. 2007. Effect of high Cu supplement on growth performance, rumen fermentation, and immune responses in goat kids. Small Ruminant Res. 69:115–123.

Terrill, T.H., J.A. Mosjidis, D.A. Moore, S.A. Shaik, J.E. Miller, J.M. Burke, J.P. Muir, and R. Wolfe. 2007. Effect of pelleting on efficacy of sericea lespedeza hay as a natural dewormer in goats. Vet. Parasitol. 146:117–122.

Terrill, T.H., S. Gelaye, E.A. Amoah, S. Miller, B. Kouakou, R.N. Gates and W.W. Hana. 1998. Protein and energy value of pearl millet grain for mature goats. J. Anim Sci. 76:1964–1969.

Tewe, O.O. 1992. Detoxification of cassava products and effects of residue toxins on consuming animals. In: Roots, tubers, plantains and bananas in animal feeding (Editors: D. Machin and Solveig Nyvold). FAO Animal Production and Health Paper No. 95: 81–95 Retrieved January 19, 2008 from http://www.fao.org/DOCREP/003/T0554E/T0554E06.htm.

Thomas, J., S. Solaiman, and Y. Dupre. 2008. Growth performance and carcass characteristics of goat kids fed diets containing sericea lespedeza. Proceedings of the American Forage and Grassland Council. Knoxville, TN.

Trodahl, S., T. Skjevdal, and T.A. Steine. 1981. Goats in cold and temperate climates. Pp. 489–513 in Goat Production, C. Gall, ed. New York: Academic Press.

Van Soest, P.J., J.B. Robertson, and B.A. Lewis. 1991. Methods for dietary fiber, neutral detergent fiber, and non-starch polysaccharides in relation to animal nutrition. J. Dairy Sci. 74:3583–3597.

Whitley, N.C., D. Cazac, B.J. Rude, D. Jackson-O'Brien, and S. Parveen. 2009. Use of commercial probiotic supplement in meat goats. J. Anim. Sci. 87:723–728.

Yanez, D.R., A.I. Martin Garcia, A. Moumen, and E. Molina Alcaide. 2004. Ruminal fermentation and degradation patterns, protozoa population and urinary purine derivatives excretion in goats and wethers fed diets based on olive leaves. J. Ani. Sci. 82:3006–3014.

# 11
# Health Management, Diseases, and Parasites

*J.E. Miller, DVM, MPVM, PhD; B.M. Olcott, DVM, MS, MBA; and G.F. Bath, BVSc*

**KEY TERMS**

Health—a state of complete physical, mental, and social well-being and not merely the absence of disease or infirmity.
Disease—an impairment of health or a condition of abnormal function caused by bacteria and viruses.
Parasite—an organism that lives in or on another and from which it obtains nourishment.
Causative (etiological) factors—factors that produces an effect.
Epidemiology—transmission and control of diseases and parasites in populations.
Clinical examination—examination reflecting normal/abnormal findings.
Clinical signs/symptoms—examination findings that reflect disease or parasite presence.
Life cycle—period involving one generation of a disease or parasitic organism through means of reproduction, whether through asexual reproduction or sexual reproduction.
Treatment—intervention to cure a disease or parasitic infection.
Prevention—intervention to prevent a disease or parasitic infection.
Health program—regularly scheduled program using an integrated approach for controlling diseases and parasites.

**OBJECTIVES**

By completing this chapter, the reader will acquire knowledge on:

- The most common diseases and parasites of goats
- The complex interaction of factors involved in transmission of diseases and parasites of goats
- What signs and symptoms are associated with these diseases and parasites of goats
- How diseases and parasites can be controlled through sound health management using integrated prevention and treatment measures

## INTRODUCTION

This chapter is not a comprehensive treatise of all diseases and parasites of goats. For that, an entire book would be necessary. Diseases included in this chapter are mainly those caused by bacteria and viruses. Diseases and parasitic infections are serious constraints affecting goat production worldwide. Economic losses are caused by decreased production, cost of prevention, cost of treatment, and possible death of infected animals.

The effects of diseases and parasitic infection are influenced by host, management, and disease/parasitic agent factors. With these factors in mind, this chapter will provide and discuss basic information on the most common diseases and parasitic infections that may be encountered with some rational integrated control measures for sustainable production.

## HEALTH AND DISEASE

Health and disease are difficult to define because although they are at opposite poles of the spectrum, they represent a continuum from one to the other. For example, a goat in "excellent" health may still have parasites (a disease). For the purpose of this chapter, health is defined as the state of being that allows sustainable, optimum productivity; disease is defined as the state that prevents optimal productivity. Thus, all factors that improve health must be promoted, and everything that inhibits it must be opposed. Health is a state of harmonious well-being, while disease is a lack of equilibrium.

Within this broad definition, the cause of disease can be divided into the *essential* causes of disease (the part that is central and is impossible to replace such as an infectious organism or nutritional deficiency), and the *sufficient causes of disease* (the contributory factors that make the disease occur, such as transmitters of diseases or management practices). Without this understanding, the management of diseases is bound to fail because both central and contributory factors are necessary to cause disease, and both must be addressed to effectively treat and prevent disease. By understanding the central and contributory factors, diseases can be tackled in a holistic and consistent manner, and simultaneously all significant matters involved can be addressed.

## PHYSICAL EXAMINATION

Physical examination includes general observation of the animal or flock from a distance followed by restraint of an individual animal and in-depth examination. Physical examination also includes the careful collection of the animal's signalment (age, sex, and breed), a historical collection of information concerning the past history of the animal and flock, and the history and clinical signs associated with the immediate medical complaint.

### Normal Values

Normal values for most vital signs for goats are presented in Table 11.1 and may vary considerably depending on animal activity and environmental conditions.

**Table 11.1** Normal values for most vital signs in goats.

| | |
|---|---|
| Rectal temperature | 39–40°C (103–104°F) |
| Pulse or heart rate | |
| Adult | 70–90 beats per minute |
| Kid or stressed animal | Double the adult heart rate |
| Respiratory rate | 12–20 breaths per minute |
| Rumen contractions | 1–2 contractions per minute |
| Rumen contractions durations | 10–15 seconds |

The FAMACHA test is a goat-side test that allows approximation of the animal's packed cell volume (PCV). It is performed by everting the lower lid of the eye and examining the color of the conjunctival mucosa (mucous membrane). In general, pale or white membranes correlate with varying degrees of anemia, which correlate well with the burden of blood-sucking parasites (*Haemonchus contortus*).

These observations are part of a minimal database that will be an important aid in establishing the nature of disease. These measurements should be made repetitively and recorded, which over time will allow determination of improvement or exacerbation of the underlying condition.

## IMMUNITY AND IMMUNIZATION

Animals can become immune (resistant to infection, disease, or unwanted biological invasion) either by *inherent* or *acquired* resistance to a disease. Inherent immunity does not require previous exposure to an infection (that is, goats are inherently immune to most equine diseases). The development of acquired immunity requires exposure to the antigen that causes a disease. Acquired immunity is of two types: *active* and *passive*. Passive immunity is acquired by the absorption of preformed antibodies. Examples would include a kid consuming colostrum, which contains antibodies produced when the doe was exposed to antigens or by injecting a goat with Tetanus Antitoxin (TAT), which contains antibodies that were produced by a horse exposed to tetanus antigen.

Passive immunity is short lived, usually lasting 2–3 months after birth; however, active immunity is much longer lasting, often several years or even lifelong. Animals develop active immunity if they survive *natural exposure* to an infectious agent. However, this can be accompanied by severe disease and production loss. They can also become immune by *vaccination* with an antigen that is killed or modified and cannot cause clinical disease.

Disease resistance is a combination of physical barriers (skin or mucous membranes), cell mediated immunity (lymphocytes and macrophages), and humoral immunity (antibodies).

The objective of vaccination is to activate one or more of these defense mechanisms in a way that facilitates the host's resistance to the pathogen or to its toxins. This is achieved by modifying the disease-causing organism (or its products like toxins) so that they cause little or no harm to the animal while stimulating sufficient immunity to prevent subsequent infection. The more common ways of modifying organisms include killing them (killed vaccines), inactivating their toxins (toxoid vaccines), weakening them (modified live vaccines), or extracting specific parts that cause immunity (subunit vaccines). Each has its advantages and disadvantages. Killed vaccines are very safe but often of short duration and dose-dependent; live vaccines are usually long-lasting but have the potential to be unsafe and are more delicate and require careful handling, while subunit vaccines are safe but may be expensive.

The system of immunization is further complicated by the existence of different strains of organisms, and a vaccine suitable for one may not protect fully or at all against another strain. In addition, some vaccines have a single disease-causing antigen (univalent vaccines) while others cover several disease-causing antigens (multivalent vaccines). Cross-protection between organisms or strains present in a vaccine may also occur.

The degree of immunity produced by vaccination is affected by many factors, including nutrition status, age, breed, production, and the presence of other diseases (like parasites). No vaccine confers complete protection, and none will protect all animals equally. Immunity is a powerful weapon against many diseases, and vaccination should be a part of any herd health program. However, the choice of vaccines will vary with the circumstances on a farm. Vaccination must not be confused with treatment. The latter applies to drugs of various types used against disease-causing organisms. Examples are antibiotics for bacterial infections, anthelmintics (dewormers), and sprays or dips used against external parasites.

## IMPORTANT VETERINARY DRUGS

There are several classes of treatments that are often used on goat farms, and livestock owners and veterinarians need to know these for routine use or emergencies.

In the U.S., there are very few drugs that are approved and labeled for use in goats. It should be noted that the use of any other drug for a goat in the U.S. is either Extra-Label or Illegal. Only veterinarians are allowed to use or prescribe drugs in an Extra-Label manner. Veterinarians are not allowed to prescribe Extra-Label Use Drugs (ELUD) unless a valid Veterinarian-Client-Patient Relationship (VCPR) has been established. Every country has its own veterinary drug use laws that must be followed.

Remember to use drugs only under supervision of a veterinarian or as specified on the label, and do not use them after the expiration date. Laypersons should consult with a knowledgeable local veterinarian for advice on the best products to have on hand in a given environment. Always observe mandatory withdrawal periods for milk or meat to ensure the safety of these products after animals have been treated. Goat meat and goat milk eventually end up in the human food chain. Failing to follow drug withdrawal times can cause drug residues to occur, which can cause serious detrimental effects in the humans that consume the product (see Appendix C).

### Antibacterials

The purpose of antibacterials is to treat or prevent bacterial infections. Antibacterials must be given at the correct dose via the correct route for the correct period of time to be effective. Failure to give the correct dose for the correct duration can result in disease relapse and in bacterial resistance to the drug.

Tetracycline and penicillin are the most commonly used antibiotics due to their low cost and spectrum of activity. Note that neither of these is approved for use on goats in the U.S. without a veterinarian's prescription. In the U.S., Ceftiofur is the only approved for use parenteral antibacterial, and it is available only by prescription.

### Anthelmintics (Dewormers)

There are three different classes of anthelmintic drugs available in the U.S., each with their advantages, uses, and problems. Availability varies between countries because of registration requirements. For more information, refer to the Goat Parasites section of this chapter.

### Ectoparasiticides

Insecticides are used for the treatment of insects such as lice, flies, gnats, etc.; and acaracides are used for the treatment of mites and ticks. Further information can be found in the Ectoparasite Control section of this chapter.

#### Nonsteroidal Anti-inflammatory Drugs (NSAID)

NSAIDs are used in the prevention of inflammation and in the relief of pain. As a species, goats are apparently very susceptible to pain and are very vocal when they perceive pain. The relief of inflammation allows a more rapid return to a normal state. NSAIDs would include aspirin, phenylbutazone, flunixin meglumine, ketoprofen, and others.

#### Hormones

Hormones are used in goats primarily for reproductive purposes (prostaglandin and progesterone). Dexamethasone is used as a steroidal anti-inflammatory. Oxytocin is used as an aid in parturition and for milk letdown.

#### Electrolytes and Glucose

Electrolyte solutions (sodium, potassium, chlorine, and bicarbonate) are used to replace fluid and electrolytes in animals with diarrhea and for animals that are not eating or drinking. Calcium solutions are used in the treatment of milk fever. Glucose solutions are used in the treatment of hypoglycemia in kids.

#### Antiseptics (Disinfectants)

Disinfectants are useful for killing bacteria, viruses, fungi, and some parasites. They are useful for disinfecting premises, equipment, and wounds. The classes of disinfectants include alcohols, aldehydes, oxidizing agents such as chlorine, iodine, hydrogen peroxide, phenols, and quaternary ammonium.

In addition to chemical disinfectants, instruments and tools can be sterilized by boiling in water.

#### Vaccines

Vaccine use should be based on prevalence or likelihood of specific diseases occurring on the farm. Vaccines should be administered on a planned basis as part of the herd health program.

#### Natural Products

Some goat owners tend to be holistic or naturalistic and use organic and herbal products. Be aware that many of these remedies do have pharmacological effects, and if they are given in large doses, they may prove to be fatal.

Some herbal and other products may be of use but many may not work. In most countries, all remedies are required to be registered after being tested for safety, efficacy, and toxicity. Unsubstantiated claims and unregistered products in many countries are illegal. Diatomaceous Earth is highly touted as a safe and effective deworming product. However, research has demonstrated that it has no effect on parasites.

## HERD HEALTH PROGRAMS

To protect the health of a flock, optimally a well-designed health program must be in place. The health program should be fully integrated with other management programs such as nutrition, grazing, and breeding programs. Health procedures should coincide with other farm procedures such as ear tagging, castration, and weighing. This will help to ensure that everything necessary for health is done at the right time and to the right class of animals. Herd health is part of good herd management and should be arranged to fit in with other procedures or activities on the farm. The golden rule is that this aspect of farming is best organized by a knowledgeable veterinarian based on the feeding and breeding programs to be followed. Otherwise health measures are often disorganized, inconvenient, and even forgotten.

Timing of the forage season provides the timing for kidding season, which provides the timing for the breeding season, pregnancy, lactation, and weaning. After these events and dates are decided, the best timings for vaccination, dipping, spraying, and deworming can be established. Apart from fixed vaccination times, other treatments (for example, for parasites) must be flexible and will depend on prevailing environmental conditions.

Essential health measures, desirable measures, and those that are purely optional (they depend on the disease situation on a farm) must be identified. Although general health programs are often requested, the circumstances on each farm will vary, and it is unwise to attempt to construct a single program to be used on all farms. A general program can only serve as a starting point for the process of drawing up something tailored to the specific farm.

## IMPORTANT INFECTIOUS DISEASES

For convenience, these diseases are dealt with in this section by clinical signs. Some diseases related to nutrition and reproduction may also be dealt with in other sections or chapters of this book, but they may form part of the presenting group of signs that follow. Space precludes an extensive description of any of the diseases. The reader is advised to consult books and articles that deal at length with these problems, and a selection is appended in the bibliography.

#### Subcutaneous Swellings

Goats are famous for having "lumps and bumps." The main reason for this is that goats get a highly contagious disease called Caseous Lymphadenitis (CL). However, lumps and bumps can be caused by a variety of other causes.

CL is an extremely common disease of goats worldwide. It is usually ranked by goat owners as the most important disease they have in their herds. The disease is characterized by one or more abscesses involving lymph nodes. CL is caused by Corynebacterium pseudotuberculosis. Infection occurs by ingestion of the bacteria or by direct contact between bacteria and skin. The bacteria can pass through intact skin and is found in the regional lymph nodes within 3 days of skin exposure. Abscesses most commonly involve the lymph nodes of the head and neck. Some goats within a herd appear to be very resistant while others are very susceptible. Occasionally the organism will involve internal lymph nodes and result in a wasting syndrome (Thin Doe Syndrome).

Clinical diagnosis is based on the characteristic abscessation of a lymph node located usually in lymph nodes that drain the head and neck. There is usually a herd history of abscesses. The disease is confirmed by culture. Culturing abscesses is important in that there are other bacteriologic causes of abscesses, the most important being *Arcanobacterium pyogenes*, *Staphylococcus aureus*, and in tropical countries *Burkholderia pseudomallei* (Meloidosis). Serologic tests are also available.

Treatment with antibiotics is rarely successful in the treatment of this disease. Surgical drainage of external abscesses is the treatment of choice; however, it only treats the abscess and not the disease. It is important to remember that all exudates from the abscess are highly contagious and should be collected and disposed of by burning or burying. The affected doe should be isolated from the herd until the opened abscess is completely healed over.

The best prevention for CL is to maintain a closed herd or to carefully screen new additions to the herd for the presence of abscesses or scars from old abscesses. All affected animals should be isolated from the herd. Keyhole feeders have a great deal of advantages in preventing fecal oral spread of disease, however, they serve as one of the major fomites for this disease.

There is a vaccine available for sheep through Colorado Serum (Case-Bac) that will decrease the incidence and severity of the disease. It is a killed vaccine that requires two doses initially and an annual booster. The vaccine has no efficacy in animals that are already infected. There is a vaccine (Glanvac) available in Canada and Australia. Although not available in the U.S., it is apparently effective in goats.

VACCINE REACTIONS AND INJECTION SITE ABSCESSES

Some vaccines and many medications are highly irritating and when injected can cause local inflammation and swelling. There may be an immediate swelling or it may take days for the swelling to appear. Lack of sanitation and refrigeration can result in the injection of live bacterial contaminants into the host.

MISCELLANEOUS SWELLINGS OF THE HEAD

The following conditions cause swelling of the head:

Salivary glands cysts: Fluctuant fluid filled cysts on the jaw or head, which are filled with clear viscous fluid (saliva).

Retained cud: Cud that is trapped between the dental arcade and the lip and is caused by problems with dentition or loss of a molar.

Goiter: Swelling of the thyroid glands in response to inadequate levels of iodine.

Bottle jaw: Edematous swelling of the area between the mandibles, which is caused by hypoproteinemia and usually a result of parasitism.

Grass awn abscesses: Usually present around the lips and cheeks and caused by thorns, briars, and grass seeds (awns).

Hematomas and seromas: Accumulations of blood or serum, respectively, usually associated with trauma but may be a result of clotting disorders.

Hernias: Umbilical hernias and inguinal hernias may occur. They contain abdominal organs and are usually reducible.

Umbilical abscesses: Occur when the umbilicus is exposed to moisture and bacterial contaminants. In kids that fail to ingest colostrum, this may result in neonatal septicemia or "navel ill."

### Diseases of the Eye

Keratoconjunctivitis is inflammation of the cornea (keratitis) and inflammation of the conjunctiva (conjunctivitis). Keratoconjuctivitis (pinkeye) infects both the cornea, which becomes cloudy or blue, and the mucous membranes of the eye, which become red and cause a discharge. If neglected, the cornea can rupture, leading to permanent blindness. Keratoconjunctivitis can be caused by a variety of organisms, including species of *Branhamella*, *Chlamydophila*, *Moraxella*, and *Mycoplasma*. The cause should be identified by culturing for the appropriate organisms.

Treatment is with topical ophthalmic drops that must be placed in the eye several times a day. However, it can be very effective when done properly. The drug and formulation depends on the organism involved. Treatment can also be done with parenteral antibacterial injections. Systemic tetracycline has been shown to produce therapeutic drug levels in tears that persist for 2–3 days.

Vaccination may be of use, although the vaccine may not be registered for use in goats (see Important Veterinary Drugs). Other factors like dust, nutrition, management, and insects may make the animals more susceptible or aid in the spread of the condition. These factors should be addressed.

#### Other Conditions of the Eye

The following list contains other conditions of the eye:

Entropion: A congenital condition of the kid where the eyelid is rolled inward on the eye and the hair of the lid irritates the cornea. It can result in keratitis and ulcers of the cornea. Treatment is by surgically taking the "roll" out of the lid.

Listeriosis: Will cause inflammation of the eye due to the inability of the goat to blink and a decrease in tear production of the affected side. Symptoms of severe neurologic disease will also be present.

#### Diseases Characterized By Blindness without Keratitis or Conjunctivitis

Although pinkeye can cause blindness, the lesions interfering with sight are obvious. The majority of diseases causing blindness without obvious lesions of the eye are those that affect the brain or optic nerve.

Polioencephalomalacia: Caused by a deficiency of thiamine resulting in metabolic disease of the brain. Goats are blind, become recumbent, and have convulsions. Treatment is with parenteral vitamin B1. Prevention is by providing adequate thiamine in the ration.

Focal Symmetrical Encephalomalacia (FSE): This is a sequel to enterotoxemia (*C. perfringens type D*). It results in damage to the brain, which is clinically indistinguishable to that caused by lead poisoning or thiamine deficiency. Cerebral abscesses may occur anywhere within the brain. They are space-occupying lesions, and their signs will depend on location and size. Signs are progressive. Abscessation of the pituitary gland is particularly common.

Pregnancy toxemia: Caused by a deficiency of energy in late pregnancy particularly in overconditioned does. The animal becomes ketotic and eventually goes into a coma. Signs include blindness, vocalization, weakness, trembling, and death. Treatment is by supplemental energy in the form of propylene glycol per os (by mouth) or glucose parentally. It is often necessary to prematurely terminate the pregnancy either surgically or by hormonal treatment. Prevention is by keeping does in good but not excessive body condition, encouraging exercise, and supplementing energy during the third trimester of pregnancy.

Heavy metal toxicity: Lead and arsenic will cause swelling of the brain with convulsions and central blindness. Salt intoxication (water deprivation) will cause the same problem.

### Digestive Disorders

The digestive system, which extends from lips to anus, is exposed to a wide variety of disease-causing agents. Clinical signs cover a wide spectrum and include system specific signs such as bloat, regurgitation, and diarrhea, and systemic signs such as anorexia, ill-thrift, and sudden death.

For goats, the major problem of the gastrointestinal system is parasitism, which will be covered in the following section. The major clinical sign associated with infectious gastrointestinal disease is diarrhea or watery stools. This can be clinically divided into causes of diarrhea in neonates (<1 month of age) and causes of diarrhea in juveniles and adults (>1 month of age).

Diarrhea in kids is usually an infectious contagious disease. Diarrhea in the first few days of life is often caused by specific *E. coli* that secrete an enterotoxin (Enterotoxigenic *E. coli* or ETEC). These are characterized by pili, which allow adhesion to microvilli in the small intestine. The major pili types affecting kids is K99. Other causes of infectious diarrhea in the first month of life would include infection with Salmonella, Rotavirus, and Cyrptosporidia.

Diagnosis is by submission of fecal samples for bacterial culture, viral identification (usually by electron microscope), and parasite identification. Samples that culture positive for *E. coli* will require pili testing to identify them as pathogenic bacteria.

Treatment for all causes of diarrhea will require fluids and electrolytes. Young kids with diarrhea can quickly become dehydrated and acidotic. Commercial rehydration preparations are available for this purpose. Antibacterials are beneficial for bacterial enteritis and may help prevent colonization of pathogenic bacteria secondary to viral enteritis. There are no effective drugs for the treatment of Cryptosporidia.

The key to prevention is sanitation, and fecal oral transmission must be prevented. Usually these diseases occur in dairy goat kids where sanitation is poor. Commercial vaccines are available for Rota, Salmonella, and ETEC and can be used when appropriate.

It should be realized that kids less than 2 weeks cannot digest complex sugars or vegetable proteins. This means that dosing the neonate with table sugar (sucrose) or using low-quality milk replacers with soybean protein added will result in dietary scours.

Kids from 1 month of age to weaning (3 months) practice grazing and as a result are being constantly exposed to parasites. This accounts for the long list of parasitic causes of diarrhea. Nematode and coccidian parasites are covered in the Goat Parasites section.

Kids in this age group have also lost most or all of their passively acquired immunity from their dam. This makes them susceptible to diseases that require circulating antibodies for protection such as with enterotoxemia.

Enterotoxemia is characterized by diarrhea with blood, severe abdominal pain with vocalization, recumbency, shock, and death. Most commonly, affected animals are found dead. Enterotoxemia is caused by the overgrowth of *Clostridium perfringens* types C and D in the small intestine. They produces exotoxins, which are absorbed and cause extensive damage. Tentative diagnosis is based on the presence of the characteristic signs, a history of digestive disturbance, and absence of vaccination. Diagnosis is confirmed by necropsy. For treatment, antitoxin for *C. perfringens D* toxin is available. In addition, the animal should be treated for shock with anti-inflammatories and fluids. For prevention, immunization with *C. perfringens C and D* toxoid is given at 4 weeks of age and repeated in 4–6 weeks. This should be given annually to adults 4–6 weeks prior to the start of the kidding season.

Kids in the 1- to 3-month age range are busy exploring their environment with all of their senses including taste and smell. This can lead them to ingest toxic materials such as pesticides and heavy metals.

Although not present in the U.S., a very important cause of infectious diarrhea in Asia and Africa would be peste de petits ruminants (PPR), which is a viral disease of goats and sheep, extending from West Africa through to Bangladesh. It is spread by livestock movements and transmitted by nasal aerosols or indirect contact. It causes fever, depression, ocular and nasal discharge, oral ulcers, diarrhea, and death. Treatment is aimed at alleviating the symptoms, hydrating the patient, and controlling secondary bacterial infections. Prevention is by vaccination. Strict quarantine and the control of stock movements help prevent its spread. Johne's disease is a contagious, chronic, and fatal bacterial disease of the intestinal tract of all ruminants caused by *Mycobacterium avium subspecies paratuberculosis*. This bacteria is very environmentally resistant and can survive in the environment for at least a year. There is no cure for goats infected with Johne's disease. Clinical signs include weight loss and ill-thrift in adult goats. Terminally, affected goats may have diarrhea. However, diarrhea is not a common sign in goats. This disease can be confused with parasitism, caseous lymphadenitis, malnutrition, or any of a number of chronic diseases. It is spread through consumption of organisms shed in the manure of an infected ruminant. Control of the disease within an infected herd is based on serial testing of all ruminants in the herd with the removal of all infected animals. Prevention is based on testing all new additions to the herd. An ELISA test is available for screening animals for Johne's disease, however, definitive diagnosis is by culturing the organism.

### Reproductive Diseases

Reproductive failure can be due to disease in both bucks and does, so the cause or causes must be properly investigated and established. Perinatal mortality is a term used to cover fetal or kid deaths shortly before, after, or at the normal time of kidding. It is probably the biggest cause of reproductive wastage in goats kept extensively, and it is often a complex problem with many factors interwoven. Postnatal deaths are defined as either early (the first 3 days after birth) or late (up to 3 weeks after birth). Often, there may be a spectrum of abortions, stillbirths, and postnatal deaths in an outbreak.

Abortions are defined as the expulsion of dead fetuses at a stage before they are viable, while stillbirths are the expulsion of dead but potentially viable kids close to or at the expected date of birth. The cause of abortion is undetermined in 70% or more of samples submitted to diagnostic laboratories.

Abortions should be treated as caused by an infectious agent until proven otherwise. Steps to take for aborted fetuses follow:

- Collect the fetus and placenta and submit for diagnosis, or store it by freezing, or destroy it by burning or burying it.
- Isolate the aborted doe until a cause for the abortion is determined or until the kidding season is completed.
- Make sure all humans are aware that almost all causes of infectious abortion in goats are zoonotic diseases. Don't allow women who are pregnant or who may

become pregnant to handle does or kids during delivery.

In the U.S., the most common cause of infectious abortion is toxoplasmosis caused by *Toxoplasma gondii*, a protozoan parasite. Toxoplasma will cause fetal infection throughout pregnancy with fetal resorption or mummification if infection occurs in the first trimester. Abortion, stillbirths, or the birth of weak kids result from infection later in pregnancy. Diagnosis is made by demonstrating the presence of the organism. There are no drugs that effectively treat this organism. Goats that become infected are infected for life. Prevention is by limiting exposure of goats to cat feces, particularly young cats and pregnant goats. A vaccine is available in the U.K.

The second most common infectious cause of abortion in the U.S. is enzootic abortion caused by *Chlamydiophilus abortus*, which is an intracellular organism containing both RNA and DNA. It will cause abortion at any time during pregnancy of goats. The doe usually remains healthy. Diagnosis is based on identifying the organism on the placenta or fetus. Treatment is with antibiotics. Tetracycline is generally the drug of choice although resistance has been reported. Prevention is by vaccination of doelings prior to breeding and annual prebreeding boosters to does. In the face of an abortion storm, administration of tetracycline in the feed or by injection may prevent abortions.

Miscellaneous infectious causes of abortion include a number of bacterial diseases such as a *Listeriosis*, *Salmonellosis*, *Leptospirosis*, *Campylobacter*, *Brucella abortus*, and *Brucella melitensis*. The last named disease is a big problem in goats in many countries, and easily transmits to humans, where it is called malta fever. In goats, it causes abortion storms and chronic (long-term) mastitis, and milk is the main source of infection for goat kids and humans. Pasteurization is thus usually deemed essential. There is no treatment, but there are good vaccines available.

The rickettsial organism *Coxiella burnetii* causes Q fever in humans and abortions in goats. Another rickettsia (*Cytoecetes phagocytophilia*) causes tick-borne fever, which results in abortion in goats. Viral causes of abortions and perinatal death include border disease (Pestivirus) and Rift Valley fever (Bunyavirus). Any severe disease, particularly those causing fever and anorexia, may result indirectly in abortions in goats.

Nutritional causes of abortions are important in goats. An energy deficiency in late pregnancy can cause an abortion storm, particularly in young does and those carrying multiple fetuses. Angora goats are particularly susceptible to stress-induced abortion (nutritional, behavioral, or hypothermic).

Excessive caloric intake causes fat does that have difficulty delivering (dystocia); conversely too little nutrition for the late pregnant doe can result in a skinny doe that gives birth to weak kids that fail to survive. Deficiencies of selenium, copper, iodine, calcium, and vitamin A may causes problems and defects that result in abortions, stillbirth, or neonatal deaths.

A small pelvic diameter may be either genetic or developmental. If such does are mated to bucks that produce high birth-weight kids, dystocia and mortality could result.

There is ample scientific evidence that poor mothering is partly genetic in sheep, the same is probably true in goats. Factors that disturb bonding between does and kids can also cause postnatal mortality. These factors include a high stocking rate, frequent stock movement, and any type of disturbance. Weather can cause large-scale deaths in newborn kids. The combination of low temperatures, wind, and rain is collectively known as the chill factor. Finally, kids will die if they cannot consume enough milk. Damaged teats or udders (mastitis) must be identified before kidding. Pendulous udders and teats get progressively larger as the doe ages. Does must get enough feed to allow for good udder development and milk supply.

The diagnosis of the type and cause of prenatal mortality is complex and best performed by a skilled veterinarian and a competent diagnostic laboratory.

Metritis (infection of the uterus) may be part of the picture seen with neonatal mortality and thus will form part of the control measures required. However, there is a particular infection seen in does that kid in unhygienic conditions. It is known as uterine gangrene and is caused by infection with clostridial organisms like *C. septicum.*, *C. novyi*, and *C. chauvoei*. Multiple births make infection more likely. Treatment with penicillin must be prompt to be successful, and vaccination may be necessary if kidding conditions cannot be improved.

In bucks or wethers, infection of the prepuce by *Corynebacterium renale* as a consequence of excessive urea in urine will result in painful urination and may result in bucks being unwilling to mate. This disease is known as balanoposthitis, or pizzle rot, and is best prevented by lowering dietary protein intake.

Epididymitis in goats is caused mainly by *Brucella melitensis*, but not *B. ovis*, which is confined to sheep only. Epididymitis may also occur as a result of infection by organisms like species of *Hemophilus*, *Pasturella*, *Actinobacillus*, and *Corynebacterium*. Diagnosis is made by the submission of samples (blood and testicles) to a diagnostic laboratory.

**Lameness**

Lameness means to move with pain or difficulty. Lameness in goats is usually associated with the foot and is usually the result of owners failing to perform regular foot trimming. The main infectious causes of lameness involving the foot are contagious foot rot caused by *Fusobacterium necrophorum* and *Dichelobacter nodosus*, interdigital dermatitis (foot scald) caused by *Fusobacterium necrophorum*, and foot abscesses caused by *Arcanobacterium pyogenes*, *F. necrophorum*, and others.

Contagious foot rot is a major problem of goats. The causative agent, *D. nodosus*, is specifically adapted to the horny layer of the hoof, which it invades and underruns. *D. nodusus* is a contagious bacterial agent that lives only on the feet of infected sheep and goats. It survives for only short periods of time on pastures. In wet, warm weather, infection can spread rapidly throughout a herd. It can cause extreme lameness depending on the strain (virulence) and the degree of underrunning of the hoof wall. Treatment requires paring away the unattached hoof wall and soaking feet in zinc sulphate, formalin, or copper sulfate. Sodium lauryl sulphate is added to these disinfectants to increase penetration into the hoof. Antibiotics may be useful. Strict isolation is necessary to prevent further spread. There are vaccines but they are of short duration and have limited effects. They are used to reduce prevalence to the point where eradication can be attempted.

Interdigital dermatitis (scald) is similar to foot rot, although it is not contagious, and is caused only by *F. necrophorum*, which is an environmental bacteria and does not cause destruction of the hoof wall. Treatment is much easier, and if goats can be placed in a dry area and their feet soaked, the disease is usually easily cured.

Foot abscesses can be a sequel to interdigital dermatitis/foot rot or penetration of the hoof wall or sole by foreign bodies. In severe cases, it can result in infection of the interphalangeal joint (the coffin joint). Severe lameness results and can last for months. Prolonged antibiotic treatment and rest in confinement may assist recovery. If infection of the distal interphalangeal joint occurs, then amputation of the affected claw may be the most practical treatment.

Caprine arthritis encephalitis (CAE) infects goats worldwide. It is present in an adult form and a juvenile form. The adult form is the most common and causes lameness associated with arthritis of joints but can also involve the udder (hard bag) and the lungs (lungers). The front knees (carpal joints) especially become thickened and painful. Many other joints may be infected. Joint fluid is often dark, yellowish, thick, or cloudy. Diagnosis depends on antibody determination, virus isolation, or histology. The juvenile form is limited to kids 2–4 months of age. Kids begin to lose neurologic control of their back legs, and as the disease ascends, eventually the back legs are paralyzed.

Infection is mainly transmitted from doe to kid via colostrum and milk. Many serologically positive goats never show clinical signs of the disease. These goats are subclinical carriers of the disease. Other than palliative treatment for degenerative arthritis, there is no treatment and no vaccine for this disease. Prevention is the key and it depends on early identification of infected goats by repeated serological testing and culling these animals from the herd. Raising kids with known seronegative mothers is another option. This disease is not a major problem in meat goat herds or fiber goat herds since the disease is mainly spread only from mother to her offspring. In dairy goat herds where milk is pooled and fed to multiple kids, an entire kid crop can be infected by one doe. Dairy goat producers routinely pasteurize milk prior to using it to feed kids to prevent transmission of this and other milk-borne diseases.

**Abnormalities of the Mammary Gland**

The most common cause of swelling of the mammary gland is mastitis, which is inflammation of the secretary tissue of the mammary gland usually caused by a bacterial infection. Udder swelling (CAE "hard bag") can also be caused by viral infection, by physical trauma, and by the development of severe edema of the gland (periparturient edema). Theilitis is inflammation of the teat and may be caused by bacterial infections, viral infections (contagious ecthyma), chemical irritants, thermal burns, and trauma.

Mastitis is mostly a problem in dairy goats and is caused by contagious organisms that are spread from goat to goat in the milking parlor. The most common organisms are *Staphylococcus epidermidis*, *Staphylococcus aureus*, and *Streptococcus agalactia*. Any bacteria, fungus, or yeast that inadvertently gains access to the mammary gland can cause mastitis. Occasional cases are caused by *E. coli*, *Klebsiella*, *Paseurella sp.*, *Candida*, *Nocardia*, and others.

Infections are classified as peracute (gangrenous), acute, subclinical, and chronic. The major cost of mastitis is the loss of secretory tissue, which reduces milk production and subsequently kid growth. Most cases of acute mastitis cause the doe to become febrile and decrease feed intake. The affected gland becomes red, hot, swollen, and painful. The milk may be watery or contain flakes and clumps. Treatment with systemic or intramammary antibacterials may result in a cure. Alternatively, the gland remains subclinically infected with occasional clinical episodes. Gangrenous mastitis is fairly common in goats and is associated with toxin-producing bacteria that cause a vasculitis leading to avascular necrosis of one or both halves of the gland. The most common bacteria is *S. aureus* with some cases caused by *P. multocida*. The gland is swollen and hard. The skin of the udder and teat becomes cyanotic, is cold to the touch, and often leaks serum. The milk is usually watery and bloody with clumps. The doe is extremely ill and often dies from septic shock. In most cases of gangrene, the affected gland will necrose and slough. Treatment for gangrenous mastitis is aimed at saving the life of the doe and should include treatment with intravenous fluids, anti-inflammatories, antibacterials, and possibly amputation of the teat or gland.

The presence of subclinical mastitis can be monitored by the use of indirect tests that measure the number of inflammatory cells in milk. A popular goat-side test is the California Mastitis Test (CMT). The etiologic cause should be identified by bacterial culture. For dairy goats, preventive practices as used for dairy cattle have been shown to be effective. These include:

- Milk goats only after properly disinfecting and drying the teats. Use disposable paper towels and disinfect your hands between goats to prevent goat-to-goat transmission.
- Examine all goats for signs of mastitis at each milking and promptly treat all cases of mastitis that are discovered with appropriate antibacterials.
- Use a disinfectant teat dip at the end of every milking.
- Make sure milking machines are properly functioning.
- Treat all goats with a long-lasting intramammary antibiotic at the time of dry off.

"Hard udder" is a uniformly firm, hot, and hard udder that usually occurs in first parity does. Milk is normal appearing, but only a small quantity is produced. Over a period of 1–2 weeks, the udder softens and milk production improves. This condition is believed to be a sequel to CAE infection.

Periparturient edema is a normal physiological event; however, it can cause extreme swelling of the mammary gland. It occurs most severely in first parity does and appears similar to "hard udder." In severe cases, treatment consists of massage, stripping, warm compresses, and diuretics (furosomide). Response to these measures should be quick.

Gynecomastia is a condition in which male goats may develop mammary tissue and secrete milk. This occasionally results in gangrenous mastitis. "Witches milk" is a secretory product found in the udders of male and female newborns. It is a result of exposure to the doe's hormones in utero. Treatment is not necessary.

**Nervous Disorders**

Neurologic disease is associated with a wide spectrum of clinical signs, including paralysis, incoordination, ataxia, convulsions, coma, circling, blindness, trembling, and stiffness. Neurologic signs are usually dramatic and immediately noticed; however, the overall incidence of diseases of the nervous system is fairly low.

There is a wide spectrum of infectious causes including viral, bacterial, parasitic, metabolic, and toxic. Viral diseases include CAE (CAEV), border disease (pestivirus), Louping Ill (flavivurus), rabies (rhabdovirus), pseudorabies (herpesvirus), and scrapie (prion associated). Bacterial diseases include Listeriosis (*L. monocytogenes*), Tetanus (*C. tetanii*), Botulism (*C. botulinum*), Focal Symmetrical Encephalomalacia (*C. perfringens*), meningoencephalitis (*Streptococcus*, *E. coli*, and others), brain or pituitary abscesses (*A. pyogenes* and others), and *Ehrlichia ruminantium* (heartwater in Africa). Parasitic diseases include *Paralaphostrongylus tenus* (moose disease), *Psoroptes sp.* (ear mites), and *Taenia multiceps* (coenuriasis).

Noninfectious diseases are more common than infectious diseases and include metabolic and toxic disorders. Metabolic diseases include pregnancy toxemia (ketosis), hypoglycemia of kids (energy deficiency), milk fever (calcium deficiency), grass tetany (magnesium deficiency), polioencephalomalacia (thiamine deficiency), and swayback (copper deficiency).

Goats are inquisitive and acrobatic animals and as such are capable of gaining access to and ingesting a wide variety of toxins. These include heavy metals (lead and arsenic), fertilizers (urea, ammonia, nitrates), and pesticides (organophosphates, carbamates). Plant poisonings occur occasionally and include nitrate accumulators (Johnson grass, annual ryegrass), and cyanide accumulators (cherry laurel, sudan grass). Some forage accumulates tremorgenic toxins such as canary grass and dallis grass.

Heartwater is a major disease of goats in Africa and a few Caribbean islands caused by *Ehrlichia ruminantium* and transmitted by ticks of the genus Amblyomma. The disease is confined mainly to warm, dry environments. The organism damages the endothelium (lining) of blood vessels, particularly of the brain, lungs, and pericardium. Signs include fever, depression, trembling, staggering, convulsions, and foam at the nose prior to death. Fluid and foam are found in the lungs after death, as well as fluid in the pericardium. Treatment with tetracycline is effective if given early in the course of the disease, and supportive measures (fluids and anti-inflammatories) may be needed. There is a vaccine for heartwater, but its use is fraught with difficulties.

### Respiratory Disease

Diseases of the respiratory tract are fairly common in goats. Clinical signs include nasal discharge, coughing, sneezing, and increases in respiratory rate, depth, and effort (dyspnea). Goats with pneumonia will exhibit systemic signs including fever, anorexia, and depression along with signs of respiratory disease.

Bacterial lung infection is often preceded by viral infection, dusty or ammonia-laden air in barns, stress, and poor nutrition. In most cases, the natural barriers and defenses of the respiratory tract are weakened or overcome, giving the opportunity for bacterial invasion. Affected animals may sneeze or cough, are depressed, and may have a fever. A nasal discharge is usually present. The main bacteria involved are *Mannheimia hemolytica*, *Pasteurella multocida*, *C. pseudotuberculosis*, and *A. pyogenes*.

Tuberculosis is not a common disease of goats but because of its zoonotic potential has to be regarded as an important disease. Goats are susceptible to *M. bovis*, *M. avium*, and *M. tuberculosis*. Clinical signs are highly variable depending on the system infected. Pulmonary infection is most common and results in coughing and ill thrift. The enteric form presents with diarrhea and ill thrift. At necropsy, affected goats will have caseated lesions in parenchyma and lymph nodes.

Testing is done by intradermal testing using the caudal tail fold. Goats commonly will respond to bovine PPD and require the comparative cervical test to allow distinguishing between bovine and avian mycobacterial. There is no practical form of treatment. Infected animals should be destroyed, and their herds of origin should be tested or slaughtered. Prevention is based on testing all new entries to the herd and not commingling goats with infected animals including cattle.

In an outbreak, the cause should be determined by culture. Antibiotics can be lifesaving if given early and selected appropriately. In the absence of culture and sensitivity, antibacterials are usually aimed at *Pasteurella sp.* and would include tetracyclines, cephalosporins, florfenicol, fluoroquinolones, and aminocyclitols. Tilmicosin (Micotil) is not recommended for use in goats as mortality has been reported following its use. In addition to antibacterials, administration of nonsteroidal anti-inflammatories (NSAIDs, for example, flunixin) is helpful.

Prevention is achieved by reducing the predisposing causes listed above. There are no Pasteurella vaccines labeled for use in goats, however, bovine products are used with some apparent success. The vaccine should contain antigens for *M. hemolyticum* as well as *P. multocida*. Viral causes of pneumonia include CAEV and PPR. Contagious caprine pleuropneumonia (*Mycoplasma* biotype F38) and other mycoplasma are capable of causing severe pneumonia.

### Diseases of the Skin

Dermatitis (inflammation of the skin) is a common but generally minor disease problem of goats. Dermatitis can cause pruritus (itching), or pruritus can be absent. The most common parasitic skin disease is infestation with lice. Parasitic dermatitis will be covered in the Goat Parasites section.

The main bacterial skin disease is dermatophilosis (rain scald), caused by *Dermatophilus congolense*. Goats are predisposed to infection by prolonged moistening of the skin from rain followed by damage to the skin often caused by ticks. The disease causes crusts and scabs on the dorsal surfaces of the goat. Other bacteria can also invade the skin if it is continuously wet.

Ringworm is the most common fungal dermatitis and is usually caused by infection of the skin by *Trichophyton verrucosum*. Lesions are crusty, raised and round, and are often found on the head and ears. It is treated with antifungicides, iodine, or thiabendazole. It may take weeks or even months to end an outbreak.

Contagious Ecthyma (Orf) is caused by a parapox virus. The virus is environmentally resistant and can live in the environment for years. In susceptible herds, infection spread quickly and causes crusts and scabs to occur on the lips, tongues, mouths, and teats of affected goats. Lesions persist for 2–4 weeks after which complete healing occurs. Recovered animals have a solid immunity that lasts for at least a year and have partial immunity for life. In endemically infected herds, infection occurs in kids after passive immunity has waned (around 3 months). Weight loss is dramatic in kids, and does may develop mastitis following infection of the teats.

Treatment of affected animals is palliative. Lotions and creams are used to soften the crusts. Antibacterials may help prevent secondary bacterial infection. Nutritional support is critical for kids. Prevention is by intradermal vaccination with a live virus vaccine. The skin on the inside of the thigh is scarified, and vaccine is applied topically. In endemic herds, it will be necessary to vaccinate kids annually prior to weaning. Humans are readily infected by handling the vaccine or lesions on goats.

Goat pox is caused by a pox virus present in countries from West Africa, North Central Africa, the Middle East, and South Asia to China. It affects sheep as well as goats and is transmitted by direct contact to susceptible (usually young) animals. The skin develops raised red lesions that turn into scabs. After the scab is removed, a star-shaped scar remains. In severe cases, lesions also occur in the lungs. Goat pox can cause severe mortality, but recovered animals are not carriers and usually have lifelong immunity. In countries where the disease is endemic, vaccination is an option, but goat pox–free countries opt for eradication by slaughter to avoid the severe trade implications of infection.

Foot and Mouth Disease (FMD) affects a wide variety of cloven-hoofed animals, including goats. It is caused by a virus that is highly contagious and can quickly spread over large areas. Its biggest economic impact is on dairies due to decreases in milk production. Countries that have FMD face severe trade restrictions on the export of products like milk and meat. Affected goats may develop blisters (vesicles), erosions, and ulcers in their mouths and on the coronary bands of their hooves. Clinically, animals are off feed, drool, and are lame. Many infections go unnoticed. Control measures are usually a matter for government officials and can vary from vaccination to quarantine and eradication.

There are many plant toxins that can cause skin lesions by photosensitization. Photosensitization can be primary from ingested plant toxins; St. John's wort and buckwheat or more commonly secondary to liver disease caused by caltrops (tribulosis), lantana, senecio, and others. Cases of photosensitization affect white portions of skin more severely than pigmented skin.

Goats are affected by neoplasia, most commonly papillomas and squamous cell carcinomas.

### Sudden Death

Sudden death includes animals that die within a period of minutes to hours (up to 24 hours). Goats are a predator-prone species and attempt to hide symptoms of disease until they are moribund. The main infectious cause is a group of intestinal bacterial organisms called *Clostridium perfringens*, of which there are four major types designated A, B, C, and D causing a group of diseases known collectively as enterotoxemia (overeating disease) characterized by sudden death, nervous symptoms, or diarrhea. The main cause of enterotoxemia is *C. perfringens* type D, which is an obligate parasite of the gastrointestinal tract. With sudden changes in quantity (overeating) or quality of carbohydrates ingested, *C. perfringens* proliferates and produces alpha and epsilon toxins that are absorbed and cause the actual disease. The epsilon toxin in particular causes a generalized toxemia with necrosis and neurotoxicosis. This can cause sudden death but sometimes also nervous signs or diarrhea. All goats should be vaccinated because the disease is common, often rapid and lethal. Vaccination is less effective in goats than in sheep and must be given more frequently. Multivalent vaccines can cover the other serotypes.

A further group of clostridial organisms causing sudden death are much less common and include *C. septicum* (malignant edema), *C. novyi* (black disease) ,*C. chavoei* (blackleg), and others. Septic wounds are often a precursor and treatment may save goats if they are identified early enough. Vaccination is far preferable and highly effective.

Anthrax (*Bacillus anthracis*) is a severe problem in many countries, partly due to its ability to survive in the soil for decades. Infection with anthrax is uniformly fatal. If anthrax is suspected, the carcass must on no account be opened, and state veterinary authorities must be informed to ensure proper investigation and carcass disposal. A very effective vaccine for this disease is available.

Rift Valley fever (a viral disease of Africa) may cause large outbreaks of disease characterized by sudden death, hemorrhages, abortions, and liver damage.

Many toxins and plants can cause sudden death, the main types being urea (used as a protein source or fertilizer), nitrates (highly fertilized pastures and nitrate accumulating plants), and cyanide (prussic acid), which is accumulated by a number of plants.

Whenever sudden death occurs, it is important to establish an accurate diagnosis. The scene of death should be carefully examined keeping in mind that trauma and electrocution are major causes of acute death. Asphyxiation by hanging is a common cause of sudden death in countries where goats are tethered on mountainsides.

### Ill Thrift

Ill thrift is the most common problem in goat management. It is characterized by animals that are slow growing, thinner than normal, rough-hair coated, and weak. The most common cause of ill thrift is nutritional and is a result of inadequate quality or quantity of available food. Caloric maintenance requirements of goats increase with increasing burdens of parasites and disease. Parasitism and disease may also cause decreases in feed intake. A common and overlooked cause of weight loss is tooth wear. Front and side teeth need to be examined to establish if this is the cause, especially in older goats.

Many diseases resulting in ill thrift cause obvious signs. For instance, a goat with foot rot will walk less and forage less due to the pain of the infection. Other diseases are less obvious, and the signs will be hidden. Organ failure (liver or kidney), chronic infections (chronic pneumonia), or damage to viscera from previous disease (intestinal damage from parasites) will all result in the symptom of ill thrift. In adult females, these are all commonly lumped into the diagnosis of "Thin Doe Syndrome."

One of the most common infectious diseases causing ill thrift in adult goats is internal abscessation due to CL (see the Subcutaneous Swellings section). CAE is less common and has obvious signs (see Lameness). Paratuberculosis (Johne's Disease) (see Digestive Disorders) is primarily a problem of dairy goats. It is a dangerous bacterial disease because it can be introduced by buying apparently normal goats that carry the disease.

In some countries, protozoan diseases like anaplasmosis, babesiosis, theileriosis, eperythrozoonosis, and trypanosomiasis can cause severe anemia, emaciation, and death. Bacteria like mycoplasmosis and leptospirosis (a bacterial disease that can affect the kidneys) would also have to be considered. It needs to be emphasized that any long-term infection or parasitism is accompanied by weight loss, and the primary cause must be identified.

## GOAT PARASITES

Parasitism, and gastrointestinal nematode parasitism in particular, is a serious constraint affecting small ruminant production worldwide. Economic losses are caused by decreased production, cost of prevention, cost of treatment, and the death of infected animals. Problems with nematode parasitism are often classified as production disease (chronic and subclinical conditions affecting productivity such as weight loss, reduced weight gain, reproductive inefficiency, etc.). This is more so the case when goats are managed as grazers or in confinement rather than browsers. The nematode of particular concern is the Barber-pole worm (*Haemonchus contortus*). The tremendous egg-laying capacity of this worm is maintained by feeding on blood by both immature and mature stages. Severe blood loss can occur, resulting in anemia, loss of appetite, depression, loss of condition, and eventual death. Other worms contribute to "production disease" because they usually do not kill but affect the animal's ability to increase and/or maintain production. External parasites, for the most part, are a nuisance and can cause production loss simply because the animal spends more time and energy combating them than feeding. Physical injury occurs when irritation and scratching result in open wounds that then can become infected or subject to infestation with fly larvae.

The effects of parasitic infection can be influenced by the nutritional status of the host. It is also true that parasites interfere with the ability of the host to use nutrients efficiently. The well-fed animal is able to tolerate increasing infection levels, but eventually a point may be reached where parasitism overwhelms the host's ability to function properly. Depending on the host's age and sex, season of the year, and exposure to various potential infectious agents (parasitic and otherwise), nutrients are partitioned for growth, breeding, pregnancy, lactation, immunity, etc. The ability of the host to maintain a proper balance of this partitioning ensures that nutrients are used appropriately. Overall, inadequate feeding will lead to loss of productivity unless the balance is restored.

## GASTROINTESTINAL NEMATODES (WORMS)

There are a number of worms found in goats, and this section will concentrate on the predominant and usually the most pathogenic ones.

### General Life Cycle of Roundworms

The life cycle of worms consists of part of their life being spent inside the animal and part of their life on the pasture (Figure 11.1). Worms mate in the host, and females lay eggs that pass out in the feces. The eggs hatch and develop to infective larvae while remaining in the feces. The infective larvae then move out of the feces onto the surrounding forage where they can be consumed during grazing thus completing the cycle. The time from ingestion of infective larvae to egg-laying adults, called the prepatent period, is about 2–3 weeks, and the time for development from egg to infective larvae can be as short as 7–10 days (especially during the summer months). Therefore, continual pasture contamination can lead to rapid transmission and reinfection. During the colder months, larval development on pasture is delayed and may take up to a month or two to reach the infective larvae stage, thus pasture contamination and reinfection is minimized. The infective larvae have a protective sheath making them relatively resistant to adverse environmental conditions, which allows them to survive for months, thus extending transmission potential. As long as temperature and moisture conditions remain

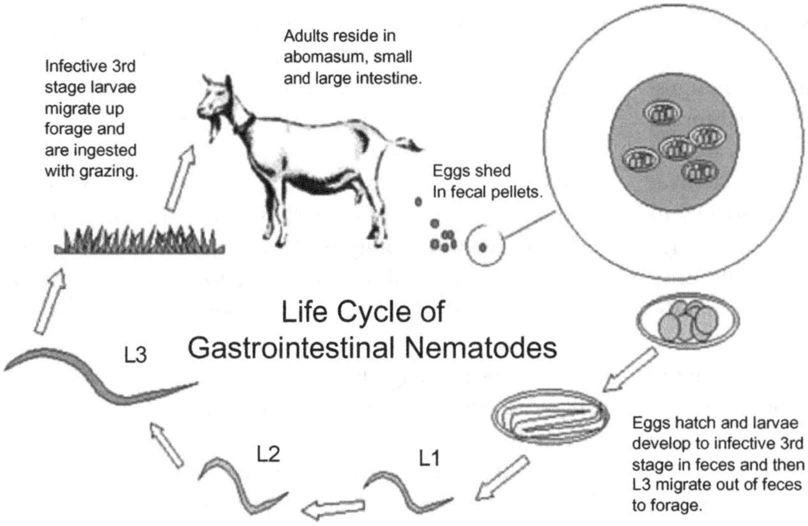


**Figure 11.1** General life cycle of roundworms (Nanci Solis and Michael Broussard, Louisiana State University).

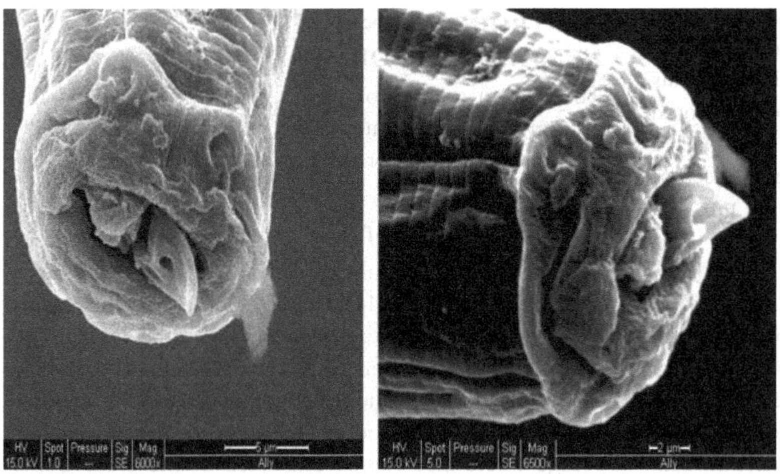

**Figure 11.2** Head end of Haemonchus contortus, showing the pointed stylet used for blood feeding (photo by Allison Moscona, Louisiana State University).

favorable, development and survival continues and pasture contamination accumulates, but if the temperature consistently gets too hot (above 35°C or 95°F) or cold (below 10°C or 50°F) and/or the moisture conditions become too dry (below 75–80% relative humidity or fecal moisture content), development and survival are hindered and pasture contamination dissipates. Transmission of parasites can be reduced (that is, breaking the life cycle) by implementing control measures to reduce the number of worms in the animal (deworming) and/or reducing pasture contamination (management).

### Abomasal Worms

BARBER POLE OR LARGE STOMACH WORM (*HAEMONCHUS CONTORTUS*)

*Haemonchus contortus* is a blood-feeding worm and gets its name from the barber-pole appearance consisting of the white ovaries that twist around the red blood-filled gut. This worm is rather large compared to other stomach and intestinal worms of goats, measuring up to 2 cm (3/4 inch) and has an anterior stylet that is used to disrupt tissue allowing blood flow (Figure 11.2). When large numbers are present, worms can readily be seen on the abomasal mucosal surface. Female worms are prolific egg layers and can contaminate the forage with a very large number of eggs. These worms thrive under hot and moist environmental conditions, which are conducive for survival and development of the free-living stages. They are found predominantly in tropical and subtropical regions of the world; and in the U.S., these conditions prevail in the Southeast. However, in the rest of the U.S. where similar environmental conditions are encountered during the summer, *H. contortus* transmission also frequently occurs.

Generally speaking, *H. contortus* transmission and infection are at the lowest level during the winter. Transmission and infection increase with the warmer temperatures and increasing moisture during the spring and then peak during the summer. Hypobiosis has not been observed to occur to any great extent in the Southeast U.S. because the life cycle can be maintained year-round, but it does occur in more northern/western temperate (cold/dry) regions of the U.S.

Animals infected with *H. contortus* show symptoms associated with blood loss (anemia), which include pale mucous membranes and bottle jaw (an accumulation of fluid under the chin). The greater the infection level the more blood is lost, and eventually the animal may die. Diarrhea may also be seen.

BROWN OR MIDDLE STOMACH WORM (*TELODORSAGIA* [*OSTERTAGIA*])

The other abomasal (*circumcincta*) worm of importance is *Telodorsagia circumcincta*, which is smaller than *H. contortus*. These worms feed mostly on nutrients in mucous and do not feed on blood, per se, but can ingest blood if present. Female worms do not produce as many eggs as *H. contortus*. Infection causes direct damage to the abomasal mucosa thereby interfering with digestion and appetite. Infection is usually considered a production disease because animals do not grow very well, but death can occur under very high infection conditions. The primary sign of infection is diarrhea and sometimes bottle jaw. This worm thrives in cooler wet environmental conditions, which are encountered in the more temperate regions. Hypobiosis occurs when environmental conditions are too cold (winter) or too dry (summer).

### Small Intestinal Worms

BANKRUPT WORM (*TRICHOSTRONGYLUS COLUBRIFORMIS*)

*Trichostrongylus colubriformis* is a very small threadlike worm and is the most predominant small intestinal worm. It thrives better under more cool and wet conditions similar to *T. circumcincta*. This worm feeds on nutrients in mucus and interferes with digestive function, resulting in diarrhea. It is called the bankrupt worm because death is seldom the end result, and animals just become poor doers leading to loss of production and income.

LONG-NECKED BANKRUPT WORM (*NEMATODIRUS* SPP.)

*Nematodirus* spp. are relatively large worms (easily seen) and are usually found in rather small numbers. If heavy infection occurs, production and income losses will result (similar to that of *T. colubriformis*). Eggs are large and football shaped.

### Large Intestinal Worms

NODULAR WORM (*OESOPHAGOSTOMUM* SPP.)

*Oesophagostomum* spp. are relatively large (easily seen) worms and are usually found in rather small numbers. These worms feed on blood and can contribute to the overall anemia in conjunction with *H. contortus*. Although this worm resides in the large intestine, the larvae are found in the mucosa of both the small and large intestine where they form nodules, thus the name nodular worm. Larvae leave these nodules and migrate to the large intestine where they become adults.

WHIPWORM (*TRICHURIS* SPP.)

*Trichuris* spp. are usually found in small numbers. The posterior end of the worm is rather large and can readily be seen. The anterior end of the worm is thread like, thus the name whipworm. These worms are also blood feeders and, like *Oesophagostomum*, contribute to the overall blood loss. Female worms produce characteristic football-shaped eggs with protruding plugs at each end.

## DIAGNOSTIC METHODS

Parasitized animals can show many signs of infection depending on the parasites present. The general signs include rough hair coat, diarrhea, depression, weight loss (or reduced weight gain), bottle jaw, and anorexia (off feed). Laboratory diagnostic findings may include anemia, increased fecal egg count (FEC), and loss of plasma protein.

### Fecal Egg Count (FEC)

FEC is used to evaluate infection level and is reported as the number of eggs per gram (EPG) of feces. While this is the best method for use with live animals, there are some difficulties associated with measurement including egg production does not always reflect the number of worms present (depends on the species); eggs cannot be completely identified by species and are grouped in various categories; how long infection has persisted; level of host immunity; fecal consistency (solid–diarrhea); and some methodologies used for EPG determination may be less precise than others. The FEC can serve as an indicator of seasonal changes in level of infection.

It is important to note that if heavy infection occurs over a short period of time (1–2 weeks) with *H. contortus*, animals may lose substantial amounts of blood without having very many eggs in the feces because late-stage larvae also are blood feeders.

The two common methods for determining FEC are the McMaster and centrifugation flotation methods. Details for these procedures can be found in goat health handbooks.

### Blood Packed Cell Volume (PCV)

Nematode parasites can affect an animal's ability to maintain erythropoiesis (making red blood cells). The PCV is the percent of the blood that is red blood cells, and normal blood is usually above 30%. When PCV drops below 20%, symptoms of anemia (lethargy, anorexia, etc.) usually start to appear. *Haemonchus contortus* is the primary nematode parasite that causes anemia and can lead to substantial acute blood loss and death. Anemia is used to support other infection response criteria, and is not necessarily a "stand-alone" diagnostic tool. The most common method for determining PCV is the use of a microhematocrit centrifuge.

### Anemia and FAMACHA

Level of anemia can be roughly evaluated by observing the color of mucous-membrane areas filled with capillaries close to the surface so that tissue color reflects blood color. Such areas are inside the lower eyelid, the gums (only where pigmentation is not present), and inside the vulva.

The FAMACHA eye color chart system was developed in South Africa (Bath et al., 1996; van Wyk and Bath, 2002) to help producers monitor and evaluate level of anemia without having to rely on laboratory testing. In this method, the lower eyelid mucous membranes are examined and compared to a laminated chart showing five colors that represent different levels of anemia: 1 (red, non-anemic); 2 (red-pink, non-anemic); 3 (pink, mild-anemic); 4 (pink-white, anemic); 5 (white, severely anemic). Since anemia is the primary pathologic effect from infection with *H. contortus,* this system can be an effective tool for identifying those animals that require treatment. It has been shown that a majority of the animals may not require treatment based on FAMACHA scores (van Wyk and Bath, 2002).

### Worm Count and Identification

The most absolute and direct method for documenting the number of worms present in an animal is to perform a necropsy and collect, identify, and count the worms present. This should be done by a properly trained veterinarian or other professional, and it might be very expensive. One can get an idea of the magnitude of *H. contortus* infection by looking for the worms that are visible on the abomasal mucosa. It should be noted that the longer the animal has been dead, the more the worms will move down the gut and won't be seen, thus giving the false impression that worms were not the problem. It is important to also note that *Telodorsagia* and *Trichostrongylus* are too small to see except under a microscope.

## DEWORMERS (ANTHELMINTICS)

Dewormers are chemicals (drugs) that have been evaluated and tested (for effectiveness and safety) for use in animals to remove worms. For the most part, pharmaceutical companies will not market a dewormer unless it is essentially 100% effective. As long as dewormers remain effective, control is relatively easy and cost effective. However, worms have developed resistance to almost all dewormers (Howell et al., 2008). Therefore, reliance on dewormers has become limited. Only FDA-approved dewormers can be used legally without restrictions in the U.S. Non-approved dewormers can be used "extra-label" (i.e., using a product other than for which it is approved) and are subject to specific regulations as delineated by the FDA. The veterinarian has to have contact with the animals and make a diagnosis that the parasite situation is potentially life threatening. The veterinarian has to establish that none of the approved dewormers will work (i.e., FEC reduction or Drenchrite® testing). Once the approved dewormers have been tested and shown not to be effective, then other dewormers can be used "extra label."

### Dewormer Classes

The three general classes of dewormers available in the U.S. are benzimidazoles, imidazothiazoles, and macrocyclic lactones. The more commonly used benzimidazole dewormers are fenbendazole (Safeguard, Panacur) and albendazole (Valbazen). Imidazothiazole dewormers are levamisole (Levisol, Tramisol) and morantel tartrate

(Rumatel), and macrocyclic lactone dewormers are ivermectin (Ivomec) and moxidectin (Cydectin). Of these, only fenbendazole and morantel tartrate are approved for use in goats. All others are non-approved and their use would be "extra-label."

### Dewormer Administration

Oral administration is preferred, and during administration of liquids, it is important to ensure the product is delivered over the base of the tongue. Thus, the dose is delivered to the rumen where it is mixed with ingesta and distributed evenly throughout the gastrointestinal tract. If the dose is delivered into the front part of the mouth, some may be spit out (wasted = reduced dose) and when swallowed, the esophageal groove reflex may be stimulated, which allows what is swallowed to bypass the rumen. The dose then goes directly into the omasum (third stomach) and moves quickly through the gastrointestinal tract, thus not allowing sufficient time to achieve full effectiveness.

The other form of oral administration is in feed products. This does not ensure that all animals will receive an effective dose because individual animals eat more/less than others due to their appetite, their place in the "pecking order," or the fact that they just may not like the formulation.

The use of injectable products, except for moxidectin, is not recommended. If used, injections are subcutaneous and best administered in an area of exposed skin so that one can see the dose being delivered.

Pour-on products are not recommended because field-use observations (unpublished) are mixed regarding whether pour-on products (approved for use in cattle only) work on small ruminants, and for the most part, they do not seem to be that effective.

### Resistance

The major problem encountered in controlling nematode parasitism in small ruminants is the resistance that many worm populations (specifically *H. contortus*) has developed to dewormers. Resistance develops primarily because deworming has been more frequent than needed. This frequent use increases the selection of more resistant worms, which will eventually result in a population of "super worms" that can't be controlled with drugs. There is no "silver bullet" that can be relied on. Resistance is genetically controlled, and once it is established, it is set in the population and those dewormers can no longer be used effectively.

## CONTROL PROGRAMS

### Smart Use of Dewormers

Conserving the effectiveness of dewormers can be achieved by using them only when infection levels dictate that intervention is necessary. The traditional concepts of treating all animals when a few show signs or all animals at regular intervals (shorter than every 3–4 months) are no longer warranted because they promote dewormer resistance.

It would be prudent to determine which dewormers are effective against a worm population. This can be achieved by conducting FEC reduction testing and should be done by a qualified professional. The concept is to have FEC done before and after (10–14 days) treatment. If the FEC after treatment is "0" (100% reduction), the dewormer is very effective. However, this should not be expected with most dewormers, and the best one (highest % reduction) should be considered for use. FEC reduction testing may seem somewhat expensive, but it will be worth the effort and expense to know which dewormers work and which don't. The worst case is not knowing which work and continuing with the use of the wrong dewormer. Another test for resistance is the Drenchrite® test. It is quite expensive, but again, may be justified. Some concepts for using dewormers smartly are:

1. Do not use the most effective dewormer exclusively unless it is the only dewormer that works. Reserve its use for deworming those animals that need it the most, and use less effective dewormers otherwise.
2. If one feels the need to rotate dewormers, do so at yearly intervals and rotate between classes using an effective one in each class.
3. Only deworm those animals that need to be dewormed and not the whole population. As a general rule, a minority of the population harbors the majority of the worm population, thus most of the animals may not need deworming, and it is not prudent to do so.
4. If resistance to multiple dewormers is present, increasing the dosage may help with some, or using combinations (from different classes) may improve effectiveness. Another concept is to take animals off feed for 24 hours before administering the dewormer. This will reduce rumen motility and the dewormer will pass through the gut slower and have more contact time with the target worms.
5. If there is drug resistance, deworming and moving animals to clean pasture (that has not had animals on it

for a long time) will more highly contaminate the new pasture with resistant worm larvae.

## Non-Drug Approach

MIXED/ALTERNATE LIVESTOCK SPECIES GRAZING

For the most part, each livestock species harbors its own parasite fauna except that small ruminants have the same parasites. Cattle and goats can be grazed together where each consumes the parasites of the other. As an alternative to co-grazing, cattle and goats can graze alternately on the same pastures. Either way, pasture contamination and re-infestation will decrease over time, thus both (or one) species should gain from this (Jordan et al., 1988).

PASTURE ROTATION

The value of pasture rotation to break the parasite cycle has been discussed for years (Levine et al., 1975; Colvin et al., 2008). The main reason to use pasture rotation is not for parasite control but to provide the most nutritious forage for animal growth and development. Many forage products reach the most nutritious stage about 30 days after animals have been removed, so rotation schemes usually have the animals returning to pastures at around 30-day intervals. Unfortunately, this 30-day interval is also about the same time necessary for newly deposited eggs to develop into the highest level of infectivity for the next grazing group. Thus, this rotation scheme may actually lead to increased worm parasite load. Rotation schemes of 2–3 months have been shown to have some effect on reducing pasture infectivity in tropical and subtropical environments; but in more temperate environments, infectivity can extend out to 8–12 months depending on the conditions. It is usually impractical to leave pastures ungrazed for such extended periods of time; therefore, being aware of the possible problems associated with whatever rotation scheme being used is important. Some success at reducing infectivity can be achieved by cutting pasture for hay between grazing periods. It should also be emphasized that when rotation schemes are used, stocking rate is usually high and the resultant increase in contamination may also make the problem worse.

COPPER OXIDE WIRE PARTICLES

Copper oxide wire particles (COWP) are marketed as a supplement for cattle being managed in copper-deficient areas. It is also known that copper has some anthelmintic activity against abomasal worms, but not other gastrointestinal worms. That makes it a very narrow spectrum product. But, in view of the potentially devastating problem of anthelmintic resistant *H. contortus*, recent work has shown that COWP effectively removes substantial numbers of *H. contortus* in goats (Burke et al., 2007). Copper toxicity, which is common in sheep, may not be much of an issue in goats because they are not as sensitive to excess copper intake. Thus, more extensive use of COWP during haemonchosis season may be useful in goats.

CONDENSED TANNIN-CONTAINING FORAGE

There is growing evidence that plants containing condensed tannins (CT) can reduce FEC and adult worms in the abomasum and small intestine. In addition, larval development in feces may be hindered. There are a number of CT-containing forage products; and in the U.S., sericea lespedeza (SL, *Lespedeza cuneata*), a perennial warm-season legume, has been shown to reduce FEC (primarily *H. contortus*) in grazing sheep and goats and when fed as hay, ground hay or pellets in confinement (Min et al., 2005; Terrill et al., 2007). Similar results have been observed with quebracho extract for both abomasal and small intestinal worms (Athanasiadou et al., 2000).

GENETIC IMPROVEMENT

Part of the ability of the host to resist worm infection is under genetic control. Resistance is based on inheritance of genes, which play a primary role in expression of host immunity. Under survival of the fittest management conditions, several sheep and goat (East African Dwarf and Saanan) breeds have been shown to be relatively resistant to infection (Preston and Allonby, 1978; Baker et al., 1998; Amarante et al., 2004). Using resistant breeds exclusively or in crossbreeding programs could lead to improved resistance to worm infection, but some level of production might be sacrificed. While this strategy may be acceptable, selection for resistant animals within a breed is also an option. Within breed, animals usually become more resistant to infection with age as their immune system matures. However, some animals do not develop this immunity as well as others and remain relatively susceptible to infection. It is well known that the majority of the worm population resides in a minority of the animal population. It would make sense, then, to encourage culling practices (based on FEC, PCV, FAMACHA, etc.) where these minority "parasitized" animals are eliminated, thus retaining more resistant stock. Using sires that throw relatively resistant offspring and/or using genetic markers to identify resistant animals would speed up this process. This approach has been used successfully (Crawford et al., 2006; Hunt et al., 2008), but it may take quite a long time to achieve satisfactory results. The real benefit to this approach is that reliance on dewormer intervention for control can be reduced, thus conserving the activity of such dewormers for when they are needed.

NEMATODE-TRAPPING FUNGI

The concept of using microfungi as a biological control agent against worms is not new (Larsen et al., 1997). These fungi occur ubiquitously in the soil throughout the world where they feed on a variety of saprophytic soil nematodes. These fungi produce sticky, loop traps on their growing hyphae, which capture and kill such nematodes. The best method of using these fungi is to concentrate their spores in feces to capture developing parasitic larvae. Of the various fungi tested, only *Duddingtonia flagrans* spores have been shown to have a high rate of survival during passage through the gastrointestinal tract of ruminants (Larsen et al., 1992). Spores deposited in the feces germinate and trap the developing larvae; thus, there are no or minimal numbers of infective larvae to migrate out of the fecal mass, which reduces pasture infectivity. This technology has been successfully applied under field conditions and is an environmentally safe biological approach for control of the fecal dwelling free-living stages of worms in small ruminants under sustainable, forage-based feeding systems (Fontenot et al., 2003).

### Vaccines

Efforts have increased in recent years to develop functional vaccines. This has been made possible by newer technologies in gene discovery and antigen identification, characterization, and production. Successful vaccines have been developed for lungworms in cattle and tapeworms in sheep (Bain and Urquhart, 1988; Lightowlers, 2006). The most promising vaccine antigen for *H. contortus* was derived from the gut of the worm, and when administered to the animal, antibodies are made. When the worm ingests blood during feeding, it also ingests these antibodies. The antibodies then attack the target gut cells of the worm and disrupt the worm's ability to process the nutrients necessary to maintain proper growth and maintenance. Thus, worms die. This vaccine has been tested successfully in sheep and goats under experimental and field conditions (Smith et al., 2001; Olcott et al., 2007). The one drawback to this vaccine is that the antigen is normally "hidden" from the host, and a number of vaccinations may be required to maintain antibody levels high enough to combat infection. Because the current process for antigen extraction is quite expensive, this will only be practical when recombinant technology produces an antigen at a lower cost.

Vaccines for other worms that do not feed on blood have focused on using antigens found in worm secretory and excretory products. These antigens do have contact with the host and should stimulate continuous antibody production. However, protection has been quite variable, and marketing such products has been a problem (Smith, 2008).

### Integrated Approaches

Traditional control of worms has relied on grazing management and treatment with dewormers. Grazing management schemes are often impractical due to the expense and the hardiness of infective larvae on pasture. Currently in the U.S., the two dewormers approved for goats are fenbendazole (Safeguard/Panacur, oral drench) and morantel tartrate (Rumatel, feed additive). Use of other dewormers or other methods of administration are not approved and constitute extra-label use. The evolution of dewormer resistance in worm populations has led to development of alternative strategies that could constitute major components in sustainable worm control programs. The most promising of these methods that are immediately applicable are smart drenching, COWP, SL, and FAMACHA.

An integrated approach using various methods should have an immediate impact on productivity and profitability of goat production systems in the Southeast U.S. and other regions where *H. contortus* and/or other worms can be a problem. Producers will be able to reduce overall dewormer usage by using smart drenching procedures and integrating an alternative (COWP and/or SL) with identification of animals in need of treatment (FAMACHA), thereby reducing cost of production while maintaining animal health and productivity. This integrated approach is essential to provide future environmentally sound worm prevention and control technologies to secure a sustainable, growing goat industry. Integration of other methodology/technology certainly will be instituted when evaluation is complete and ready for use.

## OTHER INTERNAL PARASITES

### Tapeworm (*Moniezia*)

Adult tapeworms reside in the small intestine and feed by absorbing nutrients from digested feed. They cause very little damage. Tapeworm eggs and segments are passed in the feces. Eggs are ingested by field mites and develop into the infective form. Animals grazing summer pasture ingest these mites along with forage. Many producers are concerned about tapeworms because they can see the moving white rice grainlike segments in freshly deposited feces.

There is no major concern, but growth of kids (not adults) may be somewhat reduced and intestinal blockage may rarely occur. Infection can be controlled with albendazole, fenbendazole, or oxfendazole.

### Liver Fluke (*Fasciola hepatica*)

Adult liver flukes reside in the bile ducts of the liver. The life cycle is indirect requiring an amphibious snail as an intermediate host. Eggs are passed in the feces and a larval stage called a miracidium develops inside the egg. After hatching, the miracidium infects a snail. Asexual reproduction occurs in the snail, and then the next immature larval stage called a cercaria leaves the snail and attaches to forage where it develops into a metacercaria. Animals ingest the metacercaria when grazing. The developing immature fluke migrates through the liver where damage results in unthriftiness, weight loss/reduced gains, and sometimes death. Infection can be a major problem in low-lying perennial wet areas. Snails are active mainly in the spring depending on environmental conditions, providing the source of infection (transmission). Snails burrow into the mud and become dormant the rest of the year, especially the hot summer months. Development of the adult fluke takes about 6–8 weeks. Because transmission ceases in late spring/early summer, treatment to control flukes can be divided into two periods—one period when immature and adult flukes are present (spring/early summer) and another when adults only are present (fall). Diagnosis is by using a sedimentation procedure to find eggs in feces. Clorsulon (Curatrem) is the only product that is effective against immature flukes. Clorsulon and albendazole are effective against adult flukes. Therefore, selection of either of these depends on the time of year. Another liver fluke, the deer fluke (Fascioloides magna), can kill goats by destroying the liver. Infection is rare but should be considered where deer have access to pastures. Control is difficult.

### Lungworms (*Dictyocaulus filaria, Muellerius, Protostrongylus*)

Adult lungworms live in the bronchi or tissue of lungs. The life cycle is direct with larva (not eggs) being passed in the feces where they develop to infective larvae. Grazing animals ingest the infective larvae, and after extensive migration through the body, they reach the lungs and mature to adults. Infection results in respiratory distress (chronic coughing), unthriftiness, and sometimes death. Problems with lungworm infection occur sporadically in the U.S. with transmission during the cooler months (fall/winter) of the year. Because larvae are found in feces, diagnosis is by using the Baermann procedure (Urquhart et al., 1996), which extracts the larvae from feces. Infection can be controlled with albendazole, fenbendazole, ivermectin, or oxfendazole. There are two other minor lungworms (*Muellerius capillaris* and *Protostrongylus* spp.) whose life cycles are indirect requiring land snails/slugs as intermediate hosts. Control is not as easy and fortunately pathology is minor.

### Meningeal worm (*Parelaphostrongylus tenuis*)

The meningeal worm, also known as the deer worm or meningeal deer worm, frequently infects llamas, alpacas, and sometimes goats. White-tailed deer are the natural host, so goats are at potential risk everywhere that white-tailed deer are found. Small ground dwelling slugs and snails are intermediate hosts. Goats, which are not normal hosts, can ingest the slugs/snails that harbor the infective form, and the larvae migrate into places where they don't normally reside in the deer. Migration is up the spinal nerves to the spinal cord, but then they seem to get lost, and the larvae can migrate to the brain. This causes damage to the central nervous system, which may be severe enough to result in death. Animals acquire infection in the spring/summer, but disease is usually seen in the fall/winter about 3–4 months after infection. Infected animals will show a wide variety of symptoms, which may include rear leg weakness and ataxia, paralysis, hypermetria, circling, abnormal head position, blindness, and gradual weight loss. Generally, animals with more severe symptoms have a worse prognosis.

Diagnosis is difficult in the live animal and is usually made when animals die and the larvae are found on examining the spinal cord and brain microscopically. The use of ivermectin at monthly intervals during the transmission season has been used in attempts to prevent infection, but this strategy has not been proven. However, this frequent administration interval most likely will have an effect on the development of resistance by the other resident worms.

### Coccidia (*Eimeria* spp.)

Coccidia are protozoan parasites that infect cells in the small intestine. Mature oocysts are passed in the feces and develop into infective stages (sporocysts) within the oocyst. Upon ingestion, infective stages break out of the oocysts and invade intestinal cells where they undergo asexual reproduction producing many more invasive stages. This can occur repeatedly, and eventually sexual reproduction occurs forming oocysts to complete the cycle. Infection results in damage to the intestinal lining, which can be permanent with reduced nutrient absorption capacity, thus leading to scours (sometimes bloody), unthriftiness, weight loss/reduced weight gains, and sometimes death. It is a disease associated with filth, moisture, and times of depressed immunity such as kidding, weaning, or during transportation. Devastating losses can occur quickly because of the asexual process and usually are a problem at weaning when kids are stressed. Preventing and/or controlling coccidiosis can be achieved by providing an anticoccidial product in the feed or water. There are several effective products on the market, such as amprolium, decoquinate, lasalosid, and monensin. Individual clinical cases can be treated with sulfa products. Secondary bacterial infection is common, so antibiotics may be needed. In order to reestablish healthy intestinal flora after recovery, probiotics are advised. There is no vaccine for coccidiosis. Fortunately, a solid immunity develops subsequent to infection; however, if infection was severe, stunting usually results.

## EXTERNAL PARASITES (ARTHROPODS)

External parasites can limit production. They feed on blood and organic debris. Infestation can cause discomfort, irritation, and pruritus (itching). Consequences of infestation can be skin damage, open lesions, reduced weight gains or weight loss, reduced milk production, and transmission of some diseases from sick to healthy animals.

### General Life Cycles

The life cycle of arthropods involve a series of structural changes known as metamorphosis, the actual sequence of which varies with different parasite groups. Complete metamorphosis (Figure 11.3A) begins when adults lay eggs from which larvae hatch. The larvae grow and shed their skins (moult) several times, each time to accommodate their increases in size. Eventually a hard-cased structure called a pupa is formed, which may have the capacity to survive winter. The pupa hatches into the adult parasite, the final stage of metamorphosis. Thus, there are four distinct stages: egg, larva, pupa, and adult. Incomplete metamorphosis (Figure 11.3B) involves a larva that grows and moults one or more times to become an adult-like form known as a nymph, which in turn grows and moults one or more times to become an adult. In this case larvae, nymphs, and adults all look similar but different in size.

### Flies

Flies undergo complete metamorphosis. There are a number of fly species, which are primarily a nuisance, especially under confinement conditions. The fly season is spring/summer. The constant buzzing of nuisance flies is irritating and can result in reduced foraging that may lead to production losses. Blood loss due to large numbers of blood-feeding flies (mosquitoes, black flies, horn flies, stable flies, tabanid flies, etc.) may lead to anemia, unthriftiness, and weight loss/reduced gains. Larvae (maggots) of nonblood-feeding nuisance flies can invade living and necrotic animal tissue causing a condition called myiasis.

Nasal bot flies (*Oestrus ovis*) primarily affect sheep but can be a problem for goats. They exhibit a unique quality by depositing live larvae in the nostrils. Infested animals usually have a nasal discharge that is sometimes blood tinged. Affected animals may shake their heads, grind their teeth, and be anorexic (loss of appetite). Larvae migrate

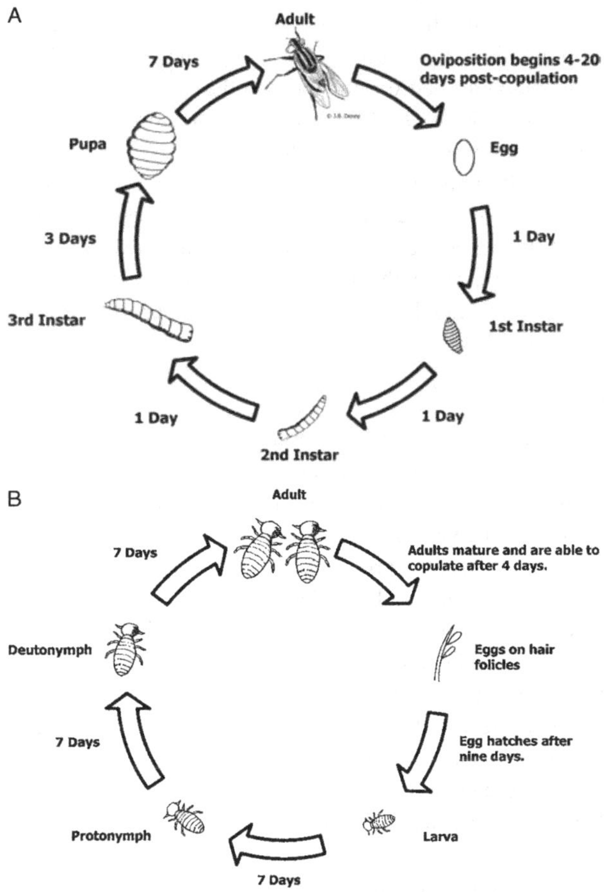


**Figure 11.3** (A) The life cycle of arthropods (complete metamorphosis) (J.J. Garza and J.B. Denny, Louisiana State University). (B) The life cycle of arthropods (incomplete metamorphosis) (J.J. Garza, Louisiana State University).

throughout the nasal passages (including into the head sinuses), and after a period of development (about 6 months), they drop from the nostrils to the ground, where they complete their development to the adult form.

Sheep keds (*Melophagus ovinus*) also primarily affect sheep, but occasionally can be found on goats. Unlike most insects, the female sheep ked gives birth to living larvae, which are nourished within her body until they reach the pupal stage, and then are expelled and glued to hairs. Adult keds emerge from the pupa and commence to feed on blood, which can damage the skin; thus buyers may downgrade the skins.

### Lice and Mites

Lice and mites undergo incomplete metamorphosis. These parasites are relatively permanent residents on the animal. They thrive and reproduce during the cooler months (fall/winter) of the year. Transmission from animal to animal is by contact usually pronounced when crowding occurs. Infestation (commonly called mange when mites are involved) may be seen as intense irritation with the animal scratching and chewing creating skin lesions that can lead to bacterial infection.

There are two kinds of lice—anoplura (suckling lice) and mallophaga (chewing lice). Suckling lice feed on blood and can cause anemia. Chewing lice scurry around feeding on organic debris leading to irritation, scratching, and skin lesions.

Mites commonly found are follicle mites (*Demodex caprae*), scabies mites (*Sarcoptes scabei*), psoroptic ear mites (*Psoroptes cuniculi*), and chorioptic scab mites (*Chorioptes bovis*).

The follicle mite causes nodules due to obstruction of the hair follicle trapping the mites and producing swelling. Nodules are usually seen on the face, neck, axillary region, and udder. These nodules may rupture, and the exuded mites can be transmitted to other animals.

Scabies mites burrow into the skin causing varying degrees of dermatitis causing crusty lesions and hair loss, usually around the muzzle, eyes, ears, inner thigh, brisket, abdomen, axillary region, and scrotum. Psoroptic ear mites cause crusty lesions and discharge from the external ear canal. Chorioptic scab mites cause mange seen primarily on the legs and feet. This is very common in goats.

### Ticks

Ticks undergo incomplete metamorphosis. They are blood feeders and are subdivided into hard and soft ticks according to structural characteristics. Adult hard ticks feed and mate on the animal. Engorged females drop to the ground and lay eggs. The eggs hatch, producing larvae, called seed ticks. The seed tick moults twice, passing through a nymphal stage before reaching maturity. A blood meal must be taken before each moult can occur. They are classified as one-, two-, or three-host ticks, depending on how many times they drop off, moult, and seek a new animal. A one-host tick remains on the animal from the seed-tick stage to maturity. A two-host tick drops off the initial host to moult from larva to nymph. The nymph seeks a second animal for the final blood meal before final moult to adult. The three-host tick drops to the ground for each moult, after which a new host is sought. The bodies of hard ticks are roughly oval, and the mouthparts protrude at the front. The mouthparts help to identify them. One mouthpart structure, the hypostome, anchors the tick to the host's skin to facilitate blood feeding. The abdomen, flattened top and bottom, can expand to several times its original size as a tick feeds. When fully engorged, females are many times greater in size. The patterns of pigmentation on the scutum of the tick also help with identification.

Soft ticks differ from hard ticks in many respects. They have a leathery outer skin rather than a hard cuticle, and both males and females engorge when feeding on the host. Their shapes vary among species and their mouthparts are located underneath the tick so it does not protrude forward.

### Fleas

Fleas occasionally can be found on goats. There are two kinds, the cat flea (*Ctenocephalides felis*) and the sticktight flea (*Echidnophaga gallinacea*). The cat flea feeds on blood and may cause anemia. The sticktight flea may cause ulcers on the head and ears.

### Ectoparasite Control

There are many insecticides/acaricides that can be used for control when necessary. These are classified as organophosphates (coumaphos, diazinon, dichlorvos, famphur, fenthion, malathion, etc.), carbamates (carbaryl and propoxur), pyrethrins/synthetic pyrethroids (cypermethrin, deltamethrin, fenvalerate, flumethrin, permethrin, etc.), macrocyclic lactones (avermectins and milbemycins), formamidines (amitraz), chloronicotinyls (Imidacloprid), Spinosyns (Spinosad) and insect growth regulators (benzoylphenyl ureas, triazine/pyrimidine derivatives, Diflubenzuron, and flufenoxuron). The most common methods of administration are dusts, sprays, pour-on products, and some oral products.

Seasonal occurrence such as flies and ticks in the warmer seasons and lice and mites in the cooler seasons, should be considered for appropriate insecticidal/acaricidal treatments. Other than insecticides, routine disposal of manure and organic materials will help control nuisance flies. It is difficult to treat for nasal bots, but macrocyclic lactones may work.

**Diagnostic Methods**

In general, most external parasites can be collected with various pieces of equipment. For flying insects, nets, traps, and aspirators are used. For crawling insects/ticks, jars, traps, combs, and forceps are used. For mites, skin scrapings are used. Most external parasites can be seen readily and identified using published descriptions and keys. However, the use of a microscope may be necessary to identify some.

## SUMMARY

This chapter briefly covers a broad spectrum of the more important disease and parasites of goats found worldwide, but with special emphasis on North America. The reader is given objectives and some key terms. Health and diseases are defined, with references to both normality and a range of factors that impact on disease. Clinical examination and normal values for findings are outlined. Various types of immunity are described, and the main types of vaccines are given, as well as reasons for apparent vaccine failure. Major classes of drugs available for treating goats are listed to assist the reader's understanding of treatment measures. Both internal and external parasites can be a major constraint to production. Parasites, unlike other infectious diseases (i.e., bacteria, viruses, etc.), are continually present, thus exposure and infection/infestation can occur year around. However, many parasite life cycles are subject to seasonal influences, which help control efforts. Since round worms are the major problem in most countries, their biology and control measures are given special attention. Sustainable production is only possible when there is a good balance between management, nutrition, and judicious use of the various control measures available.

## REFERENCES


Amarante, A.F., P.A. Bricarello, R.A. Rocha, and S.M. Gennari. 2004. Resistance of Santa Ines, Suffolk and Ile de France sheep to naturally acquired gastrointestinal nematode infections. Vet. Parasitol. 120:91–106.

Athanasiadou, S., I. Kyriazakis, F. Jackson, and R.L. Coop. 2000. Consequences of long-term feeding with condensed tannins on sheep parasitised with Trichostrongylus colubriformis. Inter. J. Parasitol. 30:1025–1033.

Bain, R.K. and G.M. Urquhart. 1988. Parenteral vaccination of calves against the cattle lungworm *Dictyocaulus viviparus*. Res. Vet. Sci. 45:270–271.

Baker, R.L., D.M. Mwamachi, J.O. Audho, E.O. Aduda, and W. Thorpe. 1998. Resistance of Galla and Small East African goats in the sub-humid tropics to gastrointestinal nematode infections and the peri-parturient rise in faecal egg counts. Vet. Parasitol. 79:53–64.

Bath, G.F., F.S. Malan, and J.A. van Wyk. 1996. The "FAMACHA" Ovine Anemia Guide to assist with the control of haemonchosis. In: Proceedings of the 7th Annual Congress of the Livestock Health and Production Group of the South African Veterinary Association, Port Elizabeth, South Africa, 5–7 June, p. 5.

Burke, J.M., T.H. Terrill, R.R. Kallu, J.E. Miller, and J. Mosjidis. 2007. Use of copper oxide wire particles to control gastrointestinal nematodes in goats. J. Anim. Sci. 85:2753–2761.

Colvin A.F., S.W. Walkden-Brown, M.R. Knox, and J.M. Scott. 2008. Intensive rotational grazing assists control of gastrointestinal nematodosis of sheep in a cool temperate environment with summer-dominant rainfall. Vet. Parasitol. 153:108–120.

Crawford, A.M., K.A. Paterson, K.G. Dodds, C. Diez Tascon, P.A. Williamson, M. Roberts Thomson, S.A. Bisset, A.E. Beattie, G.J. Greer, R.S. Green, R. Wheeler, R.J. Shaw, K. Knowler, and J.C. McEwan. 2006. Discovery of quantitative trait loci for resistance to parasitic nematode infection in sheep: I. Analysis of outcross pedigrees. BMC Genomics 18:178.

Fontenot, M.E., J.E. Miller, M.T. Peña, M. Larsen, and A. Gillespie. 2003. Efficiency of feeding *Duddingtonia flagrans* chlamydospores to grazing ewes on reducing availability of parasitic nematode larvae on pasture. Vet. Parasitol. 118:203–213.

Howell, S.B., J.M. Burk, J.E. Miller, T.H. Terrill, E. Valencia, M.J. Williams, L.H. Williamson, A.M. Zajac, and R.M. Kaplan. 2008. Prevalence of anthelmintic resistance on sheep and goat farms in the southeastern United States. J. Amer. Assoc. Vet. Med. 233:1913–1919.

Hunt, P.W., J.C. McEwan, and J.E. Miller. 2008. Future perspectives for the implementation of genetic markers for parasite resistance in sheep. Trop. Biomed. 25 (Suppl. 1):18–33.

Jordan, H.E., W.A. Phillips, R.D. Morrison, J.J. Doyle, and K. McKenzie. 1988. A 3-year study of continuous mixed grazing of cattle and sheep: parasitism of offspring. Int. J. Parasitol. 18:779–784.

Larsen, M., J. Wolstrup, S.A., Henriksen, J. Grønvold, and P. Nansen. 1992. In vivo passage through calves of nematophagous fungi selected for biocontrol of parasitic nematodes. J. Helminthol. 66:137–141.

Larsen, M., P. Nansen, J. Grϕnvold, J. Wolstrup, and S.A. Henriksen. 1997. Biological control of gastro-intestinal nematodes—facts, future, or fiction? Vet. Parasitol. 72:479–485.

Levine, N.D., D.T. Clark, R.E. Bradley, and S. Kantor. 1975. Relationship of pasture rotation to acquisition of gastro-intestinal nematodes by sheep. Amer. J. Vet. Res 36: 1459–1464.

Lightowlers, M.W. 2006. Cestode vaccines: origins, current status and future prospects. Parasitol. 133 (Suppl. S):27–42.

Min, B.R., S.P. Hart, D. Miller, G.M. Tomita, E. Loetz, and T. Sahlu. 2005. The effect of grazing forage containing condensed tannins on gastro-intestinal parasite infection and milk composition in Angora does. Vet. Parasitol. 130:105–113.

Olcott, D.D., B.M. Weeks, K. Shakya, W.D. Smith, and J.E. Miller. 2007. Effect of vaccination of goats with H-11/H-gal-GP antigens from intestinal membrane cells of *Haemonchus contortus*. J. Anim. Sci. 85 (Suppl. 2):35.

Preston, J.M. and E.W. Allonby. 1978. The influence of breed on the susceptibility of sheep and goats to a single experimental infection with *Haemonchus contortus*. Vet. Rec. 103:509–512.

Smith, W.D., J.A. van Wyk, and M.F. van Strijp. 2001. Preliminary observations on the potential of gut membrane proteins of *Haemonchus contortus* as candidate vaccine antigens in sheep on naturally infected pasture. Vet. Parasitol. 98:285–297.

Terrill, T.H., J.A. Mosjidis, D.A. Moore, S.A. Shaik, J.E. Miller, J.M. Burke, J.P. Muir, and R. Wolfe. 2007. Effect of pelleting on efficacy of sericea lespedeza hay as a natural dewormer in goats. Vet. Parasitol. 146:117–122.

Urquhart, G.M., J. Armour, J. Dunca, A.M. Dunn, and F.W. Jennings. 1996. The laboratory diagnosis of parasitism. In: Veterinary Parasitology, Blackwell Science, London, pp. 276–284.

van Wyk, J.A. and G.F. Bath. 2002. The FAMACHA system for managing haemonchosis in sheep and goats by clinically identifying individual animals for treatment. Vet. Res. 33:509–529.

## ADDITIONAL READING

Bath, G.F. and J.A. de Wet. 2000. Sheep and Goat Diseases. Tafelberg, Cape Town.

Bath, G.F., J.A. van Wyk, and K.P. Petty. 2005. Control measures for some important and unusual goat diseases in Southern Africa. Small Rum. Res. 60:127–140.

Baxendell, S.A. 1988. The diagnosis of the diseases of goats. Vade Mecum series B, No. 9. The University of Sydney Post-Graduate Foundation in Veterinary Science, Sydney.

Coetzer, J.A.W. and R.C. Tustin, Eds. 2004. Infectious Diseases of Livestock 2nd Edition. Oxford University Press, Cape Town.

Kellerman, T.S., J.A.W. Coetzer, and T.W. Naude. 1988. Plant Poisoning and Mycotoxicoses of Livestock in Southern Africa. Oxford University Press, Cape Town.

Matthews, J., 1999. Diseases of the Goat. Clarendon House Veterinary Centre, Chelmsford, UK.

Radostits, O.M., C.C. Gay, D.C. Blood, and K.W. Hinchcliff. 2007. Veterinary Medicine: A Textbook of the Diseases of Cattle Sheep, Pigs, Goats and Horses. WB Saunders Company, London.

Smith, B.P. 2008. Large Animal Internal Medicine Fourth Edition. Mosby, St. Louis.

Smith, M.C. and D.M. Sherman. 1994. Goat Medicine. Lea & Febiger, Philadelphia.

# 12
# Preferred Management Practices

*L. J. Dawson, BVSc, MS, DCAT*

**KEY TERMS**

Intramuscular—an injection directed into the muscle.
Subcutaneous—an injection directed under the skin.
Intravenous—an injection directed into a blood vessel.
Tattooing—needlelike projections into the skin for permanent identification.
Ear notching—notches made on the edges of the ear for permanent identification.
Microchips—implants or boluses transmitting radioactive waves placed underneath the skin or rumen.
Castration—making a male sterile, by either removing or damaging the testicles.
Disbudding—disrupting the blood supply to the horn buds, so they will fall.
Hoof trimming—removal of the excess growth on the hooves.
Weaning—process of removing the kid from her mother or removing a bottle-fed kid from milk.

**OBJECTIVES**

By completing this chapter, the reader will acquire knowledge on:

- How to inject a goat
- How to trim a hoof
- Best time and methods for castration
- Best time to disbud
- Different methods to dehorn
- Kidding management
- Intervention during kidding
- Colostral management
- Raising orphan kids
- Weaning management
- Controlling flies
- Keeping predators away from goats

## INTRODUCTION

The goat population in the world is approximately 800 million. A large majority of these goats (90%) are found in developing countries in Asia and Africa (FAO, 2005a; FAO, 2005b). In Central America, the goat population is around 13 million, and 10 million of them are in Mexico. In the U.S., the goat population has steadily increased and is around 3.05 million (Sahlu et al., 2009). Often, the goat is the only source of high quality protein food (milk and meat) and cash income of large segments of the population of many developing countries.

Interest in goats has mushroomed over the past 20 years in the U.S. Increased interest in goats and the value of their products requires better managing practices for optimum performance. Kid management from birth to breeding is an essential component of the goat enterprise. Kid management along with the management of the does and bucks has the greatest effect on the long-term productivity of the goat herd.

An obvious key to a successful goat operation is having a healthy, productive herd. Good animal husbandry, nutrition, vaccination, parasite control, and effective biosecurity programs for the operation could accomplish this. Good feeding and breeding will not result in maximum production if goats are not kept in good health. Effective management practices, understanding and recognizing the common diseases affecting the goats in the farm or region, and an effective herd health program is essential to a successful goat operation.

Herd health programs are usually modified to fit individual herds. These programs usually depend on the herd size, purpose of having the herd, and the production goals of the operation. Since each herd has a different situation, each manager should work with a veterinarian to create a specialized herd health plan. Keeping good records for each animal regarding medications, vaccinations, wormers, diseases, breeding, culling, etc., is key to successful goat husbandry. Use of this information will enable the producer to plan the disease prevention program. The best economic returns are realized when disease problems are at a minimum.

This chapter will discuss some of the management practices that will enhance productivity of goats. The majority of information used in this chapter is collective information gathered from four main sources:

Meat Goat Production Hand Book (2007) published by Langston University
Proceedings from the Western Veterinary Conference at Las Vegas (1998)
Sheep and Goat Medicine by Pugh (2002)
Goat Medicine by Smith and Sherman (1994)

## ADMINISTRATION OF VACCINES AND PHARMACEUTICALS

Live animals are considered unprocessed food, especially if those goats are intended for slaughter and later used in the food chain. Injection site lesions should be a major product quality concern for producers raising goats for meat. Therefore, persons involved with raising, handling, transporting, and marketing meat and milk products are encouraged to establish systems to ensure that animal drugs are used properly and to prevent illegal drug residues (Dawson, 2005).

Administering vaccines or drugs via injection is a common practice performed on goats. Injections must be done properly to reduce the chance of injury to the animal, reduce injection site defects, and prevent infections at that site.

The site chosen for giving intramuscular or subcutaneous injections is very important, especially if these animals are used for meat purposes. The three most common injection methods are subcutaneous (SC, under the skin), intramuscular (IM, in the muscle), and intravenous (IV, into a blood vessel). When a drug or vaccine lists SC as an option for injection, use the SC route. Only experienced personnel should attempt to give an intravenous injection. Intravenous injections provide the fastest absorption of a drug by the animal, and subcutaneous is the slowest (Dawson, 2007; Swize, 2007).

Injection site defects are lesions or scars found in different cuts of meat. They result from tissue irritation caused by the administration of intramuscular or sometimes subcutaneous injections. Tenderness of the meat is significantly reduced in the injected area that extends out (up to 2 inches/5 cm) in all directions from an injection site. When injection-site defects occur, the packers must trim and discard the damaged tissues. This greatly reduces the marketability and economic value of the meat (Dawson, 2005).

Commonly seen injection-site defects on goat carcasses are in the round (Figure 12.1). The round is an area on the rear leg, midway behind the hook (tuber coxae), to about 6 inches (15 cm) above the hock. The muscles in this area are the middle gluteal, gluteobiceps, semimembranosus, and semitendinosus muscles. Rounds from goats are economically important, because they are commonly processed and marketed as whole muscle products, not as ground meat. It is strongly discouraged to give intramuscular injections in the round on goats that are used for meat purposes (Dawson, 2005).

### Intramuscular Injection Sites

Intramuscular injections are commonly given in the triangular area of the neck, bounded caudally by the shoulder, dorsally by the cervical vertebra, and ventrally by the nuchal ligament of the spinous process (Figure 12.2). The volume given in the muscle should not be more than 3 mL per site. Insert the appropriate size needle (Table 12.1)

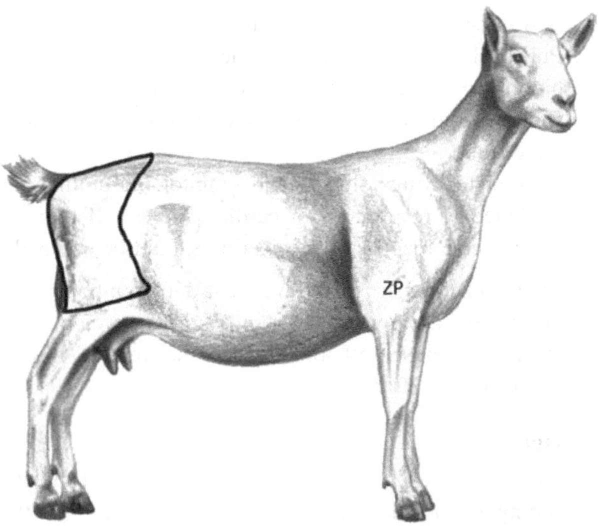


**Figure 12.1** Round is the area on the rear leg between the tuber coxae and the hock. Drawn by Zeke Proctor.

**Figure 12.2** Intramuscular injections are given in the triangular area of the neck. Drawn by Zeke Proctor.

directly into the muscle. Before injecting, draw the plunger and make sure that the needle has not entered a blood vessel. If blood enters the syringe, withdraw the needle slightly and redirect into the muscle. If the location is ideal, slowly press the plunger and administer the drug or vaccine. Massage the area after removing the needle (Dawson, 2005; Dawson, 2007a).

**Table 12.1** Needle size recommended by the author.

| Age | Gauge | Length |
|---|---|---|
| <4 weeks old | 22 or 20 | 1/2 inch (1.5 cm) |
| 4 to 16 weeks old | 20 | 5/8 to 3/4 inch (1.5–2 cm) |
| 4 to 6 months | 20 | 1 inch (2.5 cm) |
| >6 months | 20 or 18 | 1 inch (2.5 cm) |

Sources: Dawson, 2005; Swize, 2007.

**Figure 12.3** Subcutaneous injections are given behind the elbow joint or in front of the shoulders. Drawn by Zeke Proctor.

**Subcutaneous Injections**

These injections are usually given in the axillary region, behind the elbow joint, or in the triangular area in front of the shoulders (Figure 12.3). Pull a pinch of skin making a tent. Insert the appropriate size needle into the tent, making sure it does not pierce through the other side. Always insert the needle pointing toward the ground to lessen the likelihood of the drug or vaccine leaking out of the hole formed by insertion of the needle. Aspirate the plunger of the syringe and make sure there is no blood flowing back into the syringe. Depress the plunger slowly, and massage the area injected (Dawson, 2005; Dawson, 2007a).

**Intravenous Injection**

An intravenous injection requires skill to locate a vein and should always be given by a veterinarian or experienced animal health technician. Usually the jugular vein in the neck is used. Insert the needle into the vein, and ensure that the needle remains in the vein while the drug is given slowly (Figure 12.4). Animals may react quickly to drugs given in this fashion due to rapid absorption. Read the label on the bottle, and make sure the drug could be given intravenously. Intravenous injections are usually done using an 18- or 20-gauge, 1-inch (2.5-cm) hypodermic needle (Dawson, 2005; Dawson, 2007b; Swize, 2007).

To give intravenous injection or to collect blood, follow these steps:

1. Hold the goat securely. The goat's head must be up and to the side.
2. Feel for the trachea and move laterally on the neck. The area between the trachea and the neck muscles is the "jugular groove" and is where the jugular vein lies.
3. Put pressure at the bottom of the groove with the thumb, for the jugular vein to swell up with blood above your finger up to the jaw (Figure 12.4).

**Figure 12.4** Pressure applied at the bottom of the groove with your thumb. Redrawn by Zeke Proctor from the Meat Goat Handbook. Edited by T.A. Gipson, R.C. Merkel, K. Williams and T. Sahlu.

4. Use an 18- to 20-gauge needle and direct it at an angle of 45 degrees through the skin (Figure 12.4).
5. Pull back on your syringe or vacutainer and see if the blood is present. If not, adjust the depth (deep or more shallow) until blood is obtained on the syringe or vacutainer tube.
6. Administer drugs slowly, and monitor the animal for evidence of respiratory or cardiac distress. If there is any adverse reaction, the injection should be stopped.

When injecting drugs intravenously, it is important to ensure that the drug enters the vein. IV drugs given outside the vein can cause inflammation of the vein (Dawson, 2007a; Dawson, 2005).

**Dosage**

Medication dosage should be calculated and followed according to the manufacturer's recommendation on the label of the bottle. If the label does not mention goats, then the dosage and volume should be obtained from the guidelines given by a veterinarian. To use a medication not approved for goats by the manufacturer, a valid client/patient relationship needs to be established with a veterinarian. A veterinarian should make a decision about using this product, decide on the route of administration, and establish a reasonable time for milk and meat withdrawal. Also the veterinarian is responsible for any side effects of the drug administered. The veterinarian and the producer need to keep proper records on the product, if used in an extra-label manner (Dawson, 2005; Dawson, 2007b).

## ANIMAL IDENTIFICATION

Proper identification of goats is essential in order to keep good records for each animal in the herd. Record keeping is important to keep information on each animal regarding reproduction, health, management practices, milk production, etc. There are two basic types of identification: either permanent or nonpermanent identification. Permanent identification includes tattooing, ear notching, and installing microchips under the skin and in the rumen. Nonpermanent identification includes ear tags, paint, or chalk markings. Permanent identifications are not easily seen from a distance; therefore, ear tags may be necessary along with the permanent identification (Dawson, 2007a; Dawson, 2007b).

### Permanent Identification

TATTOOING

This method of identification, if properly done, stays for a long time or is permanent. Tattooing involves a pair of tattoo pliers, proper digits (alphabets or numbers), and tattoo ink or paste. Tattooing involves making needlelike projections in the goat's skin. It is usually done on the inner aspect of the ear between the ribs of the cartilage in an area free of hair. The ear is cleaned with gauze or a towel to remove dirt and wax. Tattoo ink or paste is initially applied on the area to be tattooed. The goat is adequately restrained, and the tattooing pliers is placed between the ribs of the cartilage and firmly squeezed (Figure 12.5A). Tattoo ink or paste should be reapplied on the area where the needlelike projections of the digits are present on the skin. The tattooing pliers should be placed in a bucket with disinfectant between treating animals to prevent any spread of diseases or infection between animals. The tattooing pliers and digits are thoroughly cleaned after use each day (Dawson, 2007a; Dawson, 2007b).

EAR NOTCHING

Ear notching has been commonly practiced in identifying Angora goats. Some of the meat goat associations have adapted them, with some modifications. This system has the advantage of being visible from a distance, allowing producers to identify animals without restraining the goats. A pair of ear-notching pliers is used to put V-shaped notches on the edges of the ear, and a hole-punch is used to punch holes in the middle of the ear if necessary. It is critical that the ear-notching pliers is disinfected between animals. Generally, the notches on the top of the left ear are 10, on the bottom edge are 1, on the tip of the ear are 100, and 1,000 a hole at the center of the ear (Figure 12.6). On the right ear, top of the ear are 30, bottom edge are 3, on the tip of ear are 300, and 3,000 a hole at the center (Dawson, 2007a; Dawson, 2007b).

MICROCHIPS AND RUMEN BOLUSES

This is accomplished by insertion of a microchip in the base of the ear, or a rumen microchip is placed in the rumen capable of sending radio frequency waves. A scanner is needed to identify these animals. Usually these animals will have a microchip or a rumen chip, along with an ear tag on the animal, to ensure the integrity of the microchip identification (Dawson, 2007a; Dawson, 2007b).

**Nonpermanent Identifications**

EAR TAGS

Ear tags are an easy way to identify goats in the herd. Unlike tattoos, they can be read without actually restraining the goat. Unfortunately, unlike the tattoo and ear notch, they can break or rip out of the goat's ear. There are two kinds of ear tags—plastic and metal. To attach either of

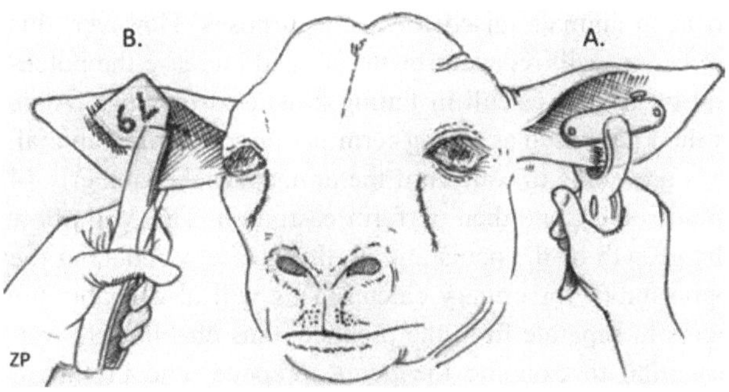


**Figure 12.5** A: Tattoo pliers is placed between the ribs of the cartilage and firmly squeezed. B: Ear tags applied with pliers and applied between the ribs of the cartilage. Drawn by Zeke Proctor.

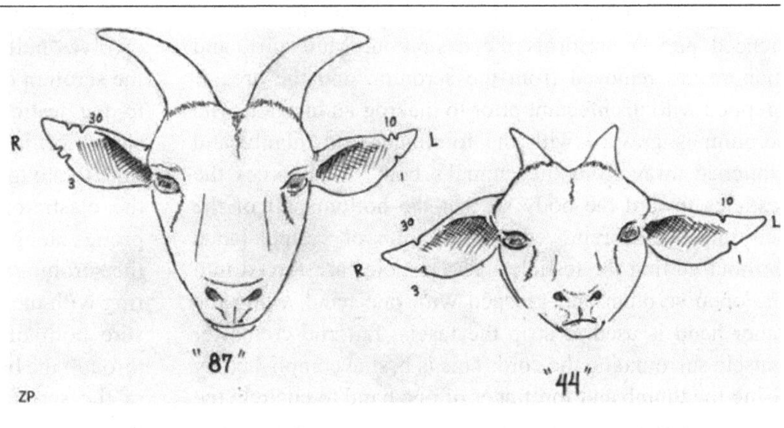


**Figure 12.6** Right ear edges 3 and 10. Left ear edges 1 and 10. Redrawn by Zeke Proctor from the Meat Goat Handbook. Edited by T.A. Gipson, R.C. Merkel, K. Williams, and T. Sahlu.

these tags, an ear tag pliers is needed (Figure 12.5B). Make sure the ear tags are inserted between the cartilage ribs of the ear (Dawson, 2007a; Dawson, 2007b).

CHALK

These are temporary markings and are used to sort animals. Chalk will not stay on the goats for a long time. Usually moisture or pressure (rubbing) will make the markings disappear.

## CASTRATION

Castration of male goats or kids is usually done during the first 7–14 days of their life. Castration of young animals produces less stress because they are more tolerant to pain at this age and have less chance of complications occurring during this procedure. Castration is usually performed to avoid undesirable odors associated with male goats, aggressive sexual behavior, accidental breeding by a buck of inferior quality, and to reduce undesirable flavors in meat in animals raised for meat purposes. However, this procedure will reduce growth rate and increase the potential of urinary calculi in fatting kids used for meat. Also, if the kid is used as a long-term pet or companion animal, it is advisable to wait until the animal reaches puberty (4 months old), and then perform castration. This will allow the growth of the penis and urethra, thereby reducing the opportunity for urinary calculi. This will also enable the penis to separate from the prepuce, thus enabling the veterinarian to examine the penis, prepuce, and urethra if needed. The two ways to castrate a goat are surgical and nonsurgical (Dawson, 2007a; Dawson, 2007b).

### Surgical Method

This method is usually performed under supervision of a veterinarian, with mild sedation under xylazine (or any approved sedative drug for goats), and pain is managed with flunixin. A handler restrains the kid, while an experienced person performs the castration. Mud, dirt, and manure are removed from the scrotum, and the area is prepped with disinfectant prior to making an incision. The scrotum is grasped with the forefinger and thumb, and stretched away from the animal's body. This forces the testicles toward the body so that the bottom half of the scrotum can be removed with a knife or scalpel blade without cutting the testicles. The testicles are forced into the open scrotum and grasped with one hand, while the other hand is used to strip the fascia, fat, and cremaster muscle surrounding the cord. This is best accomplished by using the thumb and forefinger of one hand to encircle the cord, then stripping the spermatic cord by moving your fingers up and down, while the other hand grasps the testicles and pulls it away from the animal's body. Once the cremaster muscle has been separated from the spermatic cord, the testicle is pulled or emasculated with an emasculator (Figure 12.7A), or by using the Henderson tool on a power drill to twist the cord till it breaks (Figure 12.7B). This procedure is repeated on the other testicle. Any fat or fascia hanging from the scrotum is trimmed even with the bottom of the scrotum (Dawson, 2007b; Hooper, 1998).

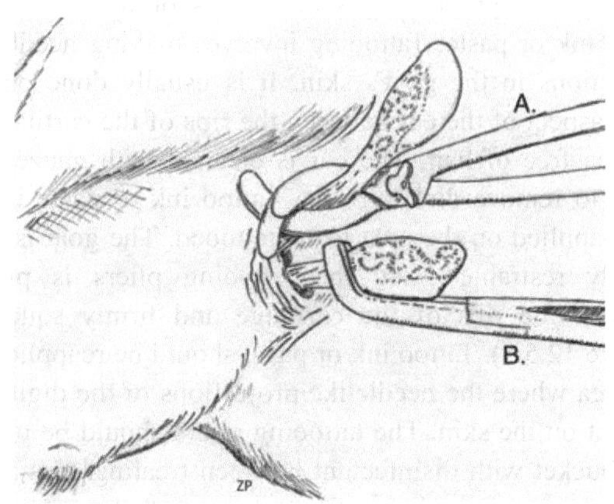

**Figure 12.7** A: Spermatic cord crushed using an emasculatome. B: Spermatic cord twisted using a Henderson tool. Drawn by Zeke Proctor.

### Nonsurgical Method

This method is usually performed with an elastrator or a Burdizzo clamp. The author prefers mild sedation if used on kids older than 2 weeks of age (Dawson, 2007a; Dawson, 2007b).

ELASTRATOR

This procedure, sometimes called bloodless castration, involves putting a heavy rubber ring around the neck of the scrotum close to the body. The ring stops blood supply to the testicles and scrotum that will later shrivel and slough off in 10–14 days. Bigger testicles may take a long time to slough off. The rubber ring is put on the prongs of the elastrator, and the handle is squeezed to open the prongs along with the band. The male kid is restrained, and the scrotum with the testicles is passed through the open ring with the prongs/band facing the kid's abdomen. Make sure both of the testicles along with the scrotum pass through the band and that the ring is positioned on the neck of the scrotum just below the penis (Figure 12.8). Kids usually show minimal discomfort for about 30 minutes to

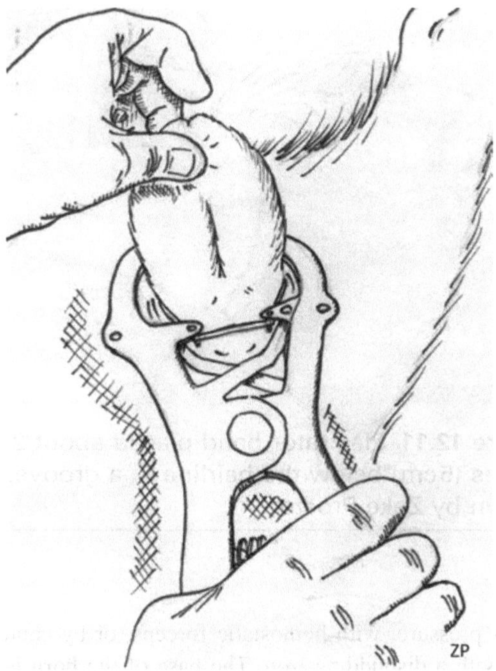


**Figure 12.8** Elastrator band applied to the neck of the scrotum. Drawn by Zeke Proctor.

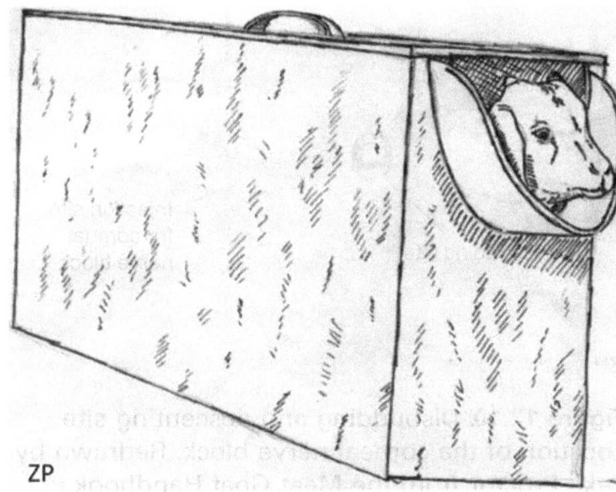


**Figure 12.9** Restraint of a kid by a disbudding box. Redrawn by Zeke Proctor from the Meat Goat Handbook. Edited by T.A. Gipson, R.C. Merkel, K. Williams, and T. Sahlu.

2 hours after banding. Kids are usually monitored for urination, appetite, and discomfort for the first 24 hours. This method has a higher risk of tetanus than other methods of castration. If the practice is to band kids at less than 3 weeks of age, make sure the does are vaccinated for tetanus 3–4 weeks before kidding. This will protect the kids from tetanus through ingestion of colostrum. Some prefer to give the kids a tetanus antigen and an injection of Procaine Penicillin (or any approved antibiotic for goats) at the time of banding, if the does were not vaccinated prior to kidding (Dawson, 2007b; Hooper, 1998).

### Burdizzo or Emasculatome

Castration can also be accomplished by using a Burdizzo clamp or emasculatome. This procedure involves clamping or crushing the spermatic cord without surgically opening the scrotum. Clamping the cord damages the blood supply to the testicle, causing the testicle to atrophy. This method is best used during the fly season because it leaves no open wounds. Goats must be between 4 weeks and 4 months of age with 2–3 months of age being ideal.

After restraining the animal under mild sedation, grasp the scrotum and manipulate one testicle to the bottom of the scrotal sac and find the spermatic cord. Place the Burdizzo clamp over the spermatic cord one-third of the way down the neck of the scrotum. Clamp down and hold for 15–20 seconds. Release the clamp, reposition it over the spermatic cord 1/2 inch (1.2 cm) below and repeat the procedure. Perform the same steps on the other side to crush the other spermatic cord.

Complications associated with the use of an emaculatome include tetanus, damage of the penis, swelling, and failure of the testicle and scrotum to atrophy and slough (Dawson, 2007b; Hooper, 1998; Smith and Sherman, 1994).

## DISBUDDING

Disbudding or dehorning is a management practice usually done for dairy goats, show goats, and goats used as companion animals. The ideal time to disbud kids is from 3 days to 3 weeks of age. A disbudding box is used to restrain these kids (Figure 12.9). Mild sedation or a cornual nerve block may be warranted to perform this procedure. If a disbudding box is not available, heavy sedation is necessary.

A hot disbudding iron is placed over the horn and pressure is applied to ensure complete contact with the skin surrounding the base of the horn. Leave the disbudding iron in place for 3 to 4 seconds at a time (Figure 12.10). The area around the horn is checked for a ring of brown

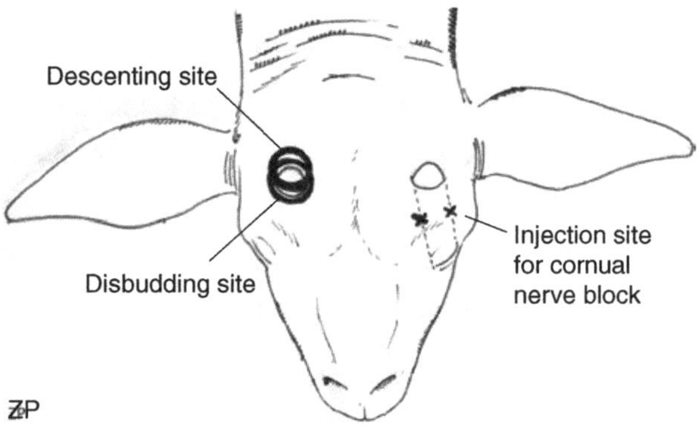


**Figure 12.10** Disbudding and descenting site. Location of the corneal nerve block. Redrawn by Zeke Proctor from the Meat Goat Handbook. Edited by T.A. Gipson, R.C. Merkel, K. Williams, and T. Sahlu.

**Figure 12.11** Elastrator band placed about 2 inches (5 cm) below the hairline in a groove. Drawn by Zeke Proctor.

or gold color encircling the horn. This indicates that there is separation of the skin to the base of the horn. This process is repeated on the other horn bud. Descenting could be done at the same time (Figure 12.10). The overzealous application of the disbudding iron can overheat the calivarium causing brain damage and death (Dawson, 2007b; Hooper, 1998; Pugh, 2002).

## DEHORNING

Kids whose buds are too large to fit into the recess of the disbudding iron will have to be dehorned surgically under heavy sedation and a cornual nerve or ring block. A veterinarian should perform this procedure. The area around the horn is clipped and prepped for surgery. Prior to the beginning of preparation for surgery, antibiotic eye ointment is put on both eyes to prevent any irritation from the antiseptic used.

Place the goat in sternal recumbency with its head up and looking forward. Make a circular incision 1–2 mm outside the junction of the skin and the horn base. Obstetric wire is placed in the incision and used to remove the horn and surrounding skin. Due to the shape of the horn, the wire should be pulled at a slightly ventral angle to ensure removal of an adequate amount of skin around the cranial aspect of the base of the horn. Repeat the procedure on the opposite horn. Check around the base of each horn to ensure that an adequate amount of skin has been removed from each horn. If there is doubt, remove more skin along the edges with a scalpel or heat the edges of the incision with a disbudding iron. Hemorrhage is controlled with direct pressure, with hemostatic forceps, or by cauterization with a disbudding iron. The base of the horn is connected with the calivarium caudally. An overzealous purchase of skin and horn could expose or damage the brain (Hooper, 1998; Pugh, 2002).

Postoperative care consists of tetanus prophylaxis and treating the surgical site like any other open wound. Cover with light bandages over the openings in the skin, especially if the sinus was entered when the horn was removed. Animals that have been recently dehorned should be kept separate from other goats until the wounds have healed. If separation is not possible, adequate pen size and feed bunk space are a must. Dehorning older animals with large-based horns is discouraged. Animal stresses, prolonged healing, persistent sinusitis, and sinocutaneous fistulas are potential complications and reasons to discourage dehorning mature goats.

Some producers prefer elastrator dehorning of adult goats from 3 months to 2 years of age. This procedure has a success rate of 25–75%. The younger the goat is, the higher the success rate will be. It takes anywhere from 3–6 weeks for the horn to fall. The key to success depends on the band placed close to the skull. The horn base should be shaved and clipped. The band is placed about 2 inches (5 cm) below the base of the horn into a groove, thus cutting off blood supply to the horn (Figure 12.11). In younger animals, the band will weaken the horn over a period of time and later the horn will fall off. In older animals, the horn becomes brittle or changes shape, and breaks off easily. Check bands weekly and add new bands immediately as required. Some goats will rub the bands until they roll off the horns or they may break. Putting duct tape over the bands may help the band to not slip off or break (Gudluck, 2008).

## HOOF TRIMMING

Hoof trimming on goats, if done on a regular basis, will keep the foot healthy and avoid any lameness associated with the feet. Many foot and leg problems are either caused by lack of routine care of the foot or improper trimming techniques. Frequency of trimming depends on many factors such as type of terrain, age, level of activity, nutrition, and genetics. Goats raised in confinement and on small acreages may require frequent trimmings. A properly trimmed hoof should look like that of a newborn kid.

The tools for hoof trimming include a set of sharp hoof shears, hoof knife, work gloves, rasp, silver nitrate sticks, and Kopertox™ (copper sulphate). Good restraint of the goat is necessary for proper trimming. A milk stand, a small ruminant restraint chute, a person holding the goat, tieing the goat to a fence, etc., are ways to restrain a goat to do an effective job of trimming the feet.

The first step in trimming is to clean the feet with a brush, then remove the dirt with a hoof knife or shear, and then trim excess hoof growth on the outside walls and toe with a hoof shear. Trimming of the outside wall and toe could be done with a shear, while the heel and sole could be removed with a hoof knife (Figure 12.12). When using a hoof knife, always cut away from the goat and yourself. Trimming of the wall and sole should be done carefully. Stop trimming the sole when you see hoof color change to pink. If there is any bleeding, you can use silver nitrate sticks. Severe bleeding is controlled with a pressure bandage for 24 hours. A properly trimmed foot will have the coronary band parallel to the weight-bearing surface of the claws. Excessive shortening of the toe by trimming will cause the animal to break down at the fetlock joint, and inadequate trimming of the toe will cause the animal to rock backward on its foot, reducing contact between the anterior surface of the sole and the ground. If the goat's feet have been neglected for some time and the toes are very long, it is usually not practical to bring them back to normal in one trimming. It is generally better to trim the feet a little at a time and gradually bring the weight-bearing surface parallel to the hoof's hairline (Dawson, 2007b; Swize, 2007).

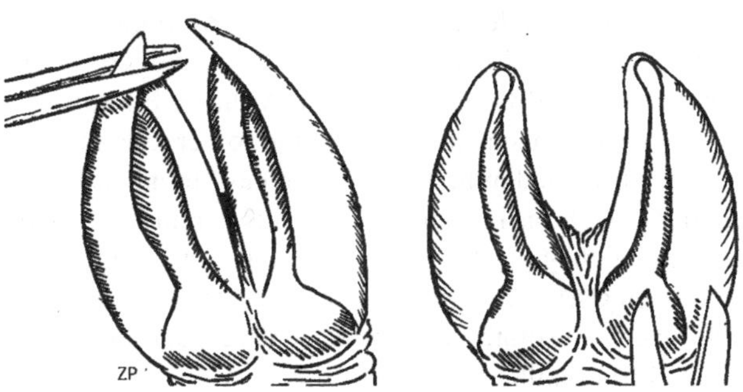


**Figure 12.12** Trimming the outer overgrown walls on the hoof and trimming the heel. Drawn by Zeke Proctor.

## KIDDING AND WEANING MANAGEMENT

A kid health and management program should actually begin prior to parturition with attention to the nutritional needs of the pregnant doe in late lactation and during the dry period. Pregnant does should be fed to have a good body condition (score of 3.0–3.5 just prior to kidding). Does should be scored in early pregnancy and again 6 weeks prior to kidding. Most fetal growth occurs in the last one-third of gestation. Feed quantity and quality may need to be increased during this time to prevent pregnancy toxemia (see Chapter 9 for more information). Clean, cool water and free choice trace-mineralized salt should also be available.

Booster vaccinations for *Clostridium perfringens* C and D and tetanus toxoid should be given not less than 3 weeks prior to kidding to produce antibodies in the colostrum. Vitamin E/selenium injections may be given during the dry period to prevent white muscle disease in kids, especially in areas where soils are selenium deficient and supplementation is inadequate. However, a nutrition program designed to provide adequate dietary selenium is preferable to injections. Provide other vaccinations or boosters for diseases causing abortion if applicable. Monitor fecal egg counts or FAMACHA score, and deworm as needed (Dawson, 2007b; Swize, 2007).

### Parturition (Kidding)

Although most meat goat does kid on pasture, there may be times when animals are brought indoors for kidding. The doe should kid in a clean environment—either a well-drained clean pasture with proper shelter, or a stall bedded with straw or other absorbent material. First-freshening does should be closely watched, especially if bred to bucks known to sire large kids.

Signs of impending kidding include udder engorgement, swelling of the vulva, restlessness, loss of appetite, frequent urination, and mucous discharge from the vulva. The ligaments in the pelvic area will relax, and the udder secretion will change from clear honey-like to yellowish and thick (colostrum). There are three stages of parturition. Stage 1 consists of uterine contractions and cervical dilation, and it usually lasts 2–6 hours. At the end of this stage, the first water bag or chorioallantois will break, and the doe will go into the second stage of parturition. During the second stage, the doe will have abdominal contractions along with the uterine contractions, and it will last from 30 minutes to 2 hours. If the doe is straining for more than an hour after the water bag has broken without any progress, or more than 30 minutes between kids, assistance is needed. Stage 3 consists of expulsion of the placenta or afterbirth, and it usually lasts 5–8 hours.

Problems during parturition may be due to an oversized kid, more than one kid entering the birth canal at a time, or postural defects (leg or head turned back, breech, that is, coming backward with both hind legs bent forward).

When assisting birth, it is important to clean the area around the vulva with a disinfectant and warm water and wear gloves to assist. There are certain diseases that can be transmitted to humans during this time; therefore, pregnant women should not assist with the kidding process. Lubricate the hand prior to entering the vagina. Feel and identify the parts of the kid. Try to ensure that all body parts felt belong to the same kid and not to a twin kid. Make sure the kid is in a normal presentation, position, and posture. Normal presentation is anterior longitudinal, dorso-sacral, and both front legs extended with the head between the legs. The fetus may have to be pushed forward toward the doe's head until the front legs and head are repositioned. Once the limbs and head are in a proper position, the kid should be gently pulled with plenty of lubrication.

Clear the mouth and nasal passages of the kid and stimulate respiration by irritating the nasal passages with straw. Never apply traction on any other presentation other than a normal presentation of two front legs and a head or a presentation of two hind legs and a tail. Pulling on an abnormal presentation, position, and posture will make it difficult to correct (Figure 12.13A).

If the anticipated kidding problems appear severe, a veterinarian must be called immediately (Dawson, 2007b; Swize, 2007).

### Kid Management at Birth

At birth, two management practices are critical to the future health and survival of the newborn kid. Dipping the cord in iodine is very important to prevent entry of organisms and also to promote rapid drying and the eventual breaking away of the cord from the navel. Another critical practice is the feeding of colostrum as soon after birth as possible. The colostrum, or first milk, contains antibodies, which the doe does not pass on to the fetus in the womb. Consumption of colostrum must occur as early as possible, ideally within 2–4 hours of birth. At 24 hours after birth, there is a rapid reduction in the absorption of colostral antibodies through the mucosa of the intestine. If a newborn kid does not or cannot nurse, the colostrum should be bottle-fed or the kid should be tube fed to ensure adequate consumption.

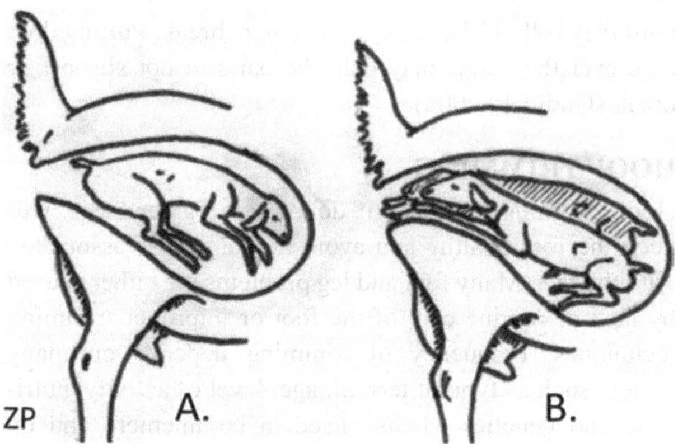


**Figure 12.13** A: Breech presentation. B: Twins. Drawn by Zeke Proctor.

Kids should receive colostrum equal to 10% of their body weight during the first 24 hours of life. For example, a 6-pound or 2.8-kg kid (96 ounces) should receive approximately 10 ounces (300 mL) of colostrum, or 1 ounce (30 mL) of colostrum per pound (about 0.5 kg body weight) within 24 hours of birth. This should be divided into at least three feedings. If fresh or frozen goat colostrum is not available, sheep or cow colostrum or colostrum substitutes could be used. Caprine Arthritis Encephalitis (CAE) virus can spread from the doe to her kids through her colostrum and milk. One of the measures advocated to control CAE virus in an infected herd is to remove the kids as soon as they are born, raise the kids on a bottle, and feed them CAE-free colostrum, heat-treated colostrum, or colostrum substitutes and milk replacer until they are weaned from milk.

Under certain conditions newborn kids may benefit from injections of vitamins A and D approximately 3–4 days after birth if they are raised away from sunlight. An iron dextran injection may be given if kids are raised in a barn or on a concrete floor or cage. A vitamin E/selenium injection may be beneficial in areas of selenium-deficient soils. These injections should be planned with an experienced veterinarian as part of the herd health management. In general, injection of vitamins and minerals is not necessary, and if supplementation is necessary, it is done more safely by dietary supplements. Realize that the fat-soluble vitamins and minerals are toxic if given in excess.

Kids should be checked carefully at birth for any physical deformities or abnormalities. Pneumonia is a major killer of young kids. A clean, dry, draft-free environment is an excellent preventative measure (Dawson, 2007b; Swize, 2007).

### Artificial Raising of Kids

Milk is the principal component of the diet of the preweaning kid. Most meat goat kids will nurse their dam until weaning. However, in dairy operations where milk is sold, or for orphaned kids and for kids of does that have lactation problems, it may be necessary to use a milk replacer. Goat milk replacers are commercially available. If necessary, a lamb milk replacer may be used to substitute for goat milk. Typical lamb milk replacers contain 22–24% protein and 28–30% fat (on a dry matter basis). If no other milk replacer is available, whole cow milk or calf milk replacers can be used. Maintaining milk replacer quality after mixing is particularly important when kids are fed ad libidum (all they can consume).

Milk can be fed by using bottles, pails, or self-feeder units. The method chosen will depend upon such factors as the size of the herd and available labor, as well as personal preference. With any system, the health of the kid, sanitation, and available labor are the major factors to consider.

Under natural suckling, kids consume small amounts of milk at very frequent intervals. Ideally, artificial rearing should mimic natural suckling, but the constraint of available labor precludes frequent feeding. Nevertheless, kids should be fed 4–5 times daily for the first and second week and 2–3 times daily thereafter. Pail or pan feeding may reduce labor somewhat, but body weight loss and need for extra "training sessions" at the beginning must be expected.

For larger herds, self-feeder units such as a "lamb bar" may successfully reduce labor. The key to use of any system is the maintenance of a low temperature of the milk (40°F) that will limit intake by the kid at any one time. Small, frequent feedings increase digestibility and decrease digestive disturbances. Rapid consumption of large quantities of milk may lead to fatal bloat due to entry of milk into the reticulorumen. Rapid passage of milk through the abomasum and small intestines can result in diarrhea or nutritional scours.

**Table 12.2** Feeding schedule and amount for nursing kids.

| Age[a] | Amount of milk/time | Feeding Schedule (3,14) |
|---|---|---|
| 1–3 days | 4 ounces (120 mL) | 5 times a day |
| 3 days–2 weeks | 8–12 ounces (240–360 mL) | 4 times a day |
| 2 weeks–3 months | 16 ounces (480 mL) | 3 times a day |
| 3–4 months | 16 ounces (480 mL) | 2 times a day |

[a]High quality hay and a grain mix (kid starter) should be provided as the kid ages.

The biggest problem with kids bottle fed milk replacer occurs with the feeding schedule. Frequently kids become "pets," and there is a tendency to feed them as much milk as they will consume each feeding. Unfortunately, this may result in bloat and sudden death due to enterotoxemia or diarrhea. It is necessary to adhere to a restricted feeding schedule and amount (Table 12.2; Dawson, 2007b; Swize, 2007).

### Dam-Raised Kids

Most meat goat kids will be raised with their dams on pasture. Although this removes the need for feeding milk replacer, these kids should not be forgotten in terms of nutritional and health needs. It must be noted that since these kids are raised in the same environment as their dams, they are also exposed to the same health, disease, management, and grazing conditions. If internal parasites are a problem in the dams, expect the same in the kids and take management steps to reduce exposure to internal parasites through pasture rotation or other means. Crowding should be avoided and, if housed at any time, clean bedding and adequate ventilation are a must. Kids are naturally curious and will begin nibbling on items in their surroundings early in life. If there are toxic substances or plants, plastic, or other harmful materials lying around, chances are some kids will eat them. If pasture is of very poor quality, kids beginning to nibble on grass or hay will not receive much nutritional benefit. This can slow down early growth.

Early access to a creep feed or creep pasture containing lush, nutritious forage will benefit kids becoming accustomed to solid feed, and will aid the development of their gastrointestinal tract in their early growth. Entry into the area containing creep feed or pasture should be restricted to kids by fencing or gates that prevent the entry of adult animals (Dawson, 2007b; Swize, 2007). Avoid overfattening of dairy goat kids.

### Weaning

In bottle-fed kids over 2 weeks of age, limiting daily milk consumption to about 48 ounces (<1.5 L) will encourage daily consumption of dry feed. No later than 3–4 weeks of age, a goat/lamb creep feed (kid starter), other suitable creep feed, or even a calf starter should be offered. As the hay and grain consumption increases, gradually reduce the milk being fed. When the kid is eating 0.25 lb or 100 g of grain per day plus some hay, and is drinking water from a bucket, it is time for weaning. Research has shown that at 2 months of age, a weaned kid has a reticuloruminal capacity five times as large as suckling kids of the same age (for more information, refer to Chapter 10).

Kids on pasture should be consuming forage such as pasture grass or hay by 2 weeks of age and grain within 4 weeks. Careful attention needs to be given to formulation of a concentrate supplement for the pre-weaning kids (kid starter). Palatability is a primary concern. Molasses at the rate of 10% of the total dry matter, corn (preferably cracked or rolled), and whole or rolled oats make up the energy "core" of a good pre-weaning diet. Balance the crude protein needs by adding cottonseed or preferably soybean meal or another high-protein source. Though few studies with kids have been done, crude protein contents of the pre-weaning ration should be within the range of 14–18%. Ground alfalfa may be added at 5% or less to provide additional stimulation for reticuloruminal development.

Several factors need to be considered when making the decision to wean. The most important consideration is whether or not the average daily consumption of concentrate and forage is adequate for growth and development to continue in the absence of milk. Fixed weaning ages are less desirable than body weight goals such as 2.0–2.5 times birth weight (Dawson, 2007b; Swize, 2007).

## FLY CONTROL

Flies can be a problem especially in a confined situation. Fly prevention should be initiated early in the season. The best method of controlling flies is good sanitation. Keeping the barn, pens, lots, and bedding around the feeders and water troughs clean and dry will prevent the larva from developing into adult flies. Preventing the accumulation of spilled feed, keeping water troughs from leaking or running over, and removing manure regularly are some of the good management practices that will aid in the reduction of files.

Integrated pest management using parasitic wasps, spiders, chickens, Muscovy ducks, etc., may be used to biologically control barn flies. Other systems such as baited traps, light traps, or stick tapes may be used. Spreading agricultural lime on the floor of the barn will also control flies. Insecticides could also be used to spray the barn. Always read and follow the label directions on any pesticide used on animals and surrounding premises (Dawson, 2007a,b; Swize, 2007).

## PREDATOR MANAGEMENT

Predator control within a goat herd is probably one of the single most important factors in maintaining profitable returns from a goat operation. During 2004, U.S. losses of goats and kids to predators totaled over 155,000 head at a value of $18.3 million. About 65% of the losses were from coyotes, 15% from dogs, 6% from mountain lions, 5% from bobcats, and the rest were from bears, foxes, and eagles (USDA, 2005).

There are management practices that keep the predators away from the prey. Predator management (Integrated Pest Management [IPM]) is an integrated approach strategy for reducing and controlling predators within the goat herd. Many methods are available for predator control ranging from passive or nonlethal to lethal control practices. Producers have a choice regarding which method to employ. However, no one method of control will completely eradicate predators from the goat herd (Grant, 2001; Pugh, 2008).

### Nonlethal Predator Management

Methods of nonlethal predator management include the following:

- Physical separation—A well-constructed fence or high-tensile electric fence can keep predators out. Coyotes prefer crawling and digging under the fence. An apron fence buried in the ground may be necessary.
- Day herding—Herdsman or a shepherd with the goats during the day can protect the animals from predators. This is a common practice in third world countries. Well-trained guardian animals (dog, llama) could be used for day herding.

- Night penning—Penning the goats at night in a properly fenced pen or barn can effectively control the predators.
- Removing dead carcasses, aborted fetuses, and afterbirth—Removing these in a timely manner can help reduce predators. Odors from these will attract predators.
- Kidding sheds—Young kids are easy prey for predators. During the kidding season, bringing the pregnant does and does and their kids away from the high-risk areas into a highly visible area, close to a barn or house will deter predators.
- Fright tactics—Loud bursts of sound on a timer, perimeter lights, motion sensors, scare crows, etc., may deter predators.
- Culling weak animals.
- Daily herd checks.
- Controlling brush around the pens and barn.
- Guardian animals (dogs, llamas, and donkeys)—Breeds of dogs including Great Pyrenees, Anatolian Shepherd, Akbash, Maremma, and Komondor are commonly used.

The deciding factor for which nonlethal method to be used depends on the producer; however, multiple methods are most effective.

### Lethal Predator Management

Lethal predator management methods include:

- Predator thinning by trapping
- Hunting
- Using toxicants, bait or gas toxicants (Grant, 2001; Pugh, 2008)

## WOUND MANAGEMENT

The most common wounds seen in goats are broken horns, lacerations, and bite wounds by predators. A broken horn is mainly caused by goats trying to establish dominance with fighting, boredom, horns trapped in the fence/feeder, or when restraining these goats by grasping the horn. When handling these wounds, hemostasis is the key. There is usually a lot of hemorrhage if there is damage to the base of the horn. Pressure is usually applied through a temporary pressure bandage. If an artery could be isolated to be hemorrhaging, it could be clamped or ligated. Usually the horn involved is removed, proper hemostasis is provided, and is treated as an open wound.

Lacerations on the body and limbs are usually handled as an open wound, and suturing may be done if the wound is fresh. Bite wounds by dogs are usually around the hind limbs or rear end of the animal. Wounds due to coyotes are usually fatal and are usually seen around the head, neck, and throat. Lacerations and bite wounds, if they are fresh and quite extensive, may require suturing and attention of a veterinarian. Occasionally drains are placed in the wounds and flushed.

Wound management due to horn damage, lacerations, and bite wounds are managed under mild or heavy sedation. Pain management is best done by using flunaxin or ketoprofen, and is administered for a few days. These animals should be placed on a broad-spectrum antibiotic to combat infection. A tetanus booster is also given at this time.

## BEHAVIOR MANAGEMENT

Goats are more sensitive animals than sheep or cattle. They are smarter, trickier, and more curious, but they are easier to train. Understanding goat behavior may reduce the stress and create a healthier environment for goats to thrive. Even herding dogs may stress goats by their noise or movements. Goats are naturally browsers and prefer open spaces, and when confined, they are more stress prone than sheep or cattle. In confinement, they are harder to manage than sheep and cattle. They are scavengers and last survivors. In confinement, this contributes to the overeating behavior, especially when management lacks a routine feeding schedule. They should be handled quietly in the pens. Excessive noise or aggression will disturb the animals.

Domesticated goats like to climb but rarely jump. When gathering the goats, they tend to drift to hilltops, if present, and will move faster than cattle or sheep. When working with goats, the more they are rushed, the longer it will take to get the job done. When hurried, they tend to be aggressive toward each other and harder to handle.

Goats exhibit natural flocking behavior, there is a definite pecking order, and they follow the leader of the pack. They move or enter a pen in family groups with older females leading the way. They act on their instinct and have a good memory. They are creatures of habit and when they get familiar with a certain sets of routines, pens, or handling procedures, they will expect to be treated the same way every time, which makes them easier to handle (Solaiman, 2007).

## SUMMARY

This chapter describes some of the common management practices followed in the U.S. for effective goat production. Some of the practices mentioned may differ in other parts of the world. The main goal is to incorporate these common practices in a herd health program that is developed to fit the needs of the farm. Information provided in the chapter can assist educators and the students in acquiring knowledge on preferred production practices for goats. Also these practices can assist experienced and inexperienced producers in making sound decisions in their management practices that will result in healthy goats and production of residue-free and unadulterated meat and milk products for human consumption.


## ACKNOWLEDGEMENT

Illustrations by Dr. Zeke Proctor, DVM. Dr. Proctor graduated from Oklahoma State University in 2009 and is currently an Associate Veterinarian in a mixed practice at Siloam Springs, Arkansas.


## REFERENCES


Dawson, L.J. 2005. Administration of injectable drugs and vaccines in goats. Proceedings of the 20th Annual Goat Field Day. American Institute of Goat Research at Langston University.

Dawson, L.J., J. Allen, and B. Alcott. 2007a. Meat goat herd health—Procedures and prevention. Meat Goat Production Handbook, edited by Gipson, T.A., R.C. Merkel, K. Williams, and T. Sahlu. Pp. 51–63, AIGR: Langston University.

Dawson, L.J., J. Allen, and B. Alcott. 2007b. Meat goat herd health—Procedures and prevention. Proceedings of the 22nd Annual Goat Field Day. Langston University.

FAO. 2005a. Selected indicators of food and agriculture development in Asia—Pacific region 1994–2004. RAP Publication, pp. 20. http://www.fao.org/docrep/008/ae937e/ae937e00.htm. Retrieved December 15, 2008.

FAO. 2005b. Goat numbers. http://faostat.fao.org. Retrieved December 15, 2008.

Grant, K. 2001. Predatory control. Proceedings of the 16th Annual Goat Field Day. Langston University.

Gundlock, D.J. Elastrator dehorning of goats. www.greatgoats.com/articles/dehorning_text.html. Retrieved December 15, 2008.

Hooper, R.N. 1998. General surgical techniques for small ruminants. Proceedings Western Veterinary Conference. Las Vegas.

Pugh, B. 2008. Predator Control. Oklahoma Basic Meat Goat Manual. Oklahoma State University, Stillwater, OK.

Pugh, D.G. 2002. Disbudding and dehorning. Sheep and Goat Medicine. Saunders Publisher, Philadelphia, PA, Pp. 221–222.

Sahlu, T., L.J. Dawson, T.A. Gipson, S.P. Hart, R.C. Merkel, R. Puchala, Z. Wang, S. Zeng, and A.L. Goetsch. 2009. ASAS Centennial Paper: Impact of Animal Science Research on U.S. Goat Production and Predictions of the Future. J. Anim. Sci. 87:400–418.

Smith, M.C. and D.M. Sherman. 1994. Goat Medicine. Dehorning and Descenting. Lea and Febiger, Malvern, PA, pp. 519–525.

Solaiman, S.G. 2007. Simply Meat Goats. Tuskegee University Publishing. Publication No. 115–1006.

Swize, R. 2007. Meat goat management. Meat Goat Production Handbook. Gipson, T.A., R.C. Merkel, K. Williams, and T. Sahlu, Eds. Pp. 25–38, AIGR: Langston University.

USDA. 2005. Sheep and goat predator loss. National Agriculture Statistics Service (NASS), Washington DC. http://usda.mannlib.cornell.edu.

# 13
# Meat Production and Quality

*K.W. McMillin, PhD*

**KEY TERMS**

Average daily gain—change in weight between two time periods divided by the days between the two time periods.

Composition—chemical nutrient or physical tissue components that comprise the entity being described.

Carcass—the base structure portion of the live animal remaining after removal of hide, head, limb extremities, and viscera during slaughter.

Conformation—the general form, shape, contour, or outline of the carcass, side, or cut that indicates the relative proportions of lean to fat and bone.

Development—progression of increased complexity of a living organism due to differential rate of change in individual body components.

Dressing percentage—proportion of the live weight that is the carcass of the animal.

Grade—a group of animals, carcasses, or meat cuts having similarly defined quality traits that estimate juiciness, tenderness, and flavor palatability or yield (cutability) traits that estimate the amount of edible meat from weight, fatness, and muscling factors.

Growth—any change in weight, size, shape, form, composition, or structure over time.

Meat—edible organs or tissues from the animal.

**OBJECTIVES**

By completing this chapter, the reader will acquire knowledge on:

- The different growth stages and patterns of tissue deposition of kid goats
- Influences on growth and body composition of kid goats before weaning
- Effects on growth and body composition of kid goats from weaning to marketing
- Appraisal of live kid goats for muscling, fatness, and structural traits
- The evaluation and characteristics of meat goat carcasses
- Meat goat traits and differences in goat meat quality
- The nutrient composition of different goat meat sources

## INTRODUCTION

The quality and quantity of goat meat is highly dependent upon production and husbandry practices for meat goats. The growth and development of goats is highly variable with genetic and environmental influences of breed, sex, nutrition, growth-promoting agents, and climate. This chapter will describe growth of goat kids, effects on growth before and after weaning, kid market goat evaluation, meat goat carcass characteristics and evaluation, goat meat traits and factors influencing goat meat quality, and nutrient composition of goat meat. A brief informative coverage of potential goat meat products is also presented in this chapter.

## GROWTH STAGES

### General Growth Changes

Growth may be defined as an increase in weight until a mature size has been reached, and maintenance is the constancy in weight or other measurement of growth. Growth is a function of the life cycle of each animal that begins with embryo fertilization and ends with death. Growth and development may be separated into two distinct processes: growth attributes and developmental changes.

Any increase in body weight, size, shape, form, composition, or structure over time may be considered to be growth. Cells are the basic unit of growth, and development is an increase in complexity of the body components that comprise the body. Usually weight is an overriding factor in determination of growth because it is more easily measured than the other body changes. Development or complexity occurs when there is a differential rate of change or distinguishing of characteristics of specific body components. In the early stages of life cycles, it may be more difficult to separate changes due to growth and changes in complexity.

Chemical components are compounds that constitute all matter. The major chemical components in the animal body are known as nutrients and are water, lipid, protein, ash (minerals), carbohydrates, and vitamins. Physical components are the tissues and organs composed of cells. Physical components of the body are often generally classified as lean, fat, bone, digestive, and neural.

The different organs and tissues grow and develop at different rates. Chronological or calendar age is measured as the number of days after parturition (birth); the physiological age is used as a guide to the relative development of the body at a given age compared with the development at a mature age. Cell types are specialized for specific functions or purposes, and the process of differentiation is the specific time when a cell or groups of cells has characteristics distinct from those of other cells or groups of cells.

Ontogeny is the progression of an individual through its life cycle. Prenatal is the time before parturition, and postnatal is from parturition to death. Puberty is the stage in the growth and development cycle when an animal is able to sexually reproduce offspring. Male goats often reach puberty at 4–8 months; female goats reach puberty in 7–10 months. The age at puberty varies between breeds, although most does may kid at 1 year of age. Some producers do not breed does during their first year, but prefer to allow the female to grow to maturity without the nutritional demands of pregnancy on the body, which may influence growth rate. It is advisable to allow the doe to be at 60–75% of her adult weight before breeding. When an animal reaches the highest degree of complexity in the body, it is considered to be mature. In a practical sense, livestock and avian species are considered to have reached maturity when a maximum body weight that remains stable without excess fat deposition has been achieved.

### Prenatal Growth

Prenatal growth occurs in three main phases:

1. The ovum phase is from fertilization to implantation on the uterine wall and is characterized by cell division through proliferative mitosis to form a zygote.
2. The embryonic phase is from implantation to tissue differentiation. After implantation, the blastocyst differentiates into the inner cell mass and the embryonic disk. The flattened plate of embryonic disk cells of the morula differentiates into the primary germ layers. The outer germ layer that surrounds the amnionic cavity is the ectoderm. The primitive streak becomes the notochord and rolls into a neural tube that differentiates into the brain, spinal cord, nerve, and epithelial structures. The middle germ layer, the mesoderm, forms somites on the sides of the neural tube that will develop into the circulatory system, mesenchyme (connective tissue and bones), and myotome muscle plates (muscles). The inner germ layer that lined the yolk sac, the endoderm, will differentiate into the digestive system and glands.
3. The fetal phase of prenatal growth is from differentiation to parturition. During this phase, tissues and organs are further differentiated, and the fetus size increases through rapid hypertrophy of cells. The greatest prenatal weight gains are in this phase and are affected by the nutritional and disease status of the doe.

### Postnatal Growth

The factors that affect postnatal growth and development are genetic potential and the influence of environment and nutrition to allow the animal to reach its genetic potential. The different body components have relative size changes during postnatal growth and development. The head is proportionately larger than other body parts at birth, so early postnatal growth is for size and shape of the body before the limbs begin to lengthen. The pattern of postnatal development follows the general pattern of growth priorities, generally in the order of nervous system, organs, digestive system, bones, muscles, and adipose (fat) tissues. The organs and tissues have priority for nutrients in that order so bone deposition precedes muscle development, and fat is deposited after bone and muscle growth is almost completed.

Postnatal growth is marked by development of neural coordination, digestive system development, and changes in bone, muscles, and fat. Long bones have a progressive ossification and growth pattern. Muscles change in size and form, with individual muscle fiber cells having an increase both in length and in sectional area (radially) during muscle development. Adipose tissue cells that are formed prenatally are influenced in postnatal growth by the nutrition of the animal and the stage of growth. The number of fat cells increases early in postnatal growth before increases in the volume or size of each fat cell, which can occur later, even after maturity.

Locations of body development are anterior to posterior and ventral to dorsal. Fat (channel fat or kidney, pelvic, and heart fat) deposition occurs around organs first, followed by intermuscular, subcutaneous, and intramuscular (marbling). Growth and development rates are governed by both genetic potentials and environmental factors, primarily availability of nutrients and climatic conditions. The composition of body parts is variable, with bone tissue containing high levels of minerals, adipose tissue containing high amounts of lipid, and muscle tissues containing high levels of moisture. Bone and muscle cells have about the same protein composition.

### Measurement of Growth

Growth in animals is usually determined by producers with easily obtained measurements, such as age, length, height, weight, width, or circumference rather than measuring specific physical and chemical components as is done by scientists. The rate of growth is expressed in terms of weight or gain per age (weight per day of age), weight gain per time period, weight change per amount of time (average daily gain), height at a specific age, time to reach a specified weight, or other easily compared measurements that provide common points of reference for growth and development of different animals. Growth is most often determined as gain in weight. The average daily gain is the change in weight between two time periods. Efficiency of growth and development is often expressed as the output divided by the input, with weight of gain per weight of feed input a common measurement.

## PRE-WEANING PERFORMANCE

Weaning is defined as the time when animals are no longer nursing from the doe. Weaning may occur when the doe no longer will allow the kid goat to nurse, but more often, it is at an earlier time because the producer will separate kid goats from the breeding herd for management or marketing purposes. High pre-weaning average daily gains reflect the genetic potential of the kid goat and the mothering ability of the doe.

Heavier body weight and faster growth rate are among the most important pre-weaning traits. Birth weight of Boer kids ranges from 6.6 to 8.8 pounds (3–4 kg) with male kid goats weighing about 1.1 pounds (0.5 kg) heavier than female kid goats. The weight of kid goats at weaning depends upon the age and methods of weaning, but it is often 40–60 pounds (18–27 kg). The growth of kid goats from birth until weaning is highly dependent upon the availability of milk from the doe and the types and amounts of feeds available when the young goat begins to eat solid materials.

Some sources indicate that goats can be successfully weaned and have desired growth when the kid goats reach 2.5 times their birth weight and are consuming at least 1 ounce (0.06 lb, 30 g) of solid feed each day. Successful weaning can also be accomplished when kid goat weight is 20 pounds (9 kg) or age is at least 8 weeks. If the goat to be weaned is not old enough to have a digestive system adapted to convert forage or grain feeds into the nutrients needed for growth and development, early weaning will result in higher mortality or slower growth rates. The physiological and other stresses when the goat is weaned are termed weaning shock. Male kid goats are more susceptible to weaning shock than females, and healthy kid goats experience less shock than those with a disease or infection such as coccidiosis. Weaning shock may be more severe for kids weaned prior to 4 or 6 weeks, but the temporary weight loss is usually overcome by subsequent compensatory growth.

### Pre-Weaning Growth Rates

It may be expected that healthy kid goats grow at least one-third to one-half pound (150–225 g) per day during the first 3 months of age. Comparisons in growth rate or of average daily gain of kid goats must consider the factors that influence growth. Kid goats from the smaller breeds will tend to grow slower than kids from the larger breeds, and kid goats from earlier maturing breeds will have a faster growth rate than kids from later maturing breeds. Buck kid goats grow 10–25% faster to weaning than doe kids (Warmington and Kirton, 1990). Kids from triplet litters will tend to grow slower than kids born as single offspring. Heavier birth weights will result in heavier kid goats even if the rate of growth is the same as for kid goats with lighter birth weights. Male kids will have heavier weight deposition than female kids. Age of the doe will influence growth. Kid goats from yearling does are expected to grow more slowly than those from older does that have more highly developed and more productive mammary systems.

Slow pre-weaning growth may be due to parasitism, worms or coccidia, insufficient milk supply, poor quality milk replacer for bottle-fed kid goats, or a bacterial or viral disease. It may be expected that after the first 3 or 4 days following birth, kid goats will consume 1–2 quarts (1–2 liters) of milk daily. Improved nutrition of lactating does can increase milk yield and thus increase weight gains in the kid goats. Growth is higher with ad libitum intake rather than limited amounts of milk.

**Supplementation during Pre-Weaning Growth**

Supplemental nutrients may compensate for a lack of nutrients by the doe or provide additional nutritional value through concentrates or grazing high quality forage. The addition of nutrient value when the doe is providing sufficient milk is known as creep feeding or, in the case of forage, creep grazing. Sometimes creep grazing, which is access to high quality pasture by kid goats and prevention of access by the does, is used to provide the kid goat with added nutrition.

Creep feeding or creep grazing usually increases growth rate pre-weaning and often improves weight gain immediately after weaning because the goat has been prepared to consume nutrients from sources other than milk and so is more prepared for the shock or stress caused by weaning.

Spanish, Boer × Angora, and Boer × Spanish kid goats were fed milk replacer twice daily until weaning at 8 weeks of age and had ad libitum access to commercial goat starter (20% CP) beginning at 3 weeks of age. The results indicated that Boer crossbred kid goats gained more body weight (0.17 pounds, 74 g) than Spanish kid goats (0.13 pounds, 60 g) from weeks 3–8 (Luo et al., 2000). Pre-weaning growth will improve by offering creep feed that is high in crude protein (14–18%) and energy and by providing access to high quality hay or pasture.

The ability of the kid goat to digest roughage or grain feeds is determined by development of the rumen. Kids that receive lower amounts of milk may have rumens that develop at an earlier age to allow them to eat on their own and compensate for a lack of milk. It is usually considered that kid goats should be weaned from milk by 10–12 weeks of age and should weigh a minimum of 30 pounds (14 kg) when weaned at this age. Kid goats receiving adequate nutrition will often have weight increases of 10–15 pounds (5–7 kg) per month for the first 3 months of age. During the pre-weaning period, goats can grow up to 0.5 pound (227 g) per day from grass with some supplementation. Average feed conversion efficiency can be from 0.1–0.125 pounds or kg of gain per pound or kg of feed for goats (Van Niekerk and Casey, 1988).

**Comparison of Pre-Weaning Growth Rates**

A standard method of comparison of pre-weaning growth rate is to compare the weight at 90 days of age. This weight can be determined by adding the birth weight to the average daily gain over the period of growth of 90 days. This allows direct comparison of growth between kid goats weaned at 6 weeks of age (42 days) and those 10 weeks old (70 days). Adjustments must be made to correct for the development that occurs at different ages of pre-weaning growth. Correction factors are based upon litter size, age of doe, and sex of the kid goat.

After the average daily gain (ADG) is determined, the 90-day weaning weight is calculated as the ADG × 90 plus the birth weight. When birth weight is not available, ADG cannot be determined over the entire pre-weaning growth period. In the absence of birth weight records, the previous equations can be replaced by 90-day weaning weight = (weaning weight/weaning age) × 90. Adjustments are made to 90-day weights because litter size and age of doe can affect weaning weight. On average, weaning weights decrease as litter size increases, and young does wean lighter kids than mature does. Multiplying the 90-day weight by the appropriate correction values (Table 13.1) will result in the adjusted 90-day weight.

Recent data showed that buck kid goats had heavier birth weights at 7.3 pounds (3.3 kg) than doe kid goats at 6.7 pounds (3.0 kg) while single, twin, and triplet kid goats had birth weights of 7.9 pounds (3.6 kg), 7.1 pounds (3.2 kg), and 6 pounds (2.7 kg), respectively. Buck kid goats gained weight faster than doe kid goats (0.4 pounds or 183 g) compared with (0.34 pounds or 154 g)). Pre-weaning ADG is higher for Boer × Kiko, and Kiko × Kiko crosses (0.40 pounds/day or 180 g/d)) and lower for Spanish × Spanish, Kiko × Spanish, and Boer × Boer (0.34 pounds or 155 g/day) crosses. Buck kid goats were heavier at weaning than doe kid goats (35 pounds or 16.1 kg compared with 30 pounds or 13.6 kg). Single, twin, and triplet kid goats had weaning weights of 39 pounds (17.7 kg), 33 pounds (15.0 kg), and 26 pounds (11.9 kg), respectively (Browning, 2007).

**Table 13.1** Correction factors for 90-day weaning weight due to litter size born and weaned, dam age, and kid goat sex.

| Effect of litter size | | |
|---|---|---|
| Number born | Number weaned | Correction factor |
| 1 | 1 | 1.00 |
| 2 | 1 | 1.04 |
| 2 | 2 | 1.18 |
| 3 | 1 | 1.08 |
| 3 | 2 | 1.23 |
| 3 | 3 | 1.27 |
| Effect of dam age, years | | |
| 1 | | 1.10 |
| 2 | | 1.09 |
| 3 and older | | 1.00 |
| Sex of kid goat | | |
| Buck | | 1.00 |
| Doe | | 1.11 |
| Wether | | 1.08 |

Source: Adapted from Browning, 2007.

These data show relative differences that a producer might expect with sex, offspring numbers, and breeds, but they should not be interpreted as absolute values for growth that all kid goats will exhibit. Many genetic and environmental factors will influence the pre-weaning weights and growth parameters. Direct heritabilities from two Boer goat herds were 0.33 and 0.36 for birth weight and 0.27 and 0.60 for weaning weight (Schoeman et al., 1997). Heterosis due to crossbreeding was observed in the growth of females from 10–210 days of age while it was not evident in males until the age of 90 days.

## POST-WEANING PERFORMANCE

Strict attention to the behavior, health, and nutrition of kid goats must be given during and immediately after weaning. Kid goats must have access to feedstuffs and potable water and be accustomed to receiving nutrition from sources other than milk to maintain desired growth and development rates. Post-weaning growth rate is less than pre-weaning growth, probably due to a lower assimilation of nutrients from solid feeds compared with milk (Warmington and Kirton, 1990). Growth after weaning depends more on the energy ingested or the level of food intake and less on the method of feeding as ad libitum, restricted grazing, intensive stall feeding, tethering, or other systems. A concentrate feed that is balanced in energy and protein is more efficient than whole grain for growing goats, but the physical form of the concentrates as whole, rolled, chopped, or other has little effect on growth and carcass traits. Deficiencies in energy from inadequate feed intake or low-quality feed may cause decreased kid growth in size or weight, delay in puberty, lowered fertility, depressed milk production, and a lowered resistance to disease and parasites. Energy requirements are different with age, body size, growth, pregnancy, lactation, weather, physical activity, and stress of meat goats.

The primary measurements of post-weaning growth are weight gain per time period and weight per a specified day of age. Post-weaning ADG is a measure of the individual animal's performance and the type of management practices in raising the goat.

The rate of tissue accumulation and the growth rate both decline as maturity is approached and are also influenced when does are pregnant or gestating. Lean tissues grow proportionately faster than fat tissues early in post-weaning growth, and fat tissues grow proportionately faster than lean tissues later in post-weaning growth closer to maturity so fat:lean and lean:bone ratios increase as goats mature (Warmington and Kirton, 1990).

### Post-Weaning Growth Restriction and Compensatory Growth

Animals may not grow to reach their genetic potential if stressor conditions such as nutritional restrictions or environmental influences interfere with normal growth rates. After removal of a nutritional deficiency or other growth-inhibiting conditions, the animal may increase the rate of growth, or compensate, to counter the previous growth limitations. This compensatory growth is often observed as accelerated growth rate with adequate nutrition following a period of feed restriction or nutrient deficiency. The increased growth may be due to increased dry matter intake (DMI), improved feed efficiency (FE), and/or enhanced endocrine and biochemical reaction functions. A restriction in growth early in the animal's life may result in a permanent stunting of growth and development, particularly if organs do not grow and develop normally or bones do not grow adequately to form the skeletal structure for muscle and fat deposition. Generally, the briefer the period of restriction and the older the animal at the time of restriction, the greater is the chance for compensatory growth to result in a similar mature weight as if no restriction had occurred.

The ADG may be misleading if there has been compensatory growth during the period that the growth of the animal is evaluated. The weight compensation after removal of the stressor may be equal to or part of the weight gain that would be observed in unstressed animals. Additional factors may affect the growth rates, compensatory growth, and mature body weights for specific animals. Post-weaning growth can also be influenced by the genetic potential of the animal and environmental conditions. Management practices that may alter growth and development include breed selection, weaning age, castration, climate control, type and availability of feedstuffs, stocking densities and stocking rates, and mixed grazing with other species.

### Post-Weaning Growth Rate with Goat Gender

At 7 months of age, bucks weigh about 88–110 pounds (40–50 kg), and doelings weigh about 77–100 pounds (35–45 kg). Yearling bucks weigh 110–155 pounds (50–70 kg) and yearling doelings weigh 100–145 pounds (45–65 kg). Mature weights for bucks and does are 200–285 pounds (90–130 kg) and 175–220 pounds (80–100 kg), respectively. Breed differences, genetics within breeds, nutrition, health and disease, breeding age and method, and management systems will cause differences in growth rates and body weights at any given time.

Buck kid goats commonly grow up to 30% faster than does during the post-weaning period. Intact male kid goats grew to heavier weights (61.3 pounds, 27.8 kg) than castrated males (48.3 pounds, 21.9 kg) and females (46.1 pounds, 20.9 kg) after feeding 0.5 pound of concentrate on Argentine Bahia grass pasture (Johnson et al., 1995a). The ADG on Marshall ryegrass pasture for 8 weeks was 0.31 pound (139 g) for buck kid goats compared with 0.15 pound (66 g) for wether kid goats (Solaiman et al., 2006). In contrast, buck, doe, and wether Spanish kid goats had ADG of 0.11 pound (49 to 51 g) when fed alfalfa and grass hay diets supplemented with concentrate at 1.5% body weight (Wildeus et al., 2007).

### Post-Weaning Growth Rate with Goat Breed

Growth rate is closely related to mature breed size (Warmington and Kirton, 1990). Boer × Spanish wether kid goats were reported to have higher DMI and 30% higher ADG than Spanish goats, but 23% lower ADG than Boer goats at 7 months of age (Cameron et al., 2001). A producer must be careful in comparing the growth of animals of different breeds under different conditions, even if the environmental differences are slight, because growth and development may often be affected by a genetic by environment effect. Heterosis or hybrid vigor will also have an influence. Examples of the types of growth of some meat goat breeds are in Table 13.2. These growth rates should not be interpreted as absolute but used as a guide to the differences that can occur with different breeds. Generally, the variations in growth rate among the different animals within a single breed are as large as the differences in average rates of growth among different breeds.

### Post-Weaning Growth Rate with Diet

#### Growth with Forage

Goats as ruminants can consume forage, browse, and concentrate feedstuff. Increased levels of poultry litter to replace part of the concentrates in 80% concentrate and 20% millet hay diets decreased ADG in each period of growth for Spanish and Spanish × Boer kid goats compared with control diets with no poultry litter (Negesse et al., 2007).

Forage with higher nutritional value will provide greater rates of growth than lesser feedstuff. Stocking rates and pasture rotation will affect quality and quantity of forage. Goats may selectively choose specific kinds and types of supplements and forage, which will affect growth rates, feed efficiencies, and feed wastage. Increased energy in the diet, particularly for goats on hay or pasture, will generally increase growth rate, but may also increase the deposition of fat and decrease lean to fat amounts, depending upon the stage of growth of the meat goat. Goats are generally not fed high-concentrate diets because of cost and composition considerations. Additionally, goat kids that exceed a market weight of 80 pounds (36 kg) will receive a price penalty at marketing because the buyers do not have markets that desire goats weighing more than 80 pounds.

Post-weaning ADG for goat kids fed warm season grass pastures (bahiagrass), mimosa browse, and cool season grass pastures (annual ryegrass) were 0.10 pound (46 g), 0.18 pound (83 g), and 0.31 pound (141 g), respectively (Solaiman et al., 2006). Condensed tannins in some forage, such as *Sericea lespedeza*, may limit consumption and thus performance of meat goats. Supplementation of lespedeza pastures with polyethylene glycol increased ADG of Boer × Spanish does 4 months of age (Merkel et al., 2003). Goats have faster growth rates on improved pastures compared with native pastures.

#### Growth with Supplementation of Forage

Crossbred Boer kid goats achieved heavier weights when supplemented with concentrates compared with range only

**Table 13.2** Some examples of different kid goat growth rates from different meat goat breeds.

| | | Initial growth phase | | Final growth phase | |
|---|---|---|---|---|---|
| Conditions | Breed | ADG, pounds (g) per day | Efficiency, weight gain per weight of feed | ADG, pounds (g) per day | Efficiency, weight gain per weight of feed |
| Wether kid goats, 4 to 5-1/2 months old, fed 16% CP pelleted feed for 4 months (Pinkerton et al., 2001) | Tennessee Meat Goat | | | 0.29 (132) | 0.13 |
| | Boer × Spanish | | | 0.46 (209) | 0.19 |
| | Boer × Nubian/ Tennessee Meat Goat | | | 0.34 (154) | 0.12 |
| | Boer × Nubian | | | 0.32 (145) | 0.11 |
| | Boer × Nubian/Alpine | | | 0.35 (159) | 0.13 |
| | Boer | | | 0.37 (168) | 0.13 |
| Wether kid goats, 4 months old, fed 75% concentrate for first and second 12-week periods (Urge et al., 2004) | Alpine | 0.15 (68) | 0.10 | 0.13 (59) | 0.08 |
| | Angora | 0.16 (72) | 0.15 | 0.11 (50) | 0.10 |
| | Boer | 0.20 (91) | 0.13 | 0.14 (64) | 0.08 |
| | Spanish | 0.14 (62) | 0.11 | 0.05 (22) | 0.04 |
| Male and female kid goats, fed 18% CP and 65% ME diets from weaning to 6 months of age and 9–13 months of age (Dzakuma et al., 2004) | Tennessee stiff-legged | 0.18 (83) | 0.17 | 0.08 (38) | 0.10 |
| | Spanish | 0.13 (58) | 0.12 | 0.07 (34) | 0.09 |

diets. ADG was higher (0.1 pound, 45.3 g) with supplementation at 2.5% of body weight for Boer cross kid goats on fescue pastures compared with no supplementation (0.03 pound, 13.6 g) (Andries et al., 2007).

Supplementation of Sudan grass for crossbred Boer × Spanish wethers with leucaena tropical legume gave slightly higher ADG than alfalfa, lablab, or desmanthus legume diets (Kanani et al., 2006). The ADG of 7-month-old Spanish does and bucks was higher with alfalfa hay and 16% CP supplement (0.14 pound, 62 g) than grass hay and the 16% CP supplement (0.08 pound, 37 g) in a 102-day trial. A repeat of this study with 10-month-old Boer and Boer-cross wether goats in an 84-day trial gave similar results, with ADG of 0.35 pound (158 g) for alfalfa hay with 16% CP supplement and 0.26 pound (119 g) for grass hay with 16% CP supplement (Wildeus et al., 2007). Spanish × Boer kid wether goats supplemented with 0.5% body weight of corn on native rangeland had 61% higher ADG than unsupplemented goats, and goats on improved legume and Amaranthus retroflexus pastures required 1% body weight supplementation with corn before showing a 31% increase in ADG compared with nonsupplemented animals (Muir and Weiss, 2006). Endophyte-infected tall fescue hay lowered growth rates by 33% over 8 weeks compared with an orchard grass hay diet (Browning et al., 2007).

The ADG and gain efficiency were greater with 75% concentrate (0.16 pound [73 g] and 0.11 pound of gain per pound of feed) compared with 50% concentrate (0.12 pound [55 g] and 0.09 pound of gain per pound of feed) in the first 12 weeks, but they were higher for 50% concentrate (0.14 pound [65 g] and 0.10 pound of gain per pound of feed) than for 75% concentrate (0.11 pound [49 g] per day and 0.08 pound of gain per pound of feed) in the second 12 weeks (Urge et al., 2004).

Table 13.3 shows some relative comparisons of the growth and efficiency of weight gain on different forage and on forage when supplemented with grain.

#### Growth with Intake Restriction

Some studies have indicated that while restricting intake to less than 90% of the National Research Council recommended values will decrease live weight gain and feed

**Table 13.3** ADG in pounds (g) of kid meat goats with different forages and supplements in the diet.

| | Dietary Treatments | | | |
|---|---|---|---|---|
| Kid goat descriptions | Millet pasture | Bermuda grass pasture | Bermuda grass pasture + 0.5 pounds (0.23 kg) corn per head per day | Bermuda grass pasture + 1 pound (0.45 kg) corn per head per day |
| 4-month-old Boer × Spanish wethers, 105-day trial | 0.16 (35) | 0.17 (37) | 0.19 (42) | 0.26 (57) |
| 4-month-old Spanish wethers, 105-day trial (Nuti et al., 2000) | 0.16 (35) | 0.15 (33) | 0.16 (35) | 0.18 (40) |

| | Complete feed supplement | Intake limited supplement | Sorghum-sudan hay (SS) | SS with supplement | SS with intake limited supplement |
|---|---|---|---|---|---|
| 5-month-old Boer × Spanish does, 63-day trial (Payne et al., 2006) | 0.34 (152) | 0.27 (123) | 0.05 (22) | 0.40 (181) | 0.38 (172) |

| | Concentrate (40% protein pellets, 40% soybean hulls, 20% Bermuda grass) hay | Bahia grass pasture with 0.33 pounds (150 g)/head/day protein pellets | Mimosa browse with 0.22 pounds (100 g)/head/day cracked corn |
|---|---|---|---|
| Boer cross wether kid goats, 14-week trial (Solaiman et al., 2006) | 0.26 (117) | 0.11 (49) | 0.20 (90) |

| | 50% concentrate | 70% concentrate | 90% concentrate |
|---|---|---|---|
| Boer cross goats, 126-day trial (Ryan et al., 2007) | 0.21 (97) | 0.23 (103) | 0.20 (90) |

| | Alfalfa hay | 18% CP concentrate | Hay first 45 days, then concentrate diet |
|---|---|---|---|
| 4-month-old Boer × Spanish intact males, 90-day feeding trial (Lee et al., 2008) | 0.09 (41) | 0.30 (134) | 0.14 (65) |

efficiencies due to a restricted nutrient intake, limit feeding to 92–98% of the National Research Council recommendation will provide the same or increased live weight gain and increased feed efficiency as 100% of the dietary recommendation. There also appears to be decreased heat production with 92–98% feeding compared with less than 90% or 100% feeding of the recommended amounts (Galyean, 1999).

#### Growth Promoters

Some trace minerals may enhance growth. Addition of 100 mg copper daily increased ADG in Boer × Spanish goats without causing copper toxicity, presumably by enhancing the immune responses (Solaiman et al., 2007).

There have been few studies on the effects of growth-promoting agents, but goats would appear to be similarly influenced as the other ruminant livestock species, sheep and cattle. Lasalocid increased weight gain by 20 g per day in kid goats after weaning at 50 days of age by increasing propionate and decreasing butyrate concentrations in the rumen (Hadjipanayiotou et al., 1988).

## CARCASS CHARACTERISTICS

### Dressing Percentage

Dressing percentage is an important determinant of the potential yield of meat from an animal. Dressing percentage is the carcass weight divided by the live weight times 100. Often the hot carcass weight, the weight taken after skinning and evisceration and before chilling, is used to calculate dressing percentage. The chilling process removes heat from the carcass but also removes moisture, so the shrinkage from carcass chilling may be 3–5% of the initial hot carcass weight. Dressing percentage will vary with the type of slaughter procedures. Leaving the cannon bones (trotters), kidney and pelvic fat, and skinned head on the hot carcass will increase the dressing percentage. Goats that are not withheld from feed before slaughter will have a lower dressing percentage than goats that only had access to water overnight before slaughter. Dressing percentages of kid and yearling male and female meat goats with different body conformations were relatively constant at 48–52% when the goats were withheld from feed overnight before sacrifice (McMillin et al., 1998).

Dressing percentages of kid goat carcasses with the head on, the trotter on, and kidney and heart fat left inside will be 50–55%. Removal of the head will result in 5–7% less dressing percentage. The dressing percentages of kid goats of most species will be 42–48%. Carcasses with more internal or external fat will have higher dressing percentages, so carcasses from older goats will usually have slightly higher dressing percentages than kid goats unless the older animals are cull breeding stock that are very thin.

Dressing percentages of Spanish doe and buck kid goats were higher after alfalfa hay and supplement diets (52.9%) than on grass hay and supplement diets (50.4%), with the same trends observed with yearling Boer and Boer × Spanish wethers (Wildeus et al., 2007).

### Carcass Composition

The physical and chemical composition of the carcass will determine the value and subsequent desirability of goat meat. Does contain less bone in their carcass than bucks and yield more edible tissue at the same carcass weight. Does also have less muscle than males at most weights because they have a greater tendency to deposit fat (Johnson et al., 1995b). Mature goats can contain 30% or more carcass fat, but usually have less fat unless on a high-energy diet. Concentrate feeding or concentrate supplementation increases carcass fat deposition compared with browsing or grazing (Warmington and Kirton, 1990). Breed, diet, age, sex, weight, growth rate, and other factors will cause differences in the chemical composition of moisture, fat, protein, and ash and the physical composition of lean, fat, and bone in meat goat carcasses (Table 13.4). Fat is deposited in visceral depots rather than carcass depots. The low proportion of subcutaneous fat on goat carcasses makes subcutaneous fat thickness measurement a poor criterion for meat yield (Webb et al., 2005). The relatively higher proportion of kidney and pelvic fat in goat carcasses makes this factor more important in determining meat yield, particularly since many goat carcasses are sold with the kidney and pelvic fat remaining with the carcass.

Data from the 9–11 rib section showed similar trends in chemical composition as chemical composition of the half carcass, except for protein content (Solaiman et al., 2006). The body chemical composition of three-fourths Boer one-fourth Saanen male goats weighing 20–35 kg could be estimated ($r^2 = 0.94$) by determining the combined chemical composition of the organs, blood, and 9–11 carcass rib section. The 9–11 rib section was accurate to estimate carcass fat percentage (Fernandes et al., 2008).

### Carcass Evaluation

#### Factors Affecting Meat Yield

Carcasses of meat animals are often evaluated to determine the relative ratios of lean, fat, and bone, and to provide an estimation of the palatability of the meat to be produced. Beef, lamb, and pork carcasses have official United States Department of Agriculture grades that segregate carcasses according to yield or cutability characteristics and quality or palatability characteristics. There has been insufficient data collection of goat carcass traits to estimate the yield or palatability of goat meat. Yield of meat, usually expressed as the amount of lean or cuts that have been closely trimmed to remove some of the fat, is highly related to the weight of the carcass, the degree of muscling relative to bone, and the amounts of subcutaneous external fat and internal kidney and pelvic (KP) fat. Heavier carcasses would be expected to result in more meat than lighter carcasses, and carcasses with heavier muscling would give more lean meat than carcasses with less muscling. Meat cuts are sold with a designated amount of fat or with all fat removed. The KP fat has no functional value for meat cuts, so greater amounts of subcutaneous fat and KP fat reduce the expected lean proportion of carcasses.

**Table 13.4** Chemical and physical composition of meat goat carcasses.

| | Chemical composition, % | | | |
|---|---|---|---|---|
| Kid goat description | Moisture | Fat | Protein | Ash |
| Boer × Spanish | 56.1 | 3.9 | 18.6 | 20.3 |
| Spanish | 57.6 | 3.0 | 16.1 | 19.8 |
| Boer × Angora | 56.4 | 3.7 | 18.0 | 20.6 |
| (Cameron et al., 2001) | | | | |
| Boer Cross Wether Kid Goats | | | | |
| Bahia grass pasture | 70.3 | 8.0 | 20.6 | 1.1 |
| Mimosa browse | 70.0 | 9.1 | 20.0 | 0.9 |
| Concentrate | 64.5 | 13.4 | 21.3 | 0.9 |
| (Solaiman et al., 2006). | | | | |
| Hay for 45 days | 77.1 | 1.3 | 20.8 | 1.3 |
| Concentrate for 45 days | 74.7 | 2.7 | 21.3 | 1.4 |
| Hay and concentrate for 45 days | 75.8 | 2.0 | 20.1 | 1.6 |
| (Lee et al., 2008) | | | | |

| | Physical composition, % | | |
|---|---|---|---|
| | Lean | Fat | Bone |
| Intact male | 71.2 | 7.4 | 21.4 |
| Wether | 66.2 | 11.0 | 22.8 |
| Doe | 67.9 | 13.0 | 19.2 |
| (Johnson et al., 1995b) | | | |
| Boer × Spanish, feedlot | 57.8 | 15.71 | 26.5 |
| Boer × Spanish, range | 55.8 | 7.34 | 36.9 |
| Spanish, feedlot | 57.6 | 13.4 | 27.6 |
| Spanish, range | 55.3 | 8.24 | 36.5 |
| (Oman et al., 1999) | | | |
| Boer × Spanish, 80% concentrate diet | 57.8 | 15.7 | 26.5 |
| Spanish, 80% concentrate diet | 58.4 | 13.6 | 27.9 |
| Spanish × Angora, 80% concentrate diet | 55.0 | 19.4 | 25.5 |
| Angora, 80% concentrate diet | 51.6 | 22.6 | 25.7 |
| (Oman et al., 2000) | | | |
| Concentrate diet | | | |
| Boer × Spanish wethers | 57.7 | 16.2 | 26.1 |
| Spanish wethers | 57.6 | 13.5 | 28.9 |
| Boer × Angora wethers | 55.7 | 16.3 | 28.0 |
| (Cameron et al., 2001) | | | |

Even though there are not official grades for goat carcasses, it is useful to classify goat carcasses into groups for marketing and pricing (McMillin and Pinkerton, 2008). Selection classifications (criteria) give an indication of the ratio of muscle to fat and bone and the relative amount of meat that may be obtained from the carcass:

- Goat carcasses in selection 1 have a superior meat conformation and give the highest meat:bone and lean meat yields. The carcass has thick leg muscling, bulging outside leg muscling, full muscling along the back and through the loin and rib, and thickness in the loin and leg junction and shoulder.

- Selection 2 carcasses have moderate muscling, with slightly thick and slightly bulging leg muscles, slightly full *Longissimus dorsi* muscling in the loins and ribs, slightly thick shoulders, and lack of muscling in the loin and leg junction.
- Selection 3 carcasses have deficient muscling and produce low yields of lean meat relative to the body size and weight. The carcasses are narrow and lack depth, with moderate leg muscling, shallow rib and loin muscling, and slightly thin shoulder muscling. The deficient muscling may contribute a shrunken appearance to the carcass and a well-defined depression along the top of the back and at the junction of the loin and leg.

The side and rear views of carcasses representing the midpoint in each selection classification are shown in Figure 13.1.

The weight of carcasses can be determined relatively easily, often before chilling to give the hot carcass weight. The back muscle (*Longissimus dorsi*) is highly related to the degree of muscling in the remainder of the carcass. The measurement of the cross-sectional area of this loineye or ribeye muscle, 12th rib in beef carcasses, forms the basis for estimating the degree of muscling. Lamb carcasses and pork carcasses are seldom split into forequarters and hindquarters commercially. The leg conformation and muscling score are subjectively evaluated to provide an indication of the relative amount of muscling on lamb and pork carcasses, respectively. Measurement of the loineye area at the 10th rib in pork carcasses is sometimes used to provide a more accurate measure of percentage muscle in the carcass. Goat carcasses are not separated into forequarters and hindquarters. The cross-sectional area of the *Longissimus dorsi* in kid goat carcasses and many adult goat carcasses is sufficiently small to result in inaccurate measurement. The Institutional Meat Purchase Specifications for Fresh Goat (USDA, 2001) use a carcass conformation score to gauge the degree of muscling in the goat carcass.

Carcass yield of goat carcasses is highly related to the amount of KPH fat and subcutaneous fat. KPH can be estimated by visual evaluation of the internal carcass fat, as shown in Figure 13.2. Deposited fat around the heart is different in goats as compared to cattle and sheep; therefore, KP is estimated in goats. The KP fat is commonly removed after evisceration in lamb and pork carcasses, but it is sold as part of beef and goat carcasses.

The measurement of subcutaneous fat covering the *Longissimus dorsi* on lamb or beef carcasses at the 12th rib (back fat) or the 10th rib for pork carcasses (rib fat) allows estimation of the proportion of fat to muscle and the relative amount of fat that will be trimmed from the carcass in making cuts. Goats do not generally deposit fat over the *Longissimus dorsi* muscle unless nutritional energy levels are much higher than the need for growth and maintenance. Fatness over the ribs and behind the shoulder is used as a relative indicator of waste fat that will be trimmed from the carcass and result in lower yields of lean meat.

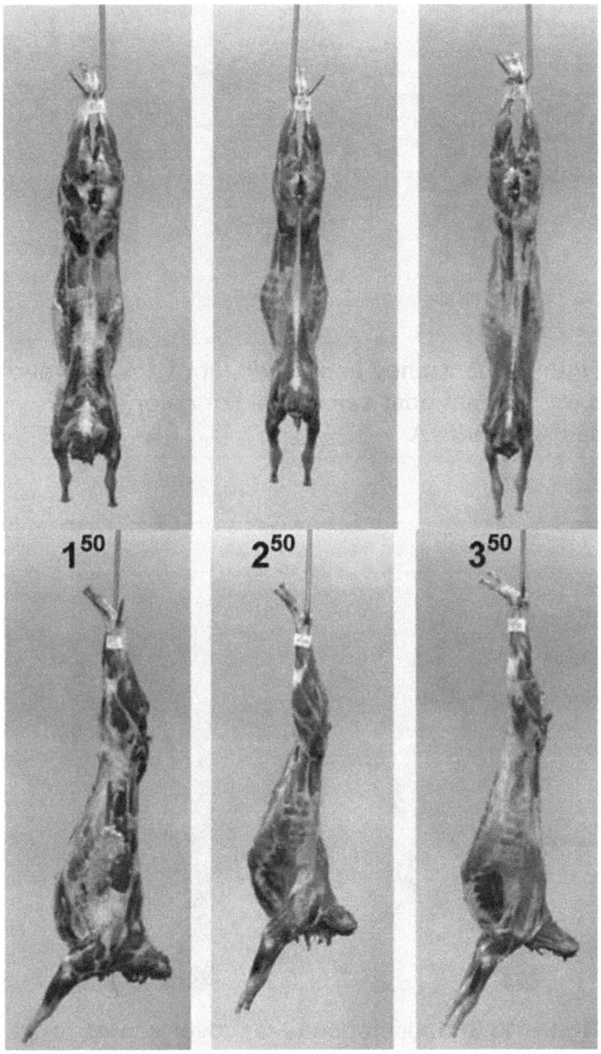


**Figure 13.1** Rear and side views of meat goat carcasses depicting the midranges of selection conformation classes 1, 2, and 3. For color detail, see Appendix A.

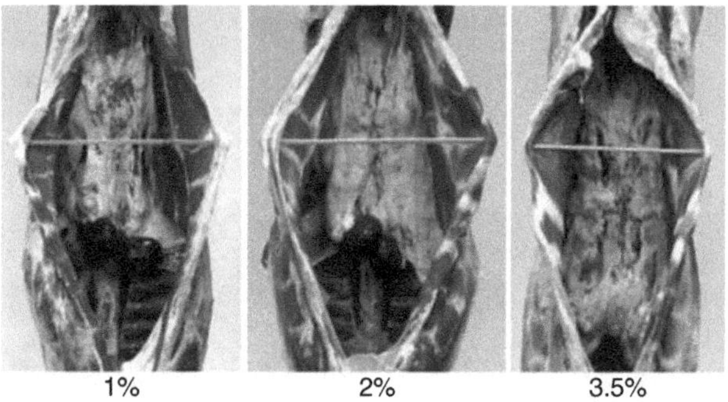


**Figure 13.2** Kidney and pelvic fat of 1%, 2%, and 3.5% in meat goat carcasses. For color detail, see Appendix A.

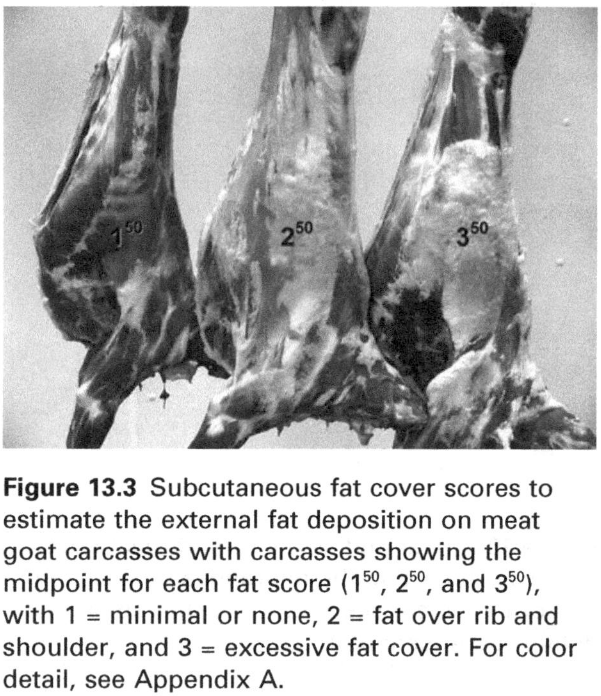


**Figure 13.3** Subcutaneous fat cover scores to estimate the external fat deposition on meat goat carcasses with carcasses showing the midpoint for each fat score ($1^{50}$, $2^{50}$, and $3^{50}$), with 1 = minimal or none, 2 = fat over rib and shoulder, and 3 = excessive fat cover. For color detail, see Appendix A.

A subjective estimation of the subcutaneous fat covering is made for goat carcasses rather than an objective linear measurement because there is little or no fat deposition over the *Longissimus dorsi* muscle on the back except when goats are undesirably over fat, the fat deposition pattern is not uniform over the body, and it is difficult to accurately measure the thickness of the fat covering. The three degrees of fat cover scores are in Figure 13.3.

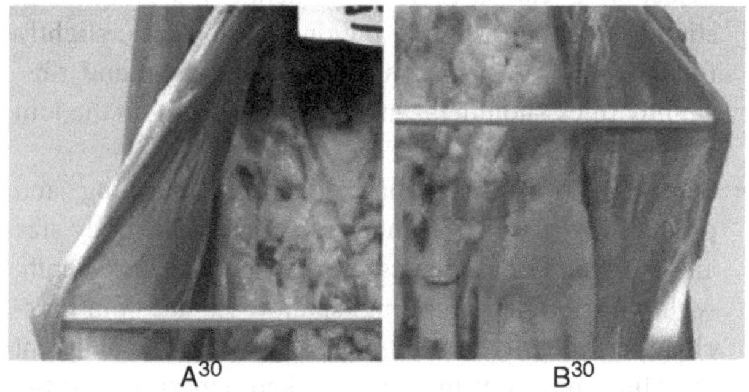


**Figure 13.4** Flank lean color indicates relative maturity of carcasses, with darker color indicating more advanced physiological maturity. Shown are scores of $A^{30}$ and $B^{30}$, with each color group ranging from 0 to 100. Color groups C, D, and E are associated with goats older than yearlings and are not shown. For color detail, see Appendix A.

#### Factors Affecting Meat Quality

Quality parameters are estimated to group carcasses by anticipated palatability. The youthfulness of beef and lamb carcasses is determined by bone structure and muscle color. Pork carcasses are evaluated for acceptable color, firmness, and texture. The relative physiological age of goats can be determined by the relative degree of ossification of the leg bones and the color of the flank muscles in goat carcasses similar to the determination of physiological maturity in lamb carcasses. The amount of intramuscular fat (marbling) is also subjectively scored in beef carcass *Longissimus dorsi* muscles to provide an indication of juiciness and flavor. Goat meat customers do not want to purchase goat carcasses with fat.

Muscles with higher amounts of pigment and darker color are associated with increased animal age. Flank lean color gives a relative indication of the physiological age of the animal. The flank muscle is one of the few muscles that can be easily observed on a goat carcass. Examples of the two lightest color scores are shown in Figure 13.4.

### Carcass Fabrication and Cuts

Primal cut proportions of carcasses are more influenced by fabrication method than by other factors (McMillin et al., 1998). Buck kid goats had higher shoulder and lower leg and loin percentages compared with doe and wether kid goats (Wildeus et al., 2007), which is representative of the

**Table 13.5** Percentage primal cut yields of meat goat carcasses.

| | Primal cut, % | | | | | | |
|---|---|---|---|---|---|---|---|
| Kid goat description | Shoulder | Breast | Rib | Loin | Leg | Shank | Flank |
| Spanish kid and yearling goats (McMillin et al., 1998) | 20–22% | | 7–9% | 6.5–8% | 23–26% | | |
| Boer × Spanish | 25.7 | 11.5 | 9.8 | 9.5 | 29.8 | 8.7 | 7.9 |
| Spanish | 24.0 | 10.8 | 10.6 | 9.5 | 30.9 | 8.9 | 4.6 |
| Boer × Angora | 24.8 | 11.3 | 9.8 | 9.9 | 31.0 | 8.6 | 5.4 |
| (Cameron et al., 2001) | | | | | | | |
| Boer cross wether kid goats | | | | | | | |
| Bahia grass pasture | 32.6 | 3.2 | 10.5 | 18.1 | 25.4 | 5.3 | |
| Mimosa browse | 28.8 | 5.1 | 10.3 | 18.7 | 20.9 | 11.1 | |
| Concentrate | 27.3 | 11.2 | 7.8 | 17.7 | 21.3 | 7.7 | |
| (Solaiman et al., 2006). | | | | | | | |
| Spanish 7-month-old | | | | | | | |
| Grass hay-based diet | 38.1 | | 17.3 | 13.6 | 31.0 | | |
| Alfalfa hay-based diet | 39.6 | | 16.4 | 12.8 | 31.2 | | |
| Boer and Boer × Spanish 10 months old | | | | | | | |
| Grass hay-based diet | 36.3 | | 18.4 | 15.7 | 29.6 | | |
| Alfalfa hay-based diet | 34.9 | | 19.0 | 17.3 | 28.8 | | |
| Spanish buck kid goats | 42.4 | | 16.6 | 12.1 | 28.9 | | |
| Spanish wether kid goats | 37.7 | | 17.0 | 13.9 | 31.4 | | |
| Spanish doe kid goats | 36.6 | | 17.0 | 13.5 | 32.9 | | |
| (Wildeus et al., 2007) | | | | | | | |

differences in body part development among the sexes. The proportions of carcass cuts may vary slightly when kid goats are fed different diets, another indication of the effect of nutrition on postnatal growth and development (Table 13.5). Most goat carcasses are often cut into pieces or cubes for stewing, currying, and braising.

Consumers of goat meat have some varying preferences, but generally, they do not want to purchase carcasses, cuts, or cubes with any adhering fat. The size of carcass desired is dependent upon the season of the year, ethnicity of the consumer, and use of the meat. Few consumers desire carcasses weighing more than 40 pounds (20 kg), so there is a substantial discount for carcasses heavier than this weight or for live goats weighing more than 80 pounds (40 kg). The heavier carcasses give larger meat portions than are desired and increased amounts of intermuscular fat that cannot be easily separated from the lean before sale to the consumer. The Institutional Meat Purchase Specifications for Fresh Goat give weight ranges for cuts from the six carcass cutting styles that were designed to produce goat cuts desirable to different retail, foodservice, and restaurant customers (USDA, 2001).

Even though carcasses are not deboned to determine lean yields, boneless cut yields and lean meat yields provide a basis for comparison among different goats and their carcasses. Spanish goats produced lower boneless cut yields and lean meat yields than heavier muscled Boer and Myotonic goats (McMillin et al., 2002). Supplementation of pasture with concentrates will also decrease boneless and lean yields due to the excess fat deposition that must be trimmed from the carcass before fabrication or the cuts before retail sale.

Yields and properties of individual muscles are important for boneless value-added product manufacture. The shoulder muscles and *Semitendinosus* were lighter in color, the *Longissimus dorsi* and hind limb muscles were heavier, the cooking yields were higher for smaller muscles such as the *Biceps brachii* and *Semitendinosus*, and tenderness was higher for *Infraspinatus and Longissumus dorsi* (McMillin and Brock, 2004).

## MEAT QUALITY AND HUMAN HEALTH

Quality of food may be considered as a collection of factors that result in the food being safe, nutritious, palatable, and acceptable to consumers. Consumers consider that meat products will be safe for consumption. Even though meat from healthy animals can be considered free of pathogenic microorganisms and diseases that would cause foodborne and other illnesses in humans, sanitation and hygienic practices during slaughter, fabrication, processing, and distribution channels are essential to maintain the wholesomeness of goat meat. The nutritional, palatability, and acceptability traits may have different ranges of expectations by consumers, but quality is directly associated with the intended usage.

### Goat Meat Palatability and Acceptability

The characteristics of water binding, color, texture/tenderness, juiciness, and flavor highly influence the purchase and palatability of meat.

#### Color

Color and other shelf-life characteristics decline with increased storage time. Goat shoulder and arm cuts were reddest, with highest chroma (color saturation) and lowest hue (redness and yellowness), compared to cuts from other carcass locations. Surface discoloration of packaged cuts occurred within 4–8 days, with case-life similar to other red meat species (Kannan et al., 2001). Color of *Longissimus dorsi* muscles from kid goats fed pasture were lighter than muscles from goats fed mimosa browse while color of muscles from kid goats fed concentrate were less yellow than muscles from goats fed pasture or browse. There were no differences in redness of muscle color due to goat diet (Solaiman et al., 2006). Hay and hay-concentrate diets produced lighter and more yellow color of loin muscles than concentrate diets for 45 days, but there were no differences in redness (Lee et al., 2008).

#### Tenderness and Juiciness

Tenderness and juiciness decline with age of goats. Tenderness is influenced by collagen and muscle fibers, with collagen content higher, solubility lower, muscle fibrils thicker, and bundles larger than sheep, giving a coarse texture (Webb et al., 2005). Proportions of cross-linking of collagen in muscle cause decreases in collagen solubility upon cooking, which decreases tenderness (Warmington and Kirton, 1990). Shear force of *Semimembranosus* in goats fed concentrates were lower than in the muscles from goats fed pasture or browse, but sensory tenderness was not different with diet, and there were no shear force or sensory tenderness differences in *Biceps femoris* from goats on browse, pasture, or concentrate diets (Solaiman et al., 2006). No differences in total collagen in loin muscles of goats fed different diets were found (Lee et al., 2008). Leg and back muscles from female goats had lower shear force and would be expected to be more tender than the same muscles from wethers and intact males (Johnson et al., 1995b). The tenderness of leg muscles from yearling goats was less for both females and males than the meat from kid goats as determined by shear force and consumer panels (McMillin et al., 1998). Different muscles will have differing amounts of connective tissue and muscle fiber types that influence tenderness. The *Longissimus dorsi* and *Biceps femoris* muscle were slightly more tender than the *Semimembranosus* and *Adductor* muscles (Johnson et al., 1995b). Shear force was less in *L. dorsi* than *Semimembranosus* and *Triceps brachii* (Kannan et al., 2001). The eleven largest muscles from the goat were characterized for linear dimensions, color, and shear force. Tenderness varied with muscle type, but was not different with moist or dry cookery methods (McMillin and Brock, 2004). Other factors of breed, muscle type, and processing conditions will influence shear force and tenderness (Webb et al., 2005).

#### Flavor

Flavor of cooked goat meat can be more intense and "goaty" with increased animal age and higher levels of nutrition, resulting in lower acceptability (Webb et al., 2005). Flavor of goat meat from kid goats was liked more than meat from yearling goats (McMillin et al., 1998). The tenderness perception of goat leg meat was not affected by panelist age, sex, or ethnicity, but palatability was lowest with the youngest consumers and consumers with the highest incomes (Dawkins et al., 2000). Leg chops were less juicy and tender than loin steaks, with higher scores given by foreign sensory panelists than domestic panelists (Griffin et al., 1992). The juiciness, flavor intensity, and off flavor of *Semimembranosus* and *Biceps femoris* from goats fed Bahia grass pasture, mimosa browse, or concentrate were not different (Solaiman et al., 2006).

#### Comparison of Goat Meat to Other Meat

Goat meat and goat meat products have been compared with meat and products from other species. Goat meat may be referred to as chevon, much the same as pig meat is called pork. The juiciness of goat meat was the same, but tenderness and overall satisfaction were less than pork, beef, and lamb at comparable maturity and fatness (Smith et al., 1974). Palatability of sheep meat was higher (Griffin et al., 1992) with higher drip loss and more juiciness than goat meat (Schönfeldt et al., 1993). Flavor and aroma are complex meat attributes that are affected by species, age, fatness, tissue type, gender, diet, and method of cooking, so reports that goat and lamb meat flavors are similar conflict with those indicating goat meat has less intense flavor, tenderness, and juiciness than lamb (Webb et al., 2005). The branched chain fatty acids have been implicated in sheep and goat flavor (Webb et al., 2005), with the "goaty" odor of goat meat attributed to 4-methyloctanoic fatty acid (hircinoic acid) (Wong et al., 1975). Goat meat was described to have more intense gamy flavor, toughness, and hardness than most of the meat species tested (beef, chicken, horse, lamb, whale, moose, ostrich, pork, rabbit, turkey, veal), except for reindeer and beaver (Rødbotten et al., 2004). These properties make goat meat desired in ethnic foods, and the acceptance of goat meat is increased with increased consumption by non-ethnic consumers.

Patties made from goat meat and lamb could be distinguished but were not different in acceptability even though both were judged to be soft and greasy. However, goat and lamb curries were found to be very acceptable by panelists (Swan et al., 1998). Patties from goat were less tender, juicy, and greasy, and more chewy than patties from lamb, with species-related goaty and muttony flavor being clearly distinguishable (Tshabalala et al., 2003).

When comparing patties with differing proportions of beef and goat meat, consumer and trained sensory panels found similar juiciness, flavor, and tenderness in any patties that contained less than 40% chevon and more than 60% beef. Increased levels of goat meat in the patties increased the cooking yield and shear force values. It was suggested that goat meat could function as a lean meat source to augment product flavor (James and Berry, 1997). Seasoned and unseasoned goat meat could be distinguished from similar beef products. Sensory scores were similar for beef and goat meat when goat meat was served before beef, but scores were lower for goat meat when beef was served before goat meat (Rhee and Myers, 2003). This may explain the lack of processed goat meat products available in mainstream food markets.

### Nutritional Composition of Goat Meat

The nutrients are chemical elements or macromolecules necessary for the proper maintenance and functioning of the body components and cell metabolism. The nutrients from feedstuff are taken into the body, digested, formed into necessary metabolite compounds, transported throughout the body to the cells, and used for biochemical reactions and to form cell organelles in the cell structure. The nutrients form the chemical composition of the edible tissues of goat meat. The six nutrient classes are water (moisture), vitamins, minerals, proteins, carbohydrates, and lipids. Meat has high amounts of protein, essential amino acids, fat-soluble vitamins, water-soluble vitamins (niacin, riboflavin, panthothenic acid, cobalomin), iron, and phosphorus. Lean muscle is also a source of conjugated linoleic acid, which has been reported to have anticarcinogenic effects, antiatherogenic properties, and antidiabetic properties, enhance immune responses, and have positive effects on energy partitioning and growth.

Goat muscle is highly nutritious and has a biological value of 60.4 compared with beef at 68.6 in rat-feeding trials with 10% protein. Goat meat was reported to be higher in thiamine and riboflavin and lower in niacin than beef, lamb, and veal (Webb et al., 2005).

Animal diet, breed, and sex will influence the fatty acid composition in goat meat and goat muscles. However, the proportions of desirable fatty acids and polyunsaturated to saturated fat ratios have been calculated to be higher in goat meat than in beef and lamb (Webb et al., 2005). Fatty acid content of *Longissimus dorsi thoracis* muscles from Boer and feral Australian male goats were primarily oleic acid (43–54%), palmitic acid (23–28%), and stearic acid (11–18%). Increased slaughter weight increased oleic and palmitic fatty acids and decreased stearic acid. Castrated goats had lower percentages of stearic, linoleic, and total saturated fatty acids (Werdi Pratiwi et al., 2006). The fatty acid profile and cholesterol content of goat meat from kid goats fed different diets are shown in Table 13.6.

## GOAT MEAT PRODUCTS

The same processing operations of tenderization, grinding, freezing, curing, smoking, emulsifying, forming, and cooking to produce convenient, portioned, and value-added meat can be used for production of goat meat products as are used in the manufacture of processed products of other species (McMillin and Brock, 2005). However, goat meat is sold primarily as whole carcasses or bone-in cubes to ethnic consumers (Kannan et al., 2001), even though there is potential to market goat meat to non-ethnic populations (Hui and McLean-Meyinsse, 1996).

Precooked goat meat developed lipid oxidation in refrigerated storage faster than cooked meat of other species (Lamikanra and Dupuy, 1990). Precooked goat leg roasts were judged by consumers to be highest in palat-

**Table 13.6** Intramuscular fatty acid composition and cholesterol of *Longissimus dorsi* from Boer cross goats fed Bahia grass pasture, mimosa browse, or 60:40 forage: concentrate.

mg per 100 g muscle

| Fatty acid | Pasture | Browse | Concentrate |
|---|---|---|---|
| 16:0 | 296$^{b}$ ±± 43.9 | 300$^{b}$ ±± 35.7 | 474$^{a}$ ±± 41.6 |
| 18:0 | 236 ±± 36.5 | 261 ± 29.7 | 337 ± 34.6 |
| 18:1 trans-9 | 2.14$^{a}$ ± 1.13 | 5.09$^{b}$ ± 0.92 | 0.11$^{a}$ ± 1.07 |
| 18:1 cis-11 | 32.8 ± 6.57 | 37.0 ± 5.34 | 45.5 ± 6.22 |
| 18:1 cis-9 | 537$^{b}$ ± 90.14 | 571$^{b}$ ± 73.1 | 1010$^{a}$ ± 85.2 |
| 18:1 cis-7 | 18.9$^{b}$ ± 2.22 | 17.8$^{b}$ ± 1.80 | 27.5$^{a}$ ± 2.10 |
| 18:2 cis-10,12 | 6.25 ± 5.64 | 1.16 ± 4.58 | 10.8 ± 5.34 |
| 18:2 cis-9,12 | 65.0 ± 8.90 | 63.1 ± 7.23 | 83.5 ± 8.43 |
| 18:3 cis-9,12,15 | 6.2$^{a}$ ± 1.13 | 16.7$^{b}$ ± 0.92 | 6.1$^{a}$ ± 1.07 |
| 18:2 cis-9, trans-11 CLA$^{d}$ | 11.6 ± 2.13 | 13.7 ± 1.73 | 11.8 ± 2.02 |
| SFA$^{d}$ | 590$^{b}$ ± 86.2 | 621$^{b}$ ± 70.0 | 878$^{a}$ ± 81.6 |
| MUFA$^{d}$ | 649$^{b}$ ± 104 | 692$^{b}$ ± 84.4 | 1,165$^{a}$ ± 98.3 |
| PUFA$^{d}$ | 171 ± 9.37 | 180 ± 7.61 | 190 ± 8.87 |
| n-6d$^{e}$ | 138$^{a,b}$ ± 7.58 | 125$^{b}$ ± 6.16 | 159$^{a}$ ± 7.18 |
| n-3d$^{ae}$ | 21.2$^{a}$ ± 1.86 | 40.8$^{b}$ ± 1.51 | 19.1$^{a}$ ± 1.76 |
| n-6:n-3 | 6.83$^{b}$ ± 0.56 | 3.21$^{c}$ ± 0.46 | 8.75$^{a}$ ± 0.53 |
| PUFA:SFA | 0.34$^{a,b}$ ± 0.04 | 0.32$^{b}$ ± 0.03 | 0.23$^{a}$ ± 0.03 |
| Total fatty acids | 1443$^{b}$ ± 201 | 1520$^{b}$ ± 163.04 | 2,279$^{a}$ ± 190.03 |
| Cholesterol, mg/100g | 66.1 ± 2.39 | 61.0 ± 1.94 | 63.3 ± 2.26 |

[a,b,c]Means within the same row with different letters differ $P < 0.05$.

[d]CLA: conjugated linoleic acid, SFA: saturated fatty acids, MUFA: monounsaturated fatty acids, PUFA: polyunsaturated fatty acids.

[e]n-6 fatty acids include 18:2c10,12, 18:2c9,12, 20:2c11,14, 20:3c8,11,14, 20:4c5,8,11,14.

Source: Solaiman et al., 2006.

ability (tenderness, juiciness, flavor, overall liking) when prepared with a prime rib spice compared with no seasonings, Italian spices, or Mexican seasonings (Hilton et al., 2006).

Fermented cabrito (young goat) snack sticks with 0 or 3.5% soy protein concentrate were not different in flavor, texture, and overall acceptability (Cosenza et al., 2003a). Smoked goat sausages with 0 and 3.5% soy protein concentrate were similar in consumer sensory studies, but the cost was more than double that of comparable smoked pork sausage (Cosenza et al., 2003b).

Cooked shredded goat meat with barbecue sauce was scored higher by consumers who previously had consumed goat meat compared with consumers who had not tasted goat meat previously. Low-fat goat meat and beef sausages were not different in tenderness, juiciness, taste, and aroma. Prior experience and willingness to purchase goat meat from a supermarket were reported to be the main factors influencing consumer preference scores for goat meat products (Kannan et al., 2005).

There are many ways in which the value of goats and goat meat can be increased, but acceptability of goat meat and processed goat meat products is highly dependent on consumer experiences and their desires for convenience, taste, and cost. Increased availability of goat meat and more convenient forms of goat meat familiar to consumers would provide increased value in the market.

## SUMMARY

Pre-weaning and post-weaning growth are often evaluated as weight gain. The ADG can be increased with higher levels of nutrition, but the ADG of individual goats will be limited by genetic potential and other production factors. Male animals grow faster than females, and goats from later maturing breeds will grow faster than goats from earlier maturing breeds. Nutrient deficiencies or other stressors will limit rate of growth, but compensatory growth may occur if the deficiencies or limitations are not too severe or for long times. Concentrate feeds will increase weight gains but will often increase fat deposition, which is undesirable to processors and consumers.

Dressing percentages, the proportion of carcass yield of the live animal, can be influenced by the goat age, weight, and diet as well as the slaughter dressing procedures. Meat goats have different patterns of muscle and fat deposition than the other species, requiring evaluation of overall conformation and external fat rather than measurements of *Longissimus dorsi* muscle area and back fat thickness. Diet, sex, and breed will have varying effects on carcass composition, primal cut yields, and meat properties.

Goat meat is judged to be less desirable than meat from other mammalian species unless the consumer has previous experience or familiarity with goat meat. Goat meat is highly nutritious and can be prepared similarly to other types of meat. Processed goat meat products can be acceptable to consumers, but the retail prices are usually higher than for comparable products from pork, beef, and lamb.

## REFERENCES


Andries, K.M., J. Bernheart, and S. Beaufont. 2007. Performance of goat kids grazing fescue alone or with supplement. J. Anim. Sci. 85(Suppl. 2):40.

Browning, R., Jr. 2007. On-farm performance testing for meat goat doe herds. MGB-0701.1. Tennessee State University, Nashville, TN, Publication TSU-07-0039(A)-6a-532199.

Browning, R., Jr., B. Donnelly, T. Payton, and M. Byars. 2007. Preweaning body weights of meat goat kids produced in a three-breed diallel managed on southeastern pastures. J. Anim. Sci. 85 (Suppl. 2):113 (Abstr.).

Cameron, M.R., J. Luo, T. Sahlu, S.P. Hart, S.W. Coleman, and A.L. Goetsch. 2001. Growth and slaughter traits of Boer × Spanish, Boer × Angora, and Spanish goats consuming a concentrate-based diet. Small Ruminant Res. 79:1423–1430.

Cosenza, G.H., S.K. Williams, D.D. Johnson, C. Sims, and C.H. McGowan. 2003a. Development and evaluation of a fermented cabrito snack stick product. Meat Sci. 64:51–57.

Cosenza, G.H., S.K. Williams, D.D. Johnson, C. Sims, and C.H. McGowan. 2003b. Development and evaluation of a cabrito smoked sausage product. Meat Sci. 64:119–124.

Dawkins, N.L., K.W. McMillin, O. Phelps, S. Gebrelul, A.J. Beyer, and A. Howard. 2000. Palatability studies as influenced by consumer demographics and chevon characteristics. J. Mus. Foods 11:45–59.

Dzakuma, J.M., E. Risch, C.O. Smith, and H.D. Blackburn. 2004. Level of feed intake on performance of two goat genotypes. So. Afr. J. Anim. Sci. 34 (Supplement 1):38–42.

Fernandes, M.H.M.R., K.T. Resende, L.O. Tedeschi, J.S. Fernandes, Jr., I.A.M.A. Teixwira, G.E. Carstens, and T.T. Berchielli. 2008. Predicting the chemical composition of the body and the carcass of 3/4 Boer × 1/4 Saanen kids using body components. Small Ruminant Res. 75: 90–98.

Galyean, M.L. 1999. Review: Restricted and programmed feeding of beef cattle—Definitions, application, and research results. Prof. Anim. Sci. 15:1–6.

Griffin, C.L., M.W. Orcutt, R.R. Riley, G.C. Smith, J.W. Savell, and M. Shelton. 1992. Evaluation of palatability of lamb, mutton, and chevon by sensory panels of various cultural backgrounds. Small Ruminant Res. 8:67–74.

Hadjipanayiotou, M., C. Papachristoforou, and S. Economides. 1988. Effects of lasalocid on growth, nutrient digestibility and rumen characteristics in Chios lambs and Damascus kids. Small Ruminant Res. 1:217–227.

Hilton, G.G., M.A. Carr, J.D. Kellermeier, and B.J. May. 2006. Development and consumer acceptance of precooked goat roasts. Sheep & Goat Res. J. 21:35–39.

Hui, J. and P.E., McLean-Meyinsse 1996. Assessing the market potential for specialty meat: Goat, rabbit, and quail. J. Intl. Food & Agribus. Marketing 8:55–68.

James, N.A. and B.W. Berry. 1997. Use of chevon in the development of low-fat meat products. J. Anim. Sci. 75:571–577.

Johnson, D.D., J.S. Eastridge, D.R. Neubauer, and C.H. McGowan. 1995a. Effect of sex class on nutrient content of meat from young goat. J. Anim. Sci. 73:296–301.

Johnson, D.D., C.H. McGowan, G. Nurse, and M.R. Anous. 1995b. Breed type and sex effects on carcass traits, composition and tenderness of young goats. Small Ruminant Res. 17:57–63.

Kanani, J., S.D. Lukefahr, and R.L. Stanko. 2006. Evaluation of tropical forage legumes (*Medicago sativa, Dolichos lablab, Leucaena leucocephala* and *Desmanthis bicornutus*) for growing goats. Small Ruminant Res. 65:1–7.

Kannan, G., B. Kouakou, and S. Gelaye. 2001. Color changes reflecting myoglobin and lipid oxidation in chevon cuts during refrigerated display. Small Ruminant Res. 42:67–75.

Kannan, G., M.C. Nelson, T.E. Hollis, T.D. Pringle, and K.W. McMillin. 2005. Demographic factors influencing consumer preference of chevon products. J. Anim. Sci. 83 (Suppl. 2):16 (Abstr. 62).

Lamikanra, V.T. and H.P. Dupuy. 1990. Analysis of volatiles related to warmed over flavor of cooked chevon. J. Food Sci. 55:861–862.

Lee, J.H., B. Kouakou, and G. Kannan. 2008. Chemical composition and quality characteristics of chevon from goats fed three different post-weaning diets. Small Ruminant Res. 75:177–184.

Luo, J., T. Sahlu, M. Cameron, and A.L. Goetsch. 2000. Growth of Spanish, Boer × Angora and Boer × Spanish goat kids fed milk replacer. Small Ruminant Res. 36:189–194.

McMillin, K.W. and A.P. Brock. 2004. Size, color, and texture of major muscles from kid goat carcasses. J. Anim. Sci. 82 (Suppl. 1):393.

McMillin, K.W. and A.P. Brock. 2005. Production practices and processing for value-added goat meat. J. Anim. Sci. 83 (E Suppl.):E57–E68.

McMillin, K.W. and F. Pinkerton. 2008. Meat goat selection, carcass evaluation & fabrication guide. Publ. 2951, Louisiana State University Agricultural Center, Baton Rouge, January, 8 p.

McMillin, K.W., L.C. Nuti, and F. Pinkerton. 2002. Carcass traits and cut yields of Boer, Spanish, and Myotonic kid goats with creep feeding and supplemented pasture finishing. J. Anim. Sci. 80 (Suppl. 2):28 (Abstr.).

McMillin, K.W., O. Phelps, N.L. Dawkins, J.M. Fernandez, A. Howard, S. Gebrelul, G. Simon, and K.E. Mellad. 1998. Carcass and meat traits of goats of different live conformations, ages, sexes and weights. Proc. Intl. Congress Meat Sci. Technol. 44:268–269.

Merkel, R.C., A.L. Goetsch, and N. Silanikove. 2003. Effects of supplementing polyethylene glycol to goat kids grazing Sericea lespedeza and early post-weaning nutritive plane upon subsequent growth. Sheep & Goat Res. J. 18:8–13.

Muir, J.P. and S.A. Weiss. 2006. Corn supplement for goats on summer rangeland or improved pasture. Sheep & Goat Res. J. 21:40–47.

Negesse, T., A.K. Patra, L.J. Dawson, A. Tolera, R.C. Merkel, T. Sahlu, and A.L. Goetsch. 2007. Performance of Spanish and Boer × Spanish doelings consuming diets with different levels of broiler litter. Small Ruminant Res. 69: 187–197.

Nuti, L., F. Pinkerton, and K.W. McMillin. 2000. Experts study benefit of corn supplement on pastured wethers. Goat Rancher, September:20–21.

Oman, J.S., D.F. Waldron, D.B. Griffin, and J.W. Savell. 1999. Effect of breed-type and feeding regime on goat carcass traits. J. Anim. Sci. 77:3215–3218.

Oman, J.S., D.F. Waldron, D.B. Griffin, and J.W. Savell. 2000. Carcass traits and retail display-life of chops from different goat breed types. J. Anim. Sci. 78:1262–1266.

Payne, B., J. Crenwelge, B.D. Lambert, and J.P. Muir. 2006. A self-limiting complete feed changes forage intake and animal performance of growing meat goats. South Afr. J. Anim. Sci. 36:257–260.

Pinkerton, F., L. Nuti, and K. McMillin. 2001. Feedlot performance and carcass characteristics of Tennessee meat goats and Boer cross wethers. Ranch and Rural Living, March, pp. 23–25.

Rhee, K.S. and C.E. Myers. 2003. Sensory properties and lipid oxidation in aerobically refrigerated cooked ground goat meat. Meat Sci. 66:189–194.

Rødbotten, M., E. Kubberød, P. Lea, and Ø. Ueland. 2004. A sensory map of the meat universe. Sensory profile of meat from 15 species. Meat Sci. 68:137–144.

Ryan, S.M., J.A. Unruh, M.E. Corrigan, J.S. Drouillard, and M. Seyfert. 2007. Effect of concentrate level on carcass traits of Boer crossbred goats. Small Ruminant Res. 73:67–76.

Schoeman, S.J., J.F. Els, and M.M. van Niekerk. 1997. Variance components of early growth traits in the Boer goat. Small Ruminant Res. 26:15–20.

Schönfeldt, H.C., R.T. Naude, W. Bok, S.M. van Heerden, L. Sowden, and E. Boshoff. 1993. Cooking- and juiciness-related quality characteristics of goat and sheep meat. Meat Sci. 34:381–394.

Smith, G.C., M.I. Pike, and Z.L. Carpenter. 1974. Comparison of the palatability of goat meat and meat from four other animal species. J. Food Sci. 39:1145–1146.

Solaiman, S.G., D. Bransby, C. Kerth, R. Noble, B. Blagburn, and C. Shoemaker. 2006. A sustainable year-round forage system for goat production in Southeastern USA. Final Report, Southern SARE Project # LS02–141.

Solaiman, S.G., T.J. Craig, Jr., G. Reddy, and C.E. Shoemaker. 2007. Effects of high levels of Cu supplement on growth performance, rumen fermentation, and immune responses in goat kids. Small Ruminant Res. 69:115–123.

Swan, J.E., C.M. Esguerra, and M.M. Farouk. 1998. Some physical, chemical and sensory properties of chevon products from three New Zealand goat breeds. Small Ruminant Res. 28:273–280.

Tshabalala, P.A., P.E. Strydom, E.C. Webb, and H.L. de Kock. 2003. Meat quality of designated South African indigenous goat and sheep breeds. Meat Sci. 65: 563–570.

Urge, M., R.C. Merkel, T. Sahlu, G. Animut, and A.L. Goetsch. 2004. Growth performance by Alpine, Angora, Boer and Spanish wether goats consuming 50 or 75% concentrate diets. Small Ruminant Res. 55:149–158.

USDA. 2001. Institutional Meat Purchase Specifications for Fresh Goat Series 11. United States Department of Agriculture Agricultural Marketing Service Livestock and Seed Program, Washington, DC. 37 pages.

Van Niekerk, W.A. and N.H. Casey. 1988. The Boer goat. II. Growth, nutrient requirements, various energy and protein levels. J. Anim. Sci. 68:1751–1759.

Warmington, B.G. and A.H. Kirton. 1990. Genetic and non-genetic influences on growth and carcass traits of goats. Small Ruminant Res. 3:147–165.

Webb, E.C., N.H. Casey, and L. Simela. 2005. Goat meat quality. Small Ruminant Res. 60:153–166.

Werdi Pratiwi, N.M., P.J. Murray, D.G. Taylor, and D. Zhang. 2006. Comparison of breed, slaughter weight and castration on fatty acid profiles in the longissimus thoracic muscle from male Boer and Australian feral goats. Small Ruminant Res. 64:94–100.

Wildeus, S., J.-M. Luginbuhl, K.E. Turner, Y.L. Nutall, and J.R. Collins. 2007. Growth and carcass characteristics in goat kids fed grass- and alfalfa-hay-based diets with limited concentrate supplementation. Sheep & Goat Res. J. 22:15–19.

Wong, E., C.B. Johnson, and L.N. Nixon. 1975. The contribution of 4-methyloctanoic (hircinoic) acid to mutton and goat meat flavour. NZ J. Agric. Res. 18:261–266.

# 14
# Milk Production

*Y.W. Park, PhD and G.F.W. Haenlein, PhD*

**KEY TERMS**

Dry period—a rest period of about 8 weeks between lactations to allow for regeneration of the mammary secretory epithelium.
Lactation curve—trend of milk production throughout the lactation, usually for 200–305 days.
Alveolus—hollow ball of milk secreting cell, forming a cluster of alveoli to join intralobular ducts.
Milk ejection—removal of milk from mammary gland by neurohormonal reflex process.
Oxytocin—milk ejection hormone by stimulating the contractile units of myoepithelial cells around the alveoli.
ACTH—adrenocorticotrophic hormone, stimulating growth of adrenal cortex and secretion of corticosteroids.
Colostrum—The first-drawn milk from the mammary gland for 0 to 5 days after parturition.
Lactose—carbohydrate in milk, which is a disaccharide composed of glucose and galactose.
Casein—major milk protein, constitutes 80% of total milk protein, precipitated by acid or coagulated by rennet enzyme.
Milk solids—solids in milk consisting of fat, protein, lactose, and ash, generally 12–13% in goat milk or cow milk.

**OBJECTIVES**

By completing this chapter, the reader will acquire knowledge on:

- Physiology of lactation
- Milk production in the tropics
- Composition of goat milk
- Factors affecting yield and composition of goat milk
- Differences in composition of milk of goats and other species
- Economics of goat versus cow milk production
- Dairy goat records
- Dairy herd improvement association
- International aspects of the efficiency of goat milk production
- Differences in goat and cow milk products (i.e., cheese, yogurt)

## INTRODUCTION

According to the FAO (2004), goats produce approximately 2% of the world's total milk supply. However, more people drink milk from goats than milk of any other animal species worldwide. Goat milk is vitally important in underdeveloped countries because it provides basic nutrition and subsistence to the majority of their populations in rural areas. Because cow milk is often unavailable, goat milk and its products are important daily food sources of protein, phosphate, and calcium for people of underdeveloped countries (Park and Haenlein, 2007). The contribution of goat milk and dairy goat products to the economy

**Table 14.1** Goat milk production contribution to countries[1] worldwide.

| | Goat milk total 1,000 MT/year | Goat milk part of all milk, % | Goat milk/ goat/year, kg[2] | Goat population 1,000 head[2] |
|---|---|---|---|---|
| India | 3,128 | 4 | 25 | 124,500 |
| Bangladesh | 1,328 | 55 | 38 | 34,500 |
| Sudan | 1,151 | 16 | 28 | 40,000 |
| Pakistan | 818 | 4 | 15 | 52,800 |
| France | 480 | 2 | 397 | 1,210 |
| Greece | 460 | 26 | 87 | 5,300 |
| Iran | 398 | 24 | 15 | 26,000 |
| Somalia | 390 | 51 | 30 | 12,700 |
| Spain | 350 | 7 | 115 | 3,050 |
| Turkey | 225 | 3 | 32 | 7,000 |
| Indonesia | 200 | 29 | 15 | 13,280 |
| Mali | 196 | 43 | 17 | 11,460 |
| Algeria | 155 | 13 | 48 | 3,200 |
| Italy | 140 | 1 | 102 | 1,375 |
| Northern Mediterranean[3] | 1,840 | 1–26 | 140 | 13,090 |
| Southern Mediterranean[4] | 618 | 1–14 | 26 | 23,540 |
| World | 12,455 | 2 | 16 | 767,930 |

[1]Countries with more than 140,000 MT annual goat milk production.

[2]Includes all goats, not just those for dairy purposes.

[3]Includes (1,000 head): Portugal (550), Spain (3,050), France (1,210), Italy (1,330), former Yugoslavia (350), Romania (680), Macedonia (20), Bulgaria (900), Greece (5,000).

[4]Includes (1,000 head): Turkey (7,000), Syria (1,000), Lebanon (380), Israel (65), Jordan (550), Egypt (3,470), Libya (1,265), Tunisia (1,400), Algeria (3,200), Morocco (5,210).

Source: Haenlein, 2006.

of Mediterranean and Middle Eastern countries is especially important, as it is in many other countries worldwide (Table 14.1).

Goat milk and its products, including cheeses and yogurt, are also valued parts of the total dairy industry in developed countries by providing gourmet foods to connoisseur consumers, and by supporting people who need alternative dairy products due to medical afflictions such as allergies and gastrointestinal disorders. Goat milk serves three types of markets around the world: (1) home consumption, (2) specialty gourmet interests, and (3) medical needs (Park and Haenlein, 2007).

## LACTATION PHYSIOLOGY

### Anatomy and Physiology of Mammary Glands

The goat udder consists of two halves, each with separate mammary glands. Appreciable mammary secretions are produced only after formation of the lobular-alveolar system. Enzymes needed for milk synthesis (lactogenesis and galactopoiesis) are present within the cells of the udder before kidding. At parturition, hormones cause a great increase in secretion of milk. Colostrum is secreted up to about 5 days after kidding and contains high concentrations of solids, especially immunoglobulins. Thereafter, normal milk is secreted with reduced solids content. Shortly before parturition, blood progesterone titers fall and estrogen, adrenocorticotrophic hormone (ACTH) and prolactin levels increase. Neural stimuli are not normally required to initiate lactation (Larson, 1985).

Lactogenesis and the milk secretion process involve intracellular synthesis of milk and subsequent passage of milk from the cytoplasm into the alveolar lumen (Cosie and Buttle, 1974). The basic anatomical component of the secretory tissue is the alveolus, which is made up of a single layer of epithelial cells that absorb precursors from the blood, synthesize, and then secrete the milk components into the lumen of the alveolus. The outer layer of the alveolus is lined with myoepithelial cells that help in contracting the alveolus and the letdown of milk. The smaller ducts within the lobules are lined with a single layer of epithelial cells and connect to the larger ducts, ending in the gland and teat cisterns. Milk is secreted by the alveolar cells through two secretory processes: synthesis and diffusion. Milk removal is accomplished by passive withdrawal of milk from teat and gland cisterns and the major ducts during nursing, hand milking, or machine milking. Active ejection of milk results from contraction of myoepithelial cells around the alveoli in response to the milk secretion hormone oxytocin elicited by the neural stimuli of the suckling kid or the milking procedure (Schmidt, 1971).

Following parturition, the increase in milk yield reaches a maximum in 4–8 weeks in milking goats and cows. The milk yield gradually declines toward the end of lactation. After maximum yield is reached, a decline in milk production can be calculated each month as a percentage of the previous month. This percentage is a measure of lactation persistency, which varies between breeds and individual animals. Although milk volume is decreased at the end of lactation, the solids content, such as protein, fat, and minerals increases, and lactose content decreases.

Milk yield declines after the onset of estrus (Peaker and Linzell, 1974), and the variation in milk production may be related to the different amounts of mammary blood flow of individual animals. Goats are frequently used as experimental animals in lactation and other physiological studies, and considerable basic knowledge of the physiology of mammals has thus been derived from goats.

### Initiation of Milk Ejection

The function of the mammary gland is under hormonal control of the pituitary gland. Milk removal from the mammary gland of most species is dependent upon the neurohormonal reflex process of milk ejection. Sensory nerve endings in the teat are stimulated, and impulses are carried via the spinal cord to the hypothalamus. This stimulus reaches the central nervous system and causes the posterior lobe of the pituitary to release oxytocin, a small peptide hormone. Oxytocin travels via the blood supply to the mammary gland, where it causes contraction of the myoepithelial cells surrounding the alveoli. The contraction process forces the milk from the alveoli into the duct system, from which it flows to the gland and teat cisterns. Adrenalin is antagonistic to oxytocin. The neurohormonal reflex of the milk ejection process is inhibited by adrenalin, resulting from fright and emotional disturbances of the animal.

## FACTORS AFFECTING MILK YIELD AND COMPOSITION

### Breed

There are six major dairy goat breeds in the U.S. The Saanen, Alpine, Toggenburg, and Oberhasli of Swiss origin are the most famous around the world for their record high milk productivity; the LaMancha are of U.S. origin; and the Nubian are of British origin. The Saanen is best known as the Holstein of the goat world, producing high milk yields with low fat levels (Haenlein and Caccese, 1984). The Nubian breed is similar to the Jersey breed of cows, but is the most numerous in the U.S. Nubians produce less milk with higher fat and protein content. The Alpine, Toggenburg, LaMancha, and Oberhasli breeds produce milk yields and composition between the levels of the Saanen and Nubians in general but with some individual exceptions (Table 14.2). There are at least 10 origi-

**Table 14.2** Individual U.S. goat breed leaders in milk production.

| Breed, goat registration no., year of official record | (Yr–mo) | Age days in milk | Lactation | | Fat, % (fat yield kg) | 4% FCM[1], yield, kg |
|---|---|---|---|---|---|---|
| | | | Milk kg | Milk/day, kg | | |
| Alpine #0177455, 1982 | 7–4 | 305 | 2,916 | 9.6 | 4.8 (140) | 3,266 |
| LaMancha #0618876, 1991 | 5–0 | 253 | 2,454 | 9.7 | 3.3 (81) | 2,197 |
| Nigerian Dwarf #1360584, 2006 | 3–10 | 305 | 782 | 2.6 | 5.4 (42) | 943 |
| Nubian #0904515, 1996 | 2–9 | 302 | 2,700 | 8.9 | 5.1 (138) | 3,150 |
| Oberhasli #0935588, 1997 | 3–0 | 304 | 2,120 | 7.0 | 5.0 (106) | 2,438 |
| Saanen #0894085, 1997 | 4–0 | 305 | 2,987 | 9.8 | 2.6 (76) | 2,335 |
| Toggenburg #0926741, 1997 | 3–1 | 305 | 3,620 | 11.9 | 3.9 (142) | 3,578 |

[1]4% Fat corrected milk = 0.4 × milk yield kg + 15 × fat yield kg
Source: ADGA, 2008

nal goat breeds with high milk-producing ability worldwide, including also the Spanish Murciana-Granadina and Canaria, and the Norwegian Nordic. There are at least 22 recognized breeds with medium milk yield genetic merit such as the Eastern Mediterranean Damascus, the Italian Maltese, and the Indian Jamnapari (Devendra and Haenlein, 2003).

The milk fat content among the different breeds varies more than the protein content. Fat and total solids content of the milk of Alpine, Saanen, and Nubian breeds exported to tropical environments have been lower than those of the same breeds raised under temperate climate conditions, which may be due to the diet and higher temperatures (Juàrez and Ramos, 1986). Indigenous goat breeds such as the West African Dwarf and Red Sokoto or the Indian Jamnapari have much richer milk composition but lower yields than the U.S. breeds, and especially the Swiss breeds.

### Age

Age is an important source of variation in milk yield and ranks second after the season of freshening (Alderson and Pollak, 1980). Milk from younger goats tends to have a higher fat content than that from older goats. Peak milk yield is attained at 4 years of age, when goats have reached mature body weights (Gall, 1981). Dairy goats are therefore not culled after their second lactation like most dairy cows. Part of the increase of milk yield with age is due to an increase in body weight, which results in a larger digestive system and a larger mammary gland for the secretion of milk. Advancing age or increased number of lactation occurrences may result in a gradual decrease in the percent of milk fat and solids-not-fat in goat milk.

### Body Size and Weight

Mature dairy goats may weigh between 30 and 80 kg. Larger does will have to produce more milk than smaller animals in order to pay for the higher maintenance costs. With high levels of milk production at the beginning of lactation, goats are usually not able to consume sufficient energy and have to draw on their body reserves for milk production (Morand-Fehr, 1991).

A positive correlation exists between milk yield and body weight. Storage of body fat during the dry period influences milk production positively at the onset of lactation. The capacity to mobilize body fat is greater in animals that have been fed liberally during the dry period due to more stored fat, but it must not cause excessive body condition to avoid ketosis after parturition. Weight losses of adult goats at the beginning of lactation are related to levels of milk yield (Chilliard et al., 1979).

### Udder Size

The productive capacity of dairy goats, like that of other dairy animals, is judged by the physical appearance, size, and quality of the udder (Haenlein and Abdellatif, 2004). Age, stage of lactation, interval since last milking, and expected lactation length must be considered, when evaluating milk production capacity of a dairy goat.

Differences in milk yield, both between and within species, are a function of the size of the mammary gland (Linzell, 1972). More milk is available for milking by expulsion of milk from the alveoli of larger udders. The decline of milk yield in later lactation is mainly due to the loss of secretory tissues and a decrease in the rate of secretion per unit of tissue capable of milk secretion. The size of the teat and gland cisterns in relation to the volume of the secretory tissues, are greater in goats than in cows.

Goats with higher milk yield tend to have pendulous udders in some breeds. A high correlation exists between linear udder measurements (circumference and udder height) and milk yield (Das and Sidhu, 1975). Between milkings, milk biosynthesis and secretion continues to increase up to 18 hours. Milk yield may be reduced to about 50% by milking once a day only, which is explained by a reduction in mammary blood flow with the increased internal mammary pressure (Pearl et al., 1973).

### Diet

Diet has similar effects on milk composition in goats as it does in dairy cows, although variations may exist (Morand-Fehr and Sauvant, 1980). An increase in energy content of the diet for high-yielding goats tends to increase milk production, and the nitrogen content of the milk may increase by 0.1–0.2% units, and the milk fat content can be reduced by 0.2–0.4% units. Water availability and intake can impact milk yield. Water deprivation for 48 hours, in high-yielding breeds results in reduction of milk yield, but it increases concentration of lactose and protein contents, mainly due to less milk yield. Indigenous goat breeds have greater adaptation for shortage of water in countries with desert conditions and are able to maintain relatively constant milk production under these environments (Shkolnik et al., 1980). The entire daily diet should contain at least 17% crude fiber to prevent serious depression in milk fat content.

Low-fat diets result in a decrease in milk fat content, but adding protected lipids increases milk fat content.

High-protein diets do not change the nitrogen content of goat milk and have no effect on yield, while a slight increase in nonprotein nitrogen content may occur. Goats can recover normal lactation after extended periods of feed restriction and low milk yield. This capacity is important under rangeland grazing conditions with greatly fluctuating feed quality and supply (Haenlein and Ramirez, 2007).

### Stage of Lactation

The lactation period of goats is normally 10 months long as in dairy cattle, but may be longer when a new pregnancy has not been initiated. The lactation period of goats is generally longer than that of dairy sheep.

Fat, protein, and mineral contents are high in the colostrum (first secretion of mammary gland after parturition) phase of early lactation, and their concentrations fall by the fifth day of lactation, reaching a low plateau by the second to fourth months in milk but increase at the end of lactation (Renner, 1982). There is a negative relationship between milk yield and the percentage composition of fat, protein, and minerals, while the relationship with lactose is positive. Lactose content in goat and cow milk is low in colostrum, rises in mid-lactation, and decreases again at the end of the lactation.

### Season

Goat milk composition and yield also vary with seasons within the year. In temperate climates in late summer, milk is often low in fat and solids-not-fat (SNF) content. A Canadian study demonstrated that cheese yields varied directly with seasonal variations in the levels of fat and protein content of the milk (Irvine, 1974).

Since natural breeding of dairy goats is seasonal, in autumn, in most dairy breeds in the Northern Hemisphere, kidding will occur in spring. Thus, in autumn most Northern dairy goats are in late lactation with decreased yields and increased milk composition, unless goats are not rebred. With artificial breeding methods and environmental light management, off-season kidding and year-round or off-season milking can be achieved.

### Length of Dry Period

This is a period when the dairy animal is pregnant but not milking. Dairy goats need this period to replenish mammary secretory tissues and body reserves and to prepare the body for the next lactation. Animals should be in good body condition but not fat. Dry periods of less than 6 weeks can cause a decrease in milk yield during the next lactation in comparison with a dry period of more than 6 weeks. Dry periods greater than 8 weeks have no significant benefit in milk yield during the next lactation period.

### Environmental Temperature

The effect of environmental temperature on milk yield and composition depends on the breed of goats, but generally cold exposure reduces goat milk secretion, partially because of reduced udder blood flow (Thompson and Thompson, 1977). At −3.9°C (25°F), mammary glucose uptake, lactose secretion, and milk yield can be as low as 30% of the values at thermal neutrality (20°C; 68°F). Low temperatures may have less effect on milk yield if extra feed is provided to cover for the extra body energy required to maintain body temperature. Within a relative humidity range of 60–80%, milk yield may be unaffected by temperature changes between 5°C (40°F) and 22°C (70°F). Milk fat content increases with temperatures below 25°C (75°F), and the SNF content follows the same pattern.

Above the range of thermal neutrality, a marked decrease in milk production occurs, because feed consumption decreases at increased temperatures while water consumption generally increases. At about 40°C (105°F), feed consumption and milk production approach zero (Figure 14.1). Milk production and feed consumption in cows are

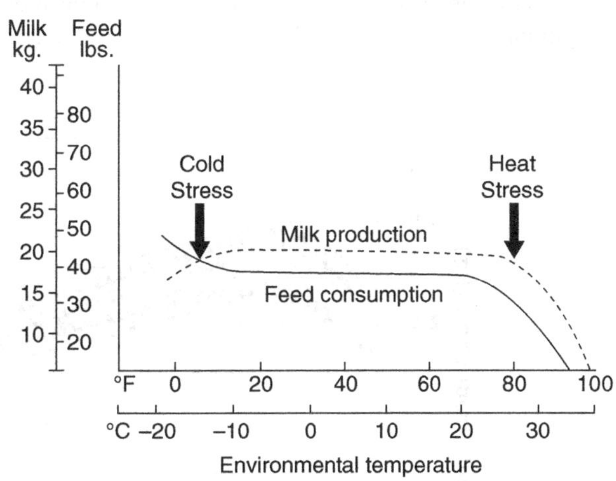


**Figure 14.1** Relationships between environmental temperature and milk production/feed consumption. When the environmental temperature is 25°C (75°F) or above, reduced feed intake is followed by a decrease in milk production. At approximately −15°C (5°F), feed consumption increases and milk production decreases (adapted from Bath et al., 1985).

reduced automatically in an effort to curtail body heat production. Respiratory rate increases about 5-fold when temperatures rise from 10–40°C (50–105°F) (Bath et al., 1985). Chloride content of milk increases and lactose decreases with high environmental temperatures.

## LACTATION LENGTH AND LACTATION CURVE

Dairy goats in the Northern Hemisphere usually start lactation after kidding in spring and get rebred for a new pregnancy in the fall (autumn), unless the farmer/owner wants to have milk in winter, avoid the winter dry period, and continue milking beyond 10 months of lactation without rebreeding in the fall. It is not unusual for some dairy goats to lactate for 2 years without much decline until the next pregnancy. Young dairy goats (doelings) can also be induced to secrete normal milk without pregnancy and kidding, after some massage stimulation of the young udder for several days. Male goats (bucks) can likewise be stimulated to secrete normal milk, which is of great advantage for the determination of genetic polymorphisms of milk proteins in both sexes for pedigree analysis and sire selection (Gall, 1981).

Length and yield of lactation may be affected by many factors including adequate feeding, maintenance of health, regular milking, proper management of the milking herd, and onset of pregnancy, besides genetics. The removal of the pituitary gland results in the immediate and complete cessation of lactation, indicating the importance of proper hormonal balance for the maintenance of milk secretion. Goats with inhibited lactation from experimental hypophysectomy restored lactation to pretreatment levels by administration of growth hormones, insulin, prolactin, thyroxin, and glucocorticoids (Cowie et al., 1964).

Globally, the length of normal goat lactations depends on breeds and environmental conditions, and can vary between 200 and 300 days. The rate of lactation decline is influenced by season of kidding, averaging about 86%, but without new pregnancy, there may not be much of a decline and the lactation curve can persist for more than 1 or 2 years.

The lactation curves of different dairy goat breeds in the U.S. are shown in Figure 14.2, where all breeds tend to have similar trends of milk production throughout the lactation period. Goats with high lactation records reach higher peaks in production than goats with low lactation records. First, lactations may have high persistency of lactation.

## MILK COMPOSITION

### General Compositional Characteristics of Goat Milk

Proximate composition of goat milk is similar to that of cow milk, although goat milk may contain more fat, protein, and minerals, and less lactose and casein than cow milk (Table 14.3). Several reports (Park, 2005) have given

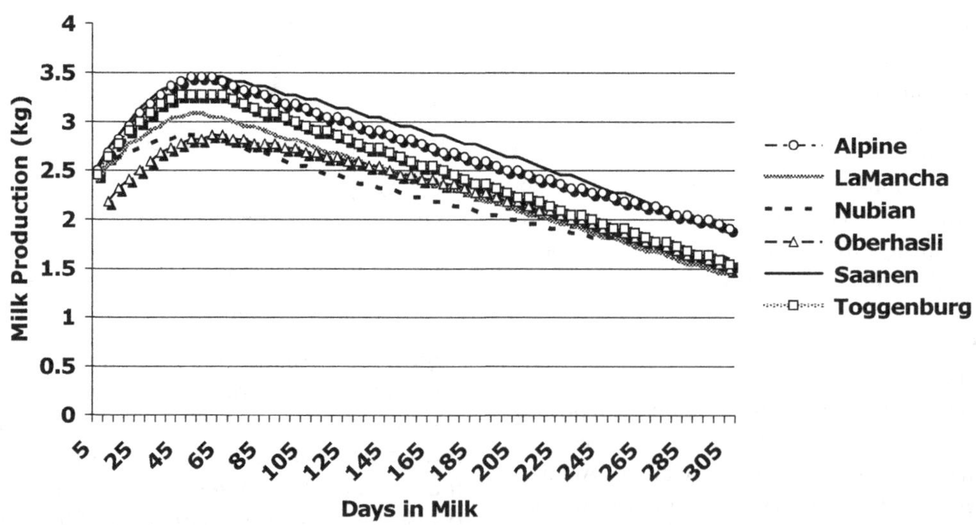


**Figure 14.2** Patterns of lactation curves for different dairy goat breeds in the U.S. (Dairy Herd Improvement Registry; ADGA, 2008). For color detail, see Appendix A.

**Table 14.3** Average concentrations of basic nutrients, minerals, and vitamins in goat milk (100 g) as compared with those in cow and human milk.

| Constituents | Goat | Cow | Human |
|---|---|---|---|
| **Basic nutrients** | | | |
| Fat (g) | 3.8 | 3.6 | 4.0 |
| Protein (g) | 3.5 | 3.3 | 1.2 |
| Lactose (g) | 4.1 | 4.6 | 6.9 |
| Minerals (ash) (g) | 0.8 | 0.7 | 0.2 |
| Total solids (g) | 12.2 | 12.3 | 123 |
| Calories (cal) | 70 | 69 | 68 |
| **Minerals** | | | |
| Ca (mg) | 134 | 122 | 33 |
| P (mg) | 141 | 119 | 43 |
| Mg (mg) | 16 | 12 | 4 |
| K (mg) | 181 | 152 | 55 |
| Na (mg) | 41 | 58 | 15 |
| Cl (mg) | 150 | 100 | 60 |
| S (mg) | 2.89 | — | — |
| Fe (mg) | 0.07 | 0.08 | 0.20 |
| Cu (mg) | 0.05 | 0.06 | 0.06 |
| Mn (mg) | 0.032 | 0.02 | 0.07 |
| Zn (mg) | 0.56 | 0.53 | 0.38 |
| I (mg) | 0.022 | 0.021 | 0.007 |
| Se (μg) | 1.33 | 0.96 | 1.52 |
| **Vitamins** | | | |
| Vitamin A (I.U.) | 185 | 126 | 190 |
| Vitamin D (I.U.) | 2.3 | 2.0 | 1.4 |
| Thiamine (mg) | 0.068 | 0.045 | 0.017 |
| Riboflavin (mg) | 0.21 | 0.16 | 0.02 |
| Niacin (mg) | 0.27 | 0.08 | 0.17 |
| Pantothenic acid (mg) | 0.31 | 0.32 | 0.20 |
| Vitamin $B_6$ (mg) | 0.046 | 0.042 | 0.011 |
| Folic acid (μg) | 1.0 | 5.0 | 5.5 |
| Biotin (μg) | 1.5 | 2.0 | 0.4 |
| Vitamin $B_{12}$ (μg) | 0.065 | 0.357 | 0.03 |
| Vitamin C (mg) | 1.29 | 0.94 | 5.00 |

Data compiled from Posati and Orr (1976), Jenness (1980), Haenlein and Caccese (1984), and Park and Chukwu (1988).

detailed chemical compositions of goat milk. As known for cow milk, composition of goat milk also varies with diet, breed, animals within breed, parity, environmental conditions, udder health, feeding and management conditions, season of year or kidding, locality, and stage of lactation. Some variations in composition can be related to different production practices in the dairy goat industry and difficulty in procuring representative milk samples from pooled bulk milk from various herds and farms (Chandan et al., 1992).

On average, goat milk contains 12–13% total solids, consisting of about 3.8% fat, 3.5% protein, 4.1% lactose, and 0.8% minerals (Table 14.3). Goat milk contains slightly less total casein, but more nonprotein nitrogen than cow milk. Goat and cow milk have 3–4 times greater levels of protein, and minerals, but much less lactose than human milk. Total solids and caloric values of goat, cow, and human milks are similar (Park, 2005). The lower casein content of goat milk compared to cow milk is closely related to slower coagulation properties and lower yield of curds during cheese manufacturing. Compositional differences in milk are highly correlated with growth rates of the young of the different species, where the newborn of goats and cows grow much faster than human infants.

### Lipids

The range of milk fat content in goat milk of different breeds and within different seasons is 2.4–7.8%. Diameters of fat globules in goat, cow, buffalo, and sheep milk average 3.5, 4.6, 5.9, and 3.3 μm, respectively (Park, 1990). The smaller fat globules of goat milk provide a better dispersed and naturally homogenized milk fat and better digestibility compared to milk with larger fat globules, because a greater relative surface area of fat globules is exposed for digestive action by lipases (Jenness, 1980).

Goat milk fat has significantly higher levels (+42% difference) of short- and medium-chain length fatty acids (MCT; C4:0—C14:0) than cow and human milk (Table 14.4). This unique difference in MCT contents in goat milk has been used for treatment of a variety of fat malabsorption syndromes in patients (Babayan, 1981). There is a great potential to provide goat milk products with beneficial properties for human nutrition and health. Goat milk has a higher content of MCT and conjugated linoleic acid (CLA) and lower levels of cholesterol than cow milk. Goat milk fat contains about 98% of free lipids (about 97% are triglycerides), and 1–3% bound lipids that are mainly in the form of membrane lipids (phospholipids) associated with fat globules (a remainder of the secretory cell membranes), of which about 47% are neutral and 53% are polar lipids (Cerbulis et al., 1985).

### Protein

Goat milk contains five principle proteins: $\alpha_{s2}$-casein ($\alpha_{s2}$-CN), β-casein (β-CN), κ-casein (κ-CN), β-lactoglobulin

**Table 14.4** Average composition of fatty acids (FA) in goat, cow, and human milk.

| Fatty acid[1] | Goat milk | Cow milk | Human milk |
|---|---|---|---|
| **Saturated FA, total** | **2.37** | **2.00** | **1.85** |
| C4:0 butyric | 0.13 (2.6) | 0.11 (3.3) | — |
| C6:0 caproic | 0.09 (2.9) | 0.06 (1.6) | — |
| C8:0 caprylic | 0.10 (2.7) | 0.04 (1.3) | — |
| C10:0 capric | 0.26 (8.4) | 0.08 (3.0) | 0.06 (1.3) |
| C12:0 lauric | 0.12 (3.3) | 0.09 (3.1) | 0.26 (3.1) |
| C14:0 myristic | 0.32 (10.3) | 0.34 (9.5) | 0.32 (9.5) |
| **C6–14:0 (MCT)[2]** | **1.02** | **0.72** | **0.64** |
| C16:0 palmitic | 0.91 (24.6) | 0.88 (26.5) | 0.92 (20.2) |
| C18:0 stearic | 0.44 (12.5) | 0.40 (14.6) | 0.29 (6.0) |
| **MUFA[3], total** | **1.06** | **0.92** | **1.65** |
| C16:1 palmitoleic | 0.08 (2.2) | 0.08 (2.3) | 0.13 (5.7) |
| C18:1 oleic | 0.98 (28.5) | 0.84 (29.8) | 1.48 (46.4) |
| C20:1 | — | trace | 0.04 |
| C22:1 | — | trace | trace |
| **PUFA[4], total** | **0.15** | **0.13** | **0.45** |
| C18:2 linoleic | 0.11 | 0.08 | 0.37 |
| C18:3 linolenic | 0.04 | 0.05 | 0.05 |
| C18:4 | — | trace | — |
| C20:4 | — | trace | 0.03 |

[1](in g/100 g fat).
[2]MCT = medium-chain triglycerides.
[3]MUFA = monounsaturated FA.
[4]PUFA = polyunsaturated FA.
Source: Posati and Orr, 1976.

(β-Lg), and α-lactalbumin (α-La). Instead of $\alpha_{s2}$-CN, the $\alpha_{s1}$-casein ($\alpha_{s2}$-CN) is present in the milk of some individual goats and breeds generally as a minor protein, while it is the major protein in cow milk (Park, 2006). β-casein is the major casein fraction in goat milk. The protein composition in goat milk is determined by genetic polymorphisms. Differences in amino acid composition between casein fractions in goat milk are much greater than differences between species (goat versus cow). Casein micelles of goat milk are less solvated, less heat stable, and lose β-casein more readily than cow milk micelles (Juàrez and Ramos, 1986).

Goat milk proteins are more digestible than those in cow milk due to the smaller, softer, and more friable curd formations of goat proteins during acidification in the stomach, providing stomach proteases easier digestive actions (Park, 2006). Nubian breed goat milk contains significantly higher levels of major buffering components, such as proteins, nonprotein N, and phosphate ($P_2O_5$) than milk from Holstein or Jersey cows (Table 14.5), which can be important in human nutrition. The use of goat milk as a hypoallergenic food substitute for cow milk has been reported in often successful treatment of patients allergic to cow milk (Walker, 1965; Park, 1994), and may be attributed to the low frequency of $\alpha_{s1}$-CN found in goat milk (Moioli et al., 1998).

The most abundant amino acids in milk of many species are glutamate (plus glutamine, 20%), proline (10%), and leucine (10%). Of these three major amino acids in goat milk, glutamate and proline are higher and leucine is lower in human milk than goat milk. Of the sulfur-containing amino acids, cystine is higher and methionine is lower in human than in goat milk (Davis et al., 1994).

### Carbohydrates

The major carbohydrate in goat milk is lactose. Lactose is a disaccharide made up of a glucose and galactose molecule and is synthesized in the mammary gland. Lactose in

**Table 14.5** Concentration of total N, NPN, and phosphate in goat and cow milk and soy-based infant formulae[1].

| Milk | $N^2$ | Total N | | NPN | | $P_2O_5$ | |
|---|---|---|---|---|---|---|---|
| | | $\bar{X}$ | SD | $\bar{X}$ | SD | $\bar{X}$ | SD |
| **Goat Milk** | | | | | | | |
| Alpine | 25 | $.390^c$ | .032 | $.048^b$ | .008 | $.166^a$ | .020 |
| Nubian | 25 | $.556^a$ | .013 | $.061^a$ | .013 | $.212^a$ | .015 |
| **Cow Milk** | | | | | | | |
| Holstein | 25 | $.392^c$ | .058 | $.033^c$ | .002 | $.173^a$ | .022 |
| Jersey | 25 | $.505^b$ | .043 | $.038^c$ | .004 | $.211^a$ | .118 |
| **Formula Milk** | | | | | | | |
| Brand A | 5 | $.227^d$ | .026 | $.020^d$ | .003 | $.211^a$ | .008 |
| Brand B | 5 | $.259^d$ | .016 | $.019^d$ | .003 | $.192^a$ | .053 |

[a,b,c,d]Means with different superscripts within a same column are significantly different ($P < .01$).
[1]Expressed in g/100 ml.
[2]Number of determinations per mean value.
Source: Park, 1991.

goat milk is about 0.4% units lower than in cow milk (Park, 2006). Goat milk has about 10 times more oligosaccharides than cow milk, which closely resembles that of human milk. This is of special interest to infant nutrition, since goat milk oligosaccharides have important functions in human nutrition. A larger variety of acidic and neutral oligosaccharides are present in goat than in cow or sheep milk.

### Minerals

Goat milk contributes importantly to the mineral nutrition of people in developing and underdeveloped countries, especially for calcium (Ca) and phosphorus (P). Goat milk contains about 134 mg Ca and 141 mg P/100 g of milk (Table 14.3). Human milk contains only one-fourth to one-sixth of these minerals as compared to goat milk. Goat milk has higher calcium, phosphorus, potassium, magnesium, and chlorine, but lower sodium and sulfur content than cow milk.

A close inverse relationship exists between lactose content and the molar sum of sodium and potassium content in milk of goat or other species (Park and Chukwu, 1988). Chloride is positively correlated with potassium and negatively with lactose. The levels of major minerals in milk do not usually fluctuate with diet, but they can vary depending on breed, individual animals, and stage of lactation, whereas trace mineral content of goat milk may be influenced by the diet and other factors. The casein micelle in goat milk has a lower degree of hydration, which is inversely related to its mineralization (Remeuf and Lenoir, 1986).

### Vitamins

Goat milk supplies adequate amounts of vitamin A and niacin, and an excess of thiamine, riboflavin and pantothenate in addition to protein, Ca, and P for human infants in relation to FAO-WHO requirements (Table 14.3). Vitamin A content of goat milk is higher than in cow milk. Goat milk is white in contrast to most cow milk (except for white Ayrshire milk), because goats convert all β-carotene (yellow color) into vitamin A (no color) in milk. Because of rumen synthesis, vitamin B content in goat and cow milk is mainly independent of diet. Goat milk, however, is deficient in folic acid and vitamin $B_{12}$ as compared to cow milk (Jenness, 1980), which contains about 5 times more folate and vitamin $B_{12}$ than goat milk. Folate is necessary for the synthesis of hemoglobin. Goat and cow milk are also deficient in pyridoxine ($B_6$), vitamin C and D, indicating a need for supplementation in human nutrition.

### Minor Constituents in Goat Milk

Bioactive components in milk of different species including goat milk have been studied by numerous researchers. These compounds can be released by enzymatic hydrolysis of milk proteins, exerting specific biological activities, such as antihypertensive, antimicrobial, opioid, antioxidant, immunomodulant, or mineral-binding capacity. Such protein fragments are formed from the precursor inactive protein during gastrointestinal digestion and/or during food processing (Korhonen and Pihlanto-Leppälä, 2003).

Goat milk contains lactoferrin, transferrin, and prolactin at comparable levels to cow milk. Human milk contains more than 2 mg lactoferrin/mL, which is 10- to 100-fold higher than in goat milk. Goat milk has high levels of folate-binding protein, which lowers the available level of folic acid in its milk (Juàrez and Ramos, 1986).

Concentrations of enzymes such as lysozyme, ribonuclease, and xanthine oxidase in goat, cow, and human milk are highly variable among and within species. Xanthine oxidase activity of goat milk is less than 10% of that in cow milk. Lipase and alkaline phosphatase levels are lower in goat milk than in cow milk (Chandan et al., 1992). Immunoglobulin (IgG) type content in goat and cow milk is much higher than in human milk, whereas human milk contains greater levels of IgA and IgM immunoglobulin than goat or cow milk.

## EFFICIENCY OF MILK PRODUCTION

Efficiency and level of milk production varies among and within breeds of goats. Marked differences in milk production levels exist between dairy and nondairy breeds in temperate and in tropical regions (Sands and McDowell, 1978). Nondairy breeds are capable of milk production above the needs of their kids, if they are fed sufficient feeds.

In European countries, the average annual milk yield of dairy goat breeds is between 200 and 500 kg, while it has exceeded 1,000 kg in some breeds in Switzerland and the U.S. (Tables 14.2 and 14.6). Milk production is usually high for registered individuals of the dairy breeds, indicating that goats fed and managed under controlled conditions can produce much higher milk volumes than under commercial conditions or in large herds or under grazing conditions with minimum feed supplementation. Under special care, astonishing official record production levels of above 2,000 kg have been recorded for individual dairy goats.

The productivity of indigenous breeds in tropical countries may be enhanced by crossbreeding with European or U.S. dairy breeds, but not all attempts have been successful because of differences in nutrition level, health care, and especially internal parasite control (Devendra, 1987) (Table 14.1). Sufficient amounts of supplementary feed concentrates are a requirement for high-producing goats, which means more cash expenses that are often difficult to cover. The dietary supply of energy from rangeland may not provide more than the energy expenditures of walking to find forage, whereby rangeland then is no more than an environment to maintain the animals. Typical rangeland at certain times of the year do not supply sufficient levels of some minerals, vitamins, and nutrients for milk production. This necessitates supplementary feeding to cover requirements, which have been published comprehensively in NRC (2007) including the extensive French, British, German, and other foreign literature.

Labor costs are a high burden in commercial goat milk production. Therefore, prices for goat milk products must be considerably higher than those for cow milk products, if dairy goat operations are to compete with dairy cow counterparts and survive (Kapture, 1991). Labor cost is not as important in small-scale family operations, where the biological value of goat milk compared to cow, sheep, or buffalo milk is of major interest. The ratio of goat milk production to digestible feed intake, under tropical conditions, is significantly better than for cows, indicating that goats can be more efficient under less favorite environments. The efficiency of milk production was investigated using native goats and Friesian and Swiss cows fed diets

**Table 14.6** U.S. official dairy goat breed lactation averages for milk, fat, and protein production[1].

| Breed | Milk (kg) | FCM[2] (kg) | Fat % (kg) | Protein % (kg) |
|---|---|---|---|---|
| Alpine | 1,083 | 958 | 3.3 (35) | 2.9 (31) |
| LaMancha | 973 | 944 | 3.9 (37) | 3.1 (30) |
| Nigerian Dwarf | 340 | 451 | 6.5 (21) | 3.9 (13) |
| Nubian | 830 | 917 | 4.7 (39) | 3.7 (30) |
| Oberhasli | 1,046 | 973 | 3.6 (37) | 2.9 (30) |
| Saanen | 1,185 | 1,044 | 3.2 (38) | 2.9 (34) |
| Toggenburg | 1,024 | 890 | 3.2 (32) | 2.7 (30) |

[1]American Dairy Goat Association Dairy Herd Improvement Registry, Spindale, NC, USA (2008).
[2]FCM = 4% Fat Corrected Milk.

with forage:concentrate ratios of 50:50, 60:40, and 80:20 under intensive Indian conditions (Sundaresan, 1978). Under high-concentrate feeding conditions, goats appear to be similar to cows in milk production efficiency. It was also evident that goats were more efficient feed converters when fed mainly coarse forage with low nutrient content.

## MILK RECORDS

The official milk-recording schemes of dairy goats are basically the same as the traditional procedures for dairy cows, consisting of monthly recordings by an official person of the quantity of milk produced during a 24-hour period; fat and protein tests in a composite proportional sample drawn at each milking; and calculation of lactation (305 day) yield or of the yield during the operational year. Fat and protein content of goat milk are determined with the Milk-Tester instrument, or the Milk-O-Scan, or Infra-Red Absorption Analyzer using the amido black method, which is simpler but equally accurate as the Kjeldahl method for protein or the Gerber method for fat. However, specific calibrations are required for either cow or goat milk. Because goat milk contains high levels of nonprotein nitrogen, direct measurements of true protein are preferable to total nitrogen (Grappin and Jeunet, 1979).

Milk yield and composition recording in goats have been limited and disadvantaged in many countries, because the cost of milk recording for goats with traditional methods is higher than for cows. This is attributable to the fact that goat herds are often small and widely dispersed, thus transportation costs are high, and the financial burden on the amount of milk produced is a limit, while recording expenses are the same for goats or cows. In addition, goat farmers receive little benefit from milk recording unless they have registered animals and are interested in selling breeding stock. Goat milk composition is required for payment from the milk processor, whose laboratory does the testing.

Keeping records, official or private, of milk production and reproduction performance of dairy goats is an absolute necessity for economic profit and for the personal satisfaction of achievements in the business of selling superior breeding animals to improve other herds both nationally and internationally. To aid in such efforts, purebred registry organizations, such as the American Dairy Goat Association (ADGA), the American Goat Society (AGS), or others in different countries, have been organized to collect official and trustworthy records that meet established standards under supervision. Records include pedigrees, type classifications, championships in show competitions, lactation milk production, and milk composition. Milk records are also computed and published by the National Dairy Improvement Laboratory of the USDA for genetic sire evaluations. Table 14.2 shows the astonishing high records that have been achieved for U.S. dairy goats under individual attention to genetic selection and superior feeding and health management. This also indicates the improvement potentials that exist in other dairy goat breeds and countries (Tables 14.2 and 14.6).

## DAIRY HERD IMPROVEMENT ASSOCIATION

For the purpose of facilitating record keeping of milk production and milk composition, each U.S. state under the guidance of the state Agricultural Extension Service has organized voluntary associations of dairy cattle and dairy goat farmers/breeders with the purpose of helping dairy herd improvement associations (DHIA). Two types of records are obtained: (1) official records with the assistance and supervision of a DHIA employee for unbiased collection of data and milk samples, and (2) unofficial records and milk samples collected by the farmer/breeder. Milk production data for each participating dairy goat from entire herds are transmitted to the National USDA computer data base monthly with an official or unofficial label and published in regional trade magazines for public information. Milk samples are also sent monthly for each participating dairy goat and entire herd to the regional laboratories for analyses of fat and protein contents. The results are also forwarded to the National USDA computer database monthly and are published together with the milk records in regional dairy trade magazines for public inspection. Rules for DHIA participation and certification of records are set forth by ADGA or AGS in the U.S. (ADGA, 2008). Similar procedures are followed in most European countries with a significant dairy goat industry, especially England, France, Germany, Italy, Spain, The Netherlands, and Norway (Haenlein, 2006).

## ECONOMICS OF IMPROVED GOAT MILK PRODUCTION

Goat milk production can be improved by genetic selection and superior management practices. The low productivity of goats in many regions of the world is mostly due to underfeeding in energy and protein sources, which can be corrected by supplementary feeding in addition to the pasturing. Supplementary feeding is profitable contrary to widespread popular opinion. See Table 14.7.

The economics of goat milk production is greatly influenced by the price of feeds, which generally comprise 50%

**Table 14.7** Economics of improved goat milk production.

| | | | | | | | | |
|---|---|---|---|---|---|---|---|---|
| **INFLUENCE OF FEEDING LEVEL** (Sands and McDowell, 1978) | | | | | | | | |
| Feeding multiples of maintenance requirements | | | | | Expected milk yield/lactation, kg | | | |
| 1.15–1.50–1.80–2.45 | | | | | 50–198–305–535 | | | |
| **INFLUENCE OF PLANE OF NUTRITION IN INDIA** (Devendra, 1987) | | | | | | | | |
| Low plane of nutrition before and after kidding: | | | | | 49 kg lactation milk yield | | | |
| Medium plane before kidding and high plane during lactation: | | | | | 109 kg lactation milk yield | | | |
| High plane before and after kidding: | | | | | 152 kg lactation milk yield | | | |
| **ECONOMICS OF GOAT HERDS IN 8 DISTRICTS OF RAJASTHAN, INDIA** (Sagar and Kanta, 1993) | | | | | | | | |
| Milk production, kg/doe/year: | 113 | 135 | 178 | 254 | 375 | 434 | 576 | 595 |
| Expenditures, Rupees/doe/year: | 29 | 23 | 102 | 48 | 164 | 80 | 160 | 174 |
| Net income, Rupees/doe/year: | 84 | 112 | 76 | 206 | 211 | 354 | 416 | 420 |
| **ECONOMICS OF GOAT HUSBANDRY IN GREECE** (Hatziminaoglou et al., 1995) | | | | | | | | |
| Gross income/doe/year | $66 | Extensive Feeding | | | $135 | | Intensive Feeding | |
| Expenses/doe/year | $58 | Extensive Feeding | | | $111 | | Intensive Feeding | |
| Net income/doe/year | $8 | Extensive Feeding | | | $24 | | Intensive Feeding | |
| **AVERAGE NET INCOME IN FOUR QUARTILES OF 120 U.S. GOAT HERDS, OFFICIAL RECORDS** (Haenlein and Abdellatif, 2004) | | | | | | | | |
| Milk yield, kg/doe/lactation: | 514 | | 701 | | 820 | | 1,050 | |
| Cost of concentrate feed/doe/year: | $83 | | 91 | | 98 | | 109 | |
| Income over feed cost/doe/year: | $183 | | 262 | | 302 | | 395 | |

Source: Park and Haenlein, 2007.

of the cost of milk production, and also by the price of goat milk payable to the producer or by the market price paid by consumers. If the supplemental feed price/kg is less than half of the price of milk/kg from the farm, it is profitable to buy concentrate feeds or grains to supplement the basic forage ration of the goats. This ratio of <0.5 : 1 also rules an improved feeding system, where 1 kg supplement should be fed for each 2–3 kg of milk produced, depending on the fluctuations of prices of milk and feed, composition in protein and energy of feed, and body condition of the doe (Park and Haenlein, 2007).

The production cost per unit of product is generally decreased, by increasing the numbers of production units. In goat milk production, the cost per kg milk produced decreases with higher levels of production per goat. The economic "break-even" price of goat milk decreases and net income increases with a higher level of milk production (Table 14.7). Currently herd sizes of at least 120 dairy goats with lactation yields of 700 kg per doe may be needed for an enterprise with an economic future (Yazman, 1984). Since goats have smaller body size and produce a lower volume of milk per animal, labor is a major constraint in management of milk production. The farm gate price for goat milk for a sustainable farm operation must be at least 1.5–2.5 times higher than for cow milk because of the higher labor costs in goat milk production. Total labor requirement in the production of 100 kg of goat milk ranges from 57–145 (average of 101) minutes compared to cows requiring only 7–31 (average of 19) minutes (Kapture, 1991). For a given labor cost per hour, the production of goat milk will require almost 5 times (101/19) that of cow milk for 100 kg milk production, explaining partially the need for the higher price for goat milk than for cow milk.

With the current prices of milk production in the U.S., including hay, buck service, veterinary costs, and milking supplies, minus the gain from sale of kids and the fertilizer value of goat manure, but not including fixed overhead, the total production cost of goat milk has been estimated to be about 2.5 times that for cow milk. At an estimated cheese yield ratio of 10 kg goat milk or cow milk:1 kg cheese, the cost for goat cheese making will also be at least 2.5 times that of cow milk cheese. According to the International Dairy Federation (IDF) (2000), average commercial prices for 100 kg of goat milk in France were currently at the equivalent of $51, in Italy $47, in Spain $44, and in Greece $36, while the U.S. price for goat milk at the farm gate varies widely from the same price as for cow milk (about $20) to about 2 times that without any officially set price.

The average milk production levels per doe in the Mediterranean region, although the premier region of goat milk production in the world, are much below those of American dairy goats, partly due to the lack of concentrates in their diets and their low genetic base. Also, low quality, unimproved pasturing used in this region, and many of their management systems, may be contributing factors. Average annual milk production per doe in France has been reported at 397 kg, in Spain at 115 kg, in Italy at 102 kg, and in Greece at 87 kg (IDF, 2000), while average official goat milk production levels in the U.S. are currently around 1,000 kg per doe (Tables 14.1 and 14.6).

If goats are raised for dual purposes of meat and milk production, a decision must be made by the goat farmer on whether it is more economical to sell most of the available milk and nurse the kids with commercial milk replacer feed, or whether it is more profitable to feed the milk to as many goat kids as possible for sale of kid meat. The decision on this premise can be made on the basis of a formula for R-value (Gall, 1990):

$$R = \frac{\text{Kid weight gain (kg)} \times \text{Kid weight price (\$)}}{\text{Amount of milk fed (kg)} \times \text{Milk price (\$)}}$$

If R is less than 1.00, selling goat milk in the market is more economical than feeding it to kids. If kid weight price is at least more than 6 times the milk price, then milk feeding for goat meat production is more profitable.

There is also a need for incentive payments for goat milk producers to meet and improve health standard requirements for fluid milk and optimum milk composition for the interests of the cheese and yogurt yields. In Europe, bonus or penalty monies are paid by using scorecards for certain desirable levels of certain parameters in goat milk contents such as coliform bacteria, listeria, fat lipolysis, bacteria counts, somatic cell counts (SCC), pH, fat, solids, protein, casein, water, and the absence of antibiotics and sediment. European limits for goat milk to be pasteurized are <1.5 million bacteria count/mL, but are <0.5 million/mL for processing of raw milk products (IDF, 2000). Current SCC limits in goat milk are set at 1.35 million/mL in France, 1.6 in Spain, 1.743 in Italy, compared to European cow milk SCC limits at 0.4 million/mL (Haenlein, 2001). The acceptable values of bacteria counts for U.S. goat milk are 1.5 million/mL and 0.5 million/mL for pasteurized and raw milk, respectively. Current SCC limits are set at 1.0 million/mL.

## GOAT MILK PRODUCTS AND THEIR CHARACTERISTICS

Compared to cow or human milk, goat milk is reported to possess unique characteristics, such as high digestibility, distinct alkalinity, high buffering capacity as well as certain therapeutic values in medicine and human nutrition (Walker, 1965; Park, 1994). Goat milk has been recommended as an ideal substitute for patients who suffer from allergies against cow milk or other food sources.

Cheese from goat milk is the major manufactured product that has significant research data available in the literature. Goat milk cheese originated in Mesopotamia and was developed especially in the Mediterranean countries of Greece, Turkey, Syria, Israel, Iraq, and also Iran (Kosikowski, 1986). The milk was probably made into soft cheese first and then into hard ripened cheeses later.

Greece became dominant in feta cheese production from sheep and goat milk. France, however, exceeds Greece in total goat cheese production with its farmstead cheese making. France today offers the best in goat milk cheeses, many of which are surface ripened. In the U.S., goat cheeses have gained popularity among ethnic groups, health food consumers, connoisseurs, and private goat farmers since early 1980. The continued shift in consumer tastes to "exotic" foreign and specialty cheeses has led to an increased volume of goat cheese importation to the U.S. The volume of imported goat cheese to the U.S. from France alone was 447 metric tons in 1988, which comprised approximately 80% of the total imported goat cheeses (Park, 1990). In 2008, it is estimated that more than 1,500 metric tons of French goat cheeses are imported to the U.S. However, domestic U.S. goat cheese production especially from artisanal goat farm producer-processors has greatly increased in competitive volumes in recent years and was much aided by the promotion of the American Cheese Society annual cheese contests.

In addition to cheeses, there are other manufactured goat milk products consumed in lesser volumes than cheeses, including yogurt, butter, kefir, ghee, sweets, powdered and condensed products, and even cosmetics.

### Diversity of Goat Cheeses Compared to Similar Cow Cheeses

Many different types of goat milk cheeses are produced and consumed worldwide, depending on locality (Park and Guo, 2006). Diversities among different varieties of goat cheeses are attributable to the high variation in seasonal milk composition, modifications of manufacturing procedures, and multitudes of aging times and conditions. Varietal differences among goat cheeses are influenced by the nature of physical and chemical changes during ripening, which are influenced by the cultures, chemicals, or flavor ingredients added to the curd during manufacturing (Kosikowski, 1977).

In contrast to the production of many semihard and hard types of cow milk cheeses, the manufacture of goat cheese is referred to as a "cottage industry." The vast majority of goat cheeses in the U.S. are of the soft-body type with one-third of them spice-added. Nutrient compositions of different varieties of goat cheeses produced in the U.S. are shown in Table 14.8. Almost all French goat cheeses are of the natural drainage type associated with slow coagulation. France is known to produce many exotic types of goat cheeses, including Crottin du Chavignol, Les Pyramides, Sainte Maure, Chabis, and Chabicou (Kosikowski, 1986). Other specific varieties of cheeses made from goat milk or a legal mixture of goat and ewe milk from other countries are Laruns, Peroil, Cabroles, Lightvan, Bryndza, Bulgarian White, Akavi, Cachcaval, Canestrano, Canniotta, Gjetost, and Feta (Kosikowski, 1977; Park and Guo, 2006).

The Agricultural Handbook No. 54 of the USDA (Sanders, 1969) describes over 400 varieties of goat cheeses and lists over 800 names of cheeses, made from goat milk or combinations of goat with other species milk such as cow, ewe, or buffalo. Most goat cheese varieties, which are consumed fresh, are set by an acid (hydrochloric, lactic, vinegar, lemon, lime, whey) coagulation process, whereas cheese varieties consumed after ripening are generally made by the enzyme (rennet, chymosin, certain herbs) setting process.

### Goat Yogurt

Goat milk yogurt is produced and consumed in the Mediterranean and many other countries. In the U.S., a few domestic commercial goat yogurts and some imported products have been available. Goat milk yogurt can be made in a similar manner to cow milk yogurt. One of the main problems in manufacturing goat milk yogurt is the weak curd and lack of consistency in curd tension and viscosity upon coagulation compared with cow milk yogurt. This is due to the difference in casein composition between the two milks (Park and Guo, 2006).

Yogurt is usually made with the symbiotic growth of two bacteria: *Streptococcus thermophilus* and *Lactobacillus delbrueckii* ssp. *bulgaricus*. These bacteria can not survive gastric passage nor colonize the intestines, while *Lactobacillus acidophilus* and *Bifidobacteria* do survive. Therefore, yogurt with these bacteria are popular due to their potential therapeutic benefits (Park and Guo, 2006).

Various techniques have been employed to improve the texture of yogurt for many centuries, including increasing solids contents and enzymatic cross-linking using different sources of proteins. The most common methods are to increase the amount of total solids in the milk by supplementation with milk powder, by condensation to reduce water content, or by adding sheep milk with its high solids content. The higher the level of solids in milk, the greater and better the viscosity and consistency of the yogurt.

### Powdered and Condensed Goat Milk Products

Significant quantities of goat milk powder are commercially produced especially in the U.S. and New Zealand. Powdered products, including whole milk, skim milk, whey, cream, ice cream mix, protein concentrates, and infant foods, are produced by several methods of drying the liquid such as spray drying, drum drying, and freeze drying.

Evaporation processing of goat milk is usually performed under reduced pressure, primarily to allow boiling at a lower temperature and thus prevent flavor damage due to heating. Evaporated goat milk is processed with similar evaporation facilities as performed for evaporated cow milk products. General composition of evaporated goat or cow milk has 7.5–9.0% fat, 17.5–22% milk solids-nonfat, and 25–31% total solids, while those of powdered goat milk contains average 28.2% fat, 65.9% solid-nonfat, and 94.1% total solids, respectively (Park and Guo, 2006).

### Other Goat Milk Products

Goat milk butter is commercially manufactured in several countries including the U.S. and Great Britain. Buttermilk is usually made from skim milk using this by-product after churning butter out of sour cream or directly with bacteria inoculation from goat skim milk (Park and Guo, 2006).

Acidophilus milk can be made from goat milk by adding *Lactobacillus acidophilus*, which can convert a high amount of the lactose to lactic acid (2%). Pasteurized milk is inoculated with *Lactobacillus acidophilus*, which destroys other competing bacteria antagonistic to man in the lower intestine (Kosikowski, 1977).

Kefir is made from pasteurized, fat-standardized, or decreamed goat milk and is an acidic, slightly foamy product. Kefir is produced by a combined acidic and alcoholic fermentation of symbiotic lactic acid bacteria and yeast "kefir grains." Kefir usually contains 0.6–0.8% lactic acid, 0.5–1.0% alcohol, and carbon dioxide.

Ghee is an Indian (and Middle East) clarified butterfat product, which is manufactured by fermenting whole goat

**Table 14.8** Nutrient profiles of selected varieties of commercial goat milk cheeses (U.S. Products)[1].

| Goat Cheese Variety | $H_2O$ | Fat | Protein | Ash | S | P | K | Mg | Ca | Na | Cl | Fe | Al | Mn | Cu | Zn |
|---|---|---|---|---|---|---|---|---|---|---|---|---|---|---|---|---|
| | g/100 g | | | | mg/100 g | | | | | | | µg/g | | | | |
| Fresh soft, plain | 59.8 | 22.5 | 19.8 | 1.74 | 3.54 | 275 | 25.8 | 14.6 | 172 | 416 | 293 | 17.8 | 14.8 | 0.96 | 7.40 | 9.05 |
| Fresh soft, garlic | 64.3 | 18.3 | 16.7 | 1.34 | 4.10 | 247 | 15.2 | 17.2 | 117 | 331 | 145 | 13.0 | 38.4 | 0.88 | 8.08 | 14.4 |
| Fresh soft, pepper | 57.3 | 22.9 | 21.6 | 1.32 | 4.13 | 236 | 27.8 | 17.9 | 107 | 298 | 201 | 28.5 | 44.5 | 1.22 | 6.50 | 8.76 |
| Fresh soft, herb | 59.1 | 21.8 | 17.3 | 1.60 | 2.99 | 225 | 35.0 | 15.3 | 112 | 336 | 165 | 17.7 | 9.1 | 1.06 | 6.68 | 7.75 |
| Feta | 52.3 | 25.3 | 25.1 | 4.30 | 3.49 | 544 | 62.2 | 17.6 | 639 | 916 | 1260 | 6.1 | 14.9 | 0.97 | 6.46 | 15.5 |
| Camembert capri | 47.3 | 28.8 | 29.1 | 1.67 | 2.37 | 264 | 7.0 | 21.8 | 229 | 279 | 301 | 8.9 | 12.7 | 0.90 | 5.38 | 16.3 |
| Blue | 25.9 | 31.8 | 20.2 | 3.32 | 8.05 | 575 | 88.8 | 34.9 | 841 | 924 | 398 | 15.2 | 36.2 | 0.93 | 5.60 | 6.89 |
| Cheddar | 41.7 | 26.6 | 30.3 | 3.60 | 7.80 | 526 | 11.0 | 42.0 | 599 | 361 | 1030 | 7.7 | 14.9 | 0.93 | 4.93 | 6.53 |
| Shepherd's hard | 31.6 | 33.1 | 28.9 | 3.93 | 4.13 | 737 | 54.4 | 61.7 | 1035 | 285 | 114 | 26.7 | 12.3 | 3.98 | 6.56 | 18.7 |
| Montasio's hard | 25.9 | 36.8 | 31.9 | 3.52 | 4.87 | 720 | 42.0 | 55.3 | 990 | 260 | 96 | 11.0 | 15.9 | 1.06 | 5.99 | 41.3[a] |
| Ancho chile | 42.3 | 29.8 | 22.2 | 4.11 | 3.17 | 688 | 53.3 | 70.5 | 939 | 464 | 162 | 24.8 | 16.1 | 1.07 | 6.44 | 9.13 |

[a]Extremely High Value.

[1]Two-thirds of the goat cheeses of this study were soft cheese varieties.

Source: Park, 1990.

or buffalo milk into curd and churning out butter, followed by heat clarification at 105–145°C. In Iran it is called Kashk or dried butter.

Sweet products are made from goat milk and are very popular in Mexico, Norway, and India. The Caheta, produced in high quantity in Mexico, is a thick liquid of caramelized milk with sugar added. It is sold as liquid or dried as small tarts. Other sweet "dulces" are made of goat milk in Latin American countries in a similar way. Brown goat cheese, "gjetost," is produced in Norway by cooking and caramelizing goat whey into hard sweet cakes. Other whey cheeses are popular in Italy and Greece.

Cosmetic products made from goat milk such as goat milk soap, hand lotion, etc., have been increasingly popular and commercially produced in high volumes in the U.S. and other countries such as Switzerland. References on goat milk soap have recently surged to more than 5,000 in Web-based literature. The number of home-based goat milk soap businesses in the U.S. has been estimated to generate multi-million dollars of annual revenues.

## SUMMARY

Goat milk production is of economic, medical, and gastronomic significance in many developed and underdeveloped countries to an extent that it is not only popular but also indispensable to millions of people. The physiology of lactation of dairy goats is similar to that of cows, but significant differences exist, which must be considered for correct and profitable management of dairy goats. The biochemistry of goat milk composition has unique differences from cow milk composition, situating goat milk as a valuable alternative to cow milk in many medical conditions, and also influencing the yogurt- and cheese-making process from goat milk considerably. The economics of goat milk production is greatly different from that of cow milk production and depends on a higher farm gate price for goat milk compared to cow milk. The farm gate price for goat milk also dictates whether it is more profitable to sell the milk or to feed it to kids for goat meat sales. Profitability of goat milk production depends very much on the economics of scale (for example, the size of the herd) and on two factors or systems: (1) low cost of feed, land, and labor under extensive grazing systems or (2) supplementary concentrate feeding under intensive management systems. It has been convincingly shown that either system can be very profitable in developed and in underdeveloped countries making it a viable part of the overall dairy industry.

## REFERENCES

ADGA. 2008. News and Events. American Dairy Goat Association Publ., Spindale, NC, U.S., Vol. 54, 4th quarter, pp. 10–13.

Alderson, P. and E.J. Pollak. 1980. Age-season adjustment factors for milk and fat of dairy goats. J. Dairy Sci. 63:148–151.

Babayan, V.K. 1981. Medium chain length fatty acid esters and their medical and nutritional applications. J. Amer. Oil Chem. Soc. 59:49A–51A.

Bath, D.L., F.N. Dickinson, H.A. Tucker, and R.D. Appleman. 1985. Biosynthesis of milk. In: Dairy Cattle: Principles, Practices, Problems, Profits. Lea & Febiger Publ., Philadelphia, U.S., pp. 291–324.

Cerbulis, J., V.P. Flanagan, and H.M. Farrell. 1985. Composition of the hydrocarbon fraction of goat milk. J. Lipid Res. 26:1438.

Chandan, R.C., R. Attaie, and K.M. Shahani. 1992. Nutritional aspects of goat milk and its products. Proc. V. Intl. Conf. Goats, New Delhi, India. Vol. II: Part II. pp. 399–420.

Chilliard, Y., P. Morand-Fehr, G. Durand, and D. Sauvant. 1979. Evolution of the metabolic activity of goat fatty tissue during the 1st month of lactation—Relation with the milk secretion. Bull. de L'Acad. Vet. de France 52:417–422.

Cosie, A.T. and H.L. Buttle. 1974. Lactation. In: Reproduction in Farm Animals. E.S.E. Hafez, Ed., Lea & Febiger Publ., Philadelphia, U.S., pp. 203–221.

Cowie, A., G. Knaggs, and J. Tindal. 1964. Complete restoration of lactation in goat after hypophysectomy. J. Endocr. 28:267–279.

Das, D. and N.S. Sidhu. 1975. Relation between udder and teat traits with milk yield in Barbari and Black Bengal breeds of goat, Capra hircus. Indian J. Hered. 7: 1–9.

Davis, T.A., H.V. Nguyen, R. Garcia-Bravo, M.L. Florotto, E.M. Jackson, D.S. Lewis, D.R. Lee, and P.J. Reeds. 1994. Amino acid composition of human milk is not unique. J. Nutr. 124:1126–1132.

Devendra, C. 1987. Strategies other than breeding for the development of small ruminants. In: Proc. Small Ruminant Production Systems in South and Southeast Asia Workshop. C. Devendra, Ed., Oct. 6–10, 1986, Bogor, Indonesia, IDRC Publ., Ottawa, Canada, pp. 332–353.

Devendra, C. and G.F.W. Haenlein. 2003. Goat breeds. In: Encyclopedia of Dairy Sciences, H. Roginski, J.W. Fuquay, P.F. Fox, Eds., Academic Press, Amsterdam, The Netherlands, Vol. 2, pp. 585–598.

FAO. 2004. Production Yearbook 2003. Food & Agriculture Publ., Rome, Italy, Vol. 57:213–215; 241–242.

Gall, C. 1981. Milk production. In: Goat Production, C. Gall, ed., Academic Press, London, U.K., pp. 309–344.

Gall, C. 1990. Potential of dual purpose goats. In: Proc. Small Ruminants Research and Development in the Near East

Workshop. A. M. Aboul-Naga, Ed., Nov. 2–4, 1988, Cairo, Egypt, IDRC Publ., Ottawa, Canada, pp. 67–73.

Grappin, R. and R. Jeunet. 1979. Routine methods for quantitative determination of butterfat and protein in goats milk. Le Lait 59:345–360.

Haenlein, G.F.W. 2001. Relationship of somatic cell counts in goat milk to mastitis and productivity. Intl. J. Anim. Sci. 16:37–51.

Haenlein, G.F.W. 2006. Production of goat milk. In: Handbook of Milk of Non-Bovine Mammals. Y.W. Park, G.F.W. Haenlein, eds., Blackwell Publ., Ames, Iowa, U.S. and Oxford, U.K., pp. 11–33.

Haenlein, G.F.W., and M.A. Abdellatif. 2004. Trends in small ruminant husbandry and nutrition and specific reference to Egypt. Small Ruminant Res. 51:185–200.

Haenlein, G.F.W. and R. Caccese. 1984. Goat milk versus cow milk. In: G.F.W. Haenlein and D.L. Ace, Eds., Extension Goat Handbook, USDA Publ., Washington, D.C. U.S., E-1, pp. 1–7.

Haenlein, G.F.W. and R.G. Ramirez. 2007. Potential mineral deficiencies on arid rangelands for small ruminants with special reference to Mexico. Small Ruminant Res. 68:35–41.

Hatziminaoglou, J., N.P. Zervas, and J. Boyazoglu. 1995. Goat production systems in the Mediterranean area: The case of Greece. In: Goat Production Systems in the Mediterranean. A. El-Aich et al., Eds., Wageningen Press, Wageningen, The Netherlands, EAAP Publ. 71:82–109.

IDF. 2000. Proceedings, Development Strategy for Sheep and Goat Dairy Sector, Intl. Symposium, April 13–14, 2000, Cyprus, Greece, Intl. Dairy Federation Publ., Brussels, Belgium, Bull. 354.

Irvine, D.M. 1974. The composition of milk as it affects the yield of cheese. Proc. 11th Annual Marshall Invitational Cheese Seminar, Marshall Div. Miles Lab. Madison, WI, U.S.

Jenness, R. 1980. Composition and characteristics of goat milk: Review 1968–1979. J. Dairy Sci. 63:1605–1630.

Juàrez, M. and M. Ramos. 1986. Physico-chemical characteristics of goat milk as distinct from those of cow milk. Intl. Dairy Bull. No. 202, pp. 54.

Kapture, J. 1991. Milking parlor efficiency: Numbers don't lie. Dairy Goat J. 9:478–479.

Korhonen, H. and A. Pihlanto-Leppälä. 2003. Food-derived bioactive peptides—opportunities for designing future foods. Curr. Pharmac. Des. 9:1297–1308.

Kosikowski, F.V. 1977. Cheese and Fermented Milk Foods. 2nd ed. Edwards Brothers, Inc. Ann Arbor, MI. U.S. pp. 90–108.

Kosikowski, F.V. 1986. Requirements for the acceptance and marketing of goat milk cheese. Dairy Goat J. 64:462–465.

Larson, B.L., Ed. 1985. Lactation. Iowa State University Press. Ames, Iowa, pp. 276.

Linzell, J.L. 1972. Milk yield, energy loss in milk, and mammary gland weight in different species. Dairy Sci. Abstr. 34:351–360.

Moioli, B., F. Pilla, A. Rando, and C. Tripaldi. 1998. Possible exploitation of milk protein polymorphisms to improve dairy traits in sheep and goats, a review. Small Ruminant Res. 27:185–195.

Morand-Fehr, P. 1991. Goat Nutrition. Pudoc Publisher, Wageningen, The Netherlands, EAAP Publication No. 46, pp. 308.

Morand-Fehr, P. and D. Sauvant. 1980. Composition and yield of goat milk as affected by nutritional manipulation. J. Dairy Sci. 63:1671–1680.

NRC. 2007. Nutrient Requirements of Small Ruminants. Nat. Res. Council, National Academy Press, Washington D.C., U.S., Animal Nutrition Series, pp. 362.

Park, Y.W. 1990. Nutrient profiles of commercial goat milk cheeses manufactured in the United States, J. Dairy Sci. 73:3059–3067.

Park, Y.W. 1991. Relative buffering capacity of goat milk, cow milk, soy-based infant formulas, and commercial non-prescription antacid drugs. J. Dairy Sci. 74:3326–3333.

Park, Y.W. 1994. Hypo-allergenic and therapeutic significance of goat milk. Small Ruminant Res. 14:151–159.

Park, Y.W. 2005. Goat Milk: Composition, Characteristics. Encyclopedia of Animal Science. W.G. Pond, N. Bell, Eds., Marcel Dekker, Inc. New York, U.S., pp. 474–477.

Park, Y.W. 2006. Goat Milk—Chemistry and Nutrition. In: Handbook of Milk of Non-Bovine Mammals. Y.W. Park, G.F.W. Haenlein, Eds., Blackwell Publishers. Ames, Iowa, U.S., and Oxford, UK, pp. 34–58.

Park, Y.W. and H. I. Chukwu. 1988. Macro-mineral concentrations in milk of two goat breeds at different stages of lactation. Small Ruminant Res. 1:157–166.

Park, Y.W. and G.F.W. Haenlein. 2007. Goat Milk, Its Products and Nutrition. In: Handbook of Food Products Manufacturing. Y.H. Hui, Ed., John Wiley & Sons, Inc., New York, U.S., pp. 447–486.

Park, Y.W. and M. R. Guo. 2006. Goat Milk Products: Processing Technology, Types and Consumption Trends. In: Handbook of Milk of Non-Bovine Mammals. Y.W. Park and G.F.W. Haenlein, Eds. Blackwell Publishers. Ames, Iowa and Oxford, England. pp. 59–106.

Peaker, M. and J.L. Linzell. 1974. Effects of estrus and exogenous estrogens on milk secretion in goat. J. Endocr. 61:213–240.

Pearl, S.L., H.F. Downey, and T.L. Lepper. 1973. Intramammary pressure and mammary blood flow in lactating goats. J. Dairy Sci. 56:1319–1323.

Posati, L.P. and M.L. Orr. 1976. Composition of foods. Agric. Handbook No. 8–1. ARS, USDA, Washington, D.C., U.S., pp. 191.

Remeuf, F. and J. Lenoir. 1986. Relationship between the physico-chemical characteristics of goat's milk and its rennetability. Intl. Dairy Bull. No. 202:68–78.

Renner, E. 1982. Milch und Milchprodudukte in der Ernährung des Menschen. Volkswirtschaftlicher Verlag, Munich, Germany, pp. 467.

Sagar, V. and A. Kanta. 1993. Economics of goat keeping. Indo-Swiss Goat Development and Fodder Production Project Publication, Jaipur, India, Studies of Goat Production and Fodder Resource Management, Rajasthan. Bull. 58, pp. 68.

Sanders, G.P. 1969. Cheese varieties and descriptions. USDA Agric. Handbook No. 54. Washington, DC.

Sands, M. and R.E. McDowell. 1978. The potential of the goat for milk production in the tropics. Cornell University, Dept. Animal Sci., Ithaca, NY, U.S., Internat. Agr. Mimeo 60, pp. 57.

Schmidt, G.L. 1971. Biology of Lactation. W.H. Freeman Co. Publ., San Francisco, CA, U.S., pp. 1–136.

Shkolnik, A., E. Malz, and S. Gordin. 1980. Desert conditions and goat milk production. J. Dairy Sci. 63:1749–1754.

Sundaresan, D. 1978. Panorama of the different races, scientific research, economic and commercial aspects of milk other than cows milk. XX International Dairy Congress, Paris, France, 6 EC, pp. 18.

Thompson, G.E. and E.M. Thompson. 1977. Effect of cold exposure on mammary circulation oxygen consumption and milk secretion in the goat. J. Physiol. (London) 272:187–196.

Walker, V.B. 1965. Therapeutic uses of goat's milk in modern medicine. Brit. Goat Society's Yearbook, pp. 24–26.

Yazman, J.A. 1984. Commercial goat milk production. In: Extension Goat Handbook. G.F.W. Haenlein and D.L. Ace, Eds., University of Delaware, Dept. Animal & Food Sci., Newark, DE, U.S., and USDA Coop. Ext. Serv. Publ., Washington, D.C., U.S., B2:1–6.

# 15
# Fiber Production

*C.J. Lupton, PhD*

**KEY TERMS**

Cashmere down—the undercoat fibers in raw cashmere having widths of 30 µm or less produced by the cashmere goat.

Classing—the act of separating whole fleeces or fleece portions by fineness, color, staple length, and strength.

Crimp—the bends or waves in textile fibers.

Kemp or objectionable fiber—a medullated animal fiber in which the diameter of the medulla is 60% or more of that of the fiber.

Med fiber—a medullated animal fiber in which the diameter of the medulla is less than 60% of the diameter of the fiber.

Medullated fiber—an animal fiber that in its original state includes a medulla, an air-filled cellular core in the center of the fiber.

Mohair—the fiber grown by the Angora goat, *Capra hircus aegagra*.

Mohair character—refers to the two-dimensional waves or crimp frequency in mohair staples.

Mohair style—refers to the three-dimensional twist or spiral formation (i.e., the type of ringlets) in mohair staples.

Skirting—the practice of manually removing from fleeces the stained or inferior fiber such as that grown on the belly and legs or that which is heavily contaminated with vegetable matter.

**OBJECTIVES**

By completing this chapter, the reader will acquire knowledge on:

- The origins and production capabilities of Angora and cashmere goats for fiber and skins
- How genetics, the environment, nutrition, age, and sex influence fiber production and properties
- The current production conditions and problems in the U.S.
- The biology of fiber growth and the physical and chemical properties of the fibers
- Which fleece and fiber characteristics are important to textile processors, which are objectively measured, and which are subjectively assessed
- How the fibers are harvested and marketed
- The names of the textile processing systems used to manufacture yarns composed of mohair or cashmere
- The main end uses into which the goat fibers and skins are transformed

## INTRODUCTION

Lustrous mohair is produced by one breed of goat, the Angora (*Capra hircus aegagrus*), whereas many different breeds produce cashmere. The word "mohair" is derived from the Arabic word "Mukhayar" translated variably as "best of selected fleece," "select choice," "silky goatskin cloth," "cloth of bright goat hair," and "hair cloth" (Hunter and Hunter, 2001). Most mohair is white although production of colored mohair for the handcraft sector is increasing, particularly in the U.S. Natural cashmere colors include white, pastel browns and grays, and black. In the context of this chapter, pashmina means simply fine quality cashmere grown in India. When dairy, meat, or cashmere goats are bred to Angora goats, a crossbred goat results that grows a fleece that is usually white and is called cashgora. Small quantities of this intermediate type of fiber and goat guard hair also find their way into commercial channels. These fibers and goatskin technology are described in short sections of this chapter.

## THE ANGORA GOAT

The Angora goat probably originated in the Asian Himalayas or highlands of Tibet and arrived in Turkey with migrating pastoral tribes during the 11th century. For the next eight centuries, the breed was maintained and developed on the Turkish plains close to Ankara, from which the name of the goat was derived. The first recorded exportation of Angora goats out of Turkey was to the Holy Roman Emperor Charles V in 1554. Later shipments followed to South Africa (1838), the U.S. (1849), Australia (1850s and 1860s), the U.K. (1881), and subsequently numerous other countries (Shelton, 1993). Mohair production flourished in South Africa (Van der Westhuysen et al., 1988) and the U.S. typically under extensive, low-input conditions. In 1909, 1.34 million Angora goats were shorn in Texas (Figure 15.1). The population increased to 4.61 million by 1965 but subsequently declined to the present day 170,000, comprising 81% of the national Angora goat herd. The year 1965 was also the peak of mohair production in Texas with 14.3 million kg being clipped. In recent years, South African mohair production peaked in 1988 with 12.2 million kg (Table 15.1). By 2007, the amount had declined to 3.0 million kg. Meanwhile, production in Turkey declined from about 4 million kg/yr in the 1970s to 0.35 million kg/yr in 2007.

The original Turkish Angora goats were described as small, refined, and delicate; annually they produced 1–2 kg of mohair in ringlets 20–25 cm in length. By contrast, present-day female Angora goats maintained under relatively harsh West Texas range conditions have body weights of 35 kg (range 30–60 kg) and produce about 2 kg (range 1.5–4 kg) of mohair every 6 months (Table 15.2). Males are often developed and maintained on a higher plane of nutrition to ensure they are in good condition at breeding time. Their body weights average about 70 kg (range 45–90 kg), and they typically produce 6 kg (range 4–9 kg) of mohair every 6 months.

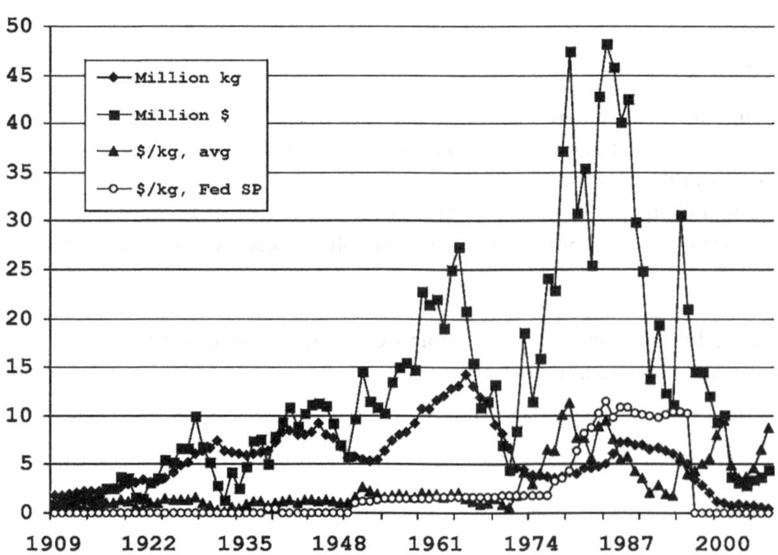


**Figure 15.1** Texas mohair production, total value, selling price, federal support payment, 1906–2006. Source: National Agricultural Statistics Service (units converted by author).

**Table 15.1** World mohair production by country, 1972–2007[1], million kg.

| Year | South Africa | Turkey | USA | Argentina | Australia | New Zealand | Lesotho | Other | Total |
|---|---|---|---|---|---|---|---|---|---|
| 1972 | 3.70 | 4.10 | 4.60 | 1.00 | — | — | 0.80 | — | 14.20 |
| 1973 | 3.40 | 4.10 | 4.50 | 1.00 | — | — | 0.60 | — | 13.60 |
| 1974 | 3.70 | 4.10 | 3.80 | 1.00 | — | — | 0.60 | — | 13.20 |
| 1975 | 3.80 | 3.90 | 3.90 | 1.00 | — | — | 0.60 | — | 13.20 |
| 1976 | 4.10 | 4.00 | 3.60 | 1.00 | — | — | 0.60 | — | 13.30 |
| 1977 | 4.60 | 4.10 | 3.60 | 1.00 | — | — | 0.40 | — | 13.70 |
| 1978 | 4.90 | 4.50 | 3.70 | 1.00 | — | — | 0.50 | — | 14.60 |
| 1979 | 5.40 | 4.50 | 4.20 | 1.00 | — | — | 0.50 | — | 15.60 |
| 1980 | 6.10 | 4.50 | 4.00 | 1.00 | — | — | 0.60 | — | 16.20 |
| 1981 | 6.90 | 4.50 | 4.50 | 1.00 | — | — | 0.60 | — | 17.50 |
| 1982 | 7.60 | 4.50 | 4.50 | 1.00 | — | — | 0.60 | — | 18.20 |
| 1983 | 7.20 | 3.80 | 4.80 | 1.10 | — | — | 0.70 | — | 17.60 |
| 1984 | 8.10 | 3.50 | 5.00 | 1.00 | — | — | 0.70 | — | 18.30 |
| 1985 | 9.20 | 3.50 | 6.00 | 1.00 | — | — | 0.80 | — | 20.50 |
| 1986 | 11.00 | 3.00 | 7.20 | 1.00 | — | — | 0.80 | — | 23.00 |
| 1987 | 11.50 | 3.00 | 7.30 | 1.00 | 1.00 | — | 0.80 | — | 24.60 |
| 1988 | 12.20 | 2.90 | 7.80 | 1.00 | 1.00 | 0.40 | 0.70 | — | 26.00 |
| 1989 | 11.70 | 2.00 | 7.80 | 1.00 | 1.20 | 0.60 | 0.60 | — | 24.90 |
| 1990 | 10.10 | 1.80 | 7.30 | 1.00 | 0.60 | 0.40 | 0.60 | — | 21.80 |
| 1991 | 7.60 | 1.20 | 7.40 | 0.90 | 0.50 | 0.30 | 0.50 | — | 18.40 |
| 1992 | 6.70 | 1.20 | 7.10 | 0.60 | 0.50 | 0.30 | 0.40 | — | 16.80 |
| 1993 | 6.00 | 0.80 | 6.50 | 0.60 | 0.40 | 0.30 | 0.40 | — | 15.00 |
| 1994 | 5.70 | 0.80 | 5.40 | 0.40 | 0.40 | 0.20 | 0.40 | — | 13.30 |
| 1995 | 5.40 | 0.60 | 4.80 | 0.50 | 0.40 | 0.20 | 0.50 | — | 12.40 |
| 1996 | 5.60 | 0.40 | 3.50 | 0.40 | 0.40 | 0.20 | 0.50 | — | 11.00 |
| 1997 | 5.20 | 0.40 | 2.50 | 0.40 | 0.30 | 0.20 | 0.40 | — | 9.40 |
| 1998 | 5.00 | 0.40 | 1.50 | 0.40 | 0.30 | 0.20 | 0.40 | — | 8.20 |
| 1999 | 4.50 | 0.40 | 1.20 | 0.25 | 0.25 | 0.20 | 0.40 | — | 7.20 |
| 2000 | 4.30 | 0.40 | 1.00 | 0.25 | 0.25 | 0.20 | 0.50 | — | 6.90 |
| 2001 | 4.20 | 0.30 | 0.80 | 0.30 | 0.25 | 0.20 | 0.50 | 0.30 | 6.85 |
| 2002 | 4.20 | 0.30 | 0.75 | 0.30 | 0.20 | 0.10 | 0.45 | 0.30 | 6.60 |
| 2003 | 3.95 | 0.25 | 0.90 | 0.30 | 0.25 | 0.15 | 0.45 | 0.30 | 6.55 |
| 2004 | 3.70 | 0.20 | 0.85 | 0.25 | 0.25 | 0.15 | 0.45 | 0.20 | 6.05 |
| 2005 | 3.60 | 0.30 | 0.80 | 0.30 | 0.20 | 0.15 | 0.60 | 0.25 | 6.20 |
| 2006 | 3.35 | 0.30 | 0.80 | 0.40 | 0.15 | 0.10 | 0.75 | 0.20 | 6.05 |
| 2007 | 3.00 | 0.35 | 0.55 | 0.45 | 0.20 | 0.10 | 0.75 | 0.20 | 5.60 |

[1]Source: Review 2007, p. 13. Mohair South Africa, Jansenville, South Africa.

### Mohair Production

Mohair production and prices have increased and fallen in response to unpredictable fashion cycles (see Figure 15.1 and Table 15.1) (Van der Westhuysen, 2005). In the 1970s, Angora goat populations in South Africa and the U.S. were increased in response to the high demand for hand-knitting yarns from European and Far Eastern countries. As the demand for hand knitting declined, starting in the late 1980s, so too did mohair production (Table 15.1). The worldwide decline in mohair production was also a response to alternative livestock operations being adopted that were expected to be more profitable, less volatile, more reliable, and less demanding in terms of labor and maintenance. In the U.S., many erstwhile Angora goat

**Table 15.2** Ranges for goat, fleece, and fiber characteristics[1].

| | Angora goat (6-month clip) | Cashmere goat | Cashmere descriptor[2] |
|---|---|---|---|
| Mature body weight, female, kg | 30–60 | 30–70 | |
| Fleece weight (mature female), kg | 1.5–4.0 | 0.08–1.0 | hair in, unscoured |
| Average fiber diameter, μm | 20–48 | 12–19 | |
| | 1yr 33–48 | 50–100 | guard hair |
| | 2yr 25–36 | | |
| Individual fiber diameter, μm | 6–60 | 5–30 | |
| | — | 30–200 | guard hair |
| Scoured yield, % | 70–93 | 65–85 | |
| Grease content, % | 1.2–8.0 | 0.7–7.2 | |
| Suint, % | 1.8–4.2 | 1.2–3.5 | |
| Mechanical yield, % of hair in, scoured | — | 15–90 | |
| Staple length, mm | 85–150 | 12–90 | |
| | — | 50–200 | guard hair |
| Medullation, % | 0–3 | 0–3 | |
| | — | 100 | guard hair |
| Color | White | White, tan, brown, gray, black | |
| S/P ratio | 6–10 : 1 | 5–7 : 1 | |
| Follicles/mm$^2$ | 16–33 | 17–46 | |
| Specific gravity | 1.31 | 1.31 | |
| Regain at 65% rh and 21°C | 14.5–17.5 | 14–17 | |
| Tenacity, cN/tex | 14.6–18.1 | 10% < wool, 40% < mohair | |
| Elongation at break, % | 38.0–45.8 | 41.2 | |
| Work at rupture, cN/tex | 2.65 | — | |
| Initial modulus, cN/tex | 384–430 | — | |
| Crimps/10 cm | 1.5–8.0 | 20–40 | |
| Curls/10 cm | 2.8–6.6 | — | |

[1]Compiled from numerous sources.
[2]Cashmere down unless otherwise specified.

breeders have turned to meat goats, hair sheep, and cattle production. Current world production of mohair is 5.7 million kg per year (greasy basis) representing less than 0.02% of total textile fiber production. Only a fraction of this mohair is processed into final textile products in the producer countries. Countries that have retained the specialized machinery and technical expertise to process mohair and manufacture products include China, France, Germany, India, Italy, Japan, South Africa, South Korea, Taiwan, Turkey, U.K., and U.S.

Although Figure 15.1 illustrates the variability of average price among years, it does not indicate the relative values among quality types or grades. These types (e.g., those used by the Mohair Council of America) are based primarily on average fiber diameter and include fine kid (24–26 μm), good kid (26–28 μm), average kid (28–30 μm), young goat (30–32 μm), fine adult (32–34 μm), and adult (>34 μm). In March 2008, Texas adult mohair sold for US$5.18/kg (greasy). Corresponding prices for the higher grades (coarse to fine) were $5.29, $7.67, $10.43, $13.60, and, $24.56/kg, respectively (Mohair Council of America Market Newsletter, March 4, 2008).

## CASHMERE GOATS

Cashmere is produced by approximately 150 breeds of goats, many of which are unimproved local types herded in flocks by pastoral nomads. China has at least five breeds of goat that have been improved over time by selection and which are considered to be superior cashmere producers (that is, annual cashmere down production greater than 300 g per head) (Teh and Gipson, 1992). These are the Liaoning, the most productive cashmere goat in China (McGregor et al., 1991); Inner Mongolian White, composed of three distinct breeds: the Alashan Downy, Albas Downy, and Erlong Mountain Downy; and the Hexi cashmere goats. The weight of cashmere removed by combing (10–50% guard hair and other impurities) from commercial Liaoning mature females (body weight [BW] = 35 kg) was reported to be in the range of 290–380 g and measuring 6.7 cm in length and 15.5 μm average diameter (McGregor et al., 1991). A more recent cashmere review (Stanton and Brown, 1997) reported on Liaoning goats that had been in a Chinese selection program for 10 years in which male and female BW were 58.5 and 49.7 kg, down weights were 675 (81% yield) and 435 g (76% yield), fiber diameters were 15.3 and 15.1 μm, and staple lengths were 5.9 and 5.8 cm, respectively. Obviously, the selection program had been successful in terms of both meat and fiber production. In contrast, the smaller Albas Downy females (BW = 27 kg) grow about 370 g of finer fiber (14 μm) having comparable length.

### Cashmere Production

Historically, cashmere was produced in the Himalayan region from China in the east to Iran in the west. Today, commercial quantities of cashmere are produced in China (including Tibet), Mongolia, Iran, Afghanistan, Turkey, Kazakhstan, Kyrgyzstan, Turkmenistan, Tajikistan, Uzbekistan, Pakistan, India, Australia, and New Zealand. Several European countries and the U.S. also produce small quantities of cashmere. In the early 1970s, Australian scientists (Smith et al., 1973) reported cashmere on feral goats, and the Australian and New Zealand cashmere industries were born. Selected Australian and New Zealand cashmere goats were imported into the U.S. in the 1980s and were generally bred to Spanish goats from Texas that had also been selected for cashmere production (Dooling and Dooling, 1996). Thus began the U.S. cashmere industry. During the Soviet-dominated era, many goats in the Central Asian republics were crossed with Russian and Angora goats, and the offspring produced the less valuable cashgora fiber. However, Millar (1986) reported that some indigenous goats of the Central Asian republics produce good quality cashmere. These goats have served as a nucleus for increased cashmere production in these countries over the past 20 years gradually replacing a dwindling sheep population. Much of the world's cashmere is still produced in traditional areas and marketed through informal trading channels with itinerant traders serving as sorters and middlemen between the producers and the Chinese, Mongolian, and European textile mills. Unlike mohair, for which reasonably accurate production records exist, cashmere production remains very much a matter for conjecture because, with the exceptions of China and Mongolia, very few official records exist (Dr. Carol Kerven, Cashmere Consultant, personal communication September 27, 2008). The consensus appears to be that annual world production is currently in the range of 11–14 million kg of raw cashmere with most of it being produced in China (60–65%) and Mongolia (20–25%).

As for mohair, cashmere prices are subject to the vagaries of fashion, weather conditions in the producing areas, and subsequent production trends. Typically, processors pay more for finer (<16.5 μm) compared to coarser (16.5–19.0 μm) cashmere of comparable cleanliness, length, hair content, luster (or lack thereof), color, style, and fiber diameter distribution. For example, in 2005, livestock owners in Mongolia sold cashmere of <15.5 μm for US$33/kg compared to $21/kg for cashmere in the range of 17.6–19 μm (Lecraw, 2005).

## CASHGORA

Cashgora is produced by goats that are the result of crossing Angoras with cashmere, feral, or dairy goats. The animals are typically vigorous and fast growing. Several serious attempts (for example, that of the French dehairer, Rene Friedlin) have been made to establish cashgora as a bona fide textile fiber, and in 1988, the International Wool Textile Organization accepted cashgora as a generic term for fibers produced by Angora × cashmere crosses. The fiber is typically white, lacking luster, with a range in average fineness from 17–23 μm, staple length of 3–9 cm (Dalton and Franck, 2001), and a relatively bold (compared to cashmere) crimp. The raw fibers contain very little grease so that the fluffy appearance of the animals gives the impression of high fiber production. In fact, females (Angora × Spanish crosses) grew only about 0.4 kg of fleece in their first year (Blakeman et al., 1990). World cashgora production was estimated at 200 tons in 1990. Today very little cashgora is produced and sold as such, and most is either being handspun or is being blended with cashmere, fine wool, or mohair thus losing its identity.

## GOAT HAIR

The dehairing process and hair removal from goatskins results in many tons of coarse (30–200 μm), heavily medullated goat hair. Much of this is disposed of in landfills but historically and still today some finds commercial end uses in such products as inexpensive felts, interlinings, special effect fibers in specialized textiles, paint brushes, mulch, and mats for soaking up oil spills.

## GOATSKIN

Goatskin is defined as the skin of the goat or leather made from it. The size of the goatskin and leather industry is large given the fact that there are currently in excess of 800 million goats in the world and at least 50% of the skins from slaughtered goats are processed. The average yield per skin is about $0.5\,m^2$. The value of an individual skin is quite low: in fact in some countries (the U.S. included), it does not always pay to try to salvage the skins and many of them are disposed of in landfills. But for the remainder, this is a classic example of value adding as the raw skins are converted into valuable leathers and end-products. During the past 30 years, goatskin processing and the leather trade in general has migrated from the U.S. and European countries to countries with less strict effluent laws, lower labor costs, and people who are willing to work in factories and conduct tasks that at best can be described as unsavory. The goatskin business is labor intensive and now provides jobs for millions of people particularly in many parts of the developing world including China, India, Pakistan, Bangladesh, and Indonesia. Naidu (2000) estimated that 0.7 million people are employed in the goatskin leather industry in India alone.

Goatskins are coproducts of the goat meat industry. This elevation from by-product is afforded to the goatskin because of its proven potential for value adding in excess of 500% by the time the leather is converted into footwear, garments, gloves, and other leather goods (Rao and Rao, 1992). Skins are collected mainly from animals processed in village and urban slaughter locations. Quality control efforts aim to minimize man-made defects such as flay cuts, bacterial damage, drag marks, and hair slip. Murthy and Ramasami (1992) reviewed the many possible antemortem, postmortem, and processing defects that can affect the value of goatskins. If the skins are close to a tanning facility, they can be tanned without salting. If they have to be stored for any length of time, they must be washed, and common salt is applied as a preservative at a rate of 500–600 g per skin. Depending upon the length of time before the skin is tanned, salting may be repeated two or three more times. These quantities of salt represent a great pollution potential, so biocides have been developed and are gradually replacing salt.

Pre-tanning, tanning, and finishing operations have evolved to produce a stable, bacteria- and enzyme-resistant matrix from the raw collagen in goatskins. Because variable technologies are being used in different parts of the world, it has been estimated that as many as 100 different chemicals are being used to produce the many forms of leather products. These include chrome salts, vegetable-tanning materials, sodium sulfide, lime, ammonium sulfate, formic and sulfuric acid, fat liquors, pigments, dyes, and lacquers. In addition to the chemical treatments, the skins are subjected to various mechanical operations to maintain flexibility. Rao and Rao (1992) outlined the multiple broad stages involved in leather production. The pre-tanning operations include soaking, liming, dehairing, fleshing, scudding, deliming, pickling, and chrome/vegetable tanning. These operations are followed by the post-tanning and finishing processes that include rechroming/semi-chroming, neutralization, retanning, dyeing, fatliquoring, and finishing (seasoning and lacquering).

Different types of leather are produced for different purposes. The most popular types of leather produced from goatskins are glaze kid, suede, nappa, aniline, and semi-aniline finished and patent leathers. The compactness, high strength, and bold grain of goat leather make it ideal for producing high quality ladies' and children's footwear. Other end uses of goat leather include garments, small leather goods like handbags, wallets, and gloves, and drum skins. Most of the leather goods are exported as products and finished leathers providing foreign exchange for the manufacturing country.

## GENETICS OF AND SELECTION FOR FIBER

Before scientific principles began to be applied to selection of Angora and cashmere goats for improved fiber production, it is difficult to estimate how much emphasis was placed on fiber versus meat production. Today with a significant amount of producer income coming from cull and excess animals, meat production is likely a component of most selection policies (Shelton, 1993). That being stated, the main focus of this section is on genetics of and selection for fiber production. Because fiber properties of Angora goats change so rapidly from birth to 18 months of age, selection based on fiber traits is not recommended until this age is achieved. Many Angora goat breeders make their selections based on the third fleece (at 18 months of age) before the females are exposed to males for the first time after which time fiber production is affected by pregnancy and later by lactation. In contrast, cashmere goats are selected initially based on their first fleece. However, as in most species, retention decisions are ongoing throughout the productive life of the goat and if the fleece becomes excessively coarse at any point, the animal would likely be culled.

Shelton (1993) noted that there are several qualitative genetic traits (that is, those controlled by a single or a few pairs of genes) such as color and developmental abnormalities, the latter having been selected against for many years but which still crop up occasionally in Angora and cashmere goats. These include peromelia (missing a leg), missing ears (and ear length in general), wattles, cryptorchidism, no horns (polled), mouth deformities, and (possibly) in the case of Angora goats, sheepy fleece. In contrast, most of the economically important traits are considered to be quantitative (that is, controlled by many sets of genes). To design an effective selection program, it is necessary to have the following knowledge for each trait being considered: historic, current, or predicted economic value, variability in the population to be selected, heritability, and phenotypic and genetic relationships between traits. Selection indexes can be constructed to produce many different outcomes, but whatever the goals of the breeder, it must be understood that the more traits included in the index, the slower will be the progress in any particular trait. Thus, rather than (or in addition to) including a particular trait in a selection index, it may be more efficient to establish independent culling or retention values for that trait.

## Angora Goats

In the Texas AgriLife Research Angora Goat Selection Index used to evaluate yearling Angora males following a 112-d central performance test, the following traits are included: clean fleece weight, average daily body weight gain, final body weight, average staple length, average fiber diameter, face cover score, character score, and neck cover score. This index was empirically derived and was designed to reward animals with above-average clean fleece weight, body gain, final weight, lock length, character, and neck cover scores, and those with below-average fiber diameter and face cover score. In addition, independent culling criteria and minimum performance standards were established for med and kemp content, final weight, clean fleece weight, and average fiber diameter. A sift committee is also charged with disqualifying animals that are deficient in breed character or conformation (special attention being paid to visible kemp and abnormal hind legs), or have horns that are too close together, a sheepy fleece, scrotal irregularities, or any other discernible health issues. This index has generally resulted in the type of animals desired by most breeders and commercial operators in the U.S. However, criticisms that have been leveled at this and similar central performance test programs (for example, those conducted in South Africa and Canada) are that this type of test does not address fertility or ability to raise kids. Further, the potential exists to produce animals that cannot perform well in the range environments in which they and their offspring will be maintained without the provision of supplementary feed. This latter consideration has led to the adoption in South Africa (and much discussion in the U.S.) of a performance test conducted solely on rangeland recognizing that the full genetic potentials to grow and produce fiber will likely not be realized but also recognizing the need to identify animals that can produce most efficiently in a range environment.

Another approach being suggested by researchers uses genetic evaluations in which pedigrees and all production records are maintained by breeders and submitted on a regular basis to a central organization that uses the data to calculate "expected progeny differences" for each of the measured traits for each contemporary animal. Although this system has been used successfully in other species, interest has been sparse among Angora and cashmere goat breeders, although Boer goat breeders have initiated a national genetic evaluation program for their breed in the U.S., and programs for sheep are well established in Australia, the U.S. (where they are still underutilized), and elsewhere. The main objection appears to be that this type of program involves more work for the breeder, and in the case of fiber-producing goats, numerous traits need to be monitored whereas in some of the species where this type of selection program has been almost universally adopted (such as dairy cattle and pigs), relatively few traits are being considered. This is one of many examples where excellent technology has been developed and made available by researchers but for very practical reasons, producers are unwilling or unable to take advantage of it.

Heritability values reported for Angora goat traits in the literature are highly variable (Table 15.3) but were categorized by Shelton (1993) as high (>25%), moderate (15 to 25%) and low (<15%). Highly heritable traits in Texas goats include staple length, clean yield, body weight (yearling and mature), face, neck, and belly covering, S (secondary follicle) to P (primary follicle) ratio, and scrotal division. Moderately heritable traits include fleece weight, fleece density, average fiber diameter, medullation, and weaning weight. Traits with low heritability are reproductive rate, longevity, and adaptability. Shelton did not report any information for heritability of the subjectively assessed

**Table 15.3** Heritability values[1] and phenotypic correlations for Angora goat traits.

| | Trait | | | | | | | | |
|---|---|---|---|---|---|---|---|---|---|
| Trait[2] | GFW | CY | CFW | AFD | ASL | MF | KF | FCS | BW |
| Heritability range | 0.07–0.45 | 0.02–0.48 | 0.12–0.38 | 0.08–0.33 | 0.07–0.42 | 0.00–0.39 | 0.05–0.42 | 0.08–0.24 | 0.07–0.50 |
| Average heritability | 0.16 | 0.34 | 0.16 | 0.18 | 0.33 | 0.25 | 0.18 | 0.22 | 0.26 |
| | Phenotypic correlations[3] | | | | | | | | |
| GFW | — | **0.23**[4] | **0.95** | **0.50** | **0.21** | **0.15** | 0.04 | **0.25** | 0.09 |
| CY | | | 0.07 | **0.18** | **0.36** | −0.01 | **−0.17** | 0.11 | −0.03 |
| CFW | | | | **0.57** | **0.33** | **0.15** | −0.01 | **0.29** | 0.08 |
| AFD | | | | | **0.31** | **0.40** | 0.08 | −0.03 | 0.12 |
| ASL | | | | | | 0.04 | −0.05 | 0.07 | −0.03 |
| MF | | | | | | | **0.60** | −0.13 | 0.08 |
| KF | | | | | | | | −0.13 | 0.05 |
| FCS | | | | | | | | | **−0.17** |

[1]Heritability values from numerous sources.

[2]GFW = grease fleece weight, CY = clean yield, CFW = clean fleece weight, AFD = average fiber diameter, ASL = average staple length, MF = med fibers, KF = kemp fibers, FCS = face cover score, BW = body weight.

[3]For yearling males in a central performance test (Pfeiffer et al., 2004).

[4]Correlation coefficients in bold font are significant at $P < 0.001$.

traits character ($h^2 = 0.14–0.34$), style ($h^2 = 0.13–0.23$), staple type, or luster. Snyman and Olivier (1999) presented the values in parenthesis for South African Angora goats.

Pfeiffer et al. (2004) reported phenotypic correlations for yearling Angora males on performance tests (7 years data, n = 462). See Table 15.3. Most are favorable, but three important exceptions are the positive correlations among average fiber diameter and body weight (r = 0.12), clean fleece weight (r = 0.57), and staple length (r = 0.31). By selecting exceptional animals that do not exhibit these general trends, individual breeders have successfully bred goats to produce greater fleece weights containing finer mohair. Gifford et al. (1991) reported phenotypic and genetic parameters of fleece traits for Australian Angora goats and pointed out that comparable parameters were already available from U.S., Turkish, and New Zealand sources.

### Cashmere Goats

Much of the genetic data relating to cashmere production was generated in Australia, New Zealand, and China although some of the latter information was difficult to access. Quantitative traits that were included in breeding objectives published by Pattie and Restall (1992) are production of down, down average fiber diameter, and live weight. Compared to the comparable traits in Angora goats, the reported heritability values (Table 15.4) for cashmere goats are high. The phenotypic relationships between BW and fiber traits are either positive or not significant, and an unfavorable phenotypic (and genetic) relationship exists between average fiber diameter and down weight. Other traits that have been considered for inclusion in cashmere selection programs include average staple length of down, secondary follicle density, S : P follicle ratio, change in average fiber diameter with age, and cashmere style (subjectively assessed) that have heritability estimates (reported in one study, Pattie and Restall, 1992) of 0.70, 0.17, 0.29, 0.21, and 0.16, respectively. The scoring of cashmere style is subjective and inaccurate. Although considered by many breeders and fiber purchasers to incorporate an overall subjective assessment of down length, fineness, luster, and crimp, one study with U.S. cashmere (Lupton et al., 1999) demonstrated that subjectively assessed style scores can be accurately predicted with only two objectively measured traits, average fiber curvature, and standard deviation of fiber curvature. In view of the recently reported study of McGregor (2007) in which he identified 11 different forms of cashmere fiber crimp that were related to the national origin of the fibers (Australia, China, and Iran), this simple relationship between curvature, variability, and style score may not be applicable to cashmere from all sources.

**Table 15.4** Heritability values and phenotypic correlations for Cashmere goat traits.

| | Trait | | | | |
|---|---|---|---|---|---|
| Trait[1] | BW | GFW | DY | DW | DAFD |
| Heritability range[2] | 0.22–0.29 | 0.25–0.45 | 0.23–0.90 | 0.36–0.61 | 0.47–0.83 |
| Average heritability[3] | 0.26 (0.57) | 0.35 (0.36) | 0.49 | 0.45 (0.69) | 0.67 (0.42) |
| | | | Phenotypic correlations[2] | | |
| BW | — | 0.23 | −0.11 | 0.05 | 0.13 |
| GFW | | | 0.10 | 0.64 | 0.31 |
| DY | | | | 0.79 | 0.35 |
| DW | | | | | 0.46 |

[1]BW = body weight, GFW = grease fleece weight, DY = down yield, DW = down weight, and DAFD = down average fiber diameter

[2]As reported by Pattie and Restall, 1992.

[3]Average of four studies cited by Pattie and Restall, 1992 (data in parenthesis from Ma Ning et al., 1996).

## Environmental, Nutritional, Age, and Sex Influences on Fiber Production

### Angora Goat

Although photoperiod influences growth rate of mohair (and cashmere), nutrition, as influenced by climatic and grazing conditions and supplementation, usually has a much greater effect. Mohair growth is lowest in winter and highest in summer (Litherland et al., 2000). In fact some follicles produce no mohair at all during the winter period. Mohair fleeces shorn in February (grown in autumn and winter) are typically lighter in weight, and are shorter, finer, and less kempy than those shorn in August (grown predominantly in spring and summer).

Mohair production by kids and mature animals is influenced greatly by the amount and composition of the diet. In fact, nutritional variation within and between years is the most important environmental factor influencing fiber production and properties. Inadequate nutrition causes less, shorter, and finer fibers to be produced. Unless goats are on a very high plane of nutrition, mohair production and average fiber diameter invariably respond positively to protein and rumen-protected amino acid (for example, encapsulated methionine) supplementation. However, the most strategic and economically rewarding use of protein supplementation may be to assist female kids in attaining a successful mating weight by the time they are 18 months of age (Huston et al., 1993). Energy supplementation is indicated whenever the welfare of the goats is at risk. Thus, energy supplementations to pregnant and lactating does and to reduce weight loss during drought have been common and necessary practices in the major producing countries.

From birth to about 4 years of age, mohair fibers become progressively coarser, after which time, the average fiber diameter remains fairly constant. After 4 years of age, Angora does begin to produce less mohair. Some follicles stop producing altogether. In summary, in the age range 1.5–6.5 years, the effects of age are positive for BW, clean yield, average fiber diameter, and med and kemp content, and are negative for fleece weight (Lupton et al., 1996). Mature females produce 2–2.5 kg of greasy mohair (average fiber diameter approximately 34 μm) every 6 months, this being under genetic and nutritional control to varying degrees. In contrast, kids (6 and 12 months of age) typically produce less than 1 kg at their first shearing and between 1 and 1.5 kg at the second shearing (22 to 29 μm). In a 6-month period, the fiber grows to a length of 10–15 cm, this being a requirement for worsted (long-staple) processing. Age and body weight are correlated with each other and fiber diameter when nutrition is adequate. However, when nutrition is poor, body weight becomes a more accurate predictor of fiber diameter. Males having greater body weights than females tend to produce more, coarser, and longer mohair than females of similar age maintained under similar conditions. Pregnant and lactating Angora does tend to produce less and finer mohair than comparable nonpregnant does.

### Cashmere Goat

Cashmere growth rate declines linearly from midsummer to early winter when the fibers stop growing altogether and eventually shed from the follicles. By extending day lengths as winter approaches using artificial light, the fiber-growing period can be extended for cashmere goats. To date, no practical economic use has been made of this phenomenon.

Highly productive Chinese and Mongolian cashmere goats grow in excess of 200 g of cashmere down per year. In contrast, many Australian and U.S. cashmere goats grow in the range of 50–150 g of down per year. These are genetic differences since the lower-producing goats are most likely on a much better plane of nutrition than their Chinese counterparts. Nutrition effects on fiber in the lower-producing goats have been reported to be minimal or nonexistent. However, experiments with more productive Australian goats (McGregor, 1988) produced a curvilinear relationship between digestible dry matter intake and cashmere production with maximum down production at energy intakes of approximately 1.4 × maintenance. Intake was also shown to have a positive influence on average fiber diameter. Protein supplementation produced no increase in down or guard hair growth. Interestingly, goats deprived of energy were shown to divert nutrients to cashmere growth while those fed ad libitum partitioned nutrients to hair growth. This latter observation has been a common thread throughout most nutrition experiments with cashmere goats.

Li Yongjun et al. (1996) reported age and sex effects in the main economic traits of the highly productive Liaoning goats. In males, clean down production (945 g) peaked at 3 years of age whereas production of the female increased with age (350 g at 1 year of age to 486 g at 6 years of age). Males and females produced their finest cashmere as yearlings (13.2 and 12.9 µm, respectively). Males peaked at 3 years of age (16.9 µm) before starting to decline slightly while female fleeces remained fairly constant at 15.3 µm between 3 and 6 years of age. For males, body weight reached a plateau at 3 years of age (approximately 82 kg) whereas female body weight increased each year from 1–6 years of age from 24–54 kg.

## CURRENT PRODUCTION CONDITIONS AND PROBLEMS

Angora goat numbers have been in decline in the major-producing countries for many years, and it is difficult to predict if future fashion requirements or other factors will reverse this trend. Cashmere production on the other hand appears to be on the increase, and along with meat goat production in general, prospects for further expansion appear good. The remaining commentary is restricted primarily to the situation on the North American continent. A major recurring challenge in the fiber goat industry, faced annually by cashmere producers and twice a year by mohair producers, is that of potentially deadly exposure of the goats to bad weather conditions for a few weeks after shearing. Most producers that have been in the business for any length of time have shelters that can be used by the goats during the immediate post-shearing period. However, weather conditions being what they are, many freshly shorn goats succumb to fast-moving cold fronts and storms each year. Most of the other economic and production challenges faced by fiber goat producers are common to most goat- and sheep-raising enterprises.

The threats and constraints faced by U.S. animal fiber producers include the following:

- an increase in predation that has technological solutions that cannot be used because of concerns by some segments of our society
- globalization and growing competition from goat producers in foreign countries in which goat meat, mohair, cashmere, and textiles are being produced at lower cost than in the U.S.
- growing competition from aesthetically pleasing, functional, and cheaper synthetic fibers
- mislabeled cashmere and mohair products and the subsequent undermining of confidence and satisfaction of the consumer for these luxury fibers
- low per capita consumption of goat meat in the U.S.
- competition from other livestock species that may require lower inputs and may be more profitable than goats (primarily cattle)
- relatively high labor costs in North America and a shortage of labor with specific “goat skills,” especially shearers
- the increasing value of land is decreasing the opportunities for profitable goat production
- loss of infrastructure as the Angora goat and sheep industries declined
- contamination of cashmere and mohair with polypropylene, black and colored fibers from colored sheep, meat and cashmere goats, cattle, and other animals undermines the acceptability and value of U.S. fibers in foreign markets
- perception by some consumers that chemical residues are contained in meat (growth hormone and antibiotics) and animal fibers (insecticides)
- the expectation of many young people for more comfortable and prosperous lives than could reasonably be expected from raising goats
- and currently, the high and increasing cost of feed are making it extremely difficult to remain profitable in the goat business.

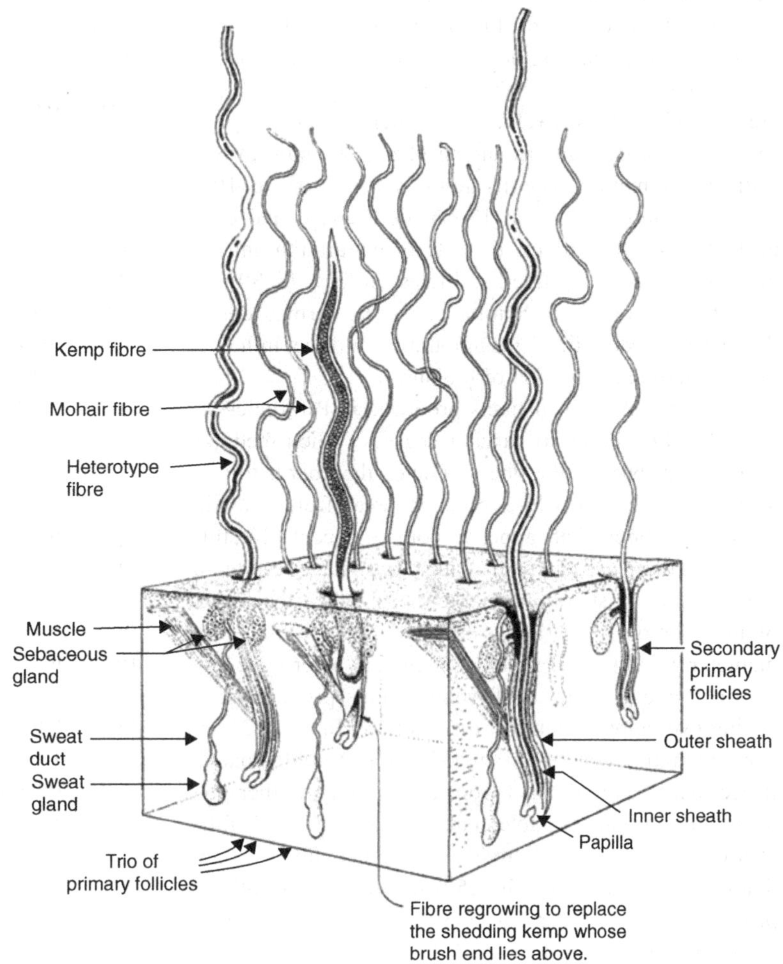


**Figure 15.2** Simplified drawing of adult Angora goat skin (from Ryder and Stephenson, 1968, courtesy of Elsevier).

Finally, it must also be emphasized that in the absence of a new family of effective anthelmintics being brought to the market, goat production systems will have to be refined or relocated to reduce losses to internal parasites. Producers are realizing that failure to be able to control parasites is making it impossible to have a profitable goat operation in some areas of the country. The business of vegetation management and targeted grazing using goats (and sheep) particularly to control fire hazards and to eat plants that are aversive to cattle could experience considerable growth in the near term (Launchbaugh et al., 2006). Furthermore, grazing toxic or undesirable plants will be made more effective by selecting for improved genetics, using mixed-livestock grazing, and providing supplements designed to ameliorate the aversive effects of chemically defended invasive or toxic species. The topics of parasite resistance to anthelmintics and prescribed or targeted grazing are discussed in greater detail elsewhere in this book.

## BIOLOGY OF FIBER GROWTH

Mohair fibers are produced by cell division in primary (P) and secondary (S) follicles in the skin of Angora goats (Figure 15.2). The two types of follicles are distinguished by their accessory structures. The P follicles each have a sebaceous gland, a sudoriferous (sweat) gland, and an arrector pili muscle. The S follicles have only a sebaceous gland. Some S follicles produce more than one fiber. The central P follicles are first observed on the fetal head about 40 days into pregnancy and spread across the body over the next 20 days. During this time, two more P follicles (laterals) appear on either side of the central P follicle thus forming a trio group. After 80 days of pregnancy, S follicles associated with each trio group begin to emerge, forming a follicle group. At birth (day 149), all P follicles are fully formed and are actively producing fiber whereas only a small but quite variable proportion of the S follicles are producing fibers. Twelve weeks after birth, most of the secondary follicles are producing fibers. The ratio of S : P follicles in mature Angora goats ranges from 6–10 : 1 (compare merino sheep at 15–25 : 1). Prenatal and early postnatal nutrition affects the rate of maturation and the ultimate number of active S follicles. In fact, 75–80% of the ultimate number of S follicles can be formed during the first 8 weeks of life. This amount has a direct influence on the lifetime production of mohair.

A variable amount of fibers produced in P follicles are medullated (hollow), to varying degrees. When medullation exceeds 60% of the fiber diameter, the fiber is termed a "kemp," a chalky white, objectionable (from a textile viewpoint) fiber that appears not to accept dyestuff. Centuries of visual selection against kemp have resulted in low levels in most commercial animals. When the degree of medullation in a fiber is less than 60% of the fiber diameter, this fiber is termed a "med" or "heterotype."

The P and S follicles of Angora goats produce fibers having overlapping diameter distribution (Table 15.2) and similar length giving rise to a nonshedding, "single-coated" fleece that is quite distinct from cashmere and other goat breeds that produce double coats. In contrast, the P follicles of cashmere goats produce individual fibers having diameters in the range 30–250 µm whereas S follicles produce fibers in the range 5–30 µm. The S : P ratios in cashmere goats range from 5–7 : 1.

## NONFIBROUS CONSTITUENTS OF THE FLEECE

These typically make up 10–30% of the weight of the raw material and are comprised of moisture (10–17%), compounds produced by the goat, other natural contaminants, and residuals introduced by man. The former group includes mohair or cashmere wax (sometimes termed grease, up to approximately 8% for mohair and cashmere) secreted by the sebaceous glands and water-soluble dried sweat (suint) produced by the sudoriferous glands. Other natural contaminants include dirt, sand, and vegetable matter (VM) such as burrs, grass, seeds, sticks, and leaves. The amount and type of VM can have a major influence on the value of the raw material. Most VM is removed during normal mechanical processing, but when amounts are excessive, carbonization with sulfuric acid and heat may be necessitated. The last group of contaminants would include such things as residual insecticides, branding paint, chalk, and dipping fluids. Excluding some types of VM, the vast majority of these nonfibrous fleece constituents are removed in normal aqueous detergent scouring.

## CHEMICAL COMPOSITION

Like other mammalian fibers, mohair and cashmere (after washing) are composed almost entirely of a family of complex proteins referred to collectively as keratin. This term is also used to describe the similar sulfur-containing proteins that make up feathers, horns, and nails. Three main fractions are referred to as low- and high-sulfur and high-tyrosine proteins that are not uniformly distributed throughout the fiber. Most of the sulfur (3–4% overall) is present in the amino acid cystine. It has been estimated that wool contains more than 170 individual proteins. There is no doubt that mohair and cashmere are equally complex. Of the 22 naturally occurring amino acids, 18 are found in animal fibers. The amino acid composition of kid mohair has been shown to be very similar to that of Merino and Lincoln wool (Ward et al., 1955). The polypeptide chains are joined together by disulfide and isopeptide covalent crosslinks, ionic bonds (between $-NH_3^+$ and $-COO^-$ groups), and by noncovalent interactions such as hydrogen bonds. The positions of the disulfide bonds are rearranged during the setting of fabrics, one of the finishing processes that uses water and heat. The ionic groups give animal fibers their amphoteric nature (capability to absorb and desorb acids and alkalis) and also control dyeing behavior.

## MORPHOLOGY AND PHYSICAL PROPERTIES

### General

The physical structure of all animal fibers is complex (Figure 15.3). Crystalline, water-impenetrable microfibrils (approximately 0.2 µm wide) and water-accessible, relatively amorphous regions (cell membrane complex [CMC]) coexist at the molecular level. At the cellular level, the bulk of the fiber (the cortex) is composed of cortical cells that are cigar-shaped having dimensions of approximately 8 × 100 µm. Animal fibers are described as a composite assembly of cuticle and cortical cells that are surrounded and held together by the CMC, the main components of which are low-sulfur proteins and lipids. Although the CMC constitutes only a few percent by weight of the fiber, it is responsible for the exceptional mechanical properties of the fibers and fabrics made from them (for example, excellent abrasion resistance). The CMC also represents the only continuous phase in the fiber and is used for diffusion of

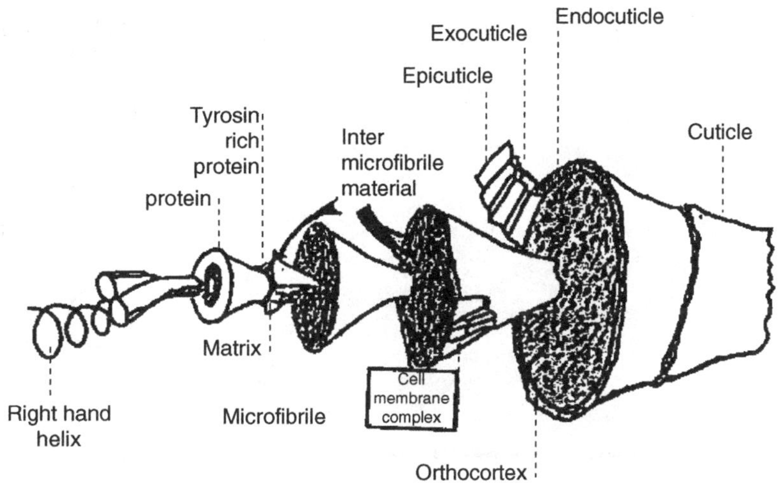


**Figure 15.3** Structure of adult mohair fiber (from Smith, 1988).

dyestuffs, other chemicals, and moisture throughout the structure. The mohair cortex is predominantly (more than 70%) orthocortex that is chemically less resistant than the remainder, paracortex. In contrast, fine wool usually contains both types arranged bilaterally in approximately equal amounts in the same fiber. This arrangement is associated with wool crimp and imparts to wool greater resistance to chemicals and heat compared with mohair. In contrast, cashmere down fibers are composed of a mixture of orthocortical and mesocortical cells, the latter having a structure intermediate between that of ortho and paracortical cells.

Cuticle cells cover the fiber surface in the form of protective, overlapping scales (Figure 15.4). The protruding edge of each scale points from the fiber root to the tip. It has been suggested that an important function of this structure is to anchor the fibers in their respective follicles. In wool, relatively thick scales (0.8–1.0 μm) protrude to such a degree that the fibers felt or mat easily when rubbed together in the presence of moisture and heat. Because the scales on mohair are thinner (0.2–0.5 μm) and lie closer to the core of the fiber, the felting propensity of mohair is lower than for cashmere and wool. Scale height has been used to estimate the composition of wool/mohair blends and to identify unknown animal fibers. Currently, DNA testing is considered to be more accurate for the latter purpose, but the analysis is only capable of differentiating species (e.g., ovine versus caprine) and not breed (e.g., Angora versus cashmere goat). The cuticle is made up of three distinct structures termed the endo-, exo-, and epicuticle, the latter being a hydrophobic, semipermeable, flattened sheath on the very outside of the fiber that may in fact be the outside layer of the CMC. As mentioned previously, a third component of the fiber, the medulla, is contained in some of the fibers produced by P follicles and consists of a continuous or fragmented central core composed of air-filled cell residues.

### Mohair

The morphology and physical properties of mohair have been reviewed in detail by Zahn (1990) and Hunter and Hunter (2001), among others. Fine mohair fibers are round in cross section. As the fibers become coarser, the cross section becomes more elliptical. Some kemp (highly medullated) fibers are collapsed and have the appearance of a flattened straw. Mohair is more crystalline than wool of comparable dimensions, which helps to explain why mohair is stronger, more wear resistant, less extensible, and stiffer than wool. Mohair is characterized by excellent whiteness, luster, durability, elasticity, resilience, resistance to soiling, soil shedding, setting, strength, abrasion resistance, draping, moisture and perspiration absorption and release, insulation, comfort and pleasing handle, and low flammability, felting, and pilling (see Table 19.2) (Hunter and Hunter, 2001). Mohair's high luster, smoothness, soil resistance, very low propensity for felting, and durability are largely a result of the structure of the scales on the fiber surface, which are thinner, less pronounced (since they hardly overlap), and longer (approximately 20 μm) than those in wool of comparable fineness.

Two main types of medullated fiber are distinguished in mohair (Figure 15.5). The objectionable kemp fibers generally have a medulla that is continuous along the fiber

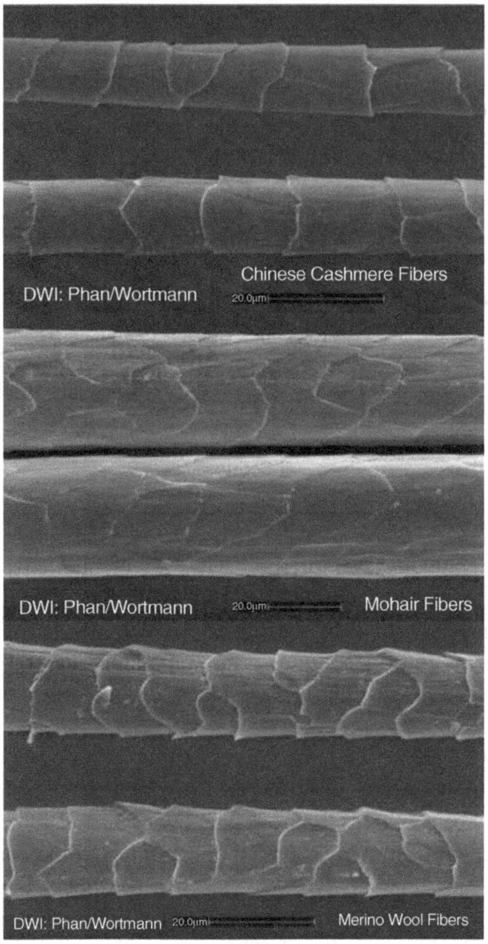


**Figure 15.4** Composite scanning electron micrograph of mohair, cashmere, and wool fibers (courtesy of International Wool Textile Organization).

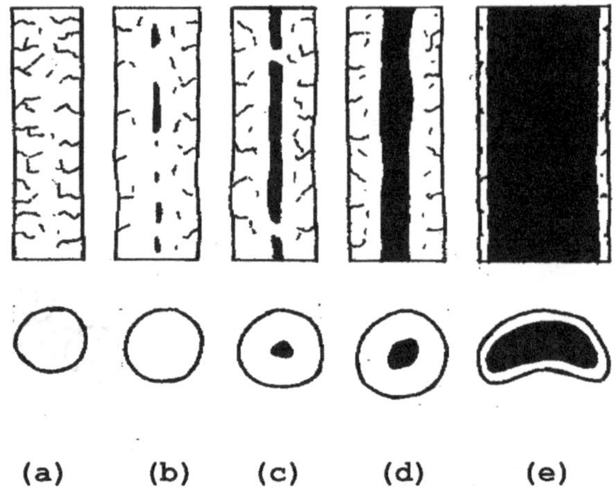

(a) (b) (c) (d) (e)

**Figure 15.5** Idealized projection microscope images of longitudinal sections of medullated mohair fibers. a) non-medullated, b) fragmented med fiber, c) interrupted med fiber, d) continuous med fiber, e) continuous kemp fiber.

length and extend to more than 60% of the width of the fiber. This much medullation produces an optical effect that causes the fibers to have a different appearance compared to unmedullated, solid fibers (that is, a chalky versus normal lustrous appearance) that becomes even more pronounced when the fibers are dyed to dark shades. The medulla itself is composed of a hollow network of cell walls. Kemp fibers are typically oval in cross section, coarser (1.8 times) than the parent population and occur as short, straight, shedded fibers as well as full-length attached fibers. Fibers with medullation less than 60% of the fiber diameter, both continuous and interrupted, cannot usually be distinguished from normal solid fibers with the naked eye and are not usually discriminated against by the textile industry. However, all medullated fibers in mohair appear to belong to a continuous population (Lupton et al., 1991). The important implication for the selection practices of breeders is that both types of medullated fibers should be selected against.

### Cashmere

Dehaired cashmere is characterized by its softness to the touch and the lightness and excellent thermal properties of fabrics composed of cashmere down. Microscopically, cashmere fibers appear to be very similar to very fine Merino wool with one or two scales encircling the fiber and six to seven scale margins per 100 µm (Figure 15.4). Distinct longitudinal striations are visible in the cuticle layer of white and pastel fibers. No medullation is present except for the continuous medulla in the few guard hair fibers that typically contaminate a dehaired cashmere sample and which are multiple times coarser than the fine down fibers. The scales on the guard hair form an irregular waved mosaic pattern with margins being close together and smooth or slightly rippled.

## OBJECTIVE MEASUREMENTS AND SUBJECTIVE ESTIMATES OF FIBER AND FLEECE

Fiber and fleece properties are used, with varying degrees of accuracy, to predict textile processing performance and yarn and fabric properties (Hunter, 1993). They also determine to a large degree both the value and the actual product into which the raw material will be converted. Consequently, instruments and standard methods (described in the Annual Book of the American Society for Testing and Materials [ASTM] Standards, Section 7, Volumes 07.01 and 07.02 [ASTM International, West Conshohocken, PA 19428-2959] and in the International Wool Textile Organization [IWTO] Specifications [the Woolmark Company, Ilkley, England]) have been developed and are in common use to accurately measure most of the value-determining properties. For mohair and cashmere down, these include average fiber diameter and its distribution; average fiber curvature and its distribution; clean yield; vegetable matter content and type (measured on most commercial lots); staple length and its distribution; staple strength and its distribution (measured on very few commercial lots); color, luster, style, and character (measured on few if any commercial lots, subjectively assessed on most). The variability inherent in these natural fibers necessitates special techniques be used when sampling commercial quantities of fibers to ensure accurate and reliable results. Typically core and grab samples are removed from individual fleeces, bags, and bales in a specified manner.

## CLEAN YIELD TESTING

A typical procedure would be that described in ASTM standard test method D 584. In summary, samples of raw fiber drawn from a commercial lot or a fleece in a specified manner are weighed, scoured, dried, and reweighed. The oven-dry, scoured fiber is then subjected to three more gravimetric tests to determine residual grease, ash content, and vegetable matter content. This permits calculation of fiber base and vegetable matter base in the original raw material. These "base" numbers are usually adjusted upward by incorporating specific amounts of "allowable" moisture, ash, and grease into the base value and calculating several commercial yield values, for example, clean mohair fiber present in the U.S.

## MECHANICAL YIELD TESTING (CASHMERE)

Modified Shirley analyzers, cards, and custom-built miniature dehairing machines have been used to estimate the amount of cashmere down that can be mechanically separated from the scoured raw material. However, a method that has likely found the most use involves measuring the fiber diameters of a representative raw sample from which the percentage of fibers 30 μm and finer can be calculated (Lupton et al., 1995). This represents a theoretical maximum yield that assumes all fibers 30 μm and finer can be separated from the guard hair by the dehairing machine (which is actually not possible).

### Average Fiber Diameter and Distribution, Medullation, Average Fiber Curvature and Distribution

Historically, determination of average fiber diameter and medullation involved sampling, conditioning, and reduction of samples to very small, short (250 μm), but still representative test specimens that were measured manually under high magnification (500×) using a projection microscope (PM). The measurements were time consuming and quite expensive. Average fiber diameter alone was estimated by an airflow technique that was faster than the PM method but was little used with mohair and cashmere. During the past 20 years, methods using automatic image analysis (such as the Optical Fiber Diameter Analyzers [OFDA] 100 and 2000 [BSC Electronics Pty. Ltd., Ardross, Australia]) and laser measurements (such as Sirolan-Laserscan, CSIRO, Geelong, Australia) have been developed and used extensively. These instruments are capable of providing very accurate estimates of average fiber diameter and distribution and average fiber curvature and distribution, and the OFDA 100 is also capable of estimating medullation (white and pastel shades only) in a short time, therefore breeders, marketers, and manufacturers now commonly use these inexpensive measurements for assisting with selection, marketing, and processing decisions, respectively.

Of particular note for the cashmere industry are the results of a recently completed study (McGregor and Butler, 2008) that showed 86% of the variability in performance and efficiency of the dehairing process was explained by measured attributes of the raw material. Raw cashmere that processed most efficiently and produced the longest dehaired cashmere tended to be white, longer, and have higher fiber curvature and lower VM than average, as well as having normal (that is, comparable in length to or slightly longer than the down fibers) length guard hair and no visible cotting. Efficiency of processing was substantially decreased with increasing average fiber diameter irrespective of color.

### Average Staple (or Fiber) Length and Distribution

A ruler was the traditional means for manually measuring staples or individual fibers that had been straightened and laid on a black felt board. Numerous instruments were developed to automate this process and remove subjectivity. However, to date, very little mohair and cashmere is measured objectively for length except by scientists for use in their selection or evaluation experiments. In the case of commercial mohair, staple length is mostly adequate (that is because it is not usually sheared until the staple length is longer than 10 cm). In the case of cashmere down, average fiber length would certainly be estimated and perhaps occasionally measured by growers and potential buyers.

### Average Staple Strength, Distribution, and Position of Break

Mohair is relatively strong and rarely exhibits weak points or "breaks" in the staple. Consequently, measuring mohair for strength is not very common in the commercial sector although instruments developed for wool testing have been used to provide accurate results for mohair and cashmere.

### Color and Dark Fiber Contamination

Though colorimeters and spectrometers are available to make objective measurements of mohair and cashmere color (especially whiteness and yellowness), they are rarely used. A standard method developed for white wools could be directly applied to mohair and white cashmere if in fact anyone required the measurement. Dark fiber contamination in mohair could sometimes be the result of bad breeding or grazing with dark fiber–shedding animals. However, the root cause is more likely to be inadequate skirting causing yellow to dark brown urine- and fecal-stained fibers to remain in the white fleece. Manual counts enhanced by special lighting and sometimes magnification are used in the industry to quantify colored and dark fibers in processed structures such as mohair sliver. In cashmere, the problem is sometimes more acute (for example, dark guard hair contaminating white down fibers) and may be a result of co-grazing white and colored animals and/or a poor job of sorting cashmere of different colors.

### Luster

Numerous attempts have been made to commercialize an objective measurement of luster that is a primary distinguishing feature of mohair. These have included goniophotometric and near-infrared spectroscopic techniques, none of which have reached commercial test method status to date. It should be observed that the exceptionally high level of luster in mohair and the absence of luster in well-bred cashmere (and also the low levels of kemp) have been arrived at in present-day populations primarily as the result of subjective assessments made by breeders.

### Style and Character

In mohair, style refers to the three-dimensional twist or ringlet structure observed in staples (sometimes referred to as locks) while character refers to the two-dimensional waves that are usually easier to observe and measure in the staple or individual fiber. When the frequency of the waves is excessive, mohair is said to have too much crimp, a condition described as "sheepiness." In the case of cashmere, these terms mean something different and are relatively ill defined. Cashmere of good style is said to have irregular crimp of relatively small magnitude and high frequency that does not lie in two dimensions but rather changes directions at irregular intervals along the length of individual fibers.

## FIBER QUALITY

Quality in mohair was described by Hunter and Hunter (2001) as being a combination of style and character, freedom from kemp, high luster, desirable handle, minimal yolk, and uniformity of length and fineness. A similar definition would be applicable to dehaired cashmere, if we replace the word "kemp" with "guard hairs" and the requirement for "high luster" with "no luster."

### Fiber Harvesting

#### Angora Goats

In extensive Texas operations (Northern Hemisphere), Angora goats are gathered and shorn in February and August. Animals are usually drafted into three main categories, kids (6 and 12 months of age), young goats (for example, 18 months of age), and adults; and they are sheared in this order (that is, the most valuable fiber first) providing about 20, 20, and 60% respectively by weight of the total clip. Professional shearers are hired to shear Angora goats. Shearing facilities range from makeshift tents with plywood boards and temporary pens to custom-built shearing sheds replete with hardwood, raised shearing floors and individual catch pens. Shearers use power-driven shears to first remove mohair from the belly and legs before tying the legs together. This essentially immobilizes the animal and facilitates removal of the rest of the fleece. The shorn mohair is then moved from the shearing floor to another location, depending on the marketing philosophy of the producer. Some owners attempt to add value to their clip at this point by removing urine- and fecal-stained portions, as well as inferior mohair (for example, that which is heavily contaminated with vegetable matter). These off-sorts are packaged separately to avoid further contamination of the clip. The remaining mohair may be further differentiated (classed) and packaged according to fiber diameter ranges, staple length, and lock structure (for example, ringlet versus flat lock), type or amount of vegetable contamination, and estimated clean yield. A trained owner or an experienced classer normally accomplishes these grading or classing operations. The skirting and classing processes are labor intensive and slow but result in bags (traditionally made of jute) or bales (made of nylon) of classed mohair that are delivered to a commission warehouse or cooperative association for sale. Some U.S. producers believe that this extra process after shearing is not cost effective and package their mohair directly from the shearing floor into burlap bags that are delivered to the warehouse as "original bag" (OB) mohair. Before processing, this mohair must be skirted and classed either at the warehouse, a custom classing firm, or at a textile mill. In South Africa, shearing is carried out using hand or power shears. Most South African growers take great personal pride in having the entire clip skirted, classed, and packaged in bales at the farm prior to delivery for sale. Both South Africa and the U.S. have official guidelines for skirting, classing, and marketing mohair.

### Cashmere Goats

Toward the end of winter, the undercoat or down of cashmere goats begins to shed. The actual time of year is quite variable throughout the world being dependent upon latitude, altitude, pregnancy status, and plane of nutrition. Generally, the colder it is, the longer the goats hold on to their insulating cashmere. In the traditional producing countries, China and Mongolia for example, the fine down is combed out of the fleece using special combs once shedding has been detected. Multiple combings are required to remove most of the cashmere. Elsewhere, whole fleeces are shorn. Combed and shorn cashmere both have to be scoured and dehaired before further textile processing, but combed cashmere normally produces much higher yields of down fiber than its shorn counterpart. Variability in coarse hair content and average fiber diameter of "raw cashmere" make meaningful price comparisons among districts, countries, or breeds very difficult unless objective measurements of these characteristics are available. In addition to producing relatively low down yields, shearing cashmere goats results in shorn animals at a time of year when exposure to cold or wet weather can produce serious stress, abortions, or even death.

## Fiber Marketing

### Mohair

At the storage facility, the different lines of mohair may be grabbed and core sampled and the samples objectively tested for the following main value-determining characteristics: clean yield, vegetable matter content, average fiber diameter (and variability), degree of medullation, and possibly staple length and staple strength. This information may or may not be made available to mohair buyers on or before sale day when they view representative opened bags or representative samples of mohair from each sale lot. Sales can be finalized by private treaty, sealed bid, or open bid. The private treaty method is quite common in the U.S., and open bid auction is more popular in South Africa. The buyer's representative then arranges for transportation to a scouring plant.

### Cashmere

Prior to the early 1980s, the extensive and remote nature of production, poor communications, informal trading, or, in the case of China, availability of raw cashmere only at central markets at prices set by the government characterized the raw cashmere market (Van der Westhuysen, 2005). Most of the raw cashmere from Mongolia and China was shipped to European countries for dehairing and conversion into luxury textiles. Since that time, the economies, marketing philosophies, and manufacturing ability of these two countries have changed to the point where export of raw material has been greatly diminished and fabrics and luxury apparel items are being produced in the countries of origin. This trend has resulted in shortages of raw materials for western manufacturers, some of whom have chosen to partner with Asian companies rather than leave the cashmere business altogether. The shortage of raw materials and subsequent elevated prices provided an impetus for numerous research projects aimed at establishing cashmere production in other countries, such as, Australia, New Zealand, U.K., U.S., and several others. To date, none of these countries have become major suppliers.

For the very small production in the U.S., much of the combed product is sold directly to hand spinners by the producers themselves using online marketing and fiber fairs. Several attempts have been made during the past 20 years to amass commercially significant amounts of classed (by color, fiber diameter, and staple length) cashmere, mainly shorn, that was then offered to buyers through cooperative organizations such as Cashmere America (founded in 1991).

### Textile Manufacturing

For in-depth information on textile manufacturing of animal fibers, the reader is referred to Volume 2 of the Wool Handbook (von Bergen, 1963). As previously mentioned, fleece and fiber properties are estimated and measured because they affect the mechanical processing behavior of the fibers, the properties of the yarns and fabrics into which they are incorporated, and they also determine to a great extent for which end-products the fibers are best suited. For a more detailed discussion of these topics, the reader is referred to the classic work of Hunter (1993) on mohair and to the excellent review by Leeder et al. (1992) that includes information on both mohair and cashmere.

#### Mohair

Mohair is washed in scouring plants to remove grease and dirt. Most mohair (being longer than 9 cm) is then mechanically processed through to the yarn stage in textile mills using the worsted system. Briefly, this system of processing involves the removal of shorter fibers so that smooth, compact yarns can be produced. Worsted yarns are woven, machine or hand knitted, or processed into rugs and carpets. Mohair may be dyed in one of several forms: loose stock, yarn, and fabric, for example. Fabrics are finished using specialized processes that are designed to produce different effects and handle (feel) in the finished product. These may include shearing and singeing (to produce a clear surface), decating (a wet process that uses heat to stabilize fabrics), fulling (a felting process to consolidate fabrics), brushing (to produce a hairy surface), and pressing. Like the dehairing process for cashmere, the specifics of many of these mechanical, dry, and wet processes for mohair are closely guarded industrial secrets.

#### Cashmere

Combed or shorn cashmere is scoured, dehaired, and processed into yarn using either the worsted or woolen system depending on staple length and intended end use. While accommodating shorter fibers, the woolen system produces softer, bulkier yarns that are required for many of the softer-handling knitwear products in which cashmere is typically used. Historically, Italy and Scotland have excelled in the production of high quality cashmere textiles. In recent times, the two major producing countries, China and Mongolia, have purchased or manufactured the necessary equipment and developed the expertise to produce fine cashmere products domestically.

## APPLICATIONS AND END USES

### Mohair

Mohair is used primarily in apparel and household textiles such as upholstery fabrics, drapes, and carpets. Compared to wool, other specialty fibers, and other natural and synthetic fibers, the main distinguishing characteristics of mohair are its luster, smoothness, and whiteness that are particularly distinctive because they lack the uniformity of appearance found in synthetics. The brightness and true color of dyestuffs are maximized when applied to the pure white mohair substrate. Mohair is particularly useful for upholstery and carpet applications because its durability and soiling resistance are exceptionally good. Because felting shrinkage is minimal, mohair can be used in products that must be washed regularly, such as dress and sport socks. In the 1980s, mohair consumption in the following broad end uses was estimated by the Mohair Council of America to be hand knitting yarns, 65%; men's suiting fabrics, 15%; women's woven accessories and rugs, 12%; and woven furnishings and velours, 8%. However, the hand-knitting component has now almost disappeared. In the U.S. particularly, the Mohair Council of America has placed a greater promotional emphasis on adult mohair for carpets and rugs. Classical end uses of mohair include lightweight tropical suitings for mens- and ladies wear; brushed articles such as sweaters, stoles, scarves, and blankets; plush and pile fabrics; upholstery and rugs; and, hand-knitting yarns (more recently, machine-knitting yarns). Often, the mohair is blended with wool or synthetic fibers that improve product performance, assist in processing, or reduce the product cost. End uses at any particular point in time are greatly influenced by fashion. Hunter (1993) documented 189 specific end uses for mohair.

### Cashmere

Most cashmere is spun into fine woolen yarns that are subsequently knitted on fully fashioned knitting machines into apparel knitwear such as cardigans and sweaters as well as accessories such as scarves, shawls, socks, and gloves. Woven fabrics are produced from both woolen and worsted yarns and are typically tailored into high quality men's and women's outerwear. Because of the high price of cashmere, it is often blended with very fine wool to obtain products that are less expensive while still having similar appeal to those made with pure cashmere to all except the most discerning wearer.

## SUMMARY

This chapter defines and describes the fibers produced by goats that have significant commercial value (that is, cashmere and mohair) in the following terms: the goats that grow them, quantities produced, geographical areas of production, genetics and selection for fiber production, effects of environment, nutrition, age, sex, current production conditions and problems, biology of fiber growth, other fleece constituents, physical and chemical properties, subjective evaluation and objective measurement, harvesting, marketing, textile manufacturing, applications, and end uses.

## REFERENCES

Blakeman, N.E., R.P. Rogers, C.J. Lupton, F.A. Pfeiffer, and M. Shelton. 1990. Comparison of fiber traits between Angora, Spanish, Angora × Spanish, and Cashmere × Spanish kids. Proc. Int Goat Prod. Symp. October 22–25. Tallahassee, FL:203–205.

Dalton, J. and R. Franck. 2001. In: Silk, mohair, cashmere and other luxury fibres. R.R. Franck, editor. Woodhead Publishing Ltd., Cambridge, England. Pp. 142–143.

Dooling, T.A. and A.R. Dooling. 1996. An overview of the status of the cashmere industry in the United States of America. Proc. 6th Int. Conf. on Goats, Beijing, China. Vol. 1:317–320.

Gifford, G.R., R.W. Ponzoni, R.J. Lampe, and J. Burr. 1991. Phenotypic and genetic parameters of fleece traits and live weight in South Australian Angora goats. Small Ruminant Res. 4:293–302.

Hunter, L. 1993. Mohair: A review of its properties, processing and applications. The CSIR Division of Textile Technology, Port Elizabeth, South Africa 6000. 278 pp. (1095 references).

Hunter, L. and E.L. Hunter. 2001. In: Silk, mohair, cashmere and other luxury fibres. R.R. Franck, editor. Woodhead Publishing Ltd., Cambridge, England. Pp. 68–132.

Huston, J.E., C.A. Taylor, C.J. Lupton, and T.D. Brooks. 1993. Effects of supplementation on intake, growth rate, and fleece production by female Angora kid goats grazing rangeland. J. Anim. Sci. 71:3124–3130.

Launchbaugh, K., J.W. Walker, and R.J. Daines, Eds. 2006. Targeted grazing: a natural approach to vegetation management and landscape enhancement. American Sheep Industry Association, Centennial, CO.

Lecraw, D. 2005. Mongolian cashmere industry value chain analysis. For the Economic Policy Reform and Competitiveness Project (USAID), Ulan Bataar, Mongolia. Chemonics International, Washington D.C.

Leeder, J.D., B.A. McGregor, and R.G. Steadman. 1992. A review and interpretation of existing research results on raw-fibre-to-end-product properties and performance of goat fibers. A Textile & Fibre Research Institute (Melbourne) Report to the Rural Industries Research and Development Corporation. Barton, Australia. 117 pp. (284 references).

Li Y., M. Ning, S. Yaqin, L. Weimin, and L. Yujie. 1996. Effects of non-genetic factors on main economic traits in Liaoning cashmere goat. Proc. 6th Int. Conf. on Goats, Beijing. Vol. 1:202–204.

Litherland, A.J., C. Toerien, T. Sahlu, P. Lee, and A.L. Goetsch. 2000. Effects of season on fleece traits of Angora does in the US. Small Ruminant Res. 38:63–70.

Lupton, C.J., J.E. Huston, J.W. Holloway, B.G. Warrington, D.F. Waldron, P.V. Thompson, F.A. Pfeiffer, and K. Qi. 1996. Animal performance and fleece characteristics of Angora goats maintained on western and southern Texas rangeland. J. Anim. Sci. 74:545–550.

Lupton, C.J., D.L. Minikheim, F.A. Pfeiffer, and J.R. Marshall. 1995. Concurrent estimation of cashmere down yield and average fiber diameter using the Optical Fiber Diameter Analyzer. Proc. 9th Int. Wool Text. Res. Conf. Biella, Italy. Vol. 2:545–554.

Lupton, C.J., F.A. Pfeiffer, and N.E. Blakeman. 1991. Medullation in mohair. Small Ruminant Res. 5:357–365.

Lupton, C.J., F.A. Pfeiffer, and A.R. Dooling. 1999. Prediction of cashmere style using objective fiber measurements. Sheep & Goat Res. J. 15:1–4.

Ma Ning, L. Yongjun, S. Yaqin, L. Weimin, and L. Yujie. 1996. Estimates of genetic parameters of main economic traits in Liaoning cashmere goats. Proc. 6th Int. Conf. on Goats, Beijing. Vol. 1:199–201.

McGregor, B.A. 1988. Effects of different nutritional regimens on the productivity of Australian cashmere goats and the partitioning of nutrients between cashmere and hair growth. Aust. J. Exp. Agric. 28:459–467.

McGregor, B.A. 2007. Cashmere fibre crimp, crimp form and fibre curvature. Int. J. Sheep & Wool Sci. 55:105–129.

McGregor, B.A., M. An, and Y. Jiang. 1991. Fleece metrology of Liaoning cashmere goats. Small Ruminant Res. 4:61–71.

McGregor, B.A. and K.L. Butler. 2008. The effects of cashmere attributes on the efficiency of dehairing and dehaired cashmere length. Text. Res. J. 78:486–496.

Millar, P. 1986. The performance of cashmere goats. Anim. Breed. Abstracts. 54,3:181–199.

Murthy, G.S.N. and T. Ramasami. 1992. Technological aspects of processing of goat skins. Proc. 5th Int. Conf. on Goats, New Delhi. Vol. 2, Part 2:534–540.

Naidu, A.S. 2000. Recent advances in the production and use of goat skins. Proc. 7th Int. Conf. on Goats, Tours, France. Vol. 2:627–630.

Pattie, W.A. and B.J. Restall. 1992. Genetical and environmental factors affecting cashmere production. Proc. 5th Int. Conf. on Goats, New Delhi. Vol. 2, Part 2:502–512.

Pfeiffer, F.A., C.J. Lupton, and D.F. Waldron. 2004. Interrelationship of traits measured on male Angora goats during a central performance test. J. Anim. Sci. 82 (Suppl. 1):59.

Rao, K.S. and G.G. Rao. 1992. Production, processing, and export of goat skins. Proc. 5th Int. Conf. on Goats, New Delhi. Vol. 2, Part 2:526–533.

Ryder, M.L. and S.K. Stephenson. 1968. Wool Growth. Academic Press, New York.

Shelton, M. 1993. Angora Goat and Mohair Production. Mohair Council of America, San Angelo, TX. 233 pages.

Smith, G.A. 1988. Rich and rare fibre differentiation. Text. Technol. Int. 22.

Smith, I.D., W.H. Clarke, and H.N. Turner. 1973. The potential of feral goats in Australia for cashmere production. J. Aust. Inst. Agr. Sci. 39:128–131.

Snyman, M.A. and J.J. Olivier. 1999. Repeatability and heritability of objective and subjective fleece traits and body weight in South African Angora goats. Small Ruminant Res. 34:103–109.

Stanton, T. and D. Brown. 1997. Worldwide cashmere situation and prospects for commercial cashmere production in the United States. Cornell University, Ithaca, New York.

Teh, T.H. and T.A. Gipson. 1992. Cashmere production in the People's Republic of China. Proc. Cashmere Prod. Amer., San Angelo, TX. 1-8.

Van der Westhuysen, J.M. 2005. Marketing goat fibers. Small Ruminant Res. 60:215–218.

Van der Westhuysen, J.M., D. Wentzel, and M.C. Grobler. 1988. Angora goats and mohair in South Africa. 3rd Ed. NMB Publisher, Port Elizabeth, South Africa. 258 pp. ISBN 0-620-12796-1.

Von Bergen, W. 1963. Wool Handbook, Part 2, 3rd Enlarged Ed. Interscience Publishers, A Div. John Wiley and Sons, New York.

Ward, W.H., C.H. Binkley, and N.S. Snell. 1955. Amino acid composition of normal wools, wool fractions, mohair, feather, and feather fractions. Text. Res. J. 25:314–325.

Zahn, H. 1990. The role of mohair in keratin research. Proc. 2nd. Int. Symp. Specialty Animal Fibers. DWI 106, 195 Aachen, Deutsches Wollforschungsinstitut.

# 16
# Environmental Enhancement

*A. Peischel, PhD*

**KEY TERMS**

Mob—a group of goats used as biological vegetation control agents.
Synergistic motion—a dynamic process including biological and environmental sequences to improve the sustainability of lands.
Biotic component—the living entity of an ecosystem.
Abiotic component—nonliving entity of an ecosystem.
Pugging—refers to intentionally walking on vegetation.
Ecosystem—a dynamic system encompassing succession, the water cycle, energy cycle, and mineral cycle.
Ladder fuel—vegetative understory in a treed area capable of ignition creating crown firing.
Land EKG—vegetation monitoring tool.

**OBJECTIVES**

By completing this chapter, the reader will acquire knowledge on:

- Use of goats as biological control agents for vegetation management
- The planning of a land enhancement project
- The synergistic motion of environmental and biological factors involved in the holistic management of lands to maintain biodiversity
- Using goats for fuel load reduction
- Using goats for forest management
- Using goats for stream bank restoration

## INTRODUCTION

The goat is an extremely agile, gregarious, and opportunistic creature. Management, through innovation and creativity, can successfully use those characteristics for the enhancement of lands. The management goal encompasses the use of all ecosystems, biological and environmental, and success depends on flexibility of management plans and the ability to re-plan. To accomplish this, biodiversity must be maintained, the physiology of plants and soil must be understood, and man must make environmentally, economically, and socially sound decisions (Savory, 1998). Goats, under control, are being used to enhance land productivity and encourage vegetative biodiversity.

## PRINCIPALS OF ENVIRONMENTAL ENHANCEMENT

### Goats as Tools

Energy, in pastoral-type agriculture, is universal and can be used, stored, concentrated, or spread; the primary source is the sun. To use the natural energy flow efficiently, it is vital to control the time of grazing/browsing, the area to be grazed/browsed, the season of grazing/browsing, the plant species to be grazed/browsed, and the goat(s) that are being used for land management applications (Smith, 1986).

The use of goats in vegetative management can take on many diverse avenues. Goats can be used for noxious weed abatement, rejuvenation of abandoned and eroded lands, edging back of woody and forb species, fire breaking and fuel load reduction, poisonous/toxic plant eradication, and enhancement of timber-producing forests through silvopasture and agroforestry techniques eliminating competition of unwanted species. Goats can stabilize stream banks and riparian areas, clean along irrigation ditches, minimize old fence lines, clear farm ponds, create flyways for ducks and geese along with landscaping around homes, and clean land in citrus orchards, nut farms, and vineyards.

The management criterion is to never underestimate the nutritional value of plants and vegetative regrowth, and to encourage a change of regression plant communities into succession plant communities. Biodiversity of vegetation provides year-round selection for goats avoiding problems such as those associated with monocultures.

Along with all of the goats' vegetative activities, they provide mankind with meat, milk, fiber, skins, and many other products (soap, cheese, creams, drums, gloves, etc.) used to enhance our lives. The goat is truly an opportunity for man to manage.

### A Synergistic Process in Motion

The integration of knowledge from separate disciplines such as ecology, plant physiology, hydrology, climatology, forestry, soil science, economics, animal science, sociology, and wildlife equals the ecosystem foundation blocks of water and mineral cycles, succession, and energy. All factors affect vegetation distribution and makeup of the various plant communities. Understanding the basic forces acting on an agricultural enterprise is important. The small amounts of energy input act as an amplification factor thereby increasing the amount of sunlight harvested and converted to a useable and marketable form.

#### Biotic Component

The biotic component of a system is that of living organisms, plants, and animals. The herbivore, through browsing and grazing, affects frequency of plants grazed or browsed, the degree of vegetation removed, plant species selected, and the quality and quantity of vegetation grazed. Other biotic factors include pollination and seed scattering by animals. Decomposition of organic matter is performed through other organisms such as earthworms, nematodes, protozoa, bacteria, or actinomycetes that consume dead material and render it useful.

#### Abiotic Components

The abiotic components of a system (that is, the nonliving environment and exchange material) affect vegetative distribution. These factors include the topography, altitude, exposure, insolation, precipitation, evaporation, evapotranspiration, and soil. The water cycle is driven by energy from the sun, and its distribution affects vegetation more than any other single environmental factor. There is a continuum between the soil, plants, animals, and the atmosphere.

Plant growth requirements include sunlight and the ability of the soil to provide moisture, support, protection, and nutrients. The type of vegetation that develops in an area is determined by soil characteristics such as texture, depth, slope, organic matter, pH, and chemical composition. These soil characteristics are determined by soil formation, which is affected by climate, vegetation, parent material, topography, time, and soil organisms.

There are many environmental factors that affect vegetation distribution in relation to pasturelands and rangelands management. These factors include topography, slope, precipitation, wind erosion, and soil mineral content. The plant community and the factors that influence those communities are important in land enhancement management decisions.

#### Soil Nutrient Redistribution and Pasture Health

Grazing/browsing management can improve soil fertility as it increases the amount of organic matter in the soil. As the percent of organic matter by weight increases, the percent of water-holding capacity by volume and the extensive root system of plants increase. If a specific nutrient in the soil is lacking, it can be added to the goats' free choice chelated loose mineral supplement, and in turn it can be deposited back into the soil. Soil receives its nutrients from the weathering of parent materials, cropping practices, rain, dust, and wind and those nutrients are recycled by plant roots in the subsoil. Livestock deposit minerals in the manure. Animals consume approximately 5–7 kg

(12–15 pounds) of mineral supplement per year, and over 90% passes through their system with dung and urine. Livestock redistribute nutrients in a browsing system depending on the variation in species browsed, the depth of the root system of the plants, and their ability to uptake different mineral elements. Therefore, using good rotation management can improve soil fertility (Gerrish and Roberts, 1999). Manure improves the physical characteristics of the soil. Soil aggregation, friability, and tilth increase water infiltration and retention, and decrease root knot nematodes and other plant root pests. Healthy pastures and healthy soil microorganisms provide high quality vegetation.

#### Vegetation Quality and Quantity

The quality and quantity of vegetation produced in a given time is dependent upon the amount of sun energy a plant captures and converts to tissue (Table 16.1). Plants need leaf area for photosynthesis, but a canopy cover of more than 30% can hinder sun-capturing ability and vegetation production. When plants are browsed, recovery time is dependent on several environmental factors such as soil fertility, soil moisture content, season of year, environmental temperature, degree of plant defoliation, time of plant removal, animal species browsing the land, and residual dry matter.

Residual dry matter is the forage dry matter remaining after a pasture/rangeland/woodland has been grazed or browsed. Different plant species (grasses, forbs, shrubs, brush, and trees) vary in recovery time, and climatic conditions can affect recovery time. When using goats in vegetation management, there are times when the available plant species are three-dimensional. Brush, shrubs, and trees are examples of three-dimensional plant species—they are plants that have more aerial canopy than basal cover. The correct amount of residual is needed for rapid plant regrowth and for yielding higher quality forage. In this scenario, when livestock numbers per hectare (acre) are increased, animal performance is improved. However, overstocking will reduce the amount of residual dry matter. Approximately 2,250 kg per hectare (2,000 pounds per acre) of residual dry matter or less is needed for optimum regrowth. High residual dry matter may slow recovery rate. With high residual dry matter, a denser canopy exists so sunlight is harder to capture; old leaves are less efficient producers of biomass than new leaves; and the higher ratio of nonphotosynthetic material to green material and the leaf to stem ratio is stressed. When the leaf to stem ratio is stressed in residual dry matter, there is more stem than leaf, the stems dry out exceptionally fast, and the nutrient quality decreases rapidly.

As the amount of mature vegetation increases in a system, digestibility decreases, crude protein decreases, and rumen turnover time increases, all having a dramatic effect on body condition of the goats. In lightly grazed paddocks with a high residual dry matter, old leaves shade the new ones and cause a net decrease in the photosynthesis rate available for new growth resulting in decreased production. Therefore, the Leaf Area Index (LAI) is a valuable tool for assessing plant health.

**Table 16.1** Chemical composition of various plants browsed by goats (%).

| Browse Type | Crude Protein | Neutral Detergent Fiber | Calcium | Phosphorus |
|---|---|---|---|---|
| Multiflora Rose<br>*Rosa multiflora* | 18.8 | 34.5 | 0.99 | 0.32 |
| Honeysuckle<br>*Lonicera* | 12.8 | 34.5 | 1.12 | 0.30 |
| Brambles<br>*Rubus* | 15.9 | 24.5 | 0.23 | 0.84 |
| Black Locust<br>*Robinia pseudoacacia* | 23.0 | 44.0 | 1.26 | 0.21 |
| Privet<br>*Ligustrum* | 18.0 | 26.8 | 0.89 | 0.34 |
| Green Briar<br>*Smilax* | 17.0 | 39.5 | 0.60 | 0.18 |
| Trumpet Creeper<br>*Campsis* | 16.7 | 43.1 | 0.42 | 0.22 |

The livestock used in a browsing regime must be under control in relation to their location, the time spent in any given location, and the animal density. The rate of plant species removal will affect the species returning to replace them in the ecosystem. Perennial vegetation should be encouraged due to depth of root zone, ability to absorb and store water, and building of organic matter. Overbrowsing plants depletes root reserves and over-resting the plants decreases the biodiversity (Table 16.2). Caution is warranted at all times to avoid erosion and the creation of hydrophobic soils. In browsing management, use of animal behavior and the herd effect allows concentrated animal energy input into a small area for a short period of time. Animals of the same physiological condition need to be foraged as a mob, and the quality of feed on offer needs to satisfy their physiological requirements.

KNOWLEDGE OF ANIMALS

The physiological and psychological effects of stress on the goat have a major impact on its production performance and health. It is therefore critical to have a basic understanding of goat behavior (social, nutritional, environmental, etc.) to enhance the welfare and well-being of the goat. The stockman needs to be observant and patient, and at the same time, understand that individual or small mob (a small group of goats) dynamics do not necessarily work with large commercial mobs; therefore, caution must be used when drawing conclusions.

When using goats for ladder fuel reduction, fire breaking, weed abatement, and riparian area restoration, there are basic behavioral patterns that can be used to enhance utilization of the goats or to curtail problems before they arise. The selection of environmentally adapted goats, goats that are in good physical condition, and structurally correct conformation are conducive to success. When placing goats into a new area, whether fenced with portable solar-powered electric polywire or netting or set/parked on horseback using herding dogs, let the livestock guardian dogs check the area first. As the dogs complete their rounds, then let the goats in. Once in the new area, the goats will circle the fence line and check out the vegetation before browsing.

In plant anatomical selection, goats eat from the tip of the plant toward the base, selecting the highest quality plants and plant parts the first time around because they will return to the same plant many times before ending consumption of the specific plant (Table 16.3). They eat the seed heads from grasses and forbs and select flowers of thistles and some brush species. Mobile lips and bipedal capabilities afford them the ability to select young buds, create a 2–2.5 m (6–8 ft) browse line, and knock down the larger diameter vegetation high in cellulose and lignin. When browsing, they are always on the move, nibbling on one plant but on the lookout for the next selection. They

**Table 16.2** Root depletion affected by grazing defoliation.

| % Leaf removal | % Root stoppage |
|---|---|
| 40 | 0 |
| 50 | 5 |
| 60 | 50 |
| 70 | 70 |
| 80 | 100 |

**Table 16.3** Browsing calendar based on the goat.

| Specie | Preferred Time | Not Preferred | Notes |
|---|---|---|---|
| Yellow Star Thistle *Centaurea solstitialis* | Leaf phase through seed head production | Cane heads are dead and empty. | All ages and classes select first at various growth phases. |
| Scotch Broom *Cytisus scoparius* | Before flowering and after fall dieback | Flowering | Caution using young doelings and pregnant does |
| Chamise *Adenostoma fasciculatum* | Fall, winter, spring | Mid- to late summer | |
| Buckeye *Aesculus californica* | Fall as leaves die back and seeds drop | Green and productive | Toxic when green |
| Tamarisk *Tamarix* | New shoots and young branches | Old decayed plants | Continuously barking and trampling |

consume vegetation from the outside perimeter toward the center of an area, being cautious not to be in a situation of being trapped. This behavior relates back to the predator/prey concept from which they evolved. Goats will turn to face an attacker and become aggressive. When herding on foot or with herding dogs, observe the amount of pressure exerted as the situation forces the animal into the attack mode and it will attempt to escape. If an escape happens, let the escapee return to the mob on its own accord. The goat is a social animal (bonded to the mob), prefers to be part of a group, and does not want to be alone. The stockman needs to investigate the cause of the escape and take corrective measures immediately to discourage others from escaping. In rolling, hilly terrain, goats must be moved uphill or at an angle across a slope when coming downhill. They naturally select the highest spot geographically for afternoon naps and night camps since it affords safety with a clear view and without obstruction.

The mob must be trained to an electric fence before they are expected to respect the electric fence line. It should be noted when using "fencing" that browsing behavior, time spent browsing, rumination time, lounging time, and flow pattern of motion will change. Fresh water needs to be provided at all times in smaller troughs with a high rate of recharge. Goats can consume more than 4–6 liters (1–1-1/2 gallons) of water a day when the weather is hot, dry, or when the vegetation is decadent, highly lignified, or fibrous. Water consumption helps keep the rumen microflora colonized and increases the digestion of fibrous materials. If night corralling is needed, it is important to have on offer a forage supplement as intake will be decreased and body condition score will decline. Goats are continuous browsers; they browse for several hours at a time, chew their cud, and return to browsing.

Goats have minimal subcutaneous fat and are susceptible to climatic changes of wet and cold. They will seek shelter during inclement weather. They need to have access to shelter such as a brush canopy, trees, or rock outcroppings, or a portable shelter. Older goats with a high body condition score can manage about 48 hours of extreme weather, but younger animals and goats in poor body condition can succumb to death in about 12 hours. They cope with heat and humidity quite well, but they change browsing patterns, plant selection, and time of browsing as the temperature soars. Trees and rock outcroppings offer some relief.

There is a dominance structure within the mob. Goats will form small familial "groups" within the larger social structure of the mob. This social structure is very crucial for behavior development of young animals. Through bonding they learn browse selection and establish browsing territories and various athletic habits; and collectively this increases survival rate (Kilgour and Dalton, 1984).

Sensory capacity of the goat comprises smell, taste, sound (above the range of human hearing), vision (enhanced peripheral vision), vocalization, body language, touch, and group size. With time and patience, goats can be taught to adapt to change. They can readily adapt to herding on horseback or bicycle, loading readily into a stock trailer, a decrease in social interacting space, and the ability to cope with an individual stress for a short period of time. Temperament is not only a heritable trait (25%), but it can be enhanced by imprinting at birth and creative insight of the stockman or herdsman. If goats are given time to assess the situation, they do make calculated decisions.

Environmental stressors such as heat or cold weather, precipitation and humidity, nutrient density, quality of feed on offer, predation, travel activity, and topography will cause economic losses. The major losses are decreased reproductive performance of both males and females, inability to maintain body condition score, decreased growth rate of offspring, increased incidences of internal parasitism, and suppressed immune system.

## Project Planning

### Initial Assessment

Before starting a land enhancement project with goats, the goals for the land must be established based on site analysis, water catchment, and erosion consequences. Soil texture should be noted because it affects rainfall infiltration and soil percolation capabilities. The final landscape goal is then described and discussed with the landowner. At this point, an experienced service provider will do an initial site survey analysis, stay in communication with the landowner, and make the final decision as to whether livestock are the best solution to attain the final landscape goals. The management goal encompasses the use of all ecosystems with the success of the project centering on the flexibility of management plans and the ability to replan.

### Business Planning

A service provider (owner of the livestock) must be successful not only as a land and livestock production manager but also as a business manager. The most important first step of any business venture is planning. A plan includes proper steps by which to proceed and accomplish the described goals. During the planning process, research is conducted and first-hand contacts are made. A financial plan for enterprise evaluation can be incorporated into the business decisions. With a financial plan, a profit and loss statement is generated indicating gross margin, gross income, and variable and fixed expenses. In the planning process, stock flow, stock allocation, and the valuation of the different classes of stock are obtained in a "before and after" inventory assessment.

The consistency and predictability of production management is the major goal, thus, a business analysis will produce the gross revenue generated for each business fraction within the plan. The analysis will show the cost of generating the revenue and the net profit for each business segment. The important factor is obtaining reasonable return for the time spent and the ability to consider the effectiveness and, if necessary, the information that might lead to the pursuit of other options. In the process of developing a business plan, the goals and objectives of the business will be developed. The goals should be realistic, attainable, and the production management land enhancement business sustainable. The business must be marketable, economically feasible, and able to produce the results for prospective clients. To successfully solicit the business, an understanding of many facets of "paid to graze/browse" is necessary (Peischel, 2006).

#### Vegetation Assessment

A major determining factor of the success of the business is the innate ability of the service provider to do a quality assessment of the environmentally adapted plant communities on site. From this information, a browse preference list will be constructed. The list will include vegetation availability and the time of year the livestock (and class of livestock) are most likely to select or prefer the vegetation in the water catchment area. In the identification of vegetation composition and succession plant communities, poisonous plants and their toxins should be identified along with the target plants. Soil texture and rainfall can cause erosion consequences if the percolation and infiltration capabilities of soil are not understood. The site must be monitored, including an initial set of monitoring points. Vegetation monitoring techniques that can be used are photo points, transects, Land EKG, green line vegetation composition measurements, water quality, etc. All methods are used to evaluate the health of land and the progress being made to increase productivity or succession. Within this monitoring process, basal and canopy cover are assessed as is ladder fuel and fuel continuity. The service provider should know the amount or total browsable/grazable biomass as part of costing out the project.

#### Land Assessment

It is the provider's responsibility to investigate the previous history of the site in question to determine previous livestock usage in reference to disease potential, and soil or water contaminants. A base map of the area and specific sites under consideration are required. On this map, the perimeter is defined, ecological constraints and exclusions are noted, and the topography detailed. This equates to an understanding of fire ecology and identification of shelter for inclement weather. In the U.S., several regulatory agencies govern land usage, and their endorsement should be considered before the commencement of the project. Wetland(s) regulations should be championed, the Environmental Protection Agency (EPA) and the National Environmental Policy Act (NEPA) requirements fulfilled, the Endangered Species Protection Act (ESP) endorsed, and Fish and Game Conservation corridors and zoning restrictions respected. Neighbors and adjacent landowners need to be fully informed as to the intentions of the landowner anticipating the use of livestock lands enhancement.

#### Animal Welfare Issues

Animal welfare and the issues related to the health and well-being of livestock are the first and foremost priorities of the provider. A health maintenance program should be in place, internal parasite assessment should be current, and no known zoonotic diseases or other transmissible diseases should be present. The body condition score (BCS) is a top priority before going into a browsing project and monitoring the condition of the goats must be done throughout the duration. The body condition of the goats at the initiation of the project should be a BCS 3–3.5 (6 out of a possible BCS 9, see Appendix D) and not below a BCS 2.5 (4 out of a possible BCS 9) at any time. When using goats in land enhancement projects, some managers prefer the BCS chart of 1–9 rather than BCS of 1–5 used for meat, dairy, and fiber goats. The wider range allows for more sensitivity for change in body condition. As goats under nutritional and environmental stress drop condition too rapidly, they have no subcutaneous fat cover, and recovery takes too long. A healthy animal is an asset to a land enhancement endeavor. It is also important for the provider to select the correct specie, breed, age, and class of livestock to be used in a prescribed herbivory venture. Animals that are adapted to the environment, climax vegetation, and topography, are assets. Those individuals or a mob that have experience working on previous projects are of great value (Provenza et al., 1992).

Livestock Guardian Dogs

Effective livestock guardians are of utmost value when working in an extensive, isolated, or predator-infested habitat. Remember that the most dangerous of all predators are domestic dogs that have joined as a pack to kill for the "thrill." The specie and breed of guardian required will be dependent upon the class of livestock to be protected, topography, and type of predator (nocturnal or diurnal). Age, level of experience, and number of guardians needed is based upon specie and aggression of predator, herd size, and management practices (portable solar-powered electric fencing, or herding). A livestock guardian dog less than 2 years of age should not be expected to guard livestock alone. The young dog needs to gain experience from a successful, competent mentor, and be used as visual backup until at least 2 years of age. As the numbers of guardians in use increase, each dog will find its niche in the working scheme of the mob. The role of each dog within the mob must be evaluated before adding or deleting a dog from a functioning group. Livestock guardian dogs should be fed a high energy and high protein food daily. The dog food should not have ruminant animal protein as an ingredient. Each dog requires its own feeder space to prevent squabbling at feeding time.

## FOREST FIRE MANAGEMENT

There are various tools that can be used to minimize the damage done by fire to grasslands, rangelands, forests, homes, and personal property. The tools used range from mechanical (bulldozers, masticator machines), hand cleaning (weedeaters, chainsaws), wildfire landscape design, herbicides (labeled for forbs, shrubs, and woody species), prescribed burns, and livestock. Each tool has a specific use and place in management. A vegetative survey analysis is conducted and an assessment made for the use of each tool. Fire protection and mitigation are factors that should be built into the management plan of ranches, farms (crops, trees, livestock), subdivisions, and secluded home sites but particularly for the urban/wild land interface. There should be a policy for interface (the I-Zone) zoning established before the layout and approval of subdivisions for structure erection (Slaughter, 1996). Self-help, neighborhood teamwork and an understanding of nature make the mitigation approachable. The consequences of fire such as erosion, floods, and biological devastations (vegetative, wildlife) should be curtailed instead of accepted as the outcome of fire. Fire prevention (or minimizing the damage) encompasses selecting the correct home site, zoning ordinances that are respected, landscape planning and design with mixed plant species, vegetative fuel discontinuity, fuel load reduction, and fire breaking. All approaches must take into consideration the climate, topography (slope, soil, elevation), vegetation type (drought-tolerant plants, plants adapted to fire, plants that can be managed to minimize fire travel), and fire frequency in the area. Goats are only one in many management scenarios that can be used to curtail the devastating loss created by fire (Peischel and Ingram, 2004).

### Fuel Load Reduction

Fuel load reduction references a reduction and lowering of ladder fuels through pruning and minimizing down fuel materials by removal or mulching. Ladder fuel is vegetation growing under or in stands of trees that will ignite, causing fire traveling up the tree and crown fire. Crown fires can generate their own weather and consume thousands of hectares (acres). They are extremely difficult to contain for fire suppression teams. The reduction of down material can be accomplished by removing "snags" out of standing timber, firewood cutting of downed trees, the physical removal of downed trees, or the use of livestock to break down and mulch-in slash material.

Goats used for fuel load reduction are managed to remove dense understory (brush, shrubs, and forbs) and lower branches to prevent "laddering." This technique takes time, and time relates to the biological control of reduction and returning to sites being reduced. On occasion, depending on availability of vegetations, it can mean the supplementation of protein, energy, or balanced minerals for the goats (depending upon the class of goat used and time of year). A free choice loose chelated mineral mix must be available. It should be balanced based upon profiles of soil, vegetation, and livestock analysis. Depending upon the vegetation present and young tree plantations or older plantations, the management can last for several years. Management includes a return maintenance program. Depending on the vegetation type, sheep or beef cattle may need to be added to the regime.

Eliminating the ladder fuel gradually prevents soil erosion, and enhances rainfall infiltration and even percolation through the soil interface. As the goats work through an area, they are also working the understory including old pine needles and leaves, breaking lower branches and splitting apart old downed branch material. Vegetation that returns can be perennial grasses, some annual grass species, evergreens, and many forbs species.

### Fuel Discontinuity

This management approach of fuels breaks the continuity of flammable cover. It is canopy related and creates a special mosaic of vegetation. The spacial distance of canopy cover between trees and shrubs is managed to keep a potential fire low and noncrowning. The reduction (and/or removal) of the lower growing species can also be minimized. The vegetation is managed to maintain a low-profile, controlled vegetation so that backfiring is easier, or kept in a growth phase, enhancing a higher water-holding capacity within the plant. Once goats have brushed an area, it can be maintained to be a "living" green belt.

### Fire Breaking

Management of vegetation in an area to establish a firebreak is intensive for both the vegetation and livestock phases. In a firebreak regime, the goal is to prevent a fire from going any farther and from crossing or jumping the break line resulting in a controllable situation. This technique requires an understanding of fire behavior to determine where the breaks need to be established. An assessment takes into consideration an evaluation of fuel flammability, heat intensity and duration, ignition temperatures (airflow and rate of heating), heat of ignition (fuel size distribution, live : dead material ratios and moisture content of the fuel), and heat transfer (interaction with topography). For example, tunnels and chimneys in canyon or saddle areas need completely different types of firebreaks than grass or rangelands.

When using goats, the goal is elimination of vegetation to slow down the oncoming fire, decrease the intensity, and allow a fire crew access to an area so they can successfully contain a fire. Surveillance of goats to maintain a positive BCS is critical. Animals' BCS cannot fall below 2.0 (3 out of possible BCS 9) when using experienced wethers. A good firebreak will help the "mop up crew" coming in after a fire. After the mop up crew, a fire site analysis must be conducted to plan erosion containment, reseed and replant vegetation, reestablish wildlife habitat, and test the water for quality and stream health.

Utilization of goats as a tool to assist in fire prevention, mitigation, and/or control is not an overnight remedy. Proper planning, site evaluation, and working goats takes time. The number one concern is maintaining the health, welfare, and body condition score of the goats.

## STREAM BANK RESTORATION

In the process of restoring a dry stream to a perennial stream, ecological phases are changed slowly over time (Peischel and Ingram, 2004). The process can take from 1 to 3 years to approach the ephemeral stream phase depending on steepness of banks and density and type of vegetative species. It may take another 4 years until the ephemeral stream begins to flow as a perennial stream. The process is "in and out" throughout the seasons and over the years. Factors taken into consideration are angle of the stream bank, soil type and structure, physical size and weight of the goats, time of the year for erosion avoidance, the plants that need to be minimized and when the goats prefer to consume them, species of vegetation returning to stabilize the stream banks, rate of water flow, temperature of the water, and turbidity.

Once the stream bank has become stabilized, it can be maintained by the management of vegetation with goats to keep the water flowing, minimize insect populations, and improve the quality of the water. It is advisable to prevent goats from consuming water from streams or ponds, and to provide goats access to fresh water daily.

## ERADICATION OF NOXIOUS VEGETATION

Before starting an eradication process, it is important to understand the physiology of the plant selected, the reason for its presence within the area, and the probable vegetative specie(s) that will take its place in the landscape mosaic. It is at this point that the class of goat becomes relevant (Peischel and Ingram, 2004).

Yellow star thistle (YST; *Centaurea solstitialis*) spreads over hundreds of thousands of hectares (acres) in California and other western states. It is an invasive plant that invites itself into overgrazed and underrested pastures. Goats and sheep can be used to readily minimize the plant, but cattle have no preference for this plant. YST cannot handle browsing pressure and does not like shading. Therefore, as the small ruminants graze it down and other vegetation species begin to return, the YST minimizes. However, it is a high quality plant, and under proper management can be used for weight gain of weaned kids and flushing does before breeding.

Dog fennel (DF; *Anthemis cotula*) grows quite aggressively in shaded areas, has a rank odor, and is preferred by young bucklings. During a growing season, it can be eliminated from along old hedgerows and within Christmas tree plantings.

Poison oak (PO; *Toxicodendron pubescens*) is very toxic to humans with an allergic reaction to the oils of the plant, whether touched or from fumes being burned. Mature goats will eat the dead leaves and growing tips of the plant whereas the young kids will eat young buds and fresh young leaves. Therefore, it is advantageous to run does with kids through an area where poison oak eradication is needed.

Goats do not select Vinca (*Vinca minor*) readily. They prefer to pass it by when cruising an area for browsing.

**Table 16.4** Diet preferences for different livestock species.

| Plant | Horse | Cattle | Sheep | Goat |
|---|---|---|---|---|
| Grass | 90 | 70 | 60 | 20 |
| Weeds (Forbs) | 4 | 20 | 30 | 20 |
| Browse (Woody) | 6 | 10 | 10 | 60 |

The plant cannot tolerate trampling and can, therefore, be pugged out (intentionally walked on) in a short period of time. If pugging is done before flowering, the process will be more successful.

Ferns (consisting of hundreds of genera and thousands of species) are very abundant in woodlands of the eastern and western regions of the U.S. Therefore, it is pertinent to identify them before engaging in land enhancement. Some ferns are toxic, others have anthelmintic properties, and some are hematologic (affect blood). Goats can be used to minimize ferns in the understory of forests but only after identification, knowledge of toxic elements, and proper time to browse the plant.

## WEED CONTROL IN PASTURE

Goats naturally have a preference for brush, forbs, and weeds (Table 16.4). Goats can be used successfully to minimize weeds in pastures and can be managed successfully when co-grazed with other species of livestock (cattle, sheep, and horses). The management of brush and weeds reduces competition for soil nutrients and moisture, and over time, restores pastures and improves the carrying capacity of the land. The most important concept to remember is that sustainability through biodiversity must persist over generations and enhance flexibility. Renewable resources are to be used wisely with pollution rate kept to a minimum. Brush, range, and pasture management are based on the physiology of the plant and the ability of man to make social, environmental, and economically sound decisions.

## SUMMARY

Goats can be successfully used to manage brush that creates a ladder fuel that feeds fires, reduce fuel load and create fuel discontinuity, establish firebreaks, restore stream banks, and control invasive vegetation. Selection and management of goats as biological and environmental agents to control weeds is different from those methods of management focused on meat production. As vegetation control agents, animals are managed intensively using high stocking density and time management in a site-specific area. In this scenario, the minimization of unwanted vegetation surpasses weight gain in importance. Animals selected for this purpose are usually wethers and trained at a young age. Use of production animals for this purpose reduces kidding rates and kid weaning weights. When using goats for land enhancement or production purposes, the management should be in compliance with all of the policies set by regulatory agencies such as the USDA, EPA, or those governing elsewhere, and should actively pursue the well-being of the animals.

## REFERENCES


Gerrish, J.A. and C.A. Roberts. 1999. Missouri Grazing Manual. Publication No. M157. University of Missouri–Columbia, MO.

Kilgour, R. and D.C. Dalton. 1984. Livestock Behavior: A Practical Guide. Westview Press, 5500 Central Ave., Boulder, CO 80301.

Peischel, A. 2006. A natural approach to vegetation management and landscape enhancement. In: Targeted Grazing. Chapter 16. www.sheepusa.org/targetedgrazing.

Peischel, A. and R. Ingram. 2004. California Browsing Academy manual. University of CA–Davis, Davis, CA.

Provenza, F.D., J.A. Pfister, and C.D. Cheney. 1992. Mechanisms of learning in diet selection with reference to phytotoxicosis in herbivores. J. Range Management. 45(1):36–45.

Savory, A. 1998. Holistic Resource Management. Island Press, POB 7, Covelo, NM.

Slaughter, R. 1996. I-Zone (Urban/Wildland Fire Prevention and Mitigation). CFESTES Bookstore, 7171 Bowling Drive, Sacramento, CA 95823-2034.

Smith, B. 1986. Intensive Grazing Management. Graziers Hui, POB 1944, Kamuela, HI 96743.

# 17
# Housing Requirements

*S.G. Solaiman, PhD, PAS*

## KEY TERMS

Housing principles—factors such as climatic data, floor design, and space allowance that must be considered for proper animal housing.

Animal requirements—factors such as animal comfort, animal welfare, and health that must be considered for animal housing.

Homeotherms—warm-blooded animals that maintain their body temperature within a predetermined range of temperatures.

Thermoneutral zone—a range of ambient temperature at which energy expenditure is minimal for regulating body temperature.

Lower critical temperature—a lower range of ambient temperature when animal productivity is negatively impacted.

Upper critical temperature—a higher range in ambient temperature that negatively impacts animal production.

Space allowance—the space required by a goat to perform its natural behavior of resting, eating, bipedal stance, and socializing.

Confinement housing—represents a maximum housing system where animals are enclosed with food and water provided for them.

Extensive housing—represents a minimum housing system where goats are outdoors and eat at their free will.

Specialized housing—represents a special requirement for housing dairy goats, including the barn, milking parlor, and milk room.

## OBJECTIVES

By completing this chapter, the reader will acquire knowledge on:

- Production systems and their housing requirements
- Extensive systems housing requirements
- Intensive systems housing requirements
- Principles for housing requirements
- Animal requirements for housing
- Physical requirements for housing
- Housing structures for goats
- Equipment required for housing
- Specialized buildings for dairy goats

## INTRODUCTION

Goats are very versatile animals and can adjust to different environmental elements. They can be a vital part of subsistence agriculture in some regions of Asia, Africa, or Latin America, with minimum to no housing, or in intensive dairy goat operations in Europe or North America. Goats in a natural setting may not require housing; however, if housing is considered, whether simple or elaborate, it must meet the animal's requirements. A few factors must be considered in the design of housing for goats. Goats should not be overcrowded if confined, and should be protected from sun, wind, and rain if possible. Environmental factors such as heat, humidity, rain, and wind can stress goats, depress their immune system, and lead to either parasitic or respiratory problems, which can be fatal. Goats can tolerate cold weather and heat as long as they live in a dry, well-ventilated, and draft-free environment. Cold and wet conditions promote respiratory problems, and heat combined with humidity or damp conditions promotes parasites, especially in young and growing kids. In addition, physiological factors such as curiosity, agility, small size, and vulnerability to predators necessitate construction of effective protective housing.

## HOUSING SYSTEMS

Housing requirements for animals managed under different production systems vary depending primarily on the availability and use of land and resources. Housing requirements can be characterized under subsistence agricultural systems, extensive grassland or range systems, and intensive or confinement systems. Research data comparing cost effectiveness of these production systems is very limited to justify elaborate housing. Lupton et al. (2008) compared a feedlot (intensive) system with pasture and elevated floor systems for meat and fiber goats. In a commercial setting, feedlot and pasture system profitability were similar, and both were more profitable compared with the more elaborate raised-floor system for mohair and goat meat production. The more elaborate housing resulted in higher quality animals; however, higher investment in housing is not feasible. Higher investment in housing is only justified when higher-priced products like goat milk and cheese are sold from the farm.

The subsistence farming system is practiced in parts of Africa, Asia, and Latin America. Under this system, the requirement for formal housing is basically minimal or none. A few animals are tethered during the day and put into a protective shelter, if available, at night.

In the extensive system, goats graze, browse, or range over large areas of marginal land not suitable for other agricultural activities. The majority of Angora goats in the U.S. are ranged with practically no shelter or housing. Animals managed extensively on improved pastures or browse (the majority of meat goats) in the U.S. are provided with portable shelters or similar structures for protection from wind, sun, and rain. This system also practices minimum housing.

In the intensive housing system, animals are confined in specialized housing for meat or milk production, and shelters are provided. This system is labor intensive and requires accurate knowledge of the goat's requirements. Feed and water is brought to the animal. This system offers the greatest protection for the animals from both predators and environmental conditions. The environment is controlled by blowing fans or exhaust fans, for air movement in summer and with heaters in the winter in colder environments. Although it may make the best use of limited land resources, this system increases labor and the capital investment. This system of housing is expensive to construct and may only be feasible in commercial settings with high output.

## HOUSING PRINCIPLES

In this section, principles governing specific requirements for climatic conditions, materials used, space allowance, and flooring for animal housing are discussed.

### Housing Design Requirements

Most goat types except dairy goats are low output commodities whose profit may not justify sophisticated housing. However, design for any housing should be according to the climatic conditions and production systems practiced in the region. Climatic data can be used as a guide for the type of housing, insulation, ventilation, and other factors that must be considered for proper animal housing. There is a wealth of information on the design, materials, and construction of livestock buildings. However, for a proper building, the requirements of animals, the performance of materials, construction standards, and housing recommendations should be considered.

Many of the new developments in building materials and designs for goat housing are aimed at lower cost. For the best outcome in animal housing, the producers, building suppliers, livestock specialists, and regulatory agencies must adopt a rational and integrated approach for goat housing design that provides for health, management and husbandry, material and construction technology, energy and pollution, and economic considerations.

Many factors must be considered for proper housing design and choice of materials. For proper lighting in a goat barn, the window area should be at least 8–10% or preferably more of the floor area. Plastic roofing can enhance lighting and reduce the need for windows. Metal roofing can be aluminum or steel, which is low in cost, but these alternatives have lower insulating properties.

Interior and exterior finish of an animal housing unit may be of different types; however, it should provide mechanical strength, moisture resistance, and easy cleaning. Metal siding is available and may last 15–20 years without refinishing but takes less abuse and is not a good insulator. Wood siding can take animal abuse and is a better insulator than metal or masonry material; however, it requires more maintenance. If the operation is seeking organic certification, treated lumber and other treated building materials will not pass organic certification standards in the U.S. Policies regarding organic certifications must be considered before purchasing treated materials. Goats can chew and rub on wood when bored, and this may lead to problems. Wood also harbors disease and is often difficult to clean and disinfect. For this reason, using metal materials such as "T" posts, welded stock panels, and combination wire pipe panels may be preferred. In any case, cost and ease of use of materials should be considered. Masonry siding requires very little maintenance but requires a high initial capital investment.

Insulation is a part of modern farm buildings, even for buildings that are considered cold structures (that do not need mechanical heating). These buildings can be minimally insulated to protect animals from summer heat and winter cold. Table 17.1 represents general recommendations for insulation of farm buildings based on a minimum R-value. As the R-value increases, the material's resistance to heat flow increases or buildings become more insulated (Parker, 2008).

**Table 17.1** R-value guidelines for animal housing.

| | Ceiling | Wall |
|---|---|---|
| Cold building operated at outside temperature | R2–R4 | No insulation |
| Buildings where animals' heat provides winter minimum temperature | R16 | R9–R12 |
| Building with supplemental heating | R24 | R13 |

Source: Parker, 2008.

In many regions of the world, low cost housing can be framed by wooden poles, and siding and roofing could be constructed simply from available weaved tree branches. In all cases, internal surfaces of all buildings and equipment to which goats have access should not have sharp edges or projections.

### Flooring Design

The floors in animal housing must be made of durable materials that are not slippery. The type of flooring can be wood, metal, and plastic. The floor should be well drained and easy to clean with a low maintenance requirement. Dirt and stone (gravel) floors are preferred and dirt (clay) mixed with sand may increase the surface area absorption. Wooden floors should be made of rough-cut hardwood. However, they can be slippery when wet, and feed and urine may be trapped in their cracks and create an environment to attract rodents. Compacted gravel can also be used for flooring that provides surface drainage. Concrete floors may be hard on feet but they are easy to clean and disinfect. Concrete pens should be bedded with 10–15 cm (5–6 inches) of straw or other absorbent materials such as poor quality hay, wood shavings, sawdust, shredded newspaper, peanut hulls, or sand. Other materials such as rubber floor mats may be used but are not recommended.

The elevated slatted false floor housing for dairy and meat goats is used in many tropical countries, and it is an effective management system practiced for half a million head of dairy goats in Taiwan (Figure 17.1) (Professor George Haenlein, personal communication). This type of flooring can be constructed from wood, corrugated metal, or plastic screens, or may be concrete (used in swine buildings). In the U.S., a durable plastic-coated metal flooring system that can be used for elevated goat housing floors is commercially available. In a study reported by Lupton et al. (2008), Angora goats were raised for meat and mohair production, using feedlot, pasture, and raised floor systems commercially or as a niche market. Goats raised on raised floors clearly had low labor requirements, a clean and healthy environment, no internal parasites, cleaner fleece, and more consistent products, either meat or fiber when compared with other systems. The high initial investment for this commercial production system failed to produce a profit in meat or fiber production. However, raised floor systems for a niche market and for a commercial goat dairy can produce higher profitability. Raised floors have a potential to protect goats from rain runoff, keep hooves dry, allow air movement, reduce feces

India Taiwan

**Figure 17.1** Elevated slatted floor design. Courtesy of Dr. George Haenlein, Professor and Dairy Goat Specialist. For color detail, see Appendix A.

**Table 17.2** Space allowance for housing goats, $m^2$ ($ft^2$).

| Type of management | Adult goat | Doe with kid | BGS[1] Adult Goat | UNH[2] Adult Goat |
|---|---|---|---|---|
| Confinement Housing | | | | |
| Individual | 1.4–1.8 (15–20) | 2.3 (25) | 4 (50) | 1.8–2.3 (20–25) |
| Group | 0.56 (6) | | 2 | |
| Exercise Pen | 2.3 (25) | 2.3 (25) | 3–4 times the pen | 4 (50) |
| Exercise Pasture Hectare (acre) | 0.065 (0.15) | | | |
| Extensive Housing | | | | |
| Group Shelter | 1.1–1.4 (12–15) | | | |

[1]British Goat Society. http://www.allgoats.com/housing.htm
[2]University of New Hampshire Cooperative Extension

and urine accumulation, and help in controlling transmission of diseases and parasites.

### Space Allowance

Space allowance recommendations for housing goats must provide for freedom of a goat's natural movements, stretching, bipedal stance (a natural goat behavior), socializing, resting, and feeding. Also goats seem to be more vulnerable to parasite infestation when kept in confinement. Providing adequate space is an important factor in health and social management of goats. Recent studies have shown that when resting space allowance was decreased, goats rested less and interacted with other goats more (Anderson and Bøe, 2007). They also reported that low ranked animals spent less time resting than others. Table 17.2 represents collective information for space allowance retrieved from searching the Internet for "housing goats" and available published goat handbooks. Generally all the U.S.–based generated data for space allowance was less than the values considered adequate for goats reported by the British Goat Society (BGS) (2002). According to the BGS, if the flooring is made of concrete and can be cleaned and disinfected on a regular basis, a space allowance of 4 $m^2$ (43 $ft^2$) per goat is adequate when goats are housed individually, and 2 $m^2$ (21 $ft^2$) is adequate when housed in a group. This area can be reduced when goats are housed on elevated slatted floors. According to the American Dairy Goat Association (ADGA) (2004), 1.4 $m^2$ (15 $ft^2$) of clean, dry, well-ventilated, and draft-free bedded area is needed per dairy goat. An additional minimum of 2.3 $m^2$ (25 $ft^2$) of well-drained and properly fenced area per animal is needed for exercise. If goats are

**Table 17.3** Trough space requirements for confinement feeding, cm (inch).

| | Concentrate | Restricted roughage | Ad libitum roughage |
|---|---|---|---|
| Adult goat >60 kg BW | 50 (20) | 25 (10) | 15 (6) |
| Growing kid <35 kg BW | 40 (16) | 20 (8) | 15 (6) |
| Weaned kid <20 kg BW | 30 (12) | 15 (6) | 10 (4) |

Source: Adapted from ADAS, 1987.

pastured, in temperate areas, at least 0.2 ha (0.5 ac) will be required for one dairy goat. More pasture area is needed in more marginal lands to avoid overgrazing.

The incidence of displacement and aggressive behavior increased as the number of goats increased per feeding space (Bøe et al., 2008), and low-ranked goats spent less time feeding and more time queuing (waiting) than goats in higher ranks. These results clearly state that for optimum performance, goats should be provided with adequate trough spaces for feeding to reduce aggression and thereby allow for lower-ranked goats to have an opportunity to eat. Table 17.3 provides a guideline for trough space requirements for goats based on size. Feeders should be reachable for goats but high enough to avoid contamination with fecal or urine matter.

## ANIMAL REQUIREMENTS FOR HOUSING

There are a number of factors that can affect productive efficiency of animals that must be considered before deciding if housing for goats is necessary. These factors include climatic requirements, animal health, animal comfort, and animal welfare (Whates and Charles, 1994). Technically goats require a shelter that is windproof, draft proof, rainproof, and is free from condensation. Meeting the animal's requirements for climatic conditions, health, and comfort can directly affect performance, whereas animal welfare is a regulatory issue for farm animals in the U.S. and around the world. Consumers are increasingly concerned about the welfare of food animals and the conditions under which their foods are produced. Providing a proper living environment for food animals will improve productivity and can be economical, if justified.

### Climatic Requirements

Climatic factors can influence animal productivity and welfare and must be considered in relation to animal housing. These factors are temperature, air supply or movement, and composition of air including ammonia, carbon dioxide, humidity or other pollutants from animal bedding, or from microbiological origin.

Animals continuously exchange heat with their environment. To maintain steady body core heat, homeotherms (all animals that maintain their body temperature within a predetermined range, including goats) gain heat from the metabolic process through oxidation of proteins, fats, and carbohydrates within the body. The rate of metabolism can depend on size, feed intake, and levels of muscular activity as well as the thermal environment. Animals can lose heat either by radiation or non-evaporative heat transfer from the body, or by evaporative heat transfer and respiration. Non-evaporative heat loss is mainly in low temperatures that can vary among animals depending on their coat or skin thickness. The majority of evaporative heat loss occurs at high temperatures, generally above 20–30°C (68–86°F) (Mount, 1979). Optimum animal productivity is achieved within the thermoneutral zone, when metabolic heat production and energy expenditure are minimums (the most efficient condition in terms of productivity). However, below the lower critical temperature (LCT), about 10°C (50°F) for most animals, and above the upper critical temperature (UCT), about 30°C (86°F) for most animals, they lose energy to regulate body temperature. The LCT, UCT, and the range of the thermoneutral zone are different among species of animals and the environments that they are in, including management practices. Skin thickness, skin wetness, level of production, and type and level of feeding may affect the animal's tolerance to extreme temperatures. Dairy goats tend to be less affected by temperatures below the LCT, compared with temperatures more than the UCT. This is evident in their presence and optimum performance in cooler regions of the world, including the U.S. Larger ruminants having less surface area to body mass are more temperature tolerant than goats, but goats seem to adjust well to different thermal zones. The ideal climatic conditions for a majority of breeds of goats are the arid and semiarid climatic conditions. In fact, according to the FAO, a majority of goats are found in these regions of the world. However, goats tend to be less tolerant to hot and humid climates as indicated by their diminished survivability in hot and humid regions of the world.

Ventilation or air exchange is also important for removal of air pollutants, such as ammonia, in animal housing to provide a stress-free environment. There are two ventilation requirements: one to prevent overheating and the other for removing carbon dioxide, ammonia, dust, and other microbial by-products while providing oxygen. Typical standards for concentration of some gasses in animal housing include below 0.3% for carbon dioxide and below 25 ppm for ammonia. The minimum ventilation requirement for confinement-housed ruminants is 0.35 $m^3$ per kg body weight per hour or between 0.43–5.66 $m^3$ per minute (15–200 $ft^3$) in summer and 0.57 $m^3$ (20 $ft^3$) per minute in winter. This value can change depending on the level of production (higher metabolic rate) and air humidity. Lactating and fast-growing animals under humid conditions may require a higher rate of air ventilation. Propeller fans are commonly used to deliver a large quantity of air for ventilation at a predetermined speed set according to the need for optimum temperature or humidity. The ventilation rate may vary depending on the ambient temperature, and as it increases, the ventilation rate increases for comfort of the animal.

### Animal Health

The housing environment, especially in confinement or the intensive systems, has a major influence on animal health and productivity. Environment in such a system refers to a collection of factors including physical, chemical, psychological, and biological factors that affects animal welfare, health, and performance. Although a number of infectious diseases have been controlled through the use of vaccines and better management, disorders of the digestive and respiratory tracts and skin diseases are prevalent in goats kept in confinement housing. In dairy goats, mastitis and lameness are typical problems. Good environmental hygiene, elevated floors and adequate space allowance for animals is desirable.

### Animal Comfort

Animal housing must create a comfortable environment for the animal by providing a reasonably clean, dry, secure area, with temperatures within the thermoneutral zone (10–30°C for ruminants), and unrestricted space for rest or sleep. The importance or priority of each factor may depend on the type of animal being used, either for milk, meat, or fiber. The level of cleanliness of housing is more important for a dairy goat to avoid mastitis than a meat or fiber goat. Also comfortable housing will provide for freedom to drink and eat, as well as allow goats to express their natural behavior of being playful, jumping, running, and fighting. Housing should be constructed with proper design and structural materials that are safe and provide animals freedom from injury. Unlike sheep, it is more challenging to provide comfortable housing for goats, because in comparison to sheep, they are aggressive, ambitious, not well insulated, and like to investigate.

### Animal Welfare

Welfare of an animal is a state of its physical and mental health. When animals are housed in confinement, although they are provided shelter, food, and water, they are no longer in their natural habitat of grazing or in the case of goats, browsing. This may impose a physical or mental hardship on animals and in turn promote parasitism and endanger their health. Goats in their natural state exhibit different behavior that must be observed and promoted when in confinement housing.

### Animal Psychology

The goat is one of the least domesticated livestock and its natural tendency is free roaming and browsing. While roaming and browsing in extensive or natural systems, where a majority of goat populations are kept, they control when to eat, drink, and seek shelter for warmth or cold. These distinct behaviors should be considered when confining goats, more so than other livestock species. Many scientists have discovered that giving the animals more control of when to eat, when to drink, or even when to be milked may be important in managing animals in groups and this may positively impact their production (Appleby and Hughes, 1993). An automated system of feeding, watering, lighting, heating, or even robotic milking gives control to animals to choose their activities. This may reduce antagonistic encounters arising from competition for access (Hunter et al., 1988), or too rare or frequent use of the resources. Individual animals' needs are different and the automated systems enable animals to predict changes in their environment and may contribute to their social well-being and productivity. With current advances in technology, many of these systems are available that may be adopted for goat production.

## PHYSICAL REQUIREMENTS AND HOUSING COMPONENTS

Careful consideration should be given when deciding which types of structures to include on the farmstead. Barns and other farm structures come in all shapes and sizes, and there is also a wide variation in costs associated with erecting these structures. Although building plans can be easily changed on paper (with little or no cost), once construction has begun they are often difficult and costly to change. In the U.S., local governments have different zoning laws, and some require building permits before any type of construction may begin on the premises.

One major point to consider when planning structures for a goat operation is whether or not structures or fabricated shelters are needed. Goats adapt readily and can endure a variety of weather conditions. In areas with heavy plant cover, such as forests in a silvopasture system, adequate cover and protection is present. However, in areas without natural cover and in climates with extreme weather conditions, supplemental shelter should be provided. Shelter may also be necessary during kidding and until young goats are environmentally adapted.

Personal preferences and goals should be taken into consideration when planning to create or modify a farmstead. While one producer may enjoy being close to his or her goats at all times, others may wish to move away from farm facilities. While one producer may take a "keep it simple" approach, others may want more elaborate or extravagant facilities. The key is to select a design that is workable and affordable. A complete detailed cost analysis is needed before construction begins to determine the most cost-effective approach while meeting housing and shelter requirements.

The type of operation is a determining factor when planning farm structures. If the operation consists of strictly meat goats, dairy, and/or fiber goats, different provisions apply. A dairy operation will need to include provisions for milking and milk storage. Fiber goats require an area for shearing and fiber collection, grading, sorting, and storage. Agritourism may be another consideration for the farm business plans. If agritourism is included, facilities should be safe and comfortable for guests visiting the farm. This can increase the cost of structures significantly. However, the potential for greater profits also increases.

If permanent building structures for housing are considered, building location is extremely important. Several factors should be considered for design of a farm building site:

- The direction of the prevailing winds to prevent damage to goats and the structure.
- The soil drainage—poor drainage leads to flooding and foot problems.
- The slope of the site, which can cause problems with water runoff during heavy rains.
- Site proximity to trees, power lines, homes, or other buildings to minimize damage during extreme weather.
- Future expansion.
- Access road to the proposed site.
- Access to water, utilities, and other services.

A list of building plans is available through different Extension offices throughout the U.S. International agencies such as Heifer International, FAO, and nongovernment aid organizations also can provide plans and advice for housing.

**Portable Structures**

When planning strictly meat goat facilities, simplicity is best. A small, low, wooden structure with open access to the outside environment works well in most cases. These structures can be constructed at low cost and will provide adequate shelter for meat goats while allowing access to browse and/or pastures. The front (open side) should face opposite of the prevailing winds. This provides greater protection for the goats from high winds and rain, as well as improves the structure's durability. Fifteen to 20 square feet (1.4–1.8 $m^2$) per mature goat should be provided (Porter, 2000). The front of the structure should be 5–6 feet (1.5–1.8 m) tall and the back should be 3–4 feet (1–1.2 m) tall (McKinney, 2000). These dimensions will allow for adequate slope on the roof for rain drainage. A gutter placed at the end of the back slope will prevent erosion along the back of the structure. These small buildings may be anchored in the ground or placed on skids or wheels, so they can be moved from paddock to paddock.

Another type of portable building is the hoop building made from welded livestock panels. These shelters are relatively simple and inexpensive to construct. Hoop buildings consist of two or more livestock panels arched to form a Quonset-shaped structure. The structure is then covered with a tarpaulin, or other weatherproof material to provide protection from the elements. Shade cloths can be used to protect goats from the sun in summer where there are no trees. The major advantages to hoop buildings are that they are relatively inexpensive to construct, they can be easily moved, and they can be disassembled and used for other purposes. The commercially available fiberglass hutches can also be used as portable shelters for goats. See Figure 17.2.

Moveable shelters are important in goat production, especially when rotational grazing/browsing systems are used. Rotational grazing/browsing is recommended for not only improvement in available pastures but for goat health reasons. Goats that are regularly rotated have fewer health problems associated with internal parasites and typically have fewer foot problems. It is both expensive and impractical to build a structure for each paddock. In some cases, old equipment such as silage and cotton wagons, tobacco wagons, boat trailers, or even old school buses can be converted into moveable shelters for goats. Before such

Portable Permanent

**Figure 17.2** Shelters for goats. Courtesy: Sandra Solaiman. For color detail, see Appendix A.

conversions are made, equipment should be thoroughly inspected to ensure that they are safe for goats or humans. Loose flooring or protruding metal parts can be potentially hazardous. Goat shelters should be considered places for the animals to escape rain, heat, and other environmental elements, not places for them to live. Goats thrive in an outdoor environment where they can browse on vegetation. Total confinement livestock systems for goats cause decreased thriftiness and lead to a general decline in health if the units are not constructed and managed properly.

### Permanent Structures

Permanent shelters can be constructed on pastures. The shelter should be constructed to provide at least the minimum space allowance for group housing (Table 17.2). It should have three closed sides with 4 feet (1.2 m) of wood/metal wall completed with 2–4 feet (0.6–1.2 m) of woven wire (half wood and half wire) for adequate ventilation. The leeward side should be open and only closed by hanging gates. The rear eave height of 1.2–1.8 m (4–6 feet) and front eave height of 1.8–2.4 m (6–8 feet) is ideal and provides for a slanted roof and moving in cleaning equipment. Metal roofing with at least a 24–36 cm (2–3 feet) overhang on the side will best protect against extreme weather conditions (Figure 17.2). As indicated earlier, a good prepared gravel bed and concrete flooring, sloped for drainage are ideal; however, they are more expensive and require bedding. But, it is easier to clean and disinfect when needed, especially to prevent disease and parasites. Dirt floors will work as well, but they may get wet and stay wet if not designed properly for drainage, and they need more frequent bedding changes. Elevated slatted floors are an excellent choice in more tropical and rainy environments.

Some producers may wish to construct a permanent barn if there are no existing agricultural structures on the farmstead. This can be costly in both time and money. However, when not in use for goat housing, such structures can be used for equipment and hay storage or as a maintenance shop. Items such as medicines or pesticides stored on the farm should not be stored in a barn near livestock and should be kept out of reach of children and unauthorized persons at all times.

### Fencing

There are two main purposes for fencing an area for goat production. The first and main purpose is to keep the animals in a desired area, and it may help minimize predators with boundary fencing. The second purpose is for controlled grazing management. Proper boundary fencing in a goat production setting is a major capital investment. However, cost will depend on the materials used. Six-strands of high tensile electric fence is less expensive than woven fence. Investigating fencing laws and consideration of adjacent landowners is important when establishing boundary fencing to avoid conflicts of interest and future disputes with neighbors. Always investigate cost-sharing opportunities. In the U.S., the Natural Resource Conservation Service (NRCS) offices will provide useful information with availability of cost share funds and tax credit availability for fencing and water delivery to livestock.

A perimeter fence should be at least 5–6 feet (1.5–1.8 m) tall depending on the intensity of four-legged or two-legged predators. A permanent 4-foot (1.2-m) woven wire with one or two strands of barbed or electric wire along the top will be a good perimeter fence. A strand of barbed wire at the bottom will provide more durability. The use of barbed wire for dairy goats should be considered carefully because of possible udder damage when goats push on it. At least five to six strands of 12.5 gauge high-tensile electric wire is also successfully used for boundary fencing, preferably energized with solar electric generators to conserve energy. The more recent eight-strand high-tensile fence with the second, third, fourth, and fifth wire electrified has proven to be an effective goat-tight fence. For kidding goats, the bottom wire should be on a cutout switch to "turn on" during kidding and off after weaning. The same number of strands or less will work for cross fencing. Electric netting or portable fences are also available for temporary fencing.

## Equipment

Equipment may include all types of feeders, waterers, corrals, and working facilities on a meat goat farm, or more specialized equipment such as milking and milk storage facilities on dairy farms. Hay, grain, and mineral feeders, and waterers and water delivery systems will be discussed here. However, corrals, working facilities, and more specialized dairy facilities will be discussed in the following sections.

Various types of equipment are available for purchase, or other designs can be built. Equipment must be easy to use and serve its specific purpose. Sanitation and cleanliness should be considered when choosing equipment. Plastic and metal materials are easier to clean and disinfect than wood. Equipment must be cleaned periodically to prevent the spread of organisms harboring disease such as listeria and coccidia. Disinfection is a must when equipment is rotated between different groups of animals, even if they are from the same herd.

### Hay, Grain, and Mineral Feeders

Feed is the largest cost in any livestock operation. Therefore, feeders (hay, grain, and mineral) that minimize waste are preferred. The best feed or mineral in the world becomes barnyard litter the moment it touches the ground because goats do not like to eat off the ground or eat soiled feed. When selecting feeders, the following factors should be considered:

- Feed wastage—when goats grasp feed or mineral.
- Trough space allowance—to avoid dominant behavior.
- Ease of removing leftover feed daily—goats do not like their feed sniffed and slobbered on.
- Durability—to handle a goat attack.
- Position of feeder—to avoid injury or feed wastage.
- Cleaning and disinfecting—to control the spread of disease.
- Moveable—for browsing systems where goats are moved often.

It is important to select feeders that are practical for both the owner and the goat. Figure 17.3 shows some examples of feeders that have been used with success.

### Water Delivery Systems

Water is the most important nutrient for goats. In fact, goats can live much longer without food than water. In

Grain Feeder

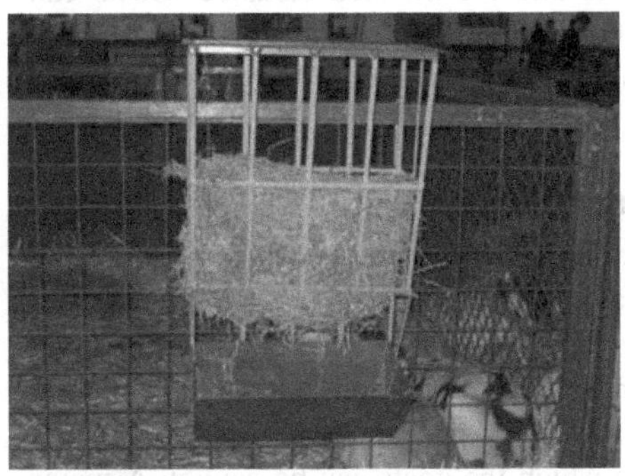

Hay Feeder

Wildwind Farm, Wetumpka, Alabama.
http://www.egwildwind.com

**Figure 17.3** Low-cost feeders. Courtesy: Sandra Solaiman. For color detail, see Appendix A.

addition to just having water, the water that is supplied should be clean and potable. Waterers come in all shapes and sizes. Everything from small pails to a 200-gallon (757 liters) tank can be used to deliver water to goats. For kids, using smaller and shallow troughs is recommended so kids can climb out of troughs easily. Allowing goats to drink from streams and ponds is not a recommended practice, especially when the quality of water is unknown. Allowing livestock to roam in streams leads to degradation of water quality downstream, contributes to soil erosion, and spreads disease.

There are two main types of watering systems for goats: automatic and non-automatic. In automatic watering systems, water is delivered automatically through floats and valves, the water is recharged frequently for freshness, and the level is maintained at a minimum. In non-automatic watering systems, water must be placed in the container. There are advantages and disadvantages to both of these systems. The main advantage of automatic waterers is that fresh water is constantly supplied to the animals with minimal effort. Disadvantages include its potential expense for installation and risk of waterline breakage or water pump malfunction. Outdoor automatic waterers in colder environments may need heating elements to prevent freezing in wintertime. The main advantage to non-automatic watering systems is low cost. These systems are typically less expensive, but more labor is involved because the water must be transported to the waterer by hauling, piping, or carrying the water to the designated location. Location of water on the pasture is another important factor to be considered. The water source should be conveniently located in an area that can serve more than one pasture at a time. The best location for water depends on size of the pasture and the traveled distance. Goats can easily travel 300–400 m (1,000–1,300 feet) to reach a water source on level land or about 250 m (800 feet) on hilly terrain.

### Working Facility

Corrals and working facilities are needed for performing functions such as treating sick animals, giving vaccinations, hoof trimming, catching, weighing, and loading goats (Schoenian, 1999). In operations with a small number of goats, welded livestock panels can be used to create a circular corral in which the goats can be caught. A major advantage to using this type of corral is that it is relatively inexpensive and can be assembled and disassembled quickly. In larger operations, a more complex handling system may be warranted.

Shelters, and/or housing and layout of the pastures, should be designed so that it is easy to isolate the animals by opening or closing the shelter gates. The width of the alleys connecting the pastures should be designed in such a way that opening the gates can be used as isolation panels to separate or herd the goats to working facilities. Many goat-working facilities consist of a gathering corral, holding pen, chute (alleyway), head catch or working box, and loading area. Facilities may also include a footbath if needed and an area for handlers to disinfect themselves before and after working the goats.

A 2–3 m (6–9 feet) working chute, 1.5 m (4–4.5 feet) high and about 30–35 cm (12–14 inches) wide will be appropriate when working on goats. For goats with horns, the chute should be designed wider at the top. Longer chutes may be used; however, they must be divided in sections with sliding gates. Also, a series of canvas flaps suspended half way down into the chute, similar to those used for sheep, may keep their heads down and may prevent goats from riding each other. Goats do not like to be confined for too long and need continual movement forward; therefore, a holding pen may be preferred. At the end of a working chute, a scale or a squeeze chute can be used to isolate individual animals or to stabilize animals and work on them. A holding pen should be half as long as a working chute and at least 4 m (12 feet) wide at the open edge of the pen and taper toward the alleyway.

Although most of the equipment mentioned in this section can be built, there are some companies in the U.S. or around the world that specialize in goat-handling equipment. As a general rule, purchased equipment is much more expensive than homemade equipment. However, purchased equipment is usually of high quality construction and easier to move and may pay for itself in the long run (Schoenian, 1999). More sophisticated working chutes and handling facilities can be purchased ready to be installed, but they may be more capital intensive (Harwell and Pinkerton, 1996).

### Quarantine Area

To keep the goat herd from contamination or infections from new arrivals, each goat farm must be equipped with a quarantine area—a separate area isolated from other animals. This area must have enough of a buffer zone to be separated from the area where the rest of the animals are kept.

Quarantine pens or pastures (more than one pen/pasture is desirable, if possible) should be outdoors preferably, with appropriate shelter, handling, and feeding facilities. New arrivals to the farm will be unloaded at the first pen, where proper vaccinations and strategic deworming practices will be performed. After 24–48 hours, the animals should be moved to the next quarantine pen or paddock, where they will be observed for several weeks (at least for 30–60 days) for any signs of abnormality or disease before joining any group of animals on the farm except the closed herd (those animals that are produced on the farm or are your seeding stock), if possible. Quarantine pastures or pens should be designed for easy cleaning after each group of animals.

As a preventive measure (not necessary), new arrivals, except the milking does, should get medicated feed containing approved coccidiostat, if possible. Goats usually are stressed when hauled, and that may cause coccidiosis. It is important to provide shelter, plenty of fresh water, and good quality hay, but to avoid grain overfeeding that may cause enterotoxemia in young kids or adults. Caretakers should tend this group of animals last and change clothes/shoes before and after for biosecurity measures.

### Sick Animal Quarters

Every goat operation should have a minimum area as a sick animal ward. The shelter and exercise area, or sick animal pasture, should be completely separated from where healthy animals are housed. The pasture(s) should have a minimum of 3–4 m (10 feet) fenced buffer zone from healthy animal pastures. Shelter, fresh water, and minerals should be provided, if not included in the complete mix ration.

### Breeding Pens

These are also either temporary pens or isolated pastures that provide for the breeding does to be exposed to bucks under a more controlled breeding management. The size of pens or pastures will depend on the operation type (meat, dairy) and the number of does breeding at one time. In a more intensive dairy setting, at least 150 $m^2$ (1,600 $ft^2$) will be sufficient for up to 30–40 does to move freely for breeding purposes. This system allows for intensive use of a buck. The majority of producers use smaller pastures with three to five goats per 0.40 hectare (1 acre) and leave bucks with does for a certain length of time (at least two estrus cycles) until all the does are exposed. The use of breeding pens or pastures along with some marking techniques will allow producers to record the breeding date for individual does for better time management. Shelter, feed, and minerals, if not provided in the complete mix ration (for dairy goats), along with fresh water should be provided.

### Kidding Pens

These are temporary pens at least 2–4 $m^2$ or more, preferably indoors in a cold environment, or under shelter in a pasture situation, protected from drafts and rain. Kidding pens improve kid survival rate, but they are not a necessity. These pens are used when animals are close to parturition and for a few days after kidding depending on management strategies. The newborn kid(s) and the doe will be closely watched, the doe will bond with kid(s), and they are protected from other adult goats on the pasture. This may seem complicated, but in a natural setting, where animals are maintained year-round on a pasture or browse, expecting does will move away from other does in the herd, find their own protected shelter, prepare a nest, and keep kids protected for 1–2 days, and then return to the herd. Shelter, feed, and minerals, if not provided in the complete mix ration, along with fresh water should be provided.

### Weaning Pens or Pastures

These pastures or pens are designed specially for the weaned kids. The feeding requirements of this group of animals are different from milking does or mature animals. This group of goats requires feed with higher digestible protein that may have higher fiber content than other grain mixes. Also kids should be separated from does to prepare does for the next breeding especially if accelerated kidding program management is practiced. Shelter, feed, and loose minerals, along with fresh water should be provided.

### Buck Pens

Separate pens with shelters and proper fencing for adequate exercise areas are needed to house herd sires. Buck pens should be constructed away from housing for does. Herd sire(s) should be kept away from breeding does at all times except for breeding time. The shelters should be constructed from durable materials that can tolerate physical abuse, and pens should be constructed with durable panels or fencing materials to keep bucks contained. Space allowance should be according to the guidelines and feed, and loose minerals and water should be provided.

## SPECIALIZED HOUSING

Housing requirement for dairy goats is more elaborate than for meat or fiber goats. The main requirements for housing dairy goats are adequate ventilation, ceilings free of condensation, dry and clean bedding areas, feeders and waterers free from contamination, and facilities that provide easy access for the animals and that are labor efficient.

Dairy goats are comfortable in temperatures ranging from 13–21°C (55–70°F), and they are less sensitive to lower temperatures −18 to +13°C (0–55°F) than temperatures of 27°C (80°F) and higher. Feed intake and milk production decreases in high environmental temperatures. As for all confinement housing, dairy buildings should be properly ventilated to remove heat, moisture, and odors. Condensation on walls and ceilings is the result of poor ventilation and improper insulation. Adequate natural lighting through windows will permit sunlight to provide vitamin D and warmth for animals, and will help in drying the facility. Windows need to be constructed in such a way that they can be closed with shades easily to ensure year-round breeding and milk supply in winter.

Dairy goats are either housed in "loose housing" where animals run loose in pens, or in "stall barns" where animals are tied in the individual stalls (Appleman and Ace, 1992). Loose housing provides adequate bedding, free movement, ample exercise area, and warmth from the pen mates, and if constructed properly, is easy to clean. However, built-up manure can release moisture in the air, and spilled water or feed can soil bedding and provide for fly breeding and contaminants for parasite buildup. These are usually cold housing facilities that are ventilated by natural air movement. Goats freely move to the feeding or exercise areas. Feeders and waterers are usually located away from bedding areas and on concrete pads to reduce contamination.

In many areas in the U.S., old dairy cattle barns are modified and used for goats. Pole barns can be constructed for dairy goats with pressure-treated posts, native lumber for siding, and metal sheets roofs (Figure 17.4). High ceiling and sidewall curtains can help in better ventilation, and open-feeding alleys can provide for mechanical feeding and cleaning. It is recommended to keep all the animals and equipment flow in a straight line by avoiding turns for a smooth and safer operation (Porter, 2009).

Fabric-covered hoop structures can be used as a low-cost dairy housing. These structures are flexible and can be moved if needed. Feeding or access alleys can be designed to keep animals away from damaging the fabric walls (Figure 17.5) (Porter, 2009).

Housing for kids and young stock is included in the housing plans but separate from the milking herd. Bucks are housed separately downwind from the milking herd in the buck pen with adequate shelter and exercise area, and feeders and waterers.

### Milking Facility or Parlor

The milking area should be separated from the bedding area. A holding area for animals before entering the

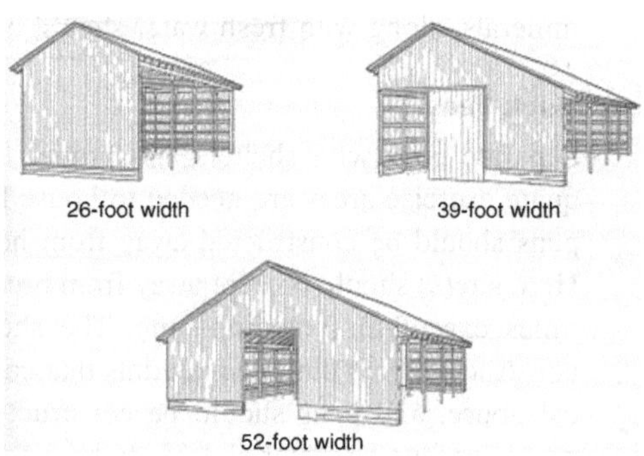


Post-frame building

NRAES publication #1

Open feeding alley

Andy and Jenny Tapper Farm, Brookfield, NH

**Figure 17.4** Post-frame building width and configurations and pen feeding alley access area. Courtesy: John C. Porter, Professor and Dairy Specialist, University of New Hampshire Cooperative Extension, www.umaine.edu/animalsci/Sheepgoats/Dairy%20Goat%20Housing.ppt. For color detail, see Appendix A.

**Figure 17.5** Low-cost hoop buildings with feeding alley adjacent to fabric wall (Ray Rodriguez, Rindge, NH). Courtesy: John C. Porter, Professor and Dairy Specialist, University of New Hampshire Cooperative Extension, www.umaine.edu/animalsci/Sheepgoats/Dairy%20Goat%20Housing.ppt. For color detail, see Appendix A.

milking parlor is recommended. Milking parlor design will depend on the construction area available, number of animals milked at a time, and number of milking staff. The most popular parlor designs for dairy goats are parallel, side opening, straight through, and herringbone. Each design has its own advantages and disadvantages. The straight-through and herringbone-style parlors handle animals in groups. The parallel and side-opening parlors will handle animals individually. When animals are handled in groups, slow milking goats can hold the rest. In the parallel and herringbone-style parlors, full view of the animal is hindered for problem diagnosis when compared with straight-through and side-opening parlors (Figure 17.6).

Milking parlors should have concrete floors for easy cleaning, and the platform must be constructed about 40–45 cm (15–18 inches) higher than the floor for easy milking. Depending on the design, the platform should provide 45 cm (18 inches) in width and 1 meter (3.5 feet) in length allowance for each animal. Extension offices in the U.S. provide more information on construction and designs of milking parlors, as do the various milking machine companies.

### Milk Stand

A portable milk stand is very popular among dairy goat breeders when taking their animals to shows, fairs, and auction sales. Goats could be milked in the stall barn on a milking stand where they live. The milk area is located in the barn, and it may be located in a screened-in facility to avoid dusty air and flies. Room size is dependent upon herd size; however, a 1.5 × 2.5 m (5 × 8 foot) area with concrete floors and a drain will be adequate for milking a small herd, but plenty of natural light or electric lighting is needed.

### Milk House or Room

The milk house should be equipped with a double sink, hot water heater, refrigerator, a table for a work place, and a rack for drying and storing utensils (Figure 17.6). At least 50% of the milk house is open space for equipment requirements. Refrigeration is needed to cool the milk. If milk produced per day exceeds 20–40 liters (5–10 gallons), it is necessary to use a water immersion cooler or a small bulk tank for cooling milk. Milk should be cooled to 4°C (40°F) and maintained at that temperature for high quality (Stevens and Ricketts, 1993). If milk is sold to the public or a processor, proper permitting is required from health authorities in the U.S. and foreign country authorities for certification. In the U.S., every state has different dairy housing and health regulations and requirements before certification.

## SUMMARY

Generally meat goats are a low output commodity; therefore, elaborate housing is not justified. Goats are versatile animals that easily adapt to different housing environments. Fiber goats are provided extensive ranging with low-cost, minimum housing facilities while meat goats may require more housing. Dairy goats require more specialized housing in a more commercial setting. This chapter describes the factors involved in designing proper housing for goats. As indicated in each design, housing requirements as well as animal requirements are addressed. In designing any animal housing, animal welfare and economics must be evaluated.

Parallel
(Courtesy: Dr. George Haenlein)

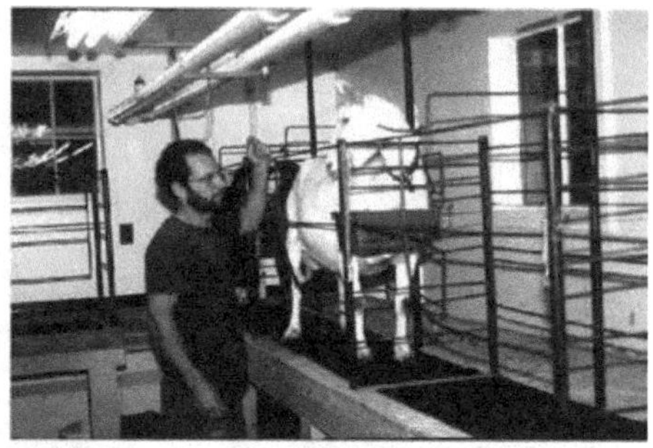

Side Opening

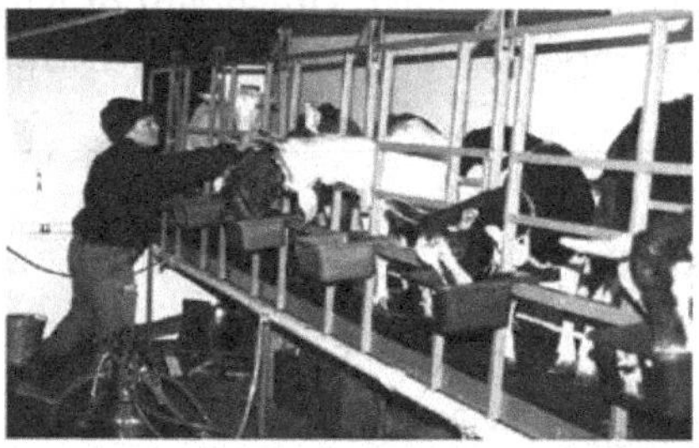

Straight Through

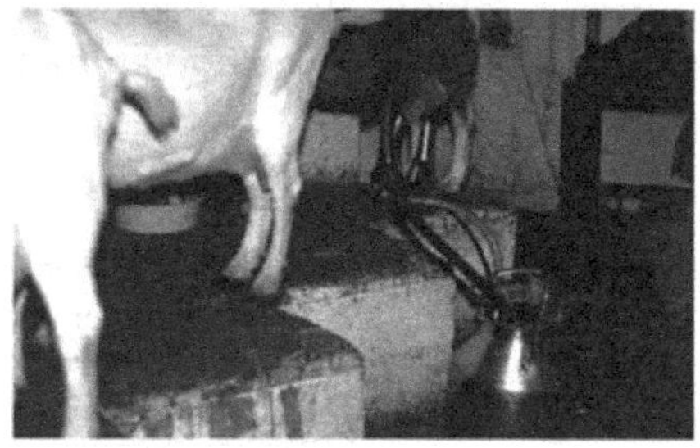

Herringbone

Andy and Jenny Tapper Farm, Brookfield, NH.

**Figure 17.6** Different types of milking parlors and milk room equipment for dairy goats. Courtesy: John C. Porter, Professor and Dairy Specialist, University of New Hampshire Cooperative Extension, www.umaine.edu/animalsci/Sheepgoats/Dairy%20Goat%20Housing.ppt. For color detail, see Appendix A.

## REFERENCES

ADAS. 1987. Cited in Slade, C.F.R. and L. Stubbings. 1994. Sheep housing. In: Livestock Housing. C.M. Wathes and D.R. Charles (Eds.). CAB International, Wallingford, UK.

American Dairy Goat Association. 2004. Did you know? http://www.adga.org/facts.htm.

Andersen, I.L. and K.E. Bøe. 2007. Resting pattern and social interactions in goats—The impact of size and organization of lying space. J. Applied Animal Behaviour Science 108 (1–2): 89–103.

Appleby, M.C. and B.O. Hughes. 1993. The future of applied ethology. J. Applied Animal Behavior Science, 35: 389–395.

Appleman, R.D. and D.L. Ace. 1992. Housing. Extension goat handbook. USDA-ES Publ., Washington D.C. Pp. 239–245.

Bøe, K.E., I.L. Andersen, G.H.M. Jørgensen. 2008. Effect of reduced feeding space for dairy goats on feed intake and social interactions. American Society of Agricultural and Biological Engineers, St. Joseph, Michigan, www.asabe.org.

British Goat Society. 2002. Housing goats. http://www.allgoats.com/housing.htm. Retrieved, Feb. 2009.

Harwell, L. and F. Pinkerton. 1996. Housing, fencing, working facilities and predators. In: Meat Goat Production and Marketing Handbook. Sponsored by Rural Economic Development Center, Raleigh, NC, and Mid-Carolina Council of Governments, Fayetteville, NC.

Hunter, E.J., D.M. Broom, S.A. Edwards, and R.M. Sibly. 1988. Social hierarchy and feeder access in a group of 20 sows using a computer-controlled feeder. Animal Production, 47:139–148.

Lupton, C.J., J.E. Huston, J.W. Hruska, B.F. Craddock, F.A. Pfeiffer, and W.L. Polk. 2008. Comparisons of three systems for concurrent production of high quality mohair and meat from Angora male kids. J. Small Ruminant Res. 74:64–71.

McKinney, T. 2000. Housing Your Goat. Langston University Agricultural Research and Extension Programs. Langston, OK. Online publication: http://www.luresext.edu/goats/library/fact_sheets/g02.htm.

Mount, L.E. 1979. Adaptation to Thermal Environment: Man and his Productive Animals. Arnold, London, UK.

Parker, R. 2008. Equine Science. Thompson Delmar Learning, Clifton Park, NY. Pp. 362.

Porter, J. 2000. Dairy goat farm planning. Langston University Agricultural Research and Extension Programs. Langston, OK. Online publication: http://www.luresext.edu/goats/library/field/dairy_goat_farm98.htm.

Porter, J. 2009. Dairy goats housing and milking facilities. University of New Hampshire Cooperative Extension. Online presentation. www.umaine.edu/animalsci/Sheepgoats/Dairy%20Goat%20Housing.ppt. Retrieved February 9, 2009.

Schoenian, S. 1999. Facilities and Equipment for Commercial Meat Goat Production. University of Maryland Cooperative Extension. http://www.sheepandgoat.com/articles/goathouse.htm.

Stevens, B. and R. Ricketts. 1993. Feeding and Housing Dairy Goats. University of Missouri Extension. http://extension.missouri.edu/explore/agguides/dairy/g03990.htm.

Whates, C.M. and D.R. Charles. 1994. Livestock Housing. CAB International, Wallingford, UK. Pp. 428.

# 18
# Business Plan, Production Enterprise, and Marketing Strategy

*S.G. Solaiman, PhD, PAS; E. Kebede, PhD; and E.M. Aviki, MBA*

**KEY TERMS**

Enterprise—a unit of economic organization or activity.

Business plan—a guide to help define business goals and how to achieve them.

Enterprise budget—a list of values including income, operation cost, and possible profit or loss associated with an enterprise.

Subsistence production—a system of production for self consumption but not for market.

Marketing—the activity and processes of communicating, delivering, and exchanging products or services that have value for customers, clients, partners, and society at large.

Marketing channels—set of interdependent units involved in the process of making a product or service available to customers.

**OBJECTIVES**

By completing this chapter, the reader will acquire knowledge on:

- Importance of a business plan
- Business records
- Management records
- Structure of enterprise budget
- Enterprise budget for different goat operations
- Nature of variable costs
- Nature of fixed costs
- Subsistence goat production
- Marketing channels for goat products

## INTRODUCTION

Goat production as a sole enterprise or specialty enterprise can be integrated into conventional small-scale operations and support sustainability of family farms in the U.S. or globally. Goats are highly adaptable animals and can be raised on a variety of forage, forbs, and browse under various environmental conditions. Given the variety of methods associated with goat production, proper planning and clear operational objectives coupled with correct management decisions can ensure successful operation. An enterprise budget analysis can be used to determine profitability and return to risk and management. It is also important to consider market structure and marketing channels for meat, milk, fiber, and skin in order to determine which products will generate profit and lead to long-term sustainability of the goat operation. In some areas of the world, measuring profit may require qualitative assessments. For example under subsistence production, profitability assessment should incorporate values such as social status and food security.

## STARTING A GOAT FARM BUSINESS

Making the decision to start a goat farm should not be taken lightly. Buying a few goats from auction and turning them loose on a pasture without proper planning can lead to failure in the goat business. Yet, so many farmers tend to do just that. Before raising goats, careful and methodical decisions should be made on what type of goats (meat, dairy, or fiber) is desired. By answering a few questions, one can clarify his/her purpose for using goats. For example, are the goats used as pets and for occasional shows? Are the goats used for clearing some land and property from unwanted weeds and brush? Or is the goal to create a profitable enterprise? Is the enterprise a commercial goat meat, milk, or fiber production? Is the enterprise concerned with maintaining the integrity of goat breeds by having a purebred farm and providing the industry with breeding stock? To be qualified to answer any of these questions, much knowledge and information is required regarding goat selection, feeding, housing, and breeding management. In the U.S., most of this information can be obtained through the local Extension office, by searching the Internet, and by visiting experienced farmers and experts in the area.

Other factors to consider are farm acreage, location, and suitability of the land for goat production. Finally, conducting a market analysis will help determine what type of goat business is likely to be most profitable given demand and supply characteristics. In many countries, including the U.S., there are various government agencies that may assist with planning efforts. Internationally, not-for-profit organizations such as Heifer International and similar agencies can also provide assistance. In the U.S., government agencies include the U.S. Natural Resource Conservation Service, various USDA programs such as Extension offices, Rural Development and Small Farm programs, and the Department of Health for dairy producers. These agencies can help with conservation planning and provide excellent resources to help start the initial goat production enterprise. In the U.S., land-grant universities are also available for support.

### Business Plan

A goat business can be defined as an enterprise in which goats are raised for the production and/or sale of meat, milk, or fiber, etc. As with all successful businesses, a sound business plan is needed. The first step to creating a business plan is to conduct a feasibility study, which will enable the producer to relate the cost of production to prices paid in the markets. Additional steps are conducted in order to gain an understanding of the competition, the market, capital and operating expenses, management and staffing needs, and meat, milk, and fiber processing costs, etc. A business plan including a feasibility study is a necessity and required by most lenders in the event that financing is needed to start a goat farming business.

### Record Keeping

Record keeping is an important aspect of animal operations. A complete and accurate record keeping system will enable the producer or farm manager to properly select superior animals and manage the operation efficiently. In addition, it can help identify the strengths and weaknesses that facilitate making proper management decisions. Detailed and accurate record keeping will result in more accurate decision making.

### Business Records

Common farm business records are based on general farm income and expenses. Operating costs may include payroll, feed, veterinary services, fertilizer, seeds and plants, chemicals, hired custom work, fuel, repair and maintenance of machinery, equipment, buildings and facilities, utilities, taxes, insurance, interest, rent/lease, and purchases. Income may include sale of crops and livestock, other product sales, equipment sales, breeding livestock sales, milk and fiber sales, and other farm incomes. There are several farm record books and software packages available for farm business record keeping. In the U.S. or elsewhere, some of them can be obtained from the Cooperative Extension offices in each state or similar agencies.

### Management Records

Accurate records on animal identification, feeding, reproduction, production, breeding, and health are essential tools to formulate proper management decisions. Animal identification number, birth date, sire, dam, and breed/cross are useful information for animal identification. The amount of feed, type of feed, frequency of feeding along with records on body weight, milk, or fiber production at different intervals can be used to determine the amount of feed needed per unit of gain in meat, milk, or fiber production (feed efficiency). If records are kept on individual animals, they can be used to select the most efficient animals for breeding purposes. Records on the number of pregnancies, number of kids born and weaned per doe, ease in breeding and in kidding, and kidding dates are important factors to be used for selecting proper breeding does. Records for breed of bucks and does used in breeding programs to produce improved crossbreeds with better genetics is important for selecting proper offspring. Additionally, health records such as problems that require veterinary attention, deworming, and other veterinary-related expenses can help select the best-fit animals needing minimum veterinary intervention. These records can help the producer/manager raise a herd that is efficient in converting feed to meat, milk, or fiber, etc.

Proper records can help select animals that are easily bred, can produce more milk and fiber, and wean large kids in a short amount of time. These records can also assist in breeding animals that are healthy with a genetic profile that is best fit for a given farm environment.

## ENTERPRISE BUDGET ANALYSIS

A goat farm enterprise selection is one of the most important planning tasks for a farm manager/owner. Selection of a meat, dairy, or fiber goat enterprise should depend on location and resources availability. The optimal location of a dairy farm is near a raw milk-hauling route, while that of a meat goat farm is near areas where meat-processing plants are located. However, fiber goat operations require warehouses for national or international trade. Resources needed for a goat enterprise include capital, land, labor, animal, feed, and veterinary supplies.

An enterprise budget is a tool used to examine both physical and economical performance of an enterprise and is a critical building block of a farm business plan and budget. Accurate record keeping can be an important input factor in the development of an enterprise budget and help the managers make informed decisions. Enterprise budgets will differ depending on the kind of business and are affected by market prices. Three samples of enterprise budgets for meat, milk, and fiber enterprises and their assumptions are provided in Tables 18.1, 18.2, and 18.3, respectively, for reference. In order to understand enterprise budgets, some economic terminology will be defined here.

### Revenue or Income

Revenue is defined as the number of items sold multiplied by the price. Market demand and therefore market prices for milk, fiber, and adult and kid goats can fluctuate throughout the year. In light of this fluctuation, operations can be structured to maximize production and sale during the times of high demand or shortage of supply. In a dairy operation, revenue can be increased by increasing production and sale of milk during months of high demand, for example, during winter in the northern hemisphere. Similarly, in the meat goat enterprise, one can increase revenue by coordinating operations in such a way that the maximum numbers of meat kids are scheduled to reach market age during specific times of year, when demand and prices are high. Currently, timing corresponds with ethnic holidays in the U.S.

### Operation Costs

There are two types of costs associated with production in any enterprise: variable and fixed costs. Variable costs are costs that can be expected to increase as the operation expands and output increases. Variable costs include costs associated with feeding, bedding, veterinary services, labor, marketing, hauling, fuel, and equipment repair. Fixed costs are those that are independent of the level of output produced. Fixed costs include buildings, land, equipment depreciation, building and fencing repairs, insurance, taxes, and interest. Note, that interest on capital investments for purchase of buildings, fencing, and equipment is usually considered a fixed cost; however, if it is used for animal purchase or equipment repairs, it can be considered a variable cost. It is also important to differentiate between cash and noncash overhead costs. Cash overhead costs are cash expenses paid out during the year that are assigned to the whole farm; these include property taxes, office expenses, interest on operating capital, liability and property insurance, equipment repairs, and management service. Noncash overhead costs are the capital recovery cost for land, equipment and other farm investments; these costs represent the depreciation in value of investments.

### Profit

Profit is defined as revenue minus expenses. An operation is profitable when total revenue exceeds total expenses. Profit and breakeven point are two useful economic indicators of the success of an enterprise (Table 18.1). Profit is usually calculated on a "per doe" basis and can be expressed as either income above variable costs or income above fixed costs. Profit can also be expressed as returns to risk and management whereby income is measured against total variable costs plus cash and noncash overhead expenses (Table 18.2). The breakeven point (BEP) is the number of units required to sell in order to cover costs calculated as total costs over sale price of one unit. BEP can be calculated using only variable costs or total cost (variable + fixed cost). There are additional economic indicators that may be useful in guiding decisions. For example, feed costs over total operational costs (including or exclud-

**Table 18.1** Meat goat budget based on annual kidding.

| Herd composition | | Production parameters | |
|---|---|---|---|
| Number of does | 100 | Kid crop raised | 170% |
| Number of bucks | 3 | Doe replacement | 20% |
| Adult death loss | 5% | Buck replacement | 33% |

| **Gross receipts** | Head | Lb/hd | Price ($/lb) | Total $ | Per doe $ |
|---|---|---|---|---|---|
| Market kids | 150 | 70 | 1.00 | 10,500 | 105.00 |
| Cull does | 15 | 90 | 0.75 | 1,013 | 10.13 |
| Cull bucks | 1 | 170 | 0.75 | 126 | 1.26 |
| **Total income** | | | | **11,639** | **116.39** |

| Operation costs | Head | Amt/Head | Unit | Cost $ | Total $ | Per doe $ |
|---|---|---|---|---|---|---|
| Feed costs | | | | | | |
| Hay | 103 | 0.25 | ton | 90.00 | 2,318 | 23.18 |
| Grain | 103 | 135 | lb | 0.09 | 1,251 | 12.51 |
| Minerals | 103 | 12.0 | lb | 0.16 | 198 | 1.98 |
| Feeds for kids | 170 | 40.0 | lb | 0.10 | 680 | 6.80 |
| Pasture[1] | 273 | 30.0 | lb | 30.00 | 900 | 9.00 |
| Health program | | | | | | |
| Deworming, adult | 103 | 3 | doses | 1.00 | 309 | 3.09 |
| Deworming, kids | 170 | 3 | doses | 0.75 | 383 | 3.83 |
| CD-T booster | 103 | 1 | doses | 0.50 | 52 | 0.52 |
| CD-T vaccination | 170 | 2 | doses | 0.50 | 170 | 1.70 |
| Other vet costs | 103 | | head | 4.00 | 412 | 4.12 |
| Buck replacement | 0.99 | | head | 250 | 248 | 2.48 |
| Bedding | 103 | | head | 3.00 | 309 | 3.09 |
| Marketing, hauling | 166 | | head | 5.00 | 830 | 8.30 |
| Supplies | 103 | | head | 3.00 | 309 | 3.09 |
| Additional costs, labor[2] | | | hours | 260 | 2,600 | 26.00 |
| Interest on total operating costs for 6 months at 5% | | | 274 | 2.74 | | |
| **Operating costs excluding labor** | | | **8,576** | **85.76** | | |
| **Total operating costs** | | | **11,241** | **112.41** | | |
| Total feed costs | | | 5,347 | | | |
| Feed costs over operating costs excluding labor | | | 63% | | | |
| Feed costs over operating costs including labor | | | 47% | | | |
| **Return to land, labor and capital (profitability)** | | | **3,062** | **30.62** | | |
| **Return to land and capital (profitability)** | | | **397** | **3.97** | | |
| **Costs per lb excluding labor** | | | | | | |
| Live weight (breakeven price) | | | | 0.71 | | |
| Carcass at 47% yield (breakeven price) | | | | 1.51 | | |
| **Costs per lb including labor** | | | | | | |
| Live weight (breakeven price) | | | | 0.96 | | |
| Carcass at 47% yield (breakeven price) | | | | 2.05 | | |

[1]Assumed 30 acres of improved pastures.

[2]Assumed hired help for 10 hours/week at $10/hour for 6 months. Owner supervision is not considered.

Source: http://www.sheepandgoat.com/articles/aboutgoats.html with modifications.

**Table 18.2** Dairy goat operation enterprise budget based on 500 head[1].

| Item | Unit | Total units | Price or cost/unit | Total value or cost | Return or cost/head |
|---|---|---|---|---|---|
| **Gross receipts** | | | | | |
| Milk | gallon | 97,500 | 3.4 | 331,500 | 663.00 |
| Kids | head | 400 | 1.00 | 400 | 0.80 |
| Small kids | head | 230 | 15 | 3,450 | 6.90 |
| Cull bucks | head | 2 | 100.00 | 200 | 0.40 |
| Cull and sale | head | 130 | 85.00 | 11,050 | 22.10 |
| Total | | | | 344,600 | 693.20 |
| **Operation costs** | | | | | |
| Rented pasture | acres | 50 | 30.00 | 1,500 | 3.00 |
| Legume hay | ton | 75 | 180.00 | 13,500 | 27.00 |
| Grass hay | ton | 75 | 120.00 | 9,000 | 18.00 |
| Grain mix | ton | 320 | 253.00 | 81,023 | 162.05 |
| Minerals | blocks | 72 | 8.20 | 590 | 1.18 |
| Loose minerals | ton | 1 | 380.00 | 456 | 0.91 |
| Kid grain | ton | 16 | 200.00 | 3,200 | 6.40 |
| Milk replacer | sack | 40 | 42.50 | 1,700 | 3.40 |
| Total feed cost | | | | 110,969 | 221.94 |
| Labor | hour | 3,650 | 9.59 | 35,004 | 70.01 |
| Labor | hour | 3,000 | 20.00 | 60,000 | 120.00 |
| Total labor costs | | | | 95,004 | 190.01 |
| Inspection | dairy | 1 | 350.00 | 350 | 0.70 |
| Veterinary costs | service/treatment | | | 11,201 | 22.40 |
| Milk transport | week | 52 | 70.00 | 3,640 | 7.28 |
| Animal transport | | 4 | 50.00 | 200 | 0.40 |
| Power | month | 12 | 666.67 | 8,000 | 16.00 |
| Bedding | bi-week | 24 | 50.00 | 1,200 | 2.40 |
| Fuel, oil, lube | dairy | 1 | 84.00 | 84 | 0.17 |
| Vehicle repair | dairy | 1 | 6,832 | 6,832 | 13.66 |
| Equipment repair | dairy | 1 | 370 | 370 | 0.74 |
| Housing improvement | | 1 | 3,973 | 3,973 | 7.95 |
| Taxes and insurance | | 1 | | 5,497 | 10.99 |
| Other expenses | month | 12 | 666.67 | 8,000 | 16.00 |
| Interest on capital | | 135,259 | 7.65% | 10,347 | 20.69 |
| Other variable operation costs | | | | 59,694 | 119.40 |
| **Total operation costs** | | | | 265,666 | 531.33 |
| **Income over operation costs** | | | | 80,934 | 161.87 |
| **Feed costs over operation costs** | | | 42% | | |
| **Feed cost over operation excluding labor** | | | 65% | | |

**Table 18.2** *Continued*

| Item | Unit | Total units | Price or cost/unit | Total value or cost | Return or cost/head |
|---|---|---|---|---|---|
| **Labor costs over total operation costs** | | | 36% | | |
| Cash overhead costs (taxes, office expenses, insurances) | | | | | |
| Interest on animals | | | | 2,226 | 4.45 |
| Office expense | | | | 4,003 | 8.06 |
| Total cash overhead costs | | | | 6,229 | 12.51 |
| Noncash overhead costs (depreciation in equipment value, etc.) | | | | | |
| Capital recovery | | | | 22,206 | 44.41 |
| Total noncash overhead costs | | | | 22,206 | 44.41 |
| Total costs | | | | 294,102 | 588.26 |
| **Return to risk and management** | | | | 52,498 | 104.94 |

[1]Assumptions: 50 acres rented @ $30/acre; 500 does and 10 bucks; Replacement rate of 20%; 130 female kids are retained to produce 100 replacements; 30 does will be culled and sold; kidding rate of 1.6; kid survival rate of 95%; most kids are sold after birth; 100% conception rate; 100% milk delivery; milking does consume around 6 lb, 3.5 lb grain; milking does produce 195 gallons or 1,677 lb (762 kg)/year; 70 hours per week hired labor; 60 hours per week owner labor.

Source: Giraud, D.D., K.M. Klonsky, and P. Livingston. 2005. Sample costs for a 500 dairy goat operation. University of California Cooperative Extension. Gt-NC-05-R. ANR Revised #7, 8-29-06.

**Table 18.3** Angora goat operation enterprise budget based on one animal unit (AU)[1,2].

| Item | Unit | Price ($) | Quantity | Amount ($) |
|---|---|---|---|---|
| **Gross receipts** | | | | |
| Mohair, adult | lb | 4.00 | 50.4 | 201.60 |
| Mohair, kid | lb | 8.00 | 7.0 | 56.00 |
| Mohair, yearling | lb | 5.75 | 12.6 | 72.45 |
| Angora kids | head | 37.35 | 2.1 | 78.44 |
| Call does | head | 41.25 | 1.4 | 57.75 |
| Total Income | | | | 466.24 |
| **Operating Costs** | | | | |
| Crop insurance | | | | |
| PRF | ac | 0.85 | 30.0 | 25.50 |
| Eradication fee | | | | |
| Predator control | ac | 0.25 | 30.0 | 7.50 |
| Custom harvest | | | | |
| Shear-goat | head | 3.00 | 17.96 | 53.89 |
| Vet. medicine | | | | |
| C/D&T | ml | 0.11 | 21.47 | 2.55 |

**Table 18.3** *Continued*

| Item | Unit | Price ($) | Quantity | Amount ($) |
|---|---|---|---|---|
| Sore mouth | dose | 0.14 | 3.5 | 0.49 |
| Deworm | ml | 0.07 | 145.84 | 11.00 |
| Others | dose | 7.00 | 0.01 | 0.07 |
| Rabies (horse) | dose | 15.00 | 0.01 | 0.15 |
| WNV | dose | 30.00 | 0.01 | 0.30 |
| Health management | | | | |
| Lice control | ml | 0.08 | 21.87 | 1.91 |
| Horse shoeing | hd | 65.00 | 0.04 | 2.60 |
| Purchased feed | | | | |
| Corn | lb | 0.13 | 180.00 | 23.40 |
| Cottonseed | lb | 0.12 | 315.00 | 39.38 |
| Hay | ton | 165.00 | 0.0066 | 1.10 |
| Oats (horse) | bu | 3.00 | 0.12 | 0.36 |
| Pasture (horse) | ac | 2.50 | 0.57 | 1.43 |
| Minerals | lb | 0.27 | 56.70 | 15.54 |
| Salt | lb | 0.11 | 56.70 | 6.41 |
| Marketing per head export | | | | |
| Sales commission | head | 3.30 | 3.5 | 11.55 |
| Goat yardage | head | 0.65 | 3.5 | 2.28 |
| Cowboy day labor | | | | |
| Special labor | day | 150.00 | 0.06 | 9.00 |
| Electricity | | | | |
| Ranch overhead | kwh | 0.15 | 23.074 | 3.46 |
| Gasoline | | | | |
| Ranch overhead | gallon | 1.90 | 21.04 | 39.98 |
| Repair and maintenance | | | | |
| Ranch overhead | AU | 10.71 | 1.00 | 10.71 |
| Ranch horse and track | AU | 0.94 | 1.00 | 0.94 |
| Interest on capital | AU | 15.84 | 1.00 | 15.84 |
| Total Direct Expenses | | | | 287.33 |
| Returns above Direct Expenses | | | | 178.99 |
| **Fixed Expenses** | | | | |
| Working dogs | each | 143.92 | 0.07 | 3.36 |
| Angora buck | each | 84.94 | 0.20 | 16.98 |
| Ranch overhead | AU | 123.91 | 1.00 | 123.91 |
| Ranch horse and track | AU | 4.86 | 1.00 | 4.86 |
| Total Fixed Expenses | | | | 149.11 |
| Total Specified Expenses | | | | 436.44 |
| **Return Above Specified Expenses** | | | | **29.80** |

[1]This sample budget is based on seven head of goats or one animal unit; pasture is owned; PRF = pasture, rangeland, forage; WNV = West Nile Virus; 1 dog is used for every 100 goats and depreciated over 5 years; 1 buck is used for every 30 does with 100% culling rate.

[2]Cost of production is based on 2007 input prices. This budget is patterned after Texas A&M University Cooperative Extension with some modifications. http://agecoext.tamu.edu/fileadmin/user_upload/Documents/Budgets/District/7/.

ing labor) and profit over total acreage both can be used in conjunction with business and management records to monitor the effects of any changes that may have occurred over time and motivate management decisions.

### Sample Enterprise Budgets

A simple budget illustration based on a 100 doe production system for meat, a 500 doe dairy enterprise, and an animal unit (AU) basis for fiber goat enterprises are shown in Tables 18.1, 18.2, and 18.3, respectively. To construct a budget, the enterprise should be described and assumptions should be made (see assumptions under each table). Kidding rate, weaning rate, price of feed, and price of market goats for meat, milk, or fiber enterprise may be different across locations and time. Changes in any of these factors can change net income. The fixed costs associated with the operation, including the cost of land, buildings, and fencing, can be accounted for as capital investments. These budgets are intended to be used as simple guides and should be adjusted to one's specific operational practices (additional goat kid, milk, or fiber sales, etc.) and marketing strategies. Management and capital investment intensities are higher for dairy than for meat and fiber operations. Therefore, more scrutiny is required in making the decision to start a dairy goat operation.

As illustrated, feed cost is one of the highest operation expenses. Due to rising energy prices for hay and grain production, feed costs can account for more than 60% of total variable costs for any given operation. The use of extensive grazing systems with natural pastureland and rangeland and/or grazing on improved pastures with proper management, where appropriate, may reduce feed costs as well as parasite burden and therefore the cost of deworming.

The second highest cost is labor, which can vary from 25–35% of operating costs. The budget analysis must include labor costs—hired help and hourly wages provided by the owner or hired management. In a majority of cases in the U.S., owners do not include the management cost in their budget and depend on the non-monetary incentives that farming provides.

With the current global economy, making profit in an agricultural enterprise, especially those involving goats can be challenging. However, through informed decisions that effectively control production costs and maximize market opportunities and revenue (producing and selling products during peak demand), goat production can be profitable. The section on tax benefits will illustrate how awareness of local, regional, and national legislation can enable producers to further enhance profitability by maximizing the benefits associated with capital investments.

### Tax Benefits

Many landowners and producers in the U.S. take advantage of the tax benefits associated with production farming. Agricultural Transfer Taxes can be waived if the intent of land purchase is for agricultural use. Agricultural Use Assessment will appraise the value of the land according to its use and not according to the current market value, which allows farms to benefit from lower taxes because of the lower value placed on agricultural land compared to nonagricultural property. In order for this to take place, the land must meet the requirement of a minimum size of land in acres and generate a minimum gross income that varies over time. It is noteworthy to mention that sales tax is waived for some farm-related purchases, equipment rental, and supplies used for agricultural production. Operating expenses from goat production, like any other agricultural enterprises, can be considered to be a tax writeoff to offset the ordinary income tax. Expenses such as education, membership in breed associations, equipment, buildings, fencing, and animal depreciations can also be deducted from income taxes. The tax incentives are more for farms that operate as businesses and have shown a profit for a certain number of years (usually 5 years) or have shown a profit in 3 of 5 years of operation. Land appreciation is another major contributing factor in increasing a farmer's net worth. Real estate values continue to increase; therefore, purchase of land to raise livestock can be a valuable investment (Schoenian, 2008).

## PARTIAL BUDGET ANALYSIS FOR SUBSISTENCE GOAT FARMING

A majority of goats are raised under subsistence farming in developing and underdeveloped countries where social values may surpass economic values. More than 90% of the world's goats are in developing countries. Goats are increasingly important in these countries as subsistence food producers. Production systems range from goats being a part of nomadic multi-species herds on arid desert rangelands or in agropastoral production systems, to goats being the primary animal enterprise in smallholder farming systems (Glimp, 1995).

Economic analysis is also a useful tool for subsistence farming, just as it is for other types of production systems; the information gained can be used for planning purposes and may lead to cost-saving changes in management practices. However, components of the budget analysis and

**Table 18.4** Partial enterprise budget for subsistence goat production.

| | |
|---|---|
| Description: | |
| Using an improved breeding buck for improved goat production | |
| **Gains** (Comparative additional gains) | |
| Costs saved | |
| Extra revenue | Higher priced offspring |
| | Food security |
| | Women prestige |
| **Losses** (Comparative additional losses) | |
| Extra costs | Payment for buck service |
| | Extra labor for animal care |
| | Extra tree fodder for buck |
| | Improved shelter for buck |
| Loss of revenue | Use fodder for fire/timber |
| | Income from selling corn used for extra feeding |
| **Extra profit** | |
| Total additional gains—total additional losses | |

Source: Andreasen and Qwist-Hoffmann, 1997. http://www.fao.org/docrep/X5676E/X5676E00.htm

objectives of the enterprise may be different from those used in developed nations. In subsistence systems, it may be difficult to assess market value or price of products, which are needed to quantify revenue and returns; and consequently, final true cost/benefit results may not be achieved. However, socioeconomic factors are often used to determine the economic value of products. These same factors are often used for informed decisions made at the household level. Under these circumstances, partial budgets can be developed for decision making and improved goat production. A simple partial budget is presented in Table 18.4 for reference.

### Structure of Partial Budget

The first step in any economic analysis, and in this case partial budget analysis, is to describe the enterprise. The next step is to assess total losses/gains and comparative additional losses/gains associated with any changes made to improve the operation. For example, in a partial budget for using an improved buck to enhance goat production, total losses and comparative additional losses would include extra costs and loss of revenue. Extra costs are the expenses associated with adopting the planned changes, which would include payment for buck service as well as any additional costs associated with the improved buck compared to a local buck. Loss of revenue is any revenue that would be lost after the change. Another example is a partial budget for the alternative uses of fodder trees for firewood or timber, and selling corn instead of feeding. In this case, total gains or comparative additional gains include costs saved as a consequence of the change and extra revenue of producing superior offspring. Extra profit is the difference between total additional gain and total additional loss. However, there are other positive outcomes such as improved food security and women prestige, given availability of veterinary services and fodder. It can be extremely challenging to estimate the values used in the calculations. Meeting this challenge requires maximum farmer involvement so that appropriate estimates can be made that accurately reflect the intangible and tangible value of gains and losses for a given subsistence farm (Andreasen and Qwist-Hoffmann, 1997).

## MARKETING

Marketing is the practice of identifying and understanding consumer needs, tailoring products to best meet those needs, and then delivering the product to achieve maximum consumer satisfaction in tandem with maximum financial return. It is a source of power for producers, which allows them to have control over prices and volume. An effective marketing plan should be developed by producers to maximize sales and profits associated with their products.

### Market Plan

The purpose of a market plan is to define the market, identify customers and competitors, outline a strategy for attracting and retaining customers, and identify and anticipate market change. The plan should include product, price, place, and promotion. A market plan should contain the strategies listed below:

- Identifying and describing customers (target market) by their age, sex, income/educational levels, profession/career, and residence. Know the customers better than anyone, including their likes, dislikes, and expectations. As business grows and the customer base expands, it may be necessary to consider modifying this section of the marketing plan and to include other customers.
- Identifying the direct competitors and the indirect competitors. Start a file on each, and identify their weaknesses and strengths. Keep files on their advertising and promotional materials and their pricing strategies. Review these files periodically to determine when and how often they advertise, sponsor promotions, and offer sales.

- Trying to describe the benefits of the goods from the customer's perspective. Emphasize its special features and selling points. Know or at least have an idea of what the customers want or expect to help build customer satisfaction and loyalty.

### Marketing Strategies

#### Pricing Strategy

Pricing strategy may be on a cost-based strategy where the price of the product is set at a specific percentage above the costs associated with production, transportation, and sale. It may alternately be a strategy of determining the economic value of the product to the target customer. If the economic value of the product exceeds the cost of production, then a price can be set that lies somewhere between the cost and the value of the product to the customer. With goat meat products, as mentioned previously, there is a significant increase in demand and value associated with goat meat, which corresponds to certain ethnic holidays. This should be accounted for in the determination of price. A study on dairy goat products, especially those sold in the health and specialty food markets in the U.K., showed that price did not have a large impact on sales (Mowlem, 2005). The relative price insensitivity of these two markets represents an area where high margins can be gained to significantly enhance the profitability of a dairy goat enterprise. Regardless of which strategy is chosen, it is important to consider the strategies used by competitors. It is important to thoroughly understand how to price a product, determine if the prices are in line with the competition, and decide what adjustments can be made to bring them in line. The key to success is constantly monitoring prices and operating costs to ensure profits.

#### Developing an Effective Promotional Strategy

Making use of various advertising media for promoting the business and identifying those that most effectively promote the business are important for success. Develop material that clearly identifies the product, its location, and price to demonstrate the ability to compete.

#### Location (Place) of Business

Describing the location of the business or where the products are sold from the customer's perspective is also important. Factors such as advantage, convenience, access to public transportation, safety of the area, etc., must be included. The location of the business should be built around the customers, should be accessible, and should provide a sense of security.

#### Developing a Marketing Budget

Operating an effective marketing plan requires money, so funds must be allocated from the operating budget to cover advertising, promotions, and all other costs associated with marketing. In developing a marketing budget, managers should account for advertising costs, the cost of collecting research data, and costs associated with monitoring shifts in the marketplace.

### Market Channels

Market channels or outlets for meat goats are slowly developing, points of origin are being better defined, and new processing plants and marketing techniques are being designed to better meet producer and consumer needs. Many areas of the U.S. have local markets for goats, but the infrastructure to market goats is still new and developing. Goat milk is used locally throughout the world and is only minimally traded globally (FAOSTAT, 2005). The dairy goat industry is more established in the U.S. and Europe compared with other goat enteprises. In the U.S., the dairy goat industry is supported through a collaborative effort of the USDA and the American Dairy Goat Association. The fiber goat industry (cashmere and mohair) is an established industry that is going through hard times due to the global economy and loss of mohair subsidies in the U.S. Goat meat is traded globally and is the sixth most traded meat after pork, chicken, beef, lamb, and turkey.

### Market Structure

The marketing structure for goat products is not well developed in the U.S. and is very complex. Figure 18.1

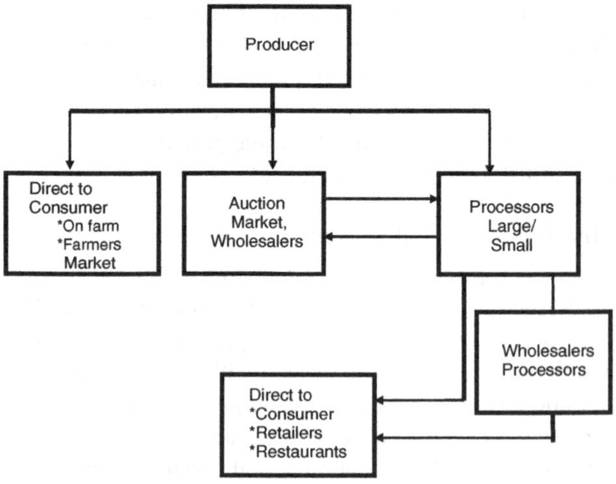


**Figure 18.1** Goat meat market structure schematic.

depicts a simplified version of the various transactions that take place as goat products (such as meat) flow through the market. It shows the route a product takes as it moves from producers to consumers. Each stage is characterized by an increase in product price. The number of times a product changes hands before reaching the end consumer is one aspect affecting the extent to which prices increase from the price received by the producer to the price paid by the end consumer. At each level of exchange, a margin is placed on the base product price until the product finally reaches the consumer. Therefore, the more steps between producer and end consumer, the greater the margin above cost of production the end consumer should expect to pay. Additionally, the more steps that exist, the less of a margin the producer is able to place on the product without significantly affecting the quantity demanded by the end consumer. Knowledge of the market structure and product flow allows a producer to design a market strategy that will allow him/her to maximize benefits from the production activity. The producer is faced with three main channels to which products may be sold: direct marketing to the local community on farm or at farmers markets, auction markets or wholesalers, and processors. There are also middlemen who sell to wholesalers and processors and wholesalers who can sell to processors. As a final step in the value chain, processors sell products to retail outlets such as local supermarkets, grocery stores, and restaurants.

On-farm or direct to consumer marketing offers the greatest profit margin on live animals or products sold by the producer because all middlemen and their fees are eliminated. It can offer a year-round marketing outlet (Alford et al., 1998). However, sanitation regulations governing milk and cheese must be considered before proceeding with this type of strategy.

Livestock auctions are convenient because they offer a year-round, local outlet for the sale of goats; however, they generally provide the lowest profit margin for producers. Prior to pursuing livestock auctions as outlets, the goat producer will have to balance the increased costs of transportation with expected revenues from auction sales and decide the minimum number of animals needed to make the longer trip profitable.

A very popular outlet used for larger dairies, which is underutilized for the sale of meat goats, is the establishment of contracts with processors. Contracting offers security in terms of pricing and quantity. Unfortunately, the local supply of goats and their products may not be large enough, thus forcing processors to seek steady supplies outside of the local market.

**Table 18.5** Meat goat marketing outlets in the Northeastern U.S. (a sample survey).

| Marketing Outlet | Total Sales | % of Total Sales |
|---|---|---|
| Marketing Cooperatives | 1 | 0.3 |
| Meat Wholesalers | 9 | 3.1 |
| Meat Retailers | 8 | 2.7 |
| Livestock Processors | 153 | 52.4 |
| Livestock Haulers | 23 | 8.0 |
| Livestock Dealers | 22 | 7.5 |
| Livestock Auctions | 40 | 13.7 |
| Live Animal Markets | 26 | 8.9 |
| Feeders | 8 | 2.7 |
| Total | 290 | 100 |

Source: http://www.sheepgoatmarketing.info/PageLoad.cfm?page=directory/Market_Directory.cfm (Accessed on March 20, 2009).

The U.S. meat goat industry is an emerging industry with a highly fragmented market. Sample data shown in Table 18.5 depict the number of meat goat market outlets primarily in the 12 northeastern states of the U.S. It shows the nine different categories involved in meat goat marketing. In this sample, livestock processors represented the largest channel and accounted for 52.4% of all meat goat purchases. Livestock auctions accounted for 13.7%; livestock hauler and dealers, 15.5%; live animal markets, 8%; and cooperatives, <1%. This sample can also be used to represent the market structure associated with goat milk and fiber outlets in the U.S. The markets for both products are small and underdeveloped, and are transferred primarily from producers to processors.

### Target Customer Groups

A driving force behind the current increased demand for meat goats is the growing ethnic population in the U.S. The Hispanic population is expected to swell to 25% of the U.S. population by 2050, and Muslim and other faith-based populations such as Buddhists have increased by more than 150% in number over the past years (Solaiman, 2007). This population has created a significant and growing demand for goat meat and other related products in the U.S. and elsewhere. Aside from use by ethnic populations, goat products are in demand for the therapeutic properties associated with goat milk for adults and infants (Walker, 1956). Cheeses produced from goat milk are gaining popularity as a specialty outlet in mainstream populations of the U.S. and other developed countries.

Many farmers, aware of these trends, have adapted and are taking advantage of these growing niche markets. However, producers, processors, and packers who lack niche market information are unaware of the various demand profiles characterizing consumers of goat meat. Currently, demand for meat and cheese surpasses supply, and the lack of sufficient market information and an organized market is evident.

## STRATEGIES TO IMPROVE GOAT MARKETING

### Cooperative Marketing

Cooperative marketing likely yields the greatest economic power for the producer due to its collective structure. Strategies for effective cooperative marketing would include gathering several producers into a collective group, ensuring allegiance to the cooperative, and collectively organizing sales directly to consumers or to a processor. Initially, cooperative marketing of goat meat can be used to meet peak demand and then be expanded to include year-round sales (Gipson, 1996). Cooperatives that exist today were created to guarantee a constant supply of goats that would otherwise be unattainable by individual farms. Among other things, cooperatives can impose a price floor in the market for goat meat, as has been used by beef markets in order to attain more attractive prices. However, market feasibility studies are important to strategically place the goat meat or other products in appropriate marketing channels so that maximum profits are obtained for members of the cooperative. The location of slaughter facilities and milk-processing facilities is also important to maximize profits due to the low returns on the sale of live goats or pre-processed goat meat or milk when compared to the returns found on the sale of processed products. Similarly, higher profit margins can be achieved by selling various cuts of goat meat or other goat products versus selling whole live goats.

### Auction Markets

Another strategy is to work with local livestock auctions in order to periodically have a goat sale where several local producers collectively market their animals. Timing of these local sales should correspond to periods of high demand and should be well advertised in advance.

### Direct Marketing

Direct marketing represents another strategy and can be conducted at farmers markets, retailers, restaurants, and community-supported agriculture events. There are several strategies that can be used to improve direct marketing. A farm can adapt a breeding schedule to ensure that goats of the desired age and weight are available for the targeted ethnic group at times when demand is high. Additionally, sales can be maximized through advertising in local newspapers, at local institutions, such as, mosques, ethnic supermarkets, ethnic restaurants, and in metropolitan areas with large ethnic populations. Producers can develop mailing lists that can be used periodically to promote awareness of product availability (Alford et al., 1998). This method can apply to other goat products such as cheese, soaps, and skin products. Another method to increase sales would be to provide transportation for animals purchased to a Custom-Exempt Slaughter Facility, which would boost sales to people who want to buy goat meat but do not have the transportation necessary to do so. If consumers buying a live animal prefer to conduct the slaughter themselves, an isolated clean area can be offered on the farm for slaughter, providing an additional market outlet.

Other exploratory avenues that market goat meat include supplying to farmers markets and local fairs as cooked or processed products. Also, contracting can be sought with various restaurants that are willing to serve gourmet dishes using goat meat. Another option is to use community-supported markets. These outlets will increase visibility of goat meat products as a healthier meat alternative and promote awareness of the health benefits of goat meat to non-ethnic populations.

### Contracting

Strategies for improving contracting include increasing local production numbers so that local purchases can fulfill the slaughter needs of the processor(s). This strategy is more focused on the long term and will require several producers to meet the quantity demanded. In the short term, efforts can be made to supply processors with the quantity necessary to fulfill times of peak demand. Because consumption of goat meat and milk in developed countries is relatively low among average consumers, breeding programs should be designed so that goats of various ages and weights, and goat milk are available year-round. This calls for clearly identifying consumers of specialty cheese and milk along with ethnic consumers who prefer goat meat, and making efforts to satisfy the periodic demand flux associated with the identified parties. Another strategy is to network and negotiate shared resource contracts with other producers to meet market objectives such as maintaining a consistent supply (McKenzie, 2007). If the producer's goal is to increase income from the enterprise, this can be achieved by moving up the value chain by including processing as a component of the overall market plan.

### Value-Added Products and Niche Marketing

A strategy of providing value-added products through niche marketing likely offers the greatest overall profit margin and allows for more consistent year-round revenue. However, the cost of equipment and the complexity associated with market development may discourage some producers (Gipson, 1996).

Agricultural producers, including goat producers, receive a much smaller portion of the end consumer's dollar when compared to food processors. Capturing the value-added dollars represents a profitable way for producers to maximize returns. Adding value to an agricultural product means changing the product and taking it to the next level, closer to a consumer product. It simply means that instead of selling 90-day-old weaned kids at auction, the producer keeps the ownership in those kids until they are processed and reach consumers, thus getting a higher price. In this system of marketing, goat producers maintain ownership of the goats, conduct the slaughtering and processing operations, and capture value through the direct-to-consumer sale of various meat cuts, sausage links, or patties. This method may also include the direct sale of value-added specialty meat products, cheeses, and possibly skin, tongue, and viscera where demand is present.

In Texas, many value-added agribusinesses have been found to average more than 60% return on product assets. Many have less than five employees and carry out less than 25,000 kg of meat sales per year (about 1,250 goats slaughtered per year or less than 25 goats processed per week). These agribusinesses are located in rural areas and cater specifically to the demands of local populations. Not surprisingly, these smaller farms have been shown to be more profitable than comparable larger farms due to their success in developing sound business plans based on market feasibility studies that they have conducted on their local markets.

A successful value-added entrepreneur will adapt to change, be open to new ideas and opportunities, and effectively manage internal and external resources (Anderson and Hall, 2009). Value-added goat meat or milk products may represent a bright future for the goat industry, provided that U.S. goat producers maintain a strong competitive advantage in areas where import markets are growing. Goat products should strive to increase their attractiveness not only in the growing ethnic consumer population, but also in the traditional American consumer markets. Producers aiming to increase penetration in traditional American markets should make efforts to structure their goat product portfolios to include items similar to those found in other meat markets, such as ground meat, sausages and roasters, or cheese, yogurt, or cream.

It is worthwhile to briefly touch on differences in the cabrito and chevon markets. The production of cabrito, or high quality goat kid meat, may draw higher prices. Alternately, chevon, meat from larger carcasses, may appeal to retailers, cruise lines, specialty food stores, and restaurants. When considering these two products, it is important to become familiar with and breed according to the cultural calendars in order to maximize sales to ethnic consumers.

Marketing outlets for value-added products targeting nontraditional consumer groups include upscale restaurants, health food outlets, delis, and supermarkets that carry specialty food for the "health conscious" consumer. Test the market for novel goat meat products to these outlets including sausages, traditional retail chops, steaks, as well as precooked roasts and loaves. On the other hand, current or new goat meat producers should consider the economics of processing and packaging operations near goat-producing areas with distribution of refrigerated and frozen goat products to distant outlets. They may also find it worthwhile to consider the option of processing and contract the growth of slaughter goats in conjunction with forward contracting of products to wholesale and retail outlets. With appropriate modifications, the current structure of the broiler and turkey industry may serve as a model for goat marketing in this manner (Ollinger et al., 2005).

### Internet Marketing

The Internet represents a tremendous opportunity to bolster sales of goat meat, cheese, or other products in the future. However, pursuing sales through this outlet imposes certain requirements that should be noted. In the U.S., Internet-based sales require that goats be slaughtered according to the USDA guidelines in federally inspected slaughterhouses. Carcasses that are not processed in a federally inspected slaughterhouse cannot cross state lines or be marketed in other states. Before pursuing this avenue, producers must ensure that facilities are readily available and cost effective. Additionally, goat milk products sold through the Internet must be processed in state or federally inspected units.

The unit producing meat or dairy products for Internet sale should be operated under USDA Hazard Analysis and

Critical Control Points (HACCP) principles and must carry liability insurance. The Internet is proving to be a profitable outlet for sale of goat products; with minimal entrepreneurial creativity, it can provide a significant source of revenue for producers.

## INTERNATIONAL EXPERIENCE IN GOAT PRODUCTS MARKETS

The largest global goat-producing and -consuming countries are found in Asia and Africa. However, export markets in these countries are limited by strict international standards and the absence of developed marketing structures. To capitalize on the growing international demand for goat products, these countries must develop strategies to meet the international standards, which currently limit the success of their operations. Countries that have adapted to existing regulations, including Australia and New Zealand, have become the leading exporters of goat meat. Australia is currently positioned as the leading exporter with goat meat exports to more than 25 countries, including the U.S., and live goat exports to 15 countries (FAO, 2009). In Asia, Africa, Australia, and New Zealand, goat meat is produced on both a large and small scale, while the U.S. holds only small-scale production facilities.

Goat milk and fiber production have a significant presence in many countries throughout the world. Although marketing strategies have been shown to vary across different markets, the objective of achieving profitability has remained universal and continues to account for the current variation in marketing strategies.

Goat milk continues to be used worldwide for production of cheese, various food products, ice cream, skin care products including lotions, and soap as well as other products. The dairy goat industry in Europe is well established with France representing one of the top global exporters of goat cheese. Australia and New Zealand carry a small but thriving goat milk industry with Australia's dairy goat industry being relatively small but geographically diverse. The industry is well positioned to service the increasing demand for exotic cheeses and goat milk as an alternative to cow milk. In response to the small size of individual production units, goat milk producers have established associations and cooperatives to promote the goat milk industry. These organizations efficiently promote compliance to regulations by providing its members with processing and marketing operations that achieve economies of scale, resulting in a homogenous end-product that consistently meets the highest quality standards.

Two of the strongest dairy goat cooperatives currently operating include the Ontario Dairy Goat Cooperative (2009) and the New Zealand Dairy Goat Cooperative (2009). The New Zealand goat cooperative is a farmer-owned cooperative that manufactures and markets products derived from goat milk. Although dairy goat associations do offer some benefit, they have not been shown to be as effective as cooperatives in increasing members' business profitability. There are several dairy goat associations in Australia, the U.S., and Europe. In these associations, farmers either sell to processors or process their own cheese and then sell directly to consumers or specialized stores. The U.S. market for goat milk products, similar to that of goat meat, is characterized by consumers who are unfamiliar with goat milk and cheese products. Consequently, aside from select supermarkets and natural food stores that carry goat cheese, the current market for dairy goat products is poorly established and grossly underdeveloped. To achieve success in this market, goat milk producers should develop market plans that focus on increasing market share by emphasizing the health benefits of goat milk as an alternative to cow milk. Goat milk has many highly marketable attributes including the fact that it is easy to digest, naturally homogenized, and hypoallergenic for infants and adults. These attributes confer a significant competitive advantage against cow milk should be leveraged in goat milk promotion efforts. Currently, the American Dairy Goat Association is working with farmer organizations to organize and expand market outlets for goat milk. Individual farmers should also prepare and be equipped to provide an adequate supply to meet the new and growing demand.

Goat fiber, namely mohair and cashmere, represent the two most expensive and luxurious fibers on the world market. Several countries including China, the U.S., and South Africa are currently producing these fibers (Jensen, 1992). In this arena, producers generally sell directly to processors at prices that are influenced by supply and demand in the world market as a whole.

Earlier in this chapter, mention was made of the variation found in goat production and marketing strategies across markets. The following section will provide further insight into this variation by comparing activities in different markets associated with goat meat, dairy, and fiber products. These examples depict the three major scenarios that currently exist in the goat industry and demonstrate how cultural, traditional, and economic factors can influence a farmer's optimal choice of production and marketing activities. For meat products, information provided by Kebede (2005) is used to demonstrate the Southeastern

U.S. experience; for milk and dairy products, information reported by Mowlem (2005) is used to depict the U.K. experience; and for fiber (mohair), information reported by Jordaan and Kristen (2008) is used to demonstrate the South African experience. These reports are included to increase awareness of the challenges faced in goat meat, milk, and fiber market situations across the globe.

### Goat Meat Marketing—Southeastern U.S. Experience

Historically, goats in the U.S. have been used principally for fiber and milk production. However, recent increases in demand for goat meat from certain ethnic groups, and to a lesser extent, from health-conscious groups have shifted some emphasis from fiber and milk production to meat production (Shelton, 1992). The numbers of goats slaughtered in federally and state inspected plants have increased in the last decade as indicated in Chapter 1. The shift in both supply and demand is an indicator of industry growth. The average price per kg of imported goat meat in the U.S. has increased from $2.30 in 1999 to $3.80 in 2006 (Solaiman, 2007) and continues to increase. Current projections show an increase in the consumption of goat meat and milk products in the U.S. that will continue in the future.

#### Goat Meat Market Channels

Goat meat marketing is not well organized and is essentially an individual venture where price varies widely. An organized marketing channel for goat meat will enable a higher farm-gate price. The market is constrained by isolated consumer demand, seasonal demand, and an uninformed public. Producers lack adequate market information on consumer preferences and they just sell to the customers they know. Southern and Southeastern states in the U.S. are major meat goat producing regions. However, the majority of goat meat is transported and consumed in the Northeast. Goat meat is produced predominantly by small producers in the U.S.

#### Goat Meat Market Study

A study was conducted that assessed the goat meat marketing channels in seven Southeastern states: Alabama, Arkansas, Georgia, Louisiana, Mississippi, North Carolina, and South Carolina, to draw conclusions about the potential strategies to organize market outlets for small producers. The results of the study indicated that about 37% of all the stores listed as selling meat and poultry products in the selected cities were also selling goat meat. About 89% of goat meat consumers were people of ethnic origin. A wide price variation was observed in the survey, which reflects that the suppliers were not well organized. Even though the majority of the suppliers had been in business for more than 5 years, they did not have an organized system of marketing, and only 26% were ready to get into a contract with other suppliers. The main marketing outlet was auction markets where price is decided by the wholesalers and where producers' benefit is minimized. The current structure of goat meat distribution and the retail market is complex, fragmented, and dominated by small enterprises revolving around consuming ethnic groups. In order to remain competitive, producers need to develop strategies to compete with new goat producers. In the long run, U.S. goat meat producers are not limited to local demand for goat meat, but they have an opportunity to meet the increasing demand for goat meat in the traditional meat-importing countries (Kebede, 2005).

In summary, this study suggests that if goat meat producers develop strategies to promote their product and identify their consumers, they can establish a sustainable market and income.

### Goat Milk Marketing—The U.K. Experience

In the U.K., the current practice of goat production is a result of changes in agriculture in the last 4 to 5 decades. A decline in farm income led to the decline in farm numbers followed by more mechanization and less labor. Many smaller farms have been incorporated with bigger units. This rationalization pushed the farmers away from the traditional production systems to more diversified production systems. Many farms now have a large recreational arrangement and a service-orientated approach. Goat farming has also developed as part of this diversification. Goat farmers in the U.K. are similar to those in the U.S.—some are goat enthusiasts who have developed their hobby into a business, and some are from other businesses or professions who have added a dairy goat enterprise to their existing farm activities (Mowlem, 2005).

#### Milk Processing

Goat milk is processed at different scales into different products for consumption. There are three distinct groups of processors: those who keep a small number of goats as a hobby and process their own milk for family consumption; small-scale producers/processors; and the large processors.

The small-scale producers/processors usually make moderate amounts of products that will be mainly sold through specialty shops, to the catering industry, and/or retailed by the processors themselves. The large-scale processors sell the majority of their output through supermarkets or retail stores.

The largest goat milk processing dairy in the U.K. started off as a small producer/processor who concentrated almost solely on pasteurizing drinking milk and yogurt production with just a few hundred milking goats. The business developed into a large and successful operation that has now acquired a processing dairy that makes the greatest volume of hard goat cheese in the U.K. About 95% of their production is sold through supermarkets and the rest through specialty shops. They process between 15 and 20 million liters of goat milk for consumption each year into cheese (60%), drinking milk (20%), yogurt (10%), and the rest for butter, cream and ice cream.

#### Marketing Outlets

Two major marketing outlets in the U.K. are health markets and specialty food markets. The market outlets depend on the scale and products they offer. The largest producers supply to the major supermarkets, and the smaller ones supply the specialty retailers such as health shops, delicatessens, specialty cheese shops, and directly to consumers.

#### Health Market

The health market created the demand for goat milk as an alternative to cow milk for people with a health problem and when goat milk is believed to be beneficial. The health benefit of goat milk is based on scientific studies reported by Walker (1956) that claimed goat milk to be therapeutic in nature and stated that a child with infantile eczema recovered when its diet was changed to goat milk instead of cow milk. There are numerous scientific reports in Chapter 14 of this book describing the superiority of goat milk to provide a convincing argument for a demand and thus market for goat milk.

#### Specialty Food Market

Goat dairy products as specialty food products are found in many countries, including the U.K. Even in developing countries, encouragement to farm goats using more productive breeds, in some cases, has resulted in milk that is surplus to the requirements of the family. This has led to the development of a small processing industry and it is now possible to find goat cheese in countries where, until comparatively recently, the goat was seen as a source of meat rather than milk and where any dairy industry even from cow's milk was virtually unknown (Peacock, 1996). The increase in the consumption of goat cheese in the U.K. is attributed to the British traveling especially to France where goat cheeses are popular (Mowlem, 2005).

#### Marketing Strategies

Marketing requires selling a volume of product with the right standard to remain profitable. So often it has been difficult for small enterprises to supply the required volume on a regular basis, and sometimes the standard of goat products has not been consistent particularly with respect to hygiene. In the U.K., a number of cooperatives have been developed among dairy goat producers and processors. They collect milk from farms and transport milk to processors, organize financial transactions, and pay farmers for the milk. In this system, the price received by farmers and a regular supply of milk to processors is guaranteed. The cooperatives could also help with generic marketing of the products by producing advertising materials. However, hygiene requirements necessary for such activities are capital intensive, therefore many small goat producers were unable to participate.

More than 60% of all dairy products are sold to supermarkets, and the rest are sold through other avenues such as farmers markets, catering, and restaurants. Goat product prices are price insensitive. They are consumed for health reasons and by those who are willing to pay the price in specialty markets.

In summary, this case study was presented to help understand the process to develop markets for goat products and how goat products can be a part of the food chain through health and specialty markets. These markets are functioning well in most of the developed countries. Developing countries can use this as a model to join the health and specialty markets.

### Goat Fiber Marketing—South African Experience

Since the 1800s, the fiber industry in South Africa has gone through periods of good fortune marked by high prices and an increased level of production, to misfortunes related to climatic change, disease outbreak, and changes in the fashion industry. The industry has assistance from the government through the establishment of a Mohair Advisory Board and Mohair Control Board that set standards. Mohair is still marketed through brokers in auctions scheduled by the Mohair Advisory Board. About 44% of South African mohair producers get more than 50% of their total farm income from their investment in Angora goats. Mohair requires relatively specialized labor and capital investments. The relationship between the producers and the processors is not well established. Producers have to identify alternative marketing structures and decision making for a successful enterprise (Jordaan and Kristen, 2008).

AUCTION MARKET

Spot market auctions currently are dominant as the primary mechanism of exchange for South African raw mohair where the price is publicly set. With the decline in prices in the 1970s, the South African Mohair Growers Association demanded an alternative marketing system. The government established a sole agent to undertake mohair preparation and handling; however, because of increased cost of clipping and the poor mohair market, growers experienced major losses.

In 1997, the Mohair Board was replaced by an independent private sector, and mohair marketing took place in the free market and producers were free to choose their outlets. Besides auctions there were farm gate sales, forward selling, contracts, electronic auctions, and tenders all emerged as possible trading platforms between producers and buyers. However, auction selling accounted for 96% (Table 18.6). The trade venues have changed to contracts and forward selling since 2001. These changes are attributed to the changes in supply and demand, and allowed mohair producers to minimize their risk.

Spot marketing may not allow optimal flow of information, goods, and returns in the supply chain, since communication between producers and processors is not easily facilitated by a spot market system. Auction markets are an uncoordinated exchange where players seek to fulfill self-interest, limited information sharing opportunities, and short-term relationships; and a relatively small amount of attention is given to product differentiation. There is limited communication between producers and processors.

NICHE MARKET

Mohair has considerable niche market appeal because of its unique composition and quality. Therefore, marketing for mohair can move from the commodity marketing to branded products with coordinated exchange and long-term relationships. Niche marketing can obtain better prices in other marketing outlets. Coordinating within the mohair supply chain, in specific circumstances, has a potential for obtaining better prices.

VERTICAL COORDINATION STRATEGY

Vertical coordination is harmonizing the successive stages of production and marketing, or an alignment of direction or control across segments of the production/marketing system. The factors that align in vertical coordination are price, quantity, quality, and terms of exchange. It can include successive activities from production to final product or any of the stages in between. The coordination of the successive stages and the selection of the strategies are dependent on the level of investment the producer has to make and the associated rewards.

An alternative supply chain that can reduce the risk and create income stability is important. The market for mohair products varies from home industry crafts to custom-tailored ladies' and men's apparel. Different types of mohair can be used for different products such as carpet, upholstery, blankets, and machine or hand knitting to name a few (Table 18.7). As a result of the diverse nature of demand for the various types of mohair, competition among the mohair products is quite diverse. Mohair exhibits its unique characteristics; therefore, a marketing structure

**Table 18.6** Volumes (%) and values (%) of mohair per trading platform in South Africa (1998–2002).

| | | Years | | | | |
|---|---|---|---|---|---|---|
| Trading platform | Volumes by mass and value | 1998 | 1999 | 2000 | 2001 | 2002 |
| Auction | Mass (%) | 98 | 99 | 97 | 92 | 94 |
| | Value (%) | 98 | 99 | 96 | 86 | 88 |
| Farm Gate Sale | Mass (%) | 2 | 1 | 3 | 2 | 3 |
| | Value (%) | 2 | 1 | 4 | 4 | 6 |
| Contracts | Mass (%) | | | | 6 | |
| | Value (%) | | | | 10 | |
| Forward Selling | Mass (%) | | | | | 3 |
| | Value (%) | | | | | 6 |

Note: Mohair Market in South Africa (n = 44).
Source: Jordaan and Kirsten (2008).

**Table 18.7** Mohair classes according to its fiber diameter and usage.

| Usage | Mohair Fiber Diameter | | | | | |
|---|---|---|---|---|---|---|
| | | 22–24 | 24–26 | 26–28 | 28–30 | |
| | 30–32 | 32–34 | 34–36 | 36–38 | 38–40 | 40+ |
| Worsted | ↔ | ↔ | ↔ | ↔ | | |
| Woolen | | ↔ | ↔ | ↔ | ↔ | |
| Hand Knitting | | | ↔ | ↔ | ↔ | ↔ |
| Machine Knitting | | ↔ | ↔ | ↔ | ↔ | |
| Blankets | | | ↔ | ↔ | ↔ | ↔ |
| Upholstery | | | | ↔ | ↔ | ↔ |
| Carpets | | | | ↔ | ↔ | ↔ |

Source: Adapted with modifications from Jordaan and Kristen, 2008.

that can convey more information on basic quality rather than prices is needed. The mohair industry or individual firms within the industry can make a strategic decision for an optimal vertical coordination strategy for each vertical exchange relationship that is executed in the process of doing business.

In summary the South African mohair market demonstrates how mohair production is affected by price, climatic change, and animal disease that affect the income of the producers. Furthermore, because of uniqueness of the mohair, it can be considered as a product rather than a commodity and can be marketed accordingly. The selection of marketing strategies can be used to minimize the risk associated with these factors.

## SUMMARY

Goat meat, milk, or fiber operations should be managed with the goal of achieving a profitable and sustainable production unit. Business planning and clear production objectives are critical to the unit's success. Enterprise budget development and good record keeping can be used to make informed management decisions and continuously increase profits from operations. Profitability in goat production need not be measured in monetary values; in subsistence agriculture, it is measured in terms of food security and social status.

Markets and channels for meat goat products remain underdeveloped, while those for goat milk and fiber products are more established. Goat products are primarily sold and consumed locally with minimal presence in global trade markets. Despite the slow uptake of goat products, particularly in the U.S., unique opportunities exist to expand market share and capture value exist in the health and specialty food market for meat and milk; in Internet-based outlets for meat and mohair; in a shift from commodity- to product-based marketing of mohair; and through forming cooperative alliances to increase overall producer efficiency and product quality.

## REFERENCES


Alford, C., J. Strickland, K. Lewis and S. Simpson. 1998. Goat Meat Production in Georgia. The Cooperative Extension Service, Georgia Extension Service Publications. Bulletin Number 1168 (August).

Anderson and Hall. 2009. Adding value to agricultural products. Texas Agricultural Extension Service, Texas A & M University, Pub. L-5361(RM1-8). Online Publication. Retrieved March 11, 2009. http://trmep.tamu.edu/cg/factsheets/rm1-8.pdf.

FAOSTAT. 2005. www.fao.org/waicent/portal/statistics.

Andreasen, L. and H. Qwist-Hoffmann. 1997. Inter-regional project for participatory upland conservation and development (Field Document 6/97)—NEPAL—A framework for a participatory economic evaluation of improved goat production by women's groups in the Bhusunde Khola watershed.

Food and Agriculture Organization (FAO). 2009. Major Food and Agricultural Commodities and Producers, Available website http://www.fao.org/es/ess/top/commodity.html?lang=en, accessed March 22, 2009.

Gipson, T.A. 1996. "Marketing Channels & Strategies." http://www.vdacs.virginia.gov/livestock/pdffiles/goatselling.pdf (accessed June 28, 2004).

Glimp, H.A. 1995. Meat goat production and marketing, Journal of Animal Science, 73:291–295.

Jensen, H.L. 1992. Cashmere—the new American challenge—1991, Leafy Spurge Symposium and Proceedings, Lincoln, NE, July 22–24, (5):11–13.

Jensen, H.L. 1992. Mohair Production, Leafy Spurge Symposium and Proceedings, Lincoln, NE, July 22–24, (5):3–5.

Jordaan, D. and J. Kirsten. 2008. Investigating alternative governance systems for the South African mohair supply chain. Agrekon. 47(2):258–284.

Kebede, E. 2005. Goat meat demand in the U.S. and goat meat marketing potential in the Southeast. George Washington Carver Agricultural Experiment Station, Tuskegee University, Alabama. Pub. No. 113–905.

McKenzie-Jakes. 2007. Getting Started in the Meat Goat Business: Trends, Development, Challenges and Opportunities in the Meat Goat Industry in Southeastern United States, Bulletin 1, Vol. 1, Florida A&M University, website www.famu.edu/goats/UserFiles/File/MeatGoatIndustry.pdf.

Mowlem, Alan. 2005. Marketing Goat Dairy Produce in the UK. Small Ruminant Research. 60:207–213.

New Zealand Dairy Goats Cooperative. 2009. History and Benefit, website http://www.dgc.co.nz/main.cfm, accessed March 24, 2009.

Ollinger, M., J.M. MacDonald, and M. Madison. 2005. Technological chain and economies of scale in U.S. poultry processing. American J. of Agric. Economics. 87 (1):116–129.

Ontario Dairy Cooperatives. 2009. Goals and Membership, website http://www.ontariodairygoat.com/member.htm, accessed March 24, 2009.

Peacock, C. 1996. Improving Goat Production in the Tropics. Oxfam/FARM-Africa.

Schoenian, S. 2008. All about meat goats. Western Maryland Research and Education Center. University of Maryland Cooperative Extension. Online publication, http://www.sheepandgoat.com/articles/aboutgoats.html.

Shelton, J.M. 1992. Meat Goat Production, Texas A&M University Research Center, San Angelo.

Solaiman, S.G. 2007. Assessing of the meat goat industry and future outlook for U.S. small farms. Online Publication. http://www.agmrc.org/commoditiesproducts/livestock/goats/meat_goats.cfm.

USDA/NASS. 2009. Livestock and Animal, website http://www.nass.usda.gov/QuickStats/indexbysubject.jsp?Passgroup=Livestock+%26+Animals, accessed on March 26, 2009.

Walker, V.B. 1956. The allergic child. S. Afr. Med. J. 30: 125–128.

# 19
# Future Needs for Teaching, Research, Extension, and Outreach

*S.G. Solaiman, PhD, PAS and G.F.W. Haenlein, PhD*

**KEY TERMS**

Teaching—activities that impart skill or knowledge using structured syllabi, assignments, and examinations.
Research—finding facts in a systematic and scientific manner.
Outreach—reaching out to promote science and technology to the community.
Extension—a coordinated framework that can lead to changes in farm practices.
Livestock production system—a subset of farming systems where animal, crop, and land interact and are unique for different regions of the world.
Search engine—an electronic program that retrieves documents and files from databases on the Internet.
Agricola—a search engine containing bibliography of the National Agricultural Library.
PubMed—a collection of bibliographic sources of the National Library of Medicine.
PubMed Central—a collection of information related to the National Institute of Health and collaborators.
Google—most popular Web-based search engine.

**OBJECTIVES**

By completing this chapter, the reader will acquire knowledge on:

- The availability or the lack of instructional materials
- The availability or the lack of solid scientific data on different basic or production information
- The availability or the lack of outreach information
- The availability or the lack of solid information collected for extension activities
- The areas lacking the most scientific data
- Importance of livestock production systems and their specialization for each region of the world
- Importance of data dissemination
- Constraints of goat production, research, and outreach

## INTRODUCTION

Goat and goat farming historically originated in the Middle East (Ganji Dareh, Iran, Mesopotamia) more than 10,000 years ago (about 8000 BC) and then spread throughout the world, east to China and India, west to Europe, south to Africa, and later to America. Although the goat was the first domesticated ruminant livestock compared to cattle or sheep, it lags behind those species in research and development as indicated throughout the chapters of this book. Several researchers during the past few years have attempted to summarize the accomplishments in research and outreach related to goats (Haenlein, 2001; Sahlu and Goetsch, 2005; Boyazoglu et al., 2005) and to demonstrate areas of research and outreach that need more attention in

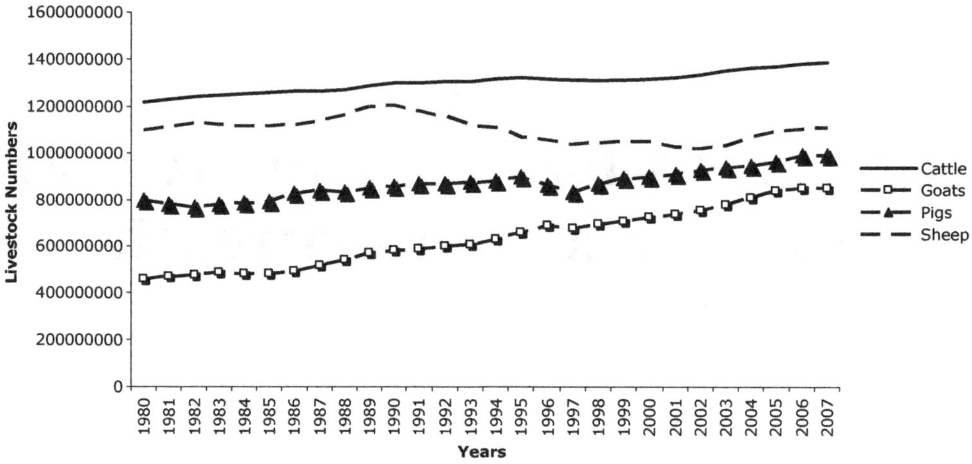


**Figure 19.1** Changes in the world's livestock numbers. Cattle, goats, pigs, and sheep numbers increased by 14.4%, 84%, 24%, and 1.4% during the last 27 years, respectively. Source: FAOSTAT, 2007, www.faostat.fao.org.

**Table 19.1** Number of records (hits) for goat, sheep, and cattle retrieved by searching National Agricultural Library, USDA (February 8, 2009).

| Items | Goat | Sheep | Cattle |
|---|---|---|---|
| Total Hits | 8,972 | >10,000 | >10,000 |
| Meat/Beef | 395 | 1,451 | >10,000 |
| Dairy | 1,315 | 1,097 | >10,000 |
| Fiber/Wool | 268 | 2,732 | — |
| Milk | 1,883 | 1,386 | 3,671 |
| Nutrition | 359 | 1,781 | 2,489 |
| Breeding | 405 | 2,291 | 4,476 |
| Genetics | 209 | 870 | 1,970 |
| Reproduction | 347 | 1,850 | 2,654 |
| Parasites | 59 | 350 | 639 |
| Management | 398 | 1,889 | 5,549 |
| Extension | 151 | 367 | 2,068 |
| Teaching | 3 | 12 | 38 |

the future. However, not even one of those reports covers the need of higher education in goat technology to provide the necessary tools for future researchers and educators of this sector of the livestock industry.

In most cases, in the U.S. for example, most of the researchers and outreach personnel are trained with cattle (dairy and beef), and working with goats is an on-the-job training activity that requires transformation and application of knowledge for goats. The limited amount of information about goats compared with sheep and cattle is evident from a simple search for published literature, records of journal articles, theses, patents, software, and technical reports related to agriculture using databases such as Agricola, PubMed, PubMed Central, and the USDA website (Table 19.1). This lack of information and proper training has been clearly stated in the literature and confirmed in the chapters of this book. The following sections will elaborate the future needs for classroom instruction, areas of research, extension programs, and outreach activities for goats.

## CLASSROOM INSTRUCTION ON GOATS

The last comprehensive pieces of information written on goats were Goat Production by C. Gall (1981), and Extension Goat Handbook by Haenlein and Ace (1984), which are more than a quarter of a century old. Recently C. Devendra (2007) released his book Goats: Biology, Production and Development in Asia, providing detailed information on development of this species in Asia. The Handbook of Milk of Non-Bovine Mammals by Park and Haenlein (2006) is another unique publication that covers goat milk.

The goat population in the world has experienced a strong increase (from 497 million in 1986 to over 850 million in 2007) during the past few decades, while most other species (except poultry) either decreased (sheep) or did not change significantly (Figure 19.1). However, scientific investment in goat science education and research remains very low (Dubeuf et al., 2004). Resources such as feed and especially water are limiting factors for livestock production in the future. Therefore, species that are more adaptable and tolerable to shortages of feed and water may be more sustainable and provide an alternative for food security in areas that may face these limitations. However, the necessary tools for higher education to train the future trainers of this animal sector are very limited. It is difficult to find literature assessing teaching capacity in higher education concentrated on educating students on goat sciences. This is based on the observation that only three articles were retrieved online when searching for goat teaching information. The number of hits recorded for teaching information for sheep and cattle were 4- and 12-fold higher than for goats, respectively. There are very few universities in the U.S. and most probably abroad that offer courses in goat production, although interest in goat production and research is increasing. After searching through all Land Grant Universities in the U.S. and seven commonwealths and territories, it was found that among more than 70 universities searched, only 20 offered either a sheep and goat production course (total of 3), small ruminant management course (1), or sheep science/sheep production course (16). It is apparent that instruction quality materials covering goat science and production are lacking and there is a need to develop such informational materials for instructional purposes.

## AREAS OF GOAT RESEARCH

In the past 20 years, a strong interest in goat research and publications was renewed by the crucial role that the International Goat Association and the journal of Small Ruminant Research played in the communication of scientific advancements in goat research. These entities enabled goat researchers to communicate their original findings and reduce duplication as well as to provide information for advancement of the industry.

The research literature or other available published materials were searched using different search engines, including PubMed (publications related to National Library of Medicine), PubMed Central (all publications related to National Institute of Health), USDA Web site, and Agricola (all agricultural literatures related to USDA National Agricultural Library). Key words related to goat type, production, nutrition, reproduction, genetics, and immunity were used for this search. Table 19.2 represents the number of records found in each category. According to the search, it is clear that within goat types, dairy goats have been investigated more than meat or fiber goats in all areas of goat production. Within the production area, more emphasis or data are available in the areas of reproduction and genetics followed by nutrition. In the area of nutrition, protein nutrition resulted in more records followed by energy nutrition. Energy nutrition of fiber goats has received more attention than other areas of their nutrition. Studies or records on goat immunity, vitamins, and minerals are very limited.

During the past 2 decades, reviewing the literature, much of the research was focused on the development of successful reproduction techniques and selection modeling. During this period, newly improved breeds were introduced—the Boer from South Africa and Kiko from New Zealand—that have revolutionized the goat industry in the world.

Several researchers have attempted to give an update on current progress and future needs for goat research in general (Sahlu and Goetsch, 2005; Sahlu et al., 2009) and in detail (Morand-Fehr, 2005). It must be noted that some research data collected in developed countries under a temperate environment may have much less direct application to subtropical or tropical environments. Also there is much to learn about differences in animal performance in intensive and extensive management systems for goats. These factors must be considered in application of all aspects of goat research.

### Goat Production and Management

Although production and management research has received less attention for goats than for sheep and cattle (Table 19.1), it has been investigated fairly as an area of emphasis within goat research (Table 19.2). Dairy and meat goats have received an equal share of research in this area as indicated in the database search. However, applicability of research conducted in temperate regions of the world with mostly intensive management practices may have little use in other regions that are less developed and are the home of a majority of the goats in the world. One of the major factors that can vary depending on regions where goats are raised is ambient condition, namely temperature and humidity. We know that ambient condition can impact energy requirement; therefore, future research should focus on determining lower critical and higher critical temperatures for goats. This could be more challenging due to the fact that goats having different genotypes vary considerably in thickness of skin and insulating fiber.

### Goat Nutrition Research

The nutrition of an animal plays an important role in its well-being, and it is a major contributing factor to the cost of production. Therefore, more research is justified in

**Table 19.2** Goat records (online hits) via several search engines (retrieved February 2009).

| Subject | PubMed[1] | PubMed Central[2] | USDA[3] | Agricola[4] | Google[5] |
|---|---|---|---|---|---|
| Goat | 27,955 | 73,475 | 23,330 | 8,972 | 40 million |
| Dairy Goat | 850 | 1,241 | 7,660 | 1,204 | 483,000 |
| Meat Goat | 561 | 1,010 | 6,900 | 622 | 357,000 |
| Fiber Goat | 248 | 4,360 | 1,320 | 242 | 1.7 million |
| Goat Production | 2,005 | 39,087 | 11,800 | 1,477 | 5.3 million |
| Dairy Goat Production | 180 | 773 | 7,340 | 243 | 189,000 |
| Meat Goat production | 105 | 596 | 6,540 | 199 | 195,000 |
| Fiber Goat Production | 32 | 2,159 | 1,120 | 73 | 122,000 |
| Goat Reproduction | 3,092 | 3,471 | 734 | 347 | 1.2 million |
| Dairy Goat Reproduction | 202 | 198 | 406 | 27 | 72,900 |
| Meat Goat Reproduction | 46 | 84 | 400 | 18 | 30,600 |
| Fiber Goat Reproduction | 7 | 215 | 290 | 4 | 107,000 |
| Goat Genetics | 3,024 | 33,731 | 1,800 | 209 | 4.3 million |
| Dairy Goat Genetics | 132 | 413 | 482 | 34 | 84,200 |
| Meat Goat Genetics | 66 | 224 | 1,440 | 11 | 81,300 |
| Fiber Goat Genetics | 14 | 2031 | 280 | 4 | 259,000 |
| Goat Immunity | 1,089 | 15,439 | 1,260 | 74 | 873,000 |
| Dairy Goat Immunity | 18 | 275 | 201 | 7 | 61,300 |
| Meat Goat Immunity | 6 | 250 | 1,070 | 1 | 50,300 |
| Fiber Goat Immunity | 3 | 641 | 102 | 0 | 78,700 |
| Goat Parasites | 959 | 3,438 | 728 | 59 | 612,000 |
| Dairy Goat Parasites | 54 | 114 | 249 | 2 | 36,400 |
| Meat Goat Parasites | 32 | 131 | 464 | 1 | 76,400 |
| Fiber Goat Parasites | 5 | 126 | 196 | 0 | 160,000 |
| Goat Nutrition | 345 | 1,340 | 1,350 | 359 | 352,000 |
| Dairy Goat Nutrition | 38 | 148 | 844 | 59 | 180,000 |
| Meat Goat Nutrition | 21 | 146 | 925 | 45 | 150,000 |
| Fiber Goat Nutrition | 17 | 86 | 555 | 26 | 98,400 |
| Goat and Protein | 12,156 | 66,217 | 3,330 | 632 | 149,000 |
| Dairy Goat and Protein | 272 | 657 | 1,100 | 145 | 52,700 |
| Meat Goat and Protein | 131 | 433 | 2,080 | 59 | 51,900 |
| Fiber Goat and Protein | 112 | 4,199 | 721 | 83 | 89,400 |
| Goat and Energy | 422 | 6,122 | 1,940 | 225 | 139,000 |
| Dairy Goat and Energy | 34 | 170 | 1,300 | 25 | 72,700 |
| Meat Goat and Energy | 14 | 128 | 1,360 | 26 | 54,300 |
| Fiber Goat and Energy | 32 | 488 | 850 | 40 | 57,800 |
| Goat and Minerals | 274 | 326 | 506 | 42 | 115,000 |
| Dairy Goat and Minerals | 14 | 43 | 273 | 11 | 293,000 |
| Meat Goat and Minerals | 14 | 34 | 323 | 5 | 26,300 |
| Fiber Goat and Minerals | 8 | 58 | 320 | 3 | 67,700 |
| Goat and Vitamins | 369 | 739 | 292 | 4 | 158,000 |
| Dairy Goat and Vitamins | 18 | 61 | 203 | 1 | 47,800 |
| Meat Goat and Vitamins | 12 | 54 | 237 | 1 | 36,500 |
| Fiber Goat and Vitamins | 6 | 51 | 193 | 1 | 77,500 |

developing alternative feeds for goats and for proper feeding practices that enhance gain or production efficiency.

The newly revised book, Nutrient Requirements of Sheep, Goats, Cervides and Camelids (NRC, 2007), has relied in part on the information from sheep and cattle, except for energy and protein requirements of goats. The Small Ruminant Committee of the National Research Council, of which the senior author was a part, faced tremendous challenges to accomplish the task due to the lack of published information on the basic science of goats, especially vitamins and minerals. It should be noted that the majority of the information may exist but it is not properly reported.

Goat nutrition research has received much less attention when compared with the amount of similar research for sheep and cattle (Table 19.1). There is a need for basic and fundamental studies to address factors influencing nutrient requirements of goats and factors such as genotype, sex, environment, and physiological processes responsible for their metabolic efficiency. Future research should address and quantify the energy and protein requirements among sexes. Energy expenditure in grazing and browsing or even in confinement of goats is not fully understood. Goats are very social animals and spend more time playing and socializing. This may justify additional energy requirements for such activities. Future research should address the impact of restricted protein and energy feeding prior to refeeding on tissue accretion and body weight gain as affected by prior body condition score of the goat.

Is the commonly acceptable minimum digestible dry matter of 65% in the diet of cattle sufficient to allow adequate intake for growth and productivity in goats? Also browse selectivity and level of intake are poorly investigated in goats. Another area that is practically ignored is information on particulate and fluid passage rates and their effects on intake and microbial fermentation in goats.

Trace mineral requirements of goats are different from other species; however, many of the goat requirements are using data from sheep and cattle. As was indicated based on literature searches, this area of nutrition has received the least attention after vitamin research. Trace minerals play an important role in an animal's metabolism, besides protein and energy metabolism and immune responses. Trace mineral supplements designed for cattle and recommended based on requirements and level of intake by cattle may be misleading when used for goats. Future research should focus on more refined prediction of intake for goats for mineral supplements and total dry matter intake. Goats have a stronger sense of taste and may have different preferences for salty or sweet feeds than cattle or sheep. Trace mineral supplements designed for cattle assume proportional intake based on salt level, but goats may have different intake. More research should be focused on taste preferences of goats as well as their requirements to enable nutritionists to design proper diets, salt blocks, and trace mineral mixes appropriate for goat intake to avoid deficiency or toxicity. Studies on vitamin requirements for goats are hard to find even if available.

### Goat Reproduction

According to the database search of reported literature, there is less research conducted on goat reproduction than for sheep or cattle. Certain areas of goat reproduction have been researched less when compared to sheep and cattle reproduction; however, this area has received more attention when compared with other areas of research in goats. With newly introduced goat breeds and composites, more research is justified to reconnect and confirm validity and applicability of the collected and reported data. Research is needed that could be used and extend as a tool in genetic improvement of goats including semen and embryo cryopreservation, embryo sexing, and the use of artificial insemination to improve production in goats.

**Table 19.2** *Continued*

[1]PubMed is a service of the U.S. National Library of Medicine that includes over 18 million citations from MEDLINE and other life science journals for biomedical articles back to 1948. PubMed includes links to full text articles and other related resources.

[2]PubMed Central (PMC) is the U.S. National Institutes of Health's (NIH) free digital archive of biomedical and life sciences journal literature.

[3]USDA covers all United States Department of Agriculture databases.

[4]Agricola is a bibliographic database of citations to the agricultural literature created by the National Agricultural Library and its cooperators.

[5]Google is the most popular search engine on the Web.

## Goat Breeding and Genetics

Although goat breeding and genetics has been one of the areas of focus and the subject of research in the past, it had received much less attention when compared to sheep or cattle (Table 19.1). Advancing molecular genetics and germplasm research enables researchers to achieve faster genetic improvement and genetic gain. Other species have taken advantage of these techniques and have made rapid, significant genetic improvement in animals. Goat science similarly can take advantage of these new techniques and produce superior animals that are adapted to different environments and are more resistant to diseases and parasites. Identifying genes that code for faster rate of gain, parasite resistance, quality of carcasses, higher milk yield, and finer fiber are necessary for progress. Developing easy-care meat, milk, and fiber goats that are the least dependent on concentrate feeding is desirable.

In the majority of developing and underdeveloped countries where goat populations reside, it is the least characterized livestock species. In these countries, goats are kept in extensive systems and are characterized as "local" or "indigenous" or "nondescriptive." Future efforts should be focused on characterizing these goat populations into distinct possible breeds and setting boundaries to maintain the breed integrity. This is a prerequisite to measuring population size and devising appropriate means for conserving endangered breeds. Besides the technical aspect, this will need socioeconomic research to make the condition conducive to the holders identifying their breeds.

## Goat Health

A profitable goat production requires a quality health program. The focus of a health program should be disease prevention and control practices rather than cure. Understanding the most common diseases of goats and how to prevent them from infecting goat herds is important. Future research should focus on production of effective vaccines for infectious diseases such as Caseous lymphadenitis, a bacterial disease that is widespread in goat herds in the U.S. and maybe elsewhere. Currently available vaccines limit the spread of infection without elimination. Future research should focus on production of DNA or whole cell vaccines that are promising (Chaplin et al., 1999; Williams, 2001). Johne's disease is another bacterial disease that invades the intestine and manifests itself as unthriftiness in goats. Although testing is available for this disease, future research should focus on production of a possible vaccine to prevent this disease. Currently used vaccine against enterotoxemia, another bacterial disease of goats, is not as effective as it is for sheep (Schoenian, 2009), and therefore, warrants further investigation. Also most information and drugs used to combat diseases are for sheep but are currently used extra-label for goats. There is an urgent need for drugs that are tested and approved for goats.

## Gene Base Technology

Goats have a larger gene pool and considerably more genetic diversity than other domesticated ruminants such as cattle and sheep. This is mainly because less selection has been practiced in this species. Gene mapping and sequencing of the goat genome will enable us to identify genes that control economically and physiologically important traits such as parasite resistance, growth rates, and immune response in goats. This information will be important in nutrition research and how nutrient metabolism interacts with genetics of the animal.

Gene-based technologies are strong tools to build new composite breeds by identifying and using biological markers that are responsible for adaptability of goats to harsh environments and their productivity under suboptimal conditions. Also these technologies should be used to select for gene markers responsible for disease and parasite resistance and other economically important traits. It is important to conserve some of the rare breeds before they become extinct.

In February 2009, the U.S. Food and Drug Administration (FDA) approved the first drug produced by milk of a genetically engineered goat. The goat was genetically altered to produce extra antithrombin—a protein that acts as a blood thinner. This may open the opportunity for future drugs made by genetically engineered biological organisms rather than by chemical synthesis. This kind of research will provide more opportunities for the use of goats to improve human health.

## Internal Parasitism Research

Internal parasites are a major drawback to raising goats in the U.S. Producers raising improved breeds of goats in the U.S. are losing the battle with parasite load in less resistant goats. There are fewer research reports from studies of parasites in goats than those reported for sheep and cattle. Also within research reported for goats, parasite research is very limited when compared with other areas of goat research. Dairy and meat goats are produced more intensively in the U.S. or elsewhere, compared to Angora goats. Therefore, more research should be directed to these breeds.

Anthelmintics currently used for goats have lost most of their effectiveness, and new ones are not available for goats; thus, use of extra-label drugs is a common practice. Therefore, there is a need for integrated management strategies to avoid high nematode burdens in goats. Co-grazing of goats with other ruminants, especially cattle, may provide a means not only to improve pastures but to reduce parasite load and anthelmintic use in goats. This area of research should be fully investigated as a holistic approach and a part of an integrated parasite management. Indigenous goats show more resistance to internal parasites. Therefore, either crossing indigenous goats with newly improved breeds to produce more resilient animals or treatment with naturally occurring substances with inherent anthelmintic activity must be researched. Plant secondary metabolites such as tannins are showing promise for reducing parasite load. These compounds can also render proteins unavailable. More research should be conducted to evaluate the optimum levels of forage that contain tannins to improve performance by reducing parasite load and nutrition of the animal. Other innovative approaches can be used, such as incorporating the use of elevated slatted false floor housing to eliminate re-infestation with internal parasite eggs. Future research should be focused on evaluating the efficacy of an integrated approach to controlling gastrointestinal parasites.

Goats, when raised intensively, seem to be more stress prone than cattle or sheep (author's personal observation); however, there is no research to document this in goats under a confinement system. High stress levels are associated with lower immune response that can directly influence resistance to diseases and parasites. More research relating to the plane and level of nutrition, environmental stressors, and immune responses is needed as a part of an integrated approach to parasite management.

The great advantages of elevated slatted false floor housings for dairy and meat goats and their impact on reducing parasite load must be investigated. This flooring system is found in many tropical countries, and it is the only management system practiced for half a million head of dairy goats in Taiwan. This system can contribute to controlling transmission of internal parasites and can be used as a part of integrated parasite management.

### Goat Housing and Environment

In the developed countries, the majority of goats are raised under intensive conditions for meat and especially milk productions. Searching literature for favorable environmental and housing conditions for goats, very limited data exist. Currently available data on favorable climatic conditions for housing design for goats are scarce and mostly are from sheep data. Future research should focus on determining thermal neutral zone, lower critical temperature, and higher critical temperature for goats. Goats being a more agile species may have a need for different sets of requirements for temperature or ventilation capacity than sheep. More consideration for housing design that accommodates goats' playful nature and space requirements for their welfare must be revisited. Automated systems of feeding and watering may be more economical and support the freewill nature of goats and may avoid agonist behavior and reduce overfeeding.

The elevated housing or lofting area that is practiced in the most tropical regions of the world is completely ignored in the U.S. The elevated slatted floor for housing goats will reduce transmission of parasites.

### Environmental Enhancement and Animal Welfare

The welfare of farm animals and the environment where they have been raised and slaughtered is increasingly becoming a major concern for consumers. Consumers are becoming increasingly concerned about the products they consume, not only in terms of fat, cholesterol, and drug residue levels, but they want to be reassured that animals have been raised under humane conditions and in a stress-free environment, and have received humane treatment during handling and slaughter.

Consumers are increasingly more conscious about the welfare of animals raised for food production under intensive systems. Also the negative impact of animals on the environment, especially in concentrated feeding facilities, is of major concern in developed countries. The unit area needed to house a goat humanely under intensive and extensive management systems must be evaluated more accurately. It appears that goats may need more surface area per animal unit than cattle and sheep to reduce parasite burden and use of anthelmintics.

Also, the positive impact of goats on enhancing the environment by reducing fire fuel load or improving pastures must be researched. Goats can be successfully used to reduce use of petrochemical-based herbicides and to avoid dangerous fires in pine plantations or other areas that are used for human habitats. However, when placing goats in these systems, the number of goats per unit area (stocking rate) and appropriate exposure time for plant growth and height, is not well defined and needs further investigation.

### Product Availability and Quality

In developed countries such as the U.S., demand for goat products, especially meat and cheese, is increasing. Consumers demand a steady supply of fresh meat products with consistent quality year-round. The efficacy of out-of-season breeding and its economic feasibility must be researched. Uniform carcasses with consistent quality are needed for successful marketing. Goats usually are low in subcutaneous fat; therefore, research should be focused on the proper cooling time and environment for reduction of carcass shrinkage. Duration of aging carcasses and environmental conditions for aging must be standardized. The current desirable weight for slaughter animals is set at about 30–35 kg in the U.S. This may be modified for production of larger carcasses resulting from new, improved meat goat breeds that mature faster and provide more appropriate retail cuts. More research is needed to predict lean to bone ratios based on more easily measurable traits similar to those for cattle. Other factors in raising slaughter kids, such as composition of gain, previous plane of nutrition, and compensatory growth, must be investigated. Also the efficacy of more than one kid crop per year and nutritional needs of the doe and kids must be defined.

Two major health problems in the U.S. are cardiovascular diseases and cancer. The fat content of goat meat is considerably lower than for beef, pork, or lamb. Goat meat, with its low cholesterol and high proportion of unsaturated to saturated fatty acids when compared to other meats, can play an important role in the diets of health conscious consumers fighting cardiovascular diseases. In addition, comparing levels of conjugated linoleic acid (CLA), a cancer-fighting fatty acid found in ruminants, to chicken, goat meat is a healthier choice. Also, goat meat is produced free of hormones, antibiotics, and other additives when compared to poultry or other red meats; however, these facts are poorly documented and researched. Consumers are more health conscious these days and, when given a choice, would probably select an alternative to the currently available red meat sources (Solaiman, 2007).

Goat milk and cheeses are also of superior quality when compared to cow milk. Goat milk has been recommended for patients that suffer allergies from cow milk or other food sources. More research is warranted to investigate other goat milk products and their nutritional quality for promoting human health.

Goat fiber, cashmere and mohair, is a high priced and very respected textile in the world. There is a need for practical and affordable objective methods for measuring luster of white mohair and color of cashmere. Such instrumental methods would be useful for selection and in marketing.

**Production System Research**

The relative importance of different production systems (see also Chapter 10) and animal species differs across various geographical regions. Small ruminants, especially goats, are important in Africa in semiarid and arid regions. Grazing systems are important for Central and South America where population density is low. Asia's livestock is found mainly in mixed farming systems. Understanding these differences in different agricultural zones and how they impact animal production systems is important. It is anticipated that more livestock will be produced under grazing systems and mixed farming systems.

Mixed or co-grazing of goats and cattle, rotational grazing, and strip grazing, and their impacts on internal parasite load should be investigated. Research relating to small ruminant productivity under different livestock production systems is lacking. Research relating the productivity changes of newly improved breeds of goats raised under a landless intensive system compared with their productivity in grazing systems is scarce. Information related to grazing of other ruminant species may not directly apply to goats, because of their selectivity for different botanical composition and types of plants. The economic value of an indigenous breed of goats versus an improved breed under different production systems must be evaluated. And finally, more research is needed to determine the most sustainable goat production systems in terms of being economically viable, environmentally friendly, and socially acceptable.

**Social Values of Goats**

Goats have a unique place in the lives of women in developing countries by giving women a sense of worth and social status. Goats are usually owned and managed by females and the children in the household. Strengthening technical capacity and improving women's agricultural expertise can enhance food security in the next century. There is a need to assess the role of women together with goats in shaping societies, not only from a socioeconomic view, but also relative to cultural and social perspectives.

In developed countries, rapid urbanization has separated people from nature, and new generations have little understanding of food and fiber. Goat shows have been gaining popularity especially in the U.S., and well-to-do families with young children are becoming more involved in these shows. The effects of goats on the lives of the young generation must be valued, and the social impacts of these activities on societies must be investigated.

**Predator Control**

In 2005 in the U.S. alone, loss from predators was more than 155,000 goats and kids. The majority of losses or 65% were attributed to coyotes, 15% to dogs, 6% to mountain lions, 5% to bobcats, and the rest were to bears, foxes, and eagles (USDA, 2005). These losses were more devastating in Angora goat herd populations in west Texas and New Mexico where they are kept extensively. Common methods used around the world for controlling animals or predators are simply housing, herding, tethering, or fencing. Guardian animals such as dogs, donkeys, or llamas have been used with different degrees of protection. Future research should focus on designing the most proper methods of animal protection with an integrated approach that is humane, sustainable, and economically feasible.

## EXTENSION PROGRAMS AND OUTREACH ACTIVITIES

The "extension" by web definition is an "act of expanding in scope; making more widely available; the spreading of something (a belief or practice) into new regions." The web definition of "outreach" is "an effort by individuals in an organization or group to connect its ideas or practices to the efforts of other organizations, groups, specific audiences, or the general public." Both of these explain connectivity and reaching out to others. Agricultural research publications in isolation are only static information, but when put to work, the outcome is dynamic. The process in which research findings are transferred and applied to real situations on a farm or in an agricultural system is the process of extension. However, system feedback on the components of the system for further research is called participatory action research that has emerged in recent years as a significant methodology for intervention, development, and change within communities and groups (Wikipedia).

Beneficiaries of extension and outreach programs are producers and farmers. Typically, there are three types of producers or farmers in the U.S. The progressive farmers take charge of their situation and impose change in their environment. The hobby farmers enjoy farming as a way of life. They take the opportunity of farming and seek success; however, they are economically independent of farm income. The limited resource farmers or producers are a group of farmers who have limited resources with technical and economic needs. In the U.S., progressive goat farmers have formed different "goat production groups" or "breed associations," and provide expertise to other producers through firsthand, on-farm research trial and error. Hobby farmers benefit from these groups and sometimes support their activities. Limited resource farmers represent farming situations often found in developing countries. Most of the goat extension work in the U.S. is geared toward this group of farmers. The lack of technical ability and capital hinders progress on these farms.

For successful outreach activity, there is a need for research output in the areas of need; research must be applicable for the given environment; it must be in harmony with available resources; and research must be sustainable. Developing regions of the world house the majority of goats; however, lack of infrastructure, supplies, and most importantly, expertise in these regions limit goat research and its application in the field. Developed countries that have less than 5% of the goat population in the world produce the majority of scientific information used by the majority of goat producers residing in a completely different physical and natural environment. However, in most cases, simply lack of extension infrastructure in these countries creates broken communication lines or connections between research and extension so that successful research output may result in failed outcomes. There is a need for more efficient dissemination of information, and the fact that the majority of work conducted on goats in the form of educational papers, minutes of extension meetings, etc., that are unpublished, makes the results difficult to distribute (Morand-Fehr and Boyazoglu, 1999).

## SUMMARY

This chapter attempted to demonstrate the relative amounts of teaching, research, and extension activities reported for goats compared to sheep and cattle using computer-based search engines. However, the needs for information related to teaching, research, and outreach were identified through reviewing published literature. Educational information for formal academics and classroom settings for goats is limited. Although the goat is the first domesticated livestock, in all aspects of science and production, it has been less researched compared to sheep and cattle. The majority of the information for goats is extrapolated from that for sheep and cattle. Therefore, extension and outreach without solid research findings for goats may have limited application. Many progressive goat producers and breeders in the U.S. and abroad have joined forces in driving this industry. However, they need more information. More than 40 million online hits can be obtained on goats using the key word "goat" alone using Google (the most popular current [2009] search engine). This compares favorably to 52.6 million hits for cattle, indicating the popularity of this species in the world and confirming the need for more information. Also there is a need for involving and encouraging interaction of this enthusiastic audience and stakeholders with researchers and educators. There is a need for greater cooperation and collaboration among research institutions within each country and among countries to reduce duplication of efforts and dissemination of information. Organizations such as the International Goat Association (IGA) and publication of the Small Ruminant Research journal by Elsevier Publishing have played important roles in communication of available research data among scientists and are commended for helping disseminate information about goats.

## REFERENCES

Boyazoglu, J., I. Hatziminaoglou, and P. Morand-Fehr. 2005. The role of the goat in society: Past, present and perspectives for the future. Small Ruminant Res. 60:13–23.

Chaplin, P.J., R. De Rose, J.S. Boyle, P. McWaters, J. Kelley, J.M. Tennent, M.L. Lew, and J.Y. Scheerlinck. 1999. Targeting improves the efficacy of a DNA vaccine against Corynebacterium pseudotuberculosis in sheep. Infect. Immun. 67:6434–6438.

Devendra, C. 2007. Goats: Biology, Production and Development in Asia. Academy of Sciences Malaysia, 902-4 Jalan Sultan Ismail, 50480 Kaula Lumpur, 246 pp.

Dubeuf, J.P., P. Morand-Fehr, and R. Rubino. 2004. Situation, changes and future of goat industry around the world. Small Ruminant Res. 51:165–173.

Gall, C. 1981. Goat Production. Academic Press, London, 619 pp.

Haenlein, G.F.W. 2001. Past, present, and future perspectives of small ruminants dairy research. J. Dairy Sci. 84: 2097–2115.

Haenlein, G.F.W. and D.L. Ace. 1984. Extension goat handbook. USDA-ES Publ., Washington D. C., 174 pp.

Morand-Fehr, P. 2005. Recent developments in goat nutrition and application: A review. Small Ruminant Res. 60: 25–43.

Morand-Fehr, P. and J. Boyazoglu. 1999. Present state and future outlook of the small ruminant sector. Small Ruminant Res. 34:175–188.

NRC. 2007. Nutrient Requirements of Small Ruminants: Sheep, Goats, Cervides, and Camelids. Academy Press. Washington, D.C., 362 pp.

Park, Y.W. and G.F.W. Haenlein (Eds.). 2006. Handbook of Milk of Non-Bovine Mammals. Blackwell Publishing, Ames, IA, USA, 449 pp.

Sahlu, T. and A.L. Goetsch. 2005. A foresight on goat research. Small Ruminant Res. 60:7–12.

Sahlu, T., L.J. Dawson, T.A. Gipson, S.P. Hart, R.C. Merkel, R. Puchala, Z. Wang, S. Zeng and A.L. Goetsch. 2009. ASAS Centennial Paper: Impact of animal science research on United States goat production and predictions for the future. J Anim. Sci. 87:400–418.

Schoenian, S. 2009. General health care of sheep and goats. http://www.sheepandgoat.com/articles/generalhealthcare.html. Accessed Feb. 10, 2009.

Solaiman, S.G. 2007. Simply Meat Goats. George Washington Carver Agricultural Experiment Station. Publication No. 115–1006, 118 pp.

USDA. 2005. Sheep and goat predator loss. National Agriculture Statistics Service (NASS), Washington DC. http://usda.mannlib.cornell.edu.

Williams, L.H. 2001. Caseous lymphadenitis in small ruminants. Vet. Clin. North Am. Food Anim. Pract. 17:359–371.

# Appendix A

**Figure 2.1** Kiko (Meat Goat).

**Figure 2.2** Boer (Meat Goat).

**Figure 2.3** Kalahari Red (Meat Goat).

**Figure 2.4** Savanna Buck (Meat Goat).

**Figure 2.5** Spanish (Meat Goat).

**Figure 2.6** Myotonic (Meat Goat).

**Figure 2.7** British Alpine (Dairy Goat).

**Figure 2.8** Barbari (Dairy Goat).

**Figure 2.9** Black Beetal (Dairy Goat).

**Figure 2.10** Damascus (Dairy Goat).

**Figure 2.11** Jamunapari (Dairy Goat).

**Figure 2.12** LaMancha (Dairy Goat).

**Figure 2.13** Nubian (Dairy Goat).

**Figure 2.14** Oberhasli (Dairy Goat).

**Figure 2.15** Saanen (Dairy Goat).

**Figure 2.16** Toggenburg (Dairy Goat).

**Figure 2.17** Angora (Fiber Goat).

**Figure 2.18** Cashmere (Fiber Goat).

**Figure 3.2** Alpine ibex (*Capra ibex ibex*) in the Gran Paradiso National Park, Italy. Photo by Marco Festa-Bianchet.

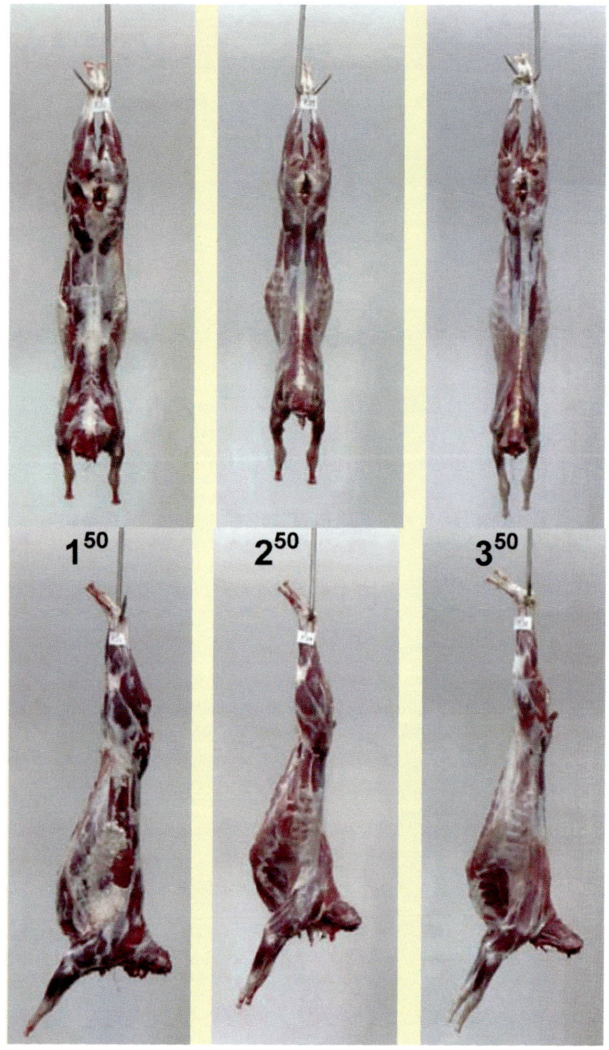


**Figure 13.1** Rear and side views of meat goat carcasses depicting the midranges of selection conformation classes 1, 2, and 3.

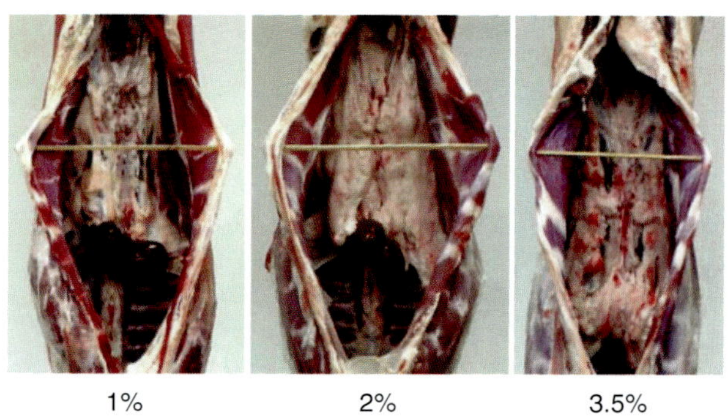


**Figure 13.2** Kidney and pelvic fat of 1%, 2%, and 3.5% in meat goat carcasses.

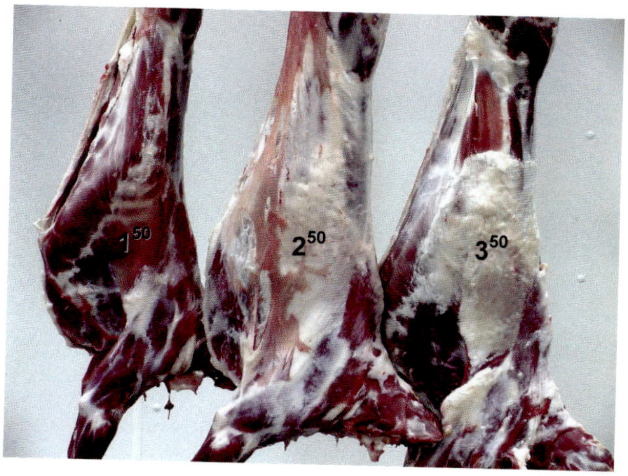


**Figure 13.3** Subcutaneous fat cover scores to estimate the external fat deposition on meat goat carcasses with carcasses showing the midpoint for each fat score ($1^{50}$, $2^{50}$, and $3^{50}$), with 1 = minimal or none, 2 = fat over rib and shoulder, and 3 = excessive fat cover.

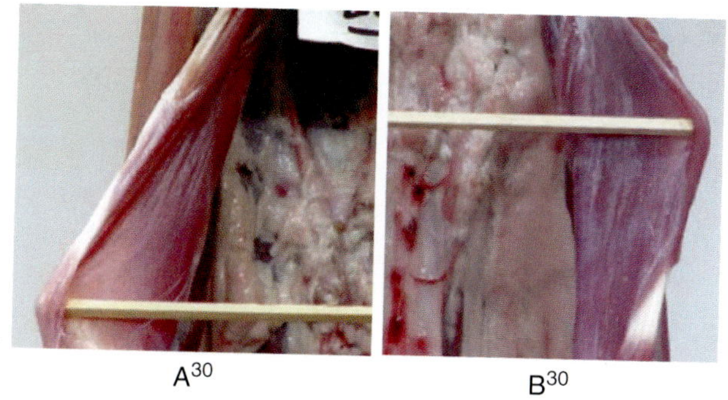


**Figure 13.4** Flank lean color indicates relative maturity of carcasses, with darker color indicating more advanced physiological maturity. Shown are scores of $A^{30}$ and $B^{30}$, with each color group ranging from 0 to 100. Color groups C, D, and E are associated with goats older than yearlings and are not shown.

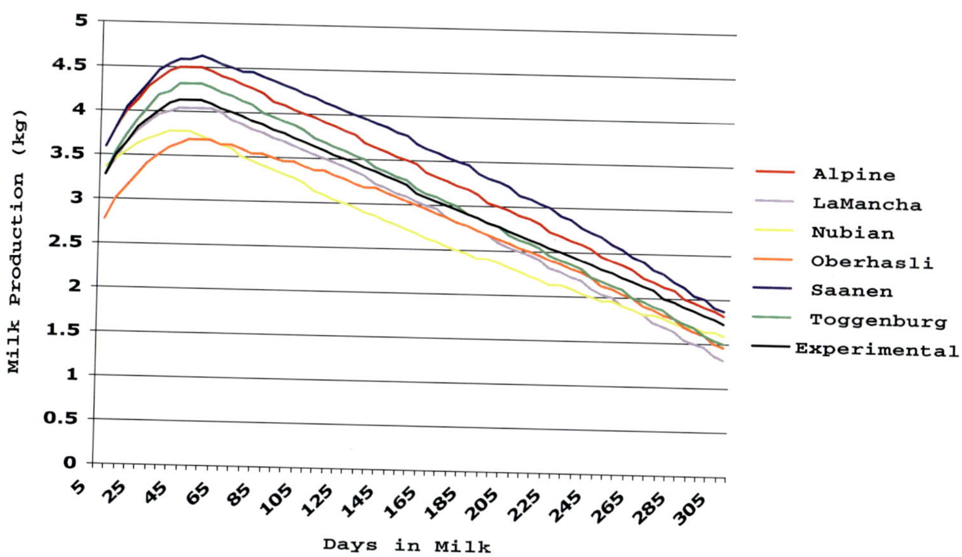


**Figure 14.2** Patterns of lactation curves for different dairy goat breeds in the U.S. (Dairy Herd Improvement Registry; ADGA, 2008).

India

Taiwan

**Figure 17.1** Elevated slatted floor design. Courtesy of Dr. George Haenlein, Professor and Dairy Goat Specialist.

Portable

Permanent

**Figure 17.2** Shelters for goats. Courtesy: Sandra Solaiman.

Grain Feeder

Hay Feeder

Wildwind Farm, Wetumpka, Alabama.
http://www.egwildwind.com

**Figure 17.3** Low-cost feeders. Courtesy: Sandra Solaiman.

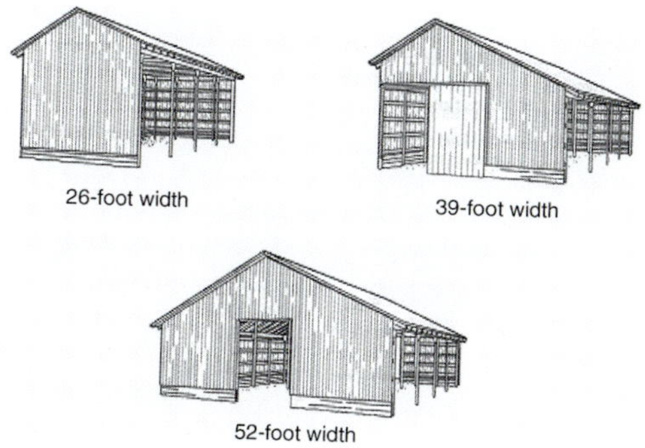

**Figure 17.4** Post-frame building width and configurations and pen feeding alley access area. Courtesy: John C. Porter, Professor and Dairy Specialist, University of New Hampshire Cooperative Extension, www.umaine.edu/animalsci/Sheepgoats/Dairy%20Goat%20Housing.ppt.

**Figure 17.5** Low-cost hoop buildings with feeding alley adjacent to fabric wall (Ray Rodriguez, Rindge, NH). Courtesy: John C. Porter, Professor and Dairy Specialist, University of New Hampshire Cooperative Extension, www.umaine.edu/animalsci/Sheepgoats/Dairy%20Goat%20Housing.ppt.

Parallel
(Courtesy: Dr. George Haenlein)

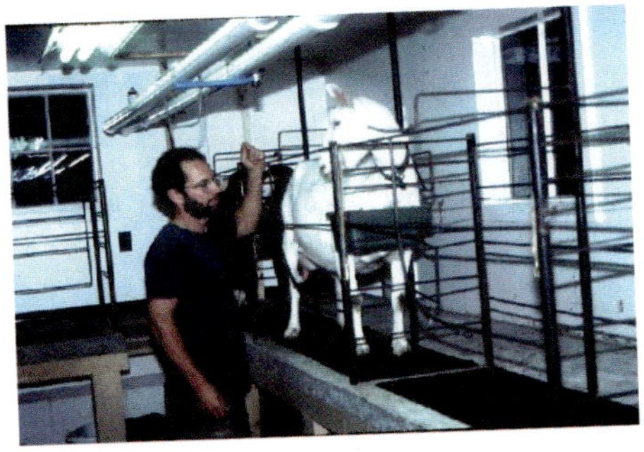

Side Opening

Straight Through

Herringbone

Andy and Jenny Tapper Farm, Brookfield, NH.

**Figure 17.6** Different types of milking parlors and milk room equipment for dairy goats. Courtesy: John C. Porter, Professor and Dairy Specialist, University of New Hampshire Cooperative Extension, www.umaine.edu/animalsci/Sheepgoats/Dairy%20Goat%20Housing.ppt.

# Appendix B

**Appendix Table B.1** Nutrient Requirements for Goats (Maintenance—Breeding).

| | Energy | | Protein | Mineral | | Vitamins | |
|---|---|---|---|---|---|---|---|
| Weight kg | TDN kg/d | ME Mcal/d | CP g/d | Ca g/d | P g/d | A RE/d | E IU/d |
| **Mature Does (Dairy) Maintenance*** | | | | | | | |
| 40 | 0.53 | 1.91 | 67 | 1.9 | 1.5 | 1256 | 212 |
| 50 | 0.62 | 2.25 | 79 | 2.1 | 1.7 | 1570 | 265 |
| 60 | 0.72 | 2.58 | 90 | 2.4 | 2.0 | 1884 | 318 |
| 70 | 0.80 | 2.90 | 101 | 2.6 | 2.2 | 2198 | 371 |
| 80 | 0.89 | 3.20 | 112 | 2.8 | 2.4 | 2512 | 424 |
| 90 | 0.97 | 3.50 | 122 | 3.0 | 2.6 | 2826 | 477 |
| **Mature Bucks (Dairy) Maintenance*** | | | | | | | |
| 50 | 0.72 | 2.59 | 86 | 2.4 | 2.0 | 1570 | 265 |
| 75 | 0.97 | 3.51 | 116 | 3.0 | 2.6 | 2355 | 398 |
| 100 | 1.21 | 4.36 | 144 | 3.7 | 3.2 | 3140 | 530 |
| 125 | 1.43 | 5.15 | 170 | 4.2 | 3.8 | 3925 | 663 |
| 150 | 1.64 | 5.91 | 195 | 4.8 | 4.3 | 4710 | 795 |
| **Mature Does (Dairy) Breeding** | | | | | | | |
| 40 | 0.58 | 2.10 | 73 | 2.0 | 1.6 | 1256 | 212 |
| 50 | 0.69 | 2.48 | 87 | 2.3 | 1.9 | 1570 | 265 |
| 60 | 0.79 | 2.84 | 99 | 2.6 | 2.1 | 1884 | 318 |
| 70 | 0.88 | 3.19 | 111 | 2.8 | 2.4 | 2198 | 371 |
| 80 | 0.98 | 3.53 | 123 | 3.1 | 2.6 | 2512 | 424 |
| 90 | 1.07 | 3.85 | 134 | 3.3 | 2.9 | 2826 | 477 |
| **Mature Bucks (Dairy) Breeding** | | | | | | | |
| 50 | 0.79 | 2.85 | 94 | 2.6 | 2.1 | 2275 | 280 |
| 75 | 1.07 | 3.86 | 128 | 3.3 | 2.9 | 3413 | 420 |
| 100 | 1.33 | 4.79 | 158 | 4.0 | 3.5 | 4550 | 560 |
| 125 | 1.57 | 5.67 | 187 | 4.6 | 4.1 | 5688 | 700 |
| 150 | 1.80 | 6.50 | 214 | 5.2 | 4.7 | 6825 | 840 |

*Feed all nondairy breeds 90% of Dairy Breed

**Appendix Table B.2** Nutrient Requirements of Growing Kids (Doelings and Wethers).*

| | | Energy | | Protein | Mineral | | Vitamins | |
|---|---|---|---|---|---|---|---|---|
| Weight kg | Daily gain g/d | TDN kg/d | ME Mcal/d | CP g/d | Ca g/d | P g/d | A RE/d | E IU/d |
| 10 | 0 | 0.20 | 0.72 | 26 | 0.9 | 0.5 | 1000 | 100 |
| 10 | 25 | 0.24 | 0.86 | 36 | 1.6 | 0.8 | 1000 | 100 |
| 10 | 100 | 0.35 | 1.27 | 69 | 3.5 | 1.6 | 1000 | 100 |
| 10 | 150 | 0.43 | 1.55 | 90 | 4.8 | 2.2 | 1000 | 100 |
| 10 | 200 | 0.51 | 1.82 | 112 | 6.2 | 2.7 | 1000 | 100 |
| 15 | 0 | 0.27 | 0.98 | 35 | 1.3 | 0.9 | 1500 | 150 |
| 15 | 25 | 0.31 | 1.11 | 46 | 1.8 | 1.0 | 1500 | 150 |
| 15 | 100 | 0.42 | 1.53 | 78 | 3.6 | 1.7 | 1500 | 150 |
| 15 | 150 | 0.50 | 1.80 | 100 | 5.0 | 2.3 | 1500 | 150 |
| 15 | 200 | 0.58 | 2.08 | 121 | 6.3 | 2.8 | 1500 | 150 |
| 20 | 0 | 0.34 | 1.21 | 43 | 1.4 | 1.0 | 2000 | 200 |
| 20 | 25 | 0.37 | 1.35 | 54 | 2.2 | 1.4 | 2000 | 200 |
| 20 | 100 | 0.49 | 1.76 | 86 | 4.0 | 2.0 | 2000 | 200 |
| 20 | 150 | 0.57 | 2.04 | 108 | 5.1 | 2.4 | 2000 | 200 |
| 20 | 200 | 0.64 | 2.31 | 130 | 6.4 | 3.0 | 2000 | 200 |
| 20 | 250 | 0.72 | 2.59 | 151 | 7.7 | 3.5 | 2000 | 200 |
| 25 | 0 | 0.40 | 1.43 | 51 | 1.6 | 1.2 | 2500 | 250 |
| 25 | 25 | 0.44 | 1.57 | 62 | 2.3 | 1.5 | 2500 | 250 |
| 25 | 100 | 0.55 | 1.95 | 94 | 4.1 | 2.2 | 2500 | 250 |
| 25 | 150 | 0.63 | 2.26 | 116 | 5.5 | 2.8 | 2500 | 250 |
| 25 | 200 | 0.70 | 2.54 | 137 | 6.5 | 3.1 | 2500 | 250 |
| 25 | 250 | 0.78 | 2.81 | 159 | 7.8 | 3.6 | 2500 | 250 |
| 30 | 0 | 0.46 | 1.64 | 59 | 1.8 | 1.4 | 3000 | 300 |
| 30 | 25 | 0.49 | 1.78 | 69 | 2.5 | 1.7 | 3000 | 300 |
| 30 | 100 | 0.61 | 2.19 | 102 | 4.2 | 2.3 | 3000 | 300 |
| 30 | 150 | 0.68 | 2.47 | 123 | 5.6 | 2.9 | 3000 | 300 |
| 30 | 200 | 0.76 | 2.74 | 145 | 7.0 | 3.5 | 3000 | 300 |
| 30 | 250 | 0.84 | 3.02 | 166 | 7.9 | 3.7 | 3000 | 300 |
| 30 | 300 | 0.91 | 3.30 | 188 | 9.3 | 4.4 | 3000 | 300 |
| 35 | 0 | 0.51 | 1.84 | 66 | 1.9 | 1.5 | 3500 | 350 |
| 35 | 25 | 0.55 | 1.98 | 77 | 2.7 | 1.9 | 3500 | 350 |
| 35 | 100 | 0.66 | 2.39 | 109 | 4.8 | 2.9 | 3500 | 350 |
| 35 | 150 | 0.74 | 2.67 | 130 | 5.7 | 3.0 | 3500 | 350 |
| 35 | 200 | 0.82 | 2.95 | 152 | 7.1 | 2.6 | 3500 | 350 |
| 35 | 250 | 0.89 | 3.22 | 174 | 8.5 | 4.2 | 3500 | 350 |
| 35 | 300 | 0.97 | 3.50 | 195 | 9.4 | 4.4 | 3500 | 350 |
| 40 | 0 | 0.56 | 2.04 | 73 | 2.1 | 1.7 | 4000 | 400 |
| 40 | 25 | 0.60 | 2.17 | 83 | 2.8 | 2.0 | 4000 | 400 |
| 40 | 100 | 0.72 | 2.59 | 116 | 5.0 | 3.0 | 4000 | 400 |
| 40 | 150 | 0.79 | 2.86 | 137 | 5.8 | 3.1 | 4000 | 400 |
| 40 | 200 | 0.87 | 3.14 | 159 | 7.2 | 3.7 | 4000 | 400 |
| 40 | 250 | 0.95 | 3.42 | 181 | 8.6 | 4.4 | 4000 | 400 |
| 40 | 300 | 1.02 | 3.69 | 202 | 10.0 | 5.0 | 4000 | 400 |

*Dairy breed requires less protein (15–20%) and more energy (5–10%) than Boer. Indigenous breed requires less protein (15–20%) and energy (5–10%) than Boer breed. Intact bucks require 5–10% more energy for growth than wethers.

**Appendix Table B.3** Nutrient Requirements for Goats (Gestation).

| | Energy | | Protein | Mineral | | Vitamins | |
|---|---|---|---|---|---|---|---|
| Weight kg | TDN kg/d | ME Mcal/d | CP g/d | Ca g/d | P g/d | A RE/d | E IU/d |
| | | **Mature Does (Dairy) Early Gestation—Single Kid** | | | | | |
| 40 | 0.64 | 2.31 | 101 | 4.2 | 2.5 | 1256 | 212 |
| 50 | 0.75 | 2.71 | 117 | 4.4 | 2.8 | 1570 | 265 |
| 60 | 0.86 | 3.09 | 132 | 4.7 | 3.1 | 1884 | 318 |
| 70 | 0.96 | 3.46 | 147 | 5.0 | 3.3 | 2198 | 371 |
| 80 | 1.06 | 3.81 | 161 | 5.2 | 3.6 | 2512 | 424 |
| 90 | 1.15 | 4.14 | 174 | 5.5 | 3.8 | 2826 | 477 |
| | | **Mature Does (Dairy) Early Gestation—Twin Kids** | | | | | |
| 40 | 0.70 | 2.52 | 117 | 5.8 | 3.2 | 1256 | 212 |
| 50 | 0.82 | 2.96 | 135 | 6.2 | 3.6 | 1570 | 265 |
| 60 | 0.94 | 3.38 | 153 | 6.5 | 3.9 | 1884 | 318 |
| 70 | 1.04 | 3.77 | 169 | 6.7 | 4.1 | 2198 | 371 |
| 80 | 1.15 | 4.16 | 187 | 7.0 | 4.4 | 2512 | 424 |
| 90 | 1.26 | 4.53 | 202 | 7.3 | 4.7 | 2826 | 477 |
| | | **Mature Does (Dairy) Early Gestation—Three or More Kids** | | | | | |
| 40 | 0.74 | 2.68 | 127 | 7.3 | 3.9 | 1256 | 212 |
| 50 | 0.87 | 3.13 | 147 | 7.6 | 4.2 | 1570 | 265 |
| 60 | 0.99 | 3.56 | 165 | 8.0 | 4.5 | 1884 | 318 |
| 70 | 1.10 | 3.97 | 183 | 8.2 | 4.8 | 2198 | 371 |
| 80 | 1.21 | 4.38 | 201 | 8.5 | 5.1 | 2512 | 424 |
| 90 | 1.32 | 4.77 | 218 | 8.8 | 5.4 | 2826 | 477 |
| | | **Mature Does (Dairy) Late Gestation—Single Kid** | | | | | |
| 40 | 0.88 | 3.18 | 141 | 4.3 | 2.7 | 1820 | 224 |
| 50 | 1.03 | 3.70 | 177 | 5.2 | 3.5 | 2275 | 280 |
| 60 | 1.17 | 4.20 | 199 | 5.5 | 3.9 | 2730 | 336 |
| 70 | 1.30 | 4.69 | 220 | 5.9 | 4.2 | 3185 | 392 |
| 80 | 1.42 | 5.14 | 239 | 6.2 | 4.5 | 3640 | 448 |
| 90 | 1.55 | 5.57 | 257 | 6.5 | 4.8 | 4095 | 504 |
| | | **Mature Does (Dairy) Late Gestation—Twin Kids** | | | | | |
| 40 | 1.05 | 3.79 | 177 | 6.2 | 3.6 | 1820 | 224 |
| 50 | 1.23 | 4.42 | 203 | 6.6 | 4.0 | 2275 | 280 |
| 60 | 1.39 | 5.02 | 229 | 6.9 | 4.3 | 2730 | 336 |
| 70 | 1.55 | 5.57 | 251 | 7.2 | 4.6 | 3185 | 392 |
| 80 | 1.70 | 6.15 | 276 | 7.6 | 4.9 | 3640 | 448 |
| 90 | 1.85 | 6.66 | 325 | 8.8 | 6.2 | 4095 | 504 |
| | | **Mature Does (Dairy) Late Gestation—Three or More Kids** | | | | | |
| 40 | 1.17 | 4.21 | 189 | 7.4 | 4.0 | 1820 | 224 |
| 50 | 1.36 | 4.89 | 215 | 7.7 | 4.3 | 2275 | 280 |
| 60 | 1.53 | 5.53 | 256 | 8.6 | 5.1 | 2730 | 336 |
| 70 | 1.71 | 6.16 | 2.82 | 8.9 | 5.5 | 3185 | 392 |
| 80 | 1.88 | 6.77 | 308 | 9.3 | 5.8 | 3640 | 448 |
| 90 | 2.04 | 7.37 | 334 | 9.6 | 6.2 | 4095 | 504 |

**Appendix Table B.4** Nutrient Requirements for Goats (Lactation).*

| | Energy | | Protein | Mineral | | Vitamins | |
|---|---|---|---|---|---|---|---|
| Weight kg | TDN kg/d | ME Mcal/d | CP g/d | Ca g/d | P g/d | A RE/d | E IU/d |
| | | **Mature Does (Dairy) Early Lactation—Single Kid** | | | | | |
| 40 | 0.89 | 3.19 | 93 | 5.9 | 3.9 | 2140 | 224 |
| 50 | 1.03 | 3.70 | 109 | 6.3 | 4.3 | 2675 | 280 |
| 60 | 1.16 | 4.20 | 124 | 6.7 | 4.6 | 3210 | 336 |
| 70 | 1.29 | 1.65 | 138 | 7.0 | 5.0 | 3745 | 392 |
| 80 | 1.41 | 5.09 | 151 | 7.3 | 5.3 | 4280 | 448 |
| 90 | 1.53 | 5.51 | 164 | 7.6 | 5.6 | 4815 | 504 |
| | | **Mature Does (Dairy) Early Lactation—Twin Kids** | | | | | |
| 40 | 1.24 | 4.48 | 120 | 10.0 | 6.4 | 2140 | 224 |
| 50 | 1.43 | 5.17 | 139 | 10.5 | 6.9 | 2675 | 280 |
| 60 | 1.61 | 5.80 | 157 | 10.9 | 7.3 | 3210 | 336 |
| 70 | 1.77 | 6.40 | 174 | 11.4 | 7.7 | 3745 | 392 |
| 80 | 1.93 | 6.97 | 190 | 11.8 | 8.1 | 4280 | 448 |
| 90 | 2.09 | 7.53 | 206 | 12.2 | 8.5 | 4815 | 504 |
| | | **Mature Does (Dairy) Early Lactation—Three or More Kids** | | | | | |
| 50 | 1.83 | 6.61 | 142 | 13.7 | 8.5 | 2675 | 280 |
| 60 | 2.05 | 7.41 | 160 | 14.1 | 8.9 | 3210 | 336 |
| 70 | 2.26 | 8.15 | 177 | 14.6 | 9.4 | 3745 | 392 |
| 80 | 2.46 | 8.85 | 229 | 16.3 | 11.0 | 4280 | 448 |
| 90 | 2.64 | 9.53 | 247 | 16.7 | 11.5 | 4815 | 504 |
| | | **Mature Does (Dairy) Late Lactation—Single Kid** | | | | | |
| 40 | 0.68 | 2.46 | 78 | 5.4 | 3.4 | 2140 | 224 |
| 50 | 0.80 | 2.88 | 92 | 5.7 | 3.7 | 2675 | 280 |
| 60 | 0.91 | 3.27 | 105 | 6.0 | 4.0 | 3210 | 336 |
| 70 | 1.01 | 3.65 | 117 | 6.3 | 4.2 | 3745 | 392 |
| 80 | 1.11 | 4.02 | 129 | 6.5 | 4.5 | 4280 | 448 |
| 90 | 1.21 | 4.36 | 140 | 6.8 | 4.8 | 4815 | 504 |
| | | **Mature Does (Dairy) Late Lactation—Twin Kids** | | | | | |
| 40 | 0.83 | 3.01 | 90 | 8.9 | 5.3 | 2140 | 224 |
| 50 | 0.97 | 3.50 | 105 | 9.3 | 5.7 | 2675 | 280 |
| 60 | 1.10 | 3.96 | 119 | 9.6 | 6.0 | 3210 | 336 |
| 70 | 1.22 | 4.40 | 132 | 9.9 | 6.3 | 3745 | 392 |
| 80 | 1.34 | 4.82 | 145 | 10.2 | 6.6 | 4280 | 448 |
| 90 | 1.45 | 5.23 | 158 | 10.5 | 6.9 | 4815 | 504 |
| | | **Mature Does (Dairy) Late Lactation—Three or More Kids** | | | | | |
| 50 | 1.14 | 4.13 | 118 | 12.8 | 7.7 | 2675 | 280 |
| 60 | 1.29 | 4.65 | 133 | 13.2 | 8.0 | 3210 | 336 |
| 70 | 1.43 | 5.15 | 148 | 13.6 | 8.4 | 3745 | 392 |
| 80 | 1.56 | 5.63 | 162 | 13.9 | 8.7 | 4280 | 448 |
| 90 | 1.69 | 6.09 | 176 | 14.2 | 9.1 | 4815 | 504 |

*All nondairy breeds require 25% (single) to 50% (triplets or more) less than dairy breeds for lactation.

**Appendix Table B.5** Nutrient Requirements for Goats (Parlor Production).

| | Energy | | Protein | Mineral | | Vitamins | |
|---|---|---|---|---|---|---|---|
| Weight kg | TDN kg/d | ME Mcal/d | CP g/d | Ca g/d | P g/d | A RE/d | E IU/d |
| **Mature Does (Dairy) Early Lactation—Parlor Production; milk yield = 4.65–6.43 kg/d** | | | | | | | |
| 50 | 2.24 | 8.06 | 144 | 16.8 | 10.1 | 2675 | 280 |
| 60 | 2.50 | 9.01 | 162 | 17.3 | 10.5 | 3210 | 336 |
| 70 | 2.74 | 9.90 | 206 | 18.7 | 11.9 | 3745 | 392 |
| 80 | 2.98 | 10.74 | 224 | 19.2 | 12.4 | 4280 | 448 |
| 90 | 3.20 | 11.54 | 242 | 19.6 | 12.8 | 4815 | 504 |
| **Mature Does (Dairy) Early Lactation—Parlor Production; milk yield = 5.82–8.04 kg/d** | | | | | | | |
| 50 | 2.64 | 9.53 | 165 | 20.7 | 12.3 | 2675 | 280 |
| 60 | 2.94 | 10.62 | 184 | 21.2 | 12.8 | 3210 | 336 |
| 70 | 3.23 | 11.65 | 203 | 21.7 | 13.3 | 3745 | 392 |
| 80 | 3.50 | 12.62 | 221 | 22.2 | 13.8 | 4280 | 448 |
| 90 | 3.76 | 13.55 | 238 | 22.6 | 14.2 | 4815 | 504 |
| **Mature Does (Dairy) Early Lactation—Parlor Production; milk yield = 6.98–9.65 kg/d** | | | | | | | |
| 50 | 3.04 | 10.98 | 185 | 24.5 | 14.5 | 2675 | 280 |
| 60 | 3.39 | 12.23 | 207 | 25.1 | 15.1 | 3210 | 336 |
| 70 | 3.72 | 13.40 | 228 | 25.6 | 15.7 | 3745 | 392 |
| 80 | 4.02 | 14.50 | 247 | 26.2 | 16.2 | 4280 | 448 |
| 90 | 4.32 | 15.56 | 266 | 26.7 | 16.7 | 4815 | 504 |
| **Mature Does (Dairy) Late Lactation—Parlor Production; milk yield = 1.99–2.76 kg/d** | | | | | | | |
| 50 | 1.31 | 4.74 | 130 | 16.4 | 9.6 | 2675 | 280 |
| 60 | 1.48 | 5.33 | 147 | 16.8 | 10.1 | 3210 | 336 |
| 70 | 1.64 | 5.90 | 164 | 17.2 | 10.5 | 3745 | 392 |
| 80 | 1.78 | 6.43 | 179 | 17.6 | 10.8 | 4280 | 448 |
| 90 | 1.93 | 6.95 | 194 | 18.0 | 11.2 | 4815 | 504 |
| **Mature Does (Dairy) Late Lactation—Parlor Production; milk yield = 2.49–3.44 kg/d** | | | | | | | |
| 50 | 1.49 | 5.37 | 143 | 16.8 | 10.1 | 2675 | 280 |
| 60 | 1.67 | 6.03 | 162 | 17.3 | 10.6 | 3210 | 336 |
| 70 | 1.84 | 6.65 | 179 | 17.8 | 11.0 | 3745 | 392 |
| 80 | 2.01 | 7.24 | 196 | 18.2 | 11.4 | 4280 | 448 |
| 90 | 2.16 | 7.80 | 211 | 18.6 | 11.8 | 4815 | 504 |
| **Mature Does (Dairy) Late Lactation—Parlor Production; milk yield = 2.99–4.13** | | | | | | | |
| 50 | 1.66 | 5.99 | 132 | 16.4 | 9.7 | 2675 | 280 |
| 60 | 1.86 | 6.72 | 176 | 17.8 | 11.0 | 3210 | 336 |
| 70 | 2.05 | 7.40 | 195 | 18.3 | 11.5 | 3745 | 392 |
| 80 | 2.23 | 8.04 | 212 | 18.8 | 12.0 | 4280 | 448 |
| 90 | 2.40 | 8.66 | 229 | 19.2 | 12.4 | 4815 | 504 |

# Appendix C

**Appendix Table C.1** Medications[1] commonly used in goats and approximate withdrawal times.

I. ANTIBACTERIALS

| Drug | Brand Name | Approval | Dosage | Route | Frequency | Withdrawal Time Meat | Withdrawal Time Milk |
|---|---|---|---|---|---|---|---|
| Ceftiofur | Naxcel®® | Approved | 0.5–1 mg/lb | IM | SID | 0 days | 0 days |
| Neomycin | Biosol®® and other products | Approved | 5 mg/lb | PO | BID | 3 days | NA |
| Amoxicillin | Amoxi-inject®® | Extra Label | 5 mg/lb | SQ | SID | 26 days | 120 hours |
| Ampicillin | Polyflex®® | Extra Label | 5 mg/lb | SQ | SID | 10 days | 72 hours |
| Benzathine Pen G | Pen BP-48®® | Extra Label | 20,000 IU/lb | SQ | Every 48 hours | 30 days | NA |
| Erythromycin | Erythro-200®® | Extra Label | 1 mg/lb | SQ | SID | 5 days | 96 hours |
| Florfenicol | Nuflor® | Extra Label | 9 mg/lb | SQ | Every 48 hours | 28 days | 120 hours |
| Oxytetracycline | LA-200® | Extra Label | 9 mg/lb | SQ | Every 48 hours | 29 days | 144 hrs |
| Procaine Pen. G | Crysticillin® | Extra Label | 10,000–20,000 IU/lb | SQ | SID | 16–21 days | 120 hours |
| Sulfadimethoxine | Albon® | Extra Label EXTRA LABEL USE IS PROHIBITED IN LACTATING DAIRY COWS. DO NOT USE IN LACTATING DAIRY DOES. | 25 mg/lb Day 1, 12.5 mg/lb Days 2–5 | PO | SID | 12 days | 5 days |
| Tylosin | Tylan®-200 | Extra Label | 10 mg/lb | IM | SID | 30 days | 96 hours |
| Chloramphenicol | Chloramphenicol | Extra Label use is prohibited. | | | | | |
| Enrofloxacin | Baytril® 100 | Extra Label use is prohibited. | | | | | |
| Furacin, nitrofurantoin | Furox® | Extra Label use is prohibited. | | | | | |
| Gentamicin | Gentocin® | Do Not Use | | | | | |
| Tilmicosin | Micotil® | Do Not Use—Toxic to Goats | | | | | |

**Appendix Table C.1** *Continued*

II. ANTI-INFLAMMATORY DRUGS

| Drug | Brand Name | Approval | Dosage | Route | Frequency | Withdrawal Time | |
|---|---|---|---|---|---|---|---|
| | | | | | | Meat | Milk |
| Aspirin | Aspirin | Extra Label | 100 mg/kg | PO | SID | 1 day | 24 hours |
| Flunixin meglumine | Banamine® | Extra Label | 1.1–2.2 mg/kg | IV or IM | SID | 10 days | 72 hours |
| Phenylbutazone | Bute | Extra Label—Do Not Use in Lactating Animals | 10–20 mg/kg | PO | SID | 60 days | DNU |
| Dipyrone | Dipyrone | Extra Label Use Is Prohibited | | | | | |

III. PREVENTION AND TREATMENT OF COCCIDIOSIS

| Drug | Brand Name | Approval | Dosage | Route | Withdrawal Time | |
|---|---|---|---|---|---|---|
| | | | | | Meat | Milk |
| Decoquinate | Deccox® | APPROVED | 13–91 gm/ton of feed | In feed | 0 days | 24 hours suggested minimum, DNU |
| Monensin | Rumensin® | APPROVED | 15–20 gms/ton of feed | In feed | 0 days | 96 hours suggested |
| Amprolium | Corid® | Extra Label | 25–50 mg/kg BW in feed or water | In feed or water | 2 days | 48 hours |
| Lasalocid | Bovatec® | Extra Label | 20–30 gms/ton of feed | In feed | 0 days | 24 hours |

IV. ANTHELMINTICS

| Drug | Brand Name | Approval | Dosage | Route | Withdrawal Time | |
|---|---|---|---|---|---|---|
| | | | | | Meat | Milk |
| IVERMECTINS | | | | | | |
| Doramectin | Dectomax® | Extra Label | 0.3 mg/kg | SQ | 56 days | 40 days |
| Eprinomectin | Eprinex® | Extra Label | 0.5 mg/kg | PO | NA | NA |
| Ivermectin | Ivomec® Drench | Extra Label | 0.3 mg/kg | PO | 14 days | 9 days |
| Ivermectin | Ivomec® 1% | Extra Label | 0.3 mg/kg | SQ | 56 days | 50 days |
| Moxidectin | Quest®, Cydectin® | Extra Label | 0.5 mg/kg | PO | 23 days | 56 days |
| Cydectin® | Drench | Extra Label | 0.3 mg/kg | PO | 14 days | NA |
| Cydectin® | Injectable | Extra Label | 0.2 mg/kg | SQ | 30 days | DNU |
| BENZIMIDAZOLES | | | | | | |
| Albendazole | Valbazen® | Extra Label | 10 mg/kg | PO | 7 days | 120 hours |
| Fenbendazole | Panacur®/Safeguard | Approved | 5mg/ kg,10 mg/kg | PO | 14 days | 120 hours |
| | | Extra Label | | PO | 14 days | 120 hours |
| Oxfendazole | Synanthic® | Extra Label | 10 mg/kg | PO | 14 days | 120 hours |
| CHOLINERGIC AGONISTS | | | | | | |
| Morantel Tartrate | Rumatel® | Approved | 10 mg/kg | PO | 30 days | 0 days |
| Levamisole | Levasole® | Extra Label | 8 mg/kg | PO | 10 days | 4 days |

**Appendix Table C.1** *Continued*

V. ANESTHETICS AND TRANQUILIZERS

| Drug | Brand Name | Approval | Dosage | Route | Withdrawal Time | |
|---|---|---|---|---|---|---|
| | | | | | Milk | Meat |
| Ketamine | Ketaset® | Extra Label | 5–10 mg/kg | IV or IM | 3 days | 48 hours |
| Lidocaine | Lidocaine | Extra Label | Variable for local anesthesia use, 1% in goats | | | |
| Thiamylal Na | Biotal | Extra Label | 10–20 mg/kg | IV | 1 day | 24 hours |
| Xylazine | Rompun® | Extra Label | 0.05–0.1 mg/kg | IV or IM | 5 days | 72 hours |
| Yohimbine | Yobin | Extra Label | 0.25 mg/kg | IV | 7 days | 72 hours |

VI. HORMONES

| Drug | Brand Name | Approval | Dosage | Route | Withdrawal Time | |
|---|---|---|---|---|---|---|
| | | | | | Milk | Meat |
| Cloprostenol | Estrumate® | Extra Label | 125 microgram | IM | 0 days | 0 days |
| Dexamethasone | Azium® | Extra Label | 20–25 mg | IM | 14 days | 4 days |
| Dinoprost | Lutalyse® | Extra Label | 5–10 mg | IM | 1 day | 24 hours |
| Oxytocin | Oxytocin | Extra Label | 10–20 IU | IM | 0 days | 0 days |

VII. ELECTROLYTES

| Drug | Brand Name | Approval | Dosage | Route | Withdrawal Time | |
|---|---|---|---|---|---|---|
| | | | | | Meat | Milk |
| Calcium | Calcium borogluconate | Extra Label | 60–100 ml of 20 to 25% Solution | IV | 0 days | 0 days |
| Calcium | Calcium gluconate | Extra Label | 50–100 ml, 10–23% calcium solution | IV | 0 days | 0 days |

NOTES: PO = oral administration; SQ = subcutaneous injection; IM = intramuscular injection; IV = intravenous injection; SID = once a day; BID = twice a day; NA = not applicable. DNU = insufficient data available to make WDI estimation; this drug is not approved for lactating goats.

[1]The drugs listed in this table are commonly used in goats. The FDA has approved only a few drugs to be used in goats. Use of drugs listed as "extra-label" is legal only if prescribed by your veterinarian in the context of a valid client-patient relationship in the U.S. The withdrawal times for various drugs were compiled from different sources. The listed dosages and withdrawal times, as well as drug status and legality of use, are subject to change. Your veterinarian will prescribe the latest, most up-to-date drugs and dosages, and provide the correct withdrawal period. Consult your veterinarian before beginning any treatment.

Courtesy of Dr. Seyedmehdi Mobini, Georgia Small Ruminant Research and Extension Center, Fort Valley State University, Fort Valley, GA. Adapted from: Gipson, T.A., R.C. Merkel, K. Williams, and T. Sahlu, Ed. 2007. Meat Goat Production Handbook. AIGR: Langston University.

# Appendix D

**Appendix Table D.1** Body Condition Score system of 1–9, used mainly for beef cattle and in special cases for goats.

| Body Condition Score | | |
|---|---|---|
| Condition | Score | Comments |
| Severely emaciated<br>Extremely poor | 1<br>(thin) | Close to death/starvation evident; outline of ribs visible and spinal processes distinct and prominent with severe depressions; physically weak; shoulder, loin and hindquarters atrophied in appearance; skin adhered to bone |
| Extremely thin<br>Poor | 2<br>(thin) | Not weak, not as emaciated as (1); skin in direct contact with bone; prominent "V" shaped cavity under tail; outline of spine and ribs are still visible; bony surface of the sternum protruding |
| Very thin<br>Frame visible | 3<br>(thin) | Wasting in appearance; ribs visible; individual spinal processes evident and depressions obvious (rib, hips) and sunken between pins and hooks; sternum is prominent |
| Slightly thin | 4<br>(moderate) | Spinous processes (dorsal/transverse) are prominent and sharp; thin flesh covering between hooks and pins; some ribs visible; definite depression between hooks |
| Frame covered<br>Balanced | 5<br>(moderate) | Spinous processes smooth; transverse processes have smooth concave curve; hooks and pins smooth; muscle becoming obvious; sternum can be palpated |
| Slightly fleshy<br>Smooth cover | 6<br>(moderate) | Spinous processes rounded; spinous to transverse processes smooth sloped; hooks and pins covered; slight depression between hooks and pins |
| Frame not visible<br>Fleshy | 7<br>(fat) | No spinous processes noticeable, ribs not visible, hooks and pins rounded with some cover; flat between hooks; palpation of sternum difficult |
| Obese | 8<br>(fat) | Edge of transverse processes barely noticeable, tailhead cavity filling with fat |
| Extremely obese<br>Severe over-condition | 9<br>(fat) | Spinous processes buried in fat; between hooks and pins rounded; between hooks rounded; tailhead cavity exhibits fat-filled folds |

# Appendix

# Index

## F